Ecological Studies, Vol. 197

Analysis and Synthesis

Ecological Studies

Volumes published since 2003 are listed at the end of this book.

Ulrich Schiewer
Editor

Ecology of Baltic Coastal Waters

Springer

Prof. Dr. Ulrich Schiewer†
University of Rostock
Germany

Cover illustration: Aerial photograph of the archipelago inside the North-Ruegenian Boddens. (Photo Hendrik Schubert)

ISBN 978-3-540-73523-6 e-ISBN 978-3-540-73523-6

Ecological Studies ISSN 0070-8356

Library of Congress Control Number: 2007934830

Cover design: WMXDesign GmbH, Heidelberg, Germany

Printed on acid-free paper

9 8 7 6 5 4 3 2 1

springer.com

Prof. em. Dr. sc. nat. Ulrich Schiewer, initiator and editor of this volume, died on the 23 May 2007 at the age of 70 years.

After growing up during the struggles of the Second World War in Pomerania and as a refugee in post-war times in Mecklenburg, Ulrich Schiewer studied Biology at the University of Rostock. He became a plant physiologist and began his career by investigating plant hormones in the group of Prof. E. Libbert. He was very active in this field, successfully investigating the mechanisms of salt acclimation of Cyanobacteria, before accepting in 1988 a call from the University of Rostock to take the Chair of Ecology, succeeding Prof. W. Schnese. Since that time, he instigated, designed and conducted a series of complex mesocosm experiments aimed at elucidating matter fluxes and the food web structure of eutrophied brackish systems.

As head of Biology, he guided the Institute through all the ups and downs of German unification, and was always able to find acceptable solutions for all the various problems that arose at this time. It is to his credit that the Institute of Biology was able not just to survive this period, but to become a widely acknowledged centre of coastal ecosystem research. Many national as well as international collaborations initiated by Prof. Schiewer, involving not only the Baltic Sea states, but also, among others, the United States, The Netherlands, France and India, still remain active today.

Prof. Schiewer mentored many dissertations and theses, and several of his former students are now working successfully in the field in which he guided their first steps.

In recognition of his achievements, Prof. Schiewer was appointed an honorary member of the Baltic Marine Biologists, an organisation in which he was active for almost three decades. He remained active in the field also after his retirement, publishing and actively attending conferences. Prof. Schiewer sat on the editorial board of several international journals and was an internationally recognised and respected scientist who published almost 200 original articles, books and other contributions.

At Rostock University, Ulrich Schiewer will be remembered not only as a highly respected scientist, but also for his boundless curiosity, kindness and generosity.

In addition to his wife, Prof. Schiewer is survived by his son and two grandchildren.

This book, which he completed shortly before his death, is not only a comprehensive overview of existing knowledge and a signpost for future development, but also serves to illustrate the enormous increase in our knowledge of brackish ecosystems gathered from all areas of the Baltic during Prof. Ulrich Schiewer's fruitful scientific career.

University of Rostock Prof. Hendrik Schubert

Preface

The Baltic Sea is one of the most investigated water bodies in the world. For decades, the many highly industrialised nations around the Baltic have financed basic and applied investigations, as well as the building and development of research stations and vessels.

After World War II, research in the Baltic Proper was intensified and investigations became much more international. The main goals of such investigations were analysis of the eutrophication and pollution of the Baltic Sea, and development of mitigating strategies (e.g. the HELCOM-Program). In contrast, research into the coastal zones was carried out mainly under national sovereignty by individual governments due to differing political regimes. Consequently, there was a lack of international collaboration and publications regarding these regions. This changed following the collapse of the former socialist governments. Nevertheless, research activities in the coastal regions still lag behind those in the Baltic Proper. A general description is further hampered by the great variety of coastal water ecosystems.

The aim of this book is to overcome this lack by presenting the important Baltic coastal zones in the form of "ecological case studies". In this way the book represents an important supplement to literature concerning the Baltic Proper.

The book begins with an overview of the Baltic Sea as reference ecosystem ("Introduction"). This is followed by a short presentation of the main characteristics of the "Baltic Coastal Zones". The ecological case studies that follow concentrate on the main types of coastal waters. They are subdivided into four groups.

"The Southern Baltic Coast" is represented by Fjörden, Boddens, Lagoons and nearshore zones. These differ in origin, in freshwater influence as well as in water exchange with the Baltic Sea. Dominance of sandy/muddy beaches is characteristic of most coastal waters in Mecklenburg-Vorpommern.

"The Eastern Baltic Coast" is characterised by great gulfs (Gulf of Gdansk, Gulf of Riga), which are in turn characterised by strong freshwater influences and coastal deltas (e.g. Gulf of Gdansk). Connected to the Baltic Proper, they are also influenced by organisms living in stable salinity.

"The Northeastern Baltic Coast" is concentrated on the Neva Estuary and the Gulf of Finland. While the Neva Estuary is dominated by the Neva River, additional small islands are characteristic of the Gulf of Finland. In contrast to the southern and eastern coast, including the Neva Estuary, hard bottom and rocky coasts are also typical here.

"The Western Baltic Coast" is, in its northern part, influenced by the decreasing water level of the Baltic Sea. The west coast is represented by typical fjords and hard bottoms in Sweden, while in Denmark coastal bights are common too. The Stockholm Archipelago and the Askö area are mostly well connected to the Baltic Proper. Odense Fjord and Kerteminde Fjord/Kertinge Nor are confronted with tidal activities and higher salinities.

Ecology of Baltic Coastal Waters represents the first science-based concept for a comparative description of the high ecological diversity of Baltic coastal waters, and demonstrates the broadly different reactions of these ecosystems and biocoenosis to natural and anthropogenic influences. The background to these activities depends not only on time and the seasonal dependencies of abiotic factors, but also the production, transport, sedimentation, transformation and degradation of organic material and nutrients. These are the ecological factors underlying the broad range of filter and buffer activities of the coastal waters. This book represents the first attempt to present data on Baltic coastal ecosystems in a clear and comprehensive way to a broad scientific audience.

Where possible, the structure of each contribution follows a common scheme: environmental characteristics; planktonic communities (structure, dynamics and productivity; nutrient cycles); benthic communities (structure and productivity; nutrient cycles); benthic–pelagic coupling and eutrophication (morphological, hydrological background; physico-chemical processes; biological processes and interactions; protection measurements); summary or conclusions.

The final chapter, "Synthesis", compiles the main data, considering common and differing characteristics and processes as well as research deficits. The chapter aims to compile general evidence for the coastal waters and outline the main differences from the Baltic Proper, leading to evaluation of both research deficits and common protection measurements.

A short summary of the main results, conclusions and the outlook for further development in coastal research round off the book.

Rostock, April 2007 Ulrich Schiewer

Contents

Part II.C Northeastern Baltic Coast

Part II.D Western Baltic Coast

Part III Synthesis

Contributors

Alimov, A.F.
Zoological Institute of the Russian Academy of Sciences, Universitetskaya Emb. 1, 199034 St. Petersburg, Russia

Chubarenko, B.V.
Atlantic Branch of P.P. Shirshov Institute of Oceanology, Russian Academy of Sciences, Prospect Mira 1, Kalingrad 236000, Russia,
e-mail: chuboris@ioran.baltnet.ru, chuboris@mail.ru

Daunys, D.
Coastal Research and Planning Institute, Klaipėda University, H. Manto 84, 92294, Klaipėda, Lithuania

Gasiūnaitė, Z.R.
Coastal Research and Planning Institute, Klaipėda University, H. Manto 84, 92294, Klaipėda, Lithuania, e-mail: zita@corpi.ku.lt

Golubkov, S.M.
Zoological Institute of the Russian Academy of Sciences, Universitetskaya Emb. 1, 199034 St. Petersburg, Russia

Herkül, K.
Estonian Marine Institute, University of Tartu, Mäealuse 10 a, 12618, Tallinn, Estonia

Hill, C.
County Administrative Board of Stockholm, P.O. Box 22067, 104 22 Stockholm, Sweden, e-mail: cathy.hill@ab.lst.se

Jaanus, A.
Estonian Marine Institute, University of Tartu, Mäealuse 10 a, 12618, Tallinn, Estonia

Jensen, M.H.
Marine, Water and Nature Division, Danish Ministry of the Environment – Environment Centre Odense, Ørbækvej 100, 5220 Odense SØ, Denmark

Kautsky, H.
Department of Systems Ecology, Stockholm University, 106 91 Stockholm, Sweden, e-mail: hassek@system.ecology.su.se

Kotta, I.
Estonian Marine Institute, University of Tartu, Mäealuse 10a, 12618, Tallinn, Estonia

Kotta, J.
Estonian Marine Institute, University of Tartu, Mäealuse 10a, 12618, Tallinn, Estonia, e-mail: jonne.kotta@sea.ee

Kruk-Dowgiałło, L.
Maritime Institute in Gdańsk, ul. Abrahama 1, 80-307 Gdańsk, Poland, e-mail: kruk@im.gda.pl

Lauringson, V.
Estonian Marine Institute, University of Tartu, Mäealuse 10a, 12618, Tallinn, Estonia

Lehtoranta, J.
Finnish Environment Institute, Box 140, 00251 Helsinki, Finland

Margoński, P.
Sea Fisheries Institute, Kollataja 1, 81-332, Gdynia, Poland, e-mail: pmargon@mir.gdynia.pl

Martin, G.
Estonian Marine Institute, University of Tartu, Mäealuse 10a, 12618, Tallinn, Estonia

Ojaveer, H.
Estonian Marine Institute, University of Tartu, Mäealuse 10a, 12618, Tallinn, Estonia

Olenin, S.
Coastal Research and Planning Institute, Klaipėda University, H. Manto 84, 92294, Klaipėda, Lithuania, e-mail: sergej@corpi.ku.lt

Peltonen, H.
Finnish Environment Institute, Box 140, 00251 Helsinki, Finland

Pitkänen, H.
Finnish Environment Institute, Box 140, 00251 Helsinki, Finland, e-mail: Heikki.Pitkanen@ymparisto.fi

Radziejewska, T.
Department of Palaeoceanology, Institute of Marine Sciences, University of Szczecin, ul. Waska 13, 71-415 Szczecin, Poland, e-mail: tera@univ.szczecin.pl

Rask, N.
Marine, Water and Nature Division, Danish Ministry of the Environment – Environment Centre Odense, Ørbækvej 100, 5220 Odense SØ, Denmark

Razinkovas, A.
Coastal Research and Planning Institute, Klaipėda University, H. Manto 84, 92294, Klaipėda, Lithuania

Riisgård, H.U.
Marine Biological Research Centre, University of Southern Denmark, Hindsholmvej 11, 5300 Kerteminde, Denmark, e-mail: hur@biology.sdu.dk

Schernewski, G.
German Baltic Sea Research Institute, Warnemünde, Seestrasse 15, 18119 Rostock, Germany, e-mail: Gerald.Schernewski@io-warnemuende.de

Schiewer, U. (deceased)
University of Rostock, Germany

Simm, M.
Estonian Marine Institute, University of Tartu, Mäealuse 10a, 12618, Tallinn, Estonia

Szaniawska, A.
Institute of Oceanography, University of Gdańsk, Al. Marszałka Piłsudskiego 46, 81–378 Gdynia, Poland, e-mail: oceasz@univ.gda.pl

Telesh, I.V.
Zoological Institute of the Russian Academy of Sciences, Universitetskaya Emb. 1, 199034 St. Petersburg, Russia, e-mail: itelesh@zin.ru

Wallström, K.
Department of Mathematics, Natural and Computer Sciences, University of Gävle, 801 76 Gävle, Sweden, e-mail: kwm@hig.se

Chapter 1
Introduction

U. Schiewer

1.1 The Baltic Sea

The Baltic Sea (Fig. 1.1) stretches from the Gulf of Finland to the Kattegat over 1,200 km in the east–west direction and from Odra Bay to Bothnian Bay near the polar circle over 1,300 km in the north–south direction. It covers an area of 415,266 km^2 and has a water volume of approximately 21,000 km^3. Traditionally, it has been divided into five main regions (Table 1.1).

- Belt Sea–Kattegat (including Sound) in the Western Baltic Sea
- Baltic Proper, with the Arkona Sea, Bornholm Sea and Gotland Sea
- Gulf of Bothnia, with the Åland Sea, the Archipelago, the Bothnian Sea and the Bothnian Bay
- Gulf of Finland
- Gulf of Riga

The catchment area of 1,745,100 km^2 comprises the nine countries bordering the Baltic Sea and a further five countries not bordering the Baltic Sea. Around 48% of the catchment area is covered by forest, 20% is arable land, 17% is open, non-productive land and 8% is marshland. Approximately 85 million people inhabit the catchment area.

The influence of the shore side is basically determined by the amount of the river inflows. Major inflows come from the rivers Neva, Vistula, Neman, Odra and Daugawa, with the Neva accounting for 20% of the freshwater inflow of the whole Baltic Sea (average annual runoff 2,500 $m^3\ s^{-1}$).

Since Baltic water can, in turn, advance far into the interior, depending on the sea water level and seasonal changes in the water regime, a basic knowledge of the structure and function of the Baltic Sea is necessary to grasp the processes occurring in the estuaries.

U. Schiewer (ed.), *Ecology of Baltic Coastal Waters. Ecological Studies 197.*

Fig. 1.1 The Baltic Sea. The main subareas of the Baltic Sea and their catchment areas (outlined in *white*) are indicated (HELCOM 2001)

Table 1.1 Characteristic data on the Baltic Sea and its five main sub-areas (HELCOM 2001)

Subarea	Sea area (km^2)	Sea volume (km^3)	Maximum depth (m)	Average depth (m)	Freshwater input (km^3 a^{-1})
Baltic Proper	211,069	13,045	459	62.1	100
Gulf of Bothnia	115,516	6,389	230	60.2	193
Gulf of Finland	29,600	1,100	123	38	100–125
Gulf of Riga	16,330	424	>60	26	18–56
Belt Sea-Kattegat	42,408	802	109	18.90	37
Baltic Sea area	415,266	459	52.3		

1.1.1 Geological Background

The geological subsoil in the north of the Baltic Sea is formed mainly by Precambrian and Palaeozoic crystalline bedrock. In the southern parts of the Baltic Sea, glacial and pre-glacial deposits predominate, overlying the old sedimentary layers of the Silure and the Tertiary (Winterhalter et al. 1981). In contrast to the crystalline bedrock, the glacial and pre-glacial deposits are very rich in calcium, which may contribute to the Calcium-anomaly in Baltic Sea water (see below).

Compared with the North Sea, which was part of the Zechsteiner shelf as early as 180 million years ago, the Baltic Sea is a very young water body. Its present shape developed only 12,000 years ago. Only the southern part of the Baltic Sea was ever a (temporary) part of the Zechsteiner Shelf Sea.

The postglacial history of the Baltic Sea began when the glacial ice crusts began to melt and a freshwater ice sea emerged. A first surface connection to the ocean appeared over central Sweden approximately 10,000 years ago. As a result of water exchange, the brackish "Yoldia Sea" was formed. Strong salinity variations were probably characteristic of that period, as there was also a second temporary connection to the White Sea. Upward movements of the earth's crust closed these early connections to the ocean at around 9,250 BC. A freshwater sea – the "Ancylus Sea" – developed, which existed from 9,250 to 7,100 BC.

As a brackish sea close to its present shape, the Baltic has existed since 7,100 BC. From 7,100 BC to 4,000 BC, the "Littorina Sea" developed, with *Littorina litorea* as the index species. During the next 4,000 years, the salinity again decreased. The freshwater mussel *Limnea ovata* (Limnea period) became the index species. This was followed by the "Mya period" (*Mya arenaria*) approximately 1,500 years ago; this index species remains characteristic of the current Baltic Sea.

A striking feature of the Baltic Sea is its topographical structure, with the Darß Sill and the adjacent large Baltic basin as characteristic features (Fig. 1.2). The Darß Sill represents the border between the Baltic Proper and the Western Baltic. With a maximum depth of 18 m the Darß Sill (and also the Drogen-Sill/Øresund with a depth of 7 m) makes the connection to the North Sea considerably more difficult. This explains why it is only through irregular bigger or smaller inflow events of North Sea water that larger amounts of water with considerably higher salinity can enter the Baltic. The deepest part of the Baltic (the Landsort Deep; 459 m) is located in the eastern Gotland Basin; its oxygen supply is completely dependent on the inflow of North Sea water.

According to the criteria formulated by Hakanson and Jansson (1983), bottom type classification differentiates between three basic area types:

- Accumulation areas prevail where fine materials can be deposited continuously
- Transportation zones appear where there is continuous deposition of fine particles/aggregates
- Erosion areas prevail where there is no deposition of fine material.

A further aspect is coastal change as a late inheritance of the ice age. Having been freed of the ice crust, which was at that time several kilometres thick, Norway,

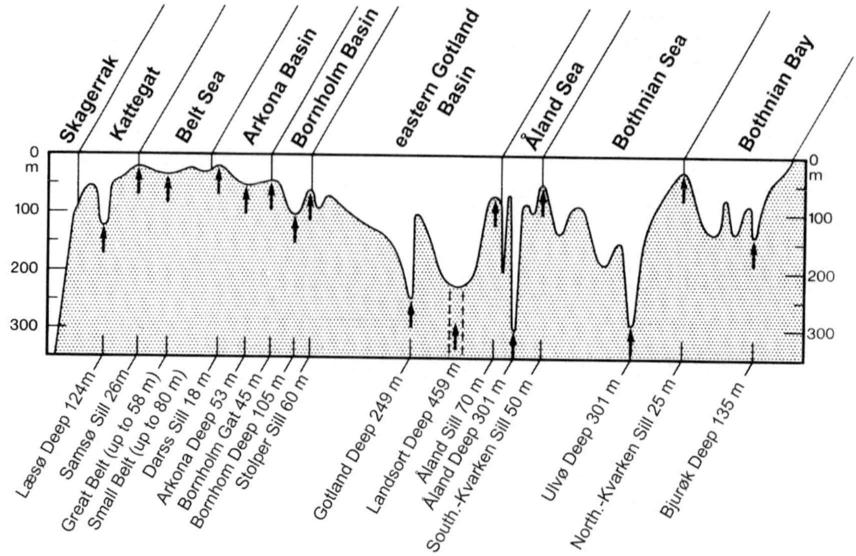

Fig. 1.2 Depth profile in longitudinal axis. Depth of basins and sills in metres (Rheinheimer 1995)

Sweden and Finland are still rising at a rate of up to 9 mm year^{-1}. As a countermove, the Southern Baltic coast sinks by 1–2 mm year^{-1} (Fig. 1.3). Over long periods this leads to clear changes in the coastlines.

1.1.2 Climate, Hydrology and Chemistry

The theoretical residence time of water in the Baltic is approximately 32 years. This is calculated from the total volume of 21×10^3 km^3 and the average value of the total inflow (approximately 16,000 m^3 s^{-1} fresh water + approximately 4,000 m^3 s^{-1} salt water at 35 psu).

Predominantly physical processes control water exchange in the Baltic Sea. In the southern part, such processes are directed mainly towards the east or north-east, with a counter current in the northern part of the Baltic. Correspondingly, we find run-off of less salty surface water along the northwest coast into the North Sea.

In the course of a year, clear differences in the surface water can be observed. During spring and summer, new fresh water coming from rivers and less salty lakes is kept on the surface by the low salinity and the developing thermal stratification. This leads to a salinity minimum at the end of the summer. Normal conditions are restored in late autumn and winter by mixing with water from the region of permanent haloclines. Mixing is guaranteed by the usual surface activities, with a shift of surface water into deeper zones via Ekman transportation.

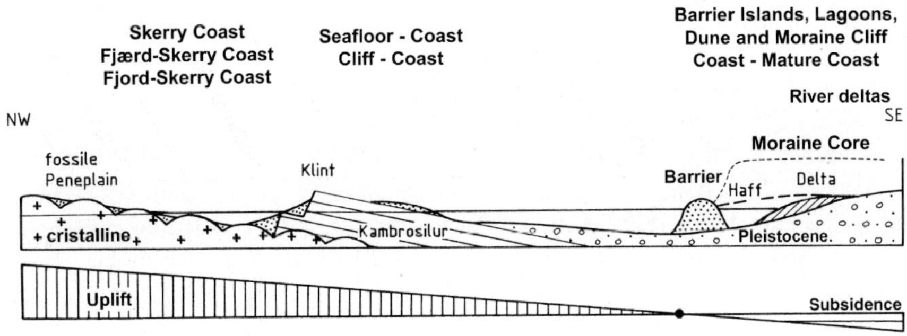

Fig. 1.3 Principal spatial arrangement and correlations of coastal types of the Baltic Sea. Structure of subsoil and isostatic compensation movements (Lampe 1996)

The Baltic Sea represents a "microtidal" system, with an average daily tidal component of 15 cm. Only the Kattegat, being under the influence of the North Sea, shows higher tide amplitudes. Changes in the sea level therefore emerge due to air pressure differences and the influence of the wind, and can reach values up to over 3 m above and 2.5 m below sea level. In the innermost Gulf of Finland, seiches can lead to changes in sea level of up to 4 m above sea level (Fonselius 1996). Tiesel (1995) describes a seasonal behaviour with high air pressure over the whole Baltic Sea in the winter (October–February, partly also in May). In the summer months it is mainly the southwest Baltic that is influenced by strong air pressure gradients.

Connected with water exchange is the transportation of sediments. Sediment transport is characterised by the shifting of sedimentary material from the shallower western or north-eastern basins into the deeper Gotland Basin. During such transport, transformation processes change the structure of the sediment considerably.

The total amounts of carbon, nitrogen and phosphorus (in tons, for the whole Baltic Sea deposition bottom area in 1–5 cm sediment) are 48.3×10^6, 5.6×10^6 and $1{,}295 \times 10^3$, respectively (Carman and Cederwall 2001). The high amount of organic carbon in the deep offshore sediments of the Baltic Proper (36.7×10^6 t) is due most likely to high primary production (Jonsson and Carman 1994) and altered decomposition efficiency.

The north-south extension of the Baltic Sea of more than 1,200 km and its integration into the North European continent results in a strong temperature gradient from south to north. Thus, the mean annual temperature near Warnemünde in the southern Baltic amounts to +8.4°C, whereas near Helsinki it is only +4.5°C. There is also an increasingly continental climatic influence on eastern and northern parts of the Baltic, explaining why regular ice coating in the winter is characteristic of the eastern and northern Baltic Sea. In severe winters it can cover almost the entire Baltic.

Also connected with the north–south extension, from the southernmost point at 54°N to the northernmost at approximately 65°N, is a clear gradient of solar irradiation. Such differences, however, are minimised during the summer due to thicker cloud coverage in the south (6/8 compared to 4/8) and longer hours of sunshine (maximum 4 h) in the north. The biggest differences are therefore found in winter.

Reduced salt water inflow, increased fresh water input and a humid climate lead to a decrease in salinity in the Baltic Proper. Whereas values of 28–30 psu can still be found in the region of the Kattegat and the Belt Sea, this goes down to 10 psu in the Western Belt Sea. East of the Darß Sill as far as the Åland islands, salinity is fairly stable at 5–7 psu. Northeast of the Åland islands there is a further reduction to 3–4 psu. The change in salinity is correlated with an ion anomaly. Positive anomalies exist in the relationship of Ca^{+} and HCO^{3-} to Cl^{-}; such anomalies result from the high water inflow of Ca^{2+} and HCO^{3-} via rivers (Kremling 1995) as well as glacial and pre-glacial deposits that are very rich in calcium. Significant differences are also found with Mg^{2+}.

Despite the structure of the basin and the differing salinity, there is a permanent isohaline in the Baltic Sea, which constantly separates the lower-salt surface water from the saltier bottom water below 60 m. This layer prevents a complete mixing of the Baltic and thus cuts the deeper basins off from the supply of atmospheric oxygen. Replacement of oxygen that has been consumed is possible only via an inflow of heavy North Sea water. Such inflows, which are irregular and vary greatly in degree, are considerably hindered by the Drogen-Sill (Øresund) with a depth of 7 m and the 18 m deep Darß Sill. This results in a constant danger of the formation of anoxic regions and regions containing H_2S below a depth of 130 m.

Apart from the permanent haloclines, there is also a temperature layering with dimictic character. The epilimnion reaches 20 m in depth, whereas the metalimnion varies between a depth of 20 and 50 m.

1.1.3 Catchment Area, Eutrophication and Pollution

Assuming a ratio between the catchment and surface area of the Baltic Sea of 4:1, a moderate load is to be expected theoretically. This is enhanced by the fact that large areas of the northern and southern catchment area are covered by forests, and large parts belong to the Scandinavian Shield and, thus, to the low-nutrient continental margin.

In contrast, there is the fact that around 85 million people currently inhabit the catchment area of the Baltic, with approximately 40 million in the immediate coastal area. The main population centres are the big cities such as Copenhagen, St. Petersburg, Stockholm, Helsinki, Riga, and the triple town of Gdansk, Sopot and Gdynia. A high population density can also be found in some coastal areas, above all at the Danish Strait coast. Since all bordering countries are developed industrial states, the whole region is strongly characterised by the presence of industry. In addition, Denmark, Germany, Sweden, Poland and the Baltic states have highly developed agricultural systems, which result in heavy pollution of the Baltic Sea.

The main problem in the Baltic Sea is eutrophication. Rönnberg and Bonsdorff (2004) have worked out a general conceptual eutrophication model for the Baltic Sea (Fig. 1.4). As far as the nutrient load is concerned, it is mainly nitrogen and phosphorus that are of interest. For the major part of the Baltic Sea, and over longer

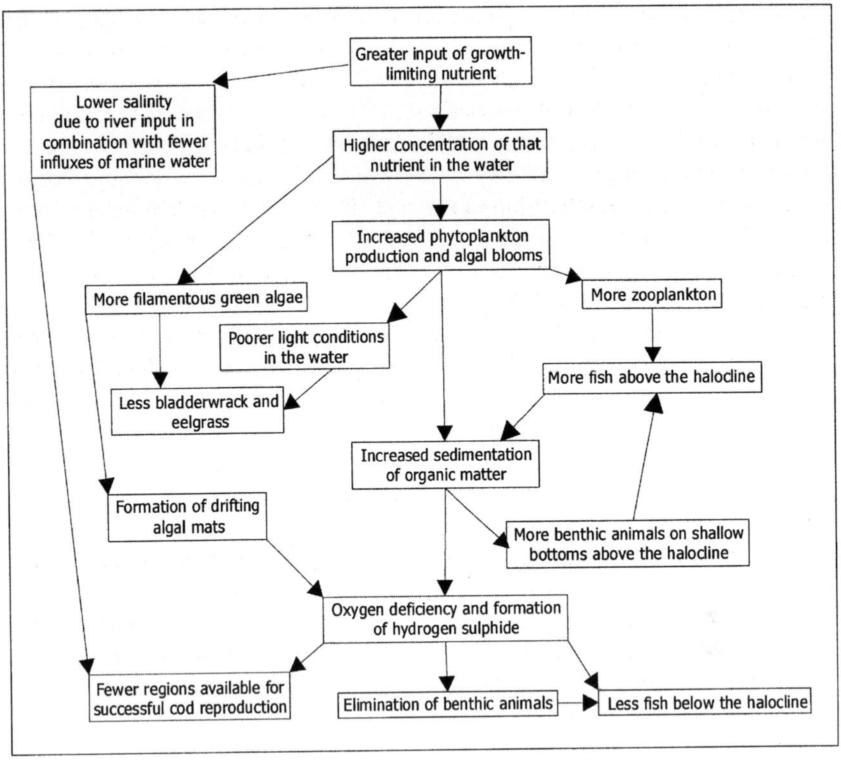

Fig. 1.4 A conceptual flow-model of eutrophication processes in the Baltic Sea (Rönnberg and Bonsdorff 2004)

times in the vegetation period, nitrogen is the limiting nutrient. At present, one-third of the nitrogen load results from direct and diffuse input from catchment areas, one-third from atmospheric input and one-third from N_2-fixation by Cyanobacteria in the Baltic itself.

The main problem is eutrophication resulting from direct and indirect nutrient input. Investigations by Wulff et al. (2001) carried out for the years 1980–1993 prove a total entry of 1,360,000 t a^{-1} total nitrogen and 59,500 t a^{-1} total phosphorus (Table 1.2). With around 48% of the nitrogen input (around 401,000 t a^{-1}) and 56% of the phosphorus input (23,000 t a^{-1}), the major contribution to these totals is made by the 11 major rivers in the eastern and southern region of the Baltic Sea. In the case of nitrogen, the rivers most involved are the Vistula, Oder, Neman, Daugawa and Neva. Most of the phosphorus comes from the rivers Neman, Vistula, Odra, Neva and Daugawa.

Using models, Wulff et al. (2001) worked out the nitrogen and phosphorus proportion of the Baltic Sea. Figure 1.5 represents the flow and deposition rates for the main regions of the Baltic Sea. A consideration of residence times and a simulation

Table 1.2 Estimated annual input of nutrients to the Baltic Sea along four major pathways (Grimvall and Stalnacke 2001)

Pathway	Study period	Nitrogen load (t a^{-1})	Phosphorus load (t a^{-1})
Riverine load	1980–1993	830,000	41,000
Coastal point sources	1990	100,000	13,000
Atmospheric deposition	1985–1989	300,000	5,000
N_2 fixation	1980	130,000	–
All pathways combined		1,360,000	59,500

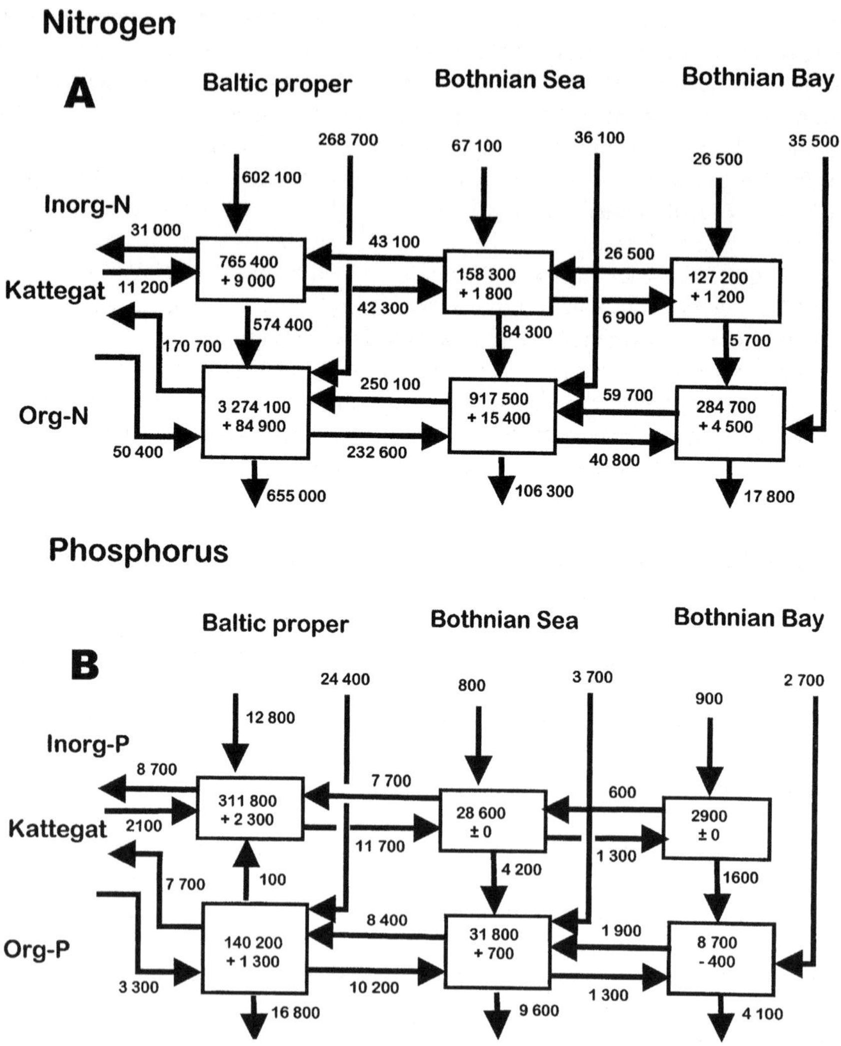

Fig. 1.5 Budgets of nitrogen (**A**) and phosphorus (**B**) for 1975–1991 in the Baltic region. Total amounts in tons. Flows and changes in pools within each basin are expressed in tons year^{-1} (Wulff et al. 2001)

of the annual proportion of nitrogen and phosphorus were made by Savchuk and Wulff (2001). According to Rönnberg and Bonsdorff (2004), an essential aspect in these calculations is the fact that, with regard to the present eutrophication, the Baltic Sea cannot be regarded as a uniform water mass. On the contrary, it is clearly differentiated into individual regions. Rönnberg and Bonsdorff suggested that the following nine sub-regions should be differentiated: Gulf of Bothnia, Archipelago region, Gulf of Finland, Gulf of Riga, Gulf of Gdansk, Swedish East coast, central Baltic, Belt Sea region and Kattegat.

The Gulf of Bothnia, with the Bothnian Sea and Bothnian Bay, represents a special case. Due to the geological composition of the catchment area, this region is limited in phosphorus. Low water temperature, low salinity (maximum around 3.5 psu) and low biodiversity are responsible for the extreme sensitivity, above all of Bothnian Bay, to human influences. The first symptoms of eutrophication are visible as algae blooms, which have been observed since the 1990s (Laine et al. 2005).

Further pollution is caused by the entry of polycyclic aromatic hydrocarbons (PAHs), polychlorinated biphenyls (PCBs), chlorinated hydrocarbons (DDT, Lindan), oil and, to some extent, heavy metals. Annual river entries are estimated at 332 kg PCBs, 2.8 kg DDT and 47.5 kg hexachlorcyclohexane (HCHs) (Agrell et al. 2002). In addition, the same proportion of PCBs enters from the air, whereas pesticide input from the air is approximately 5–7 times higher.

A permanent risk factor is the extraordinarily high, and constantly growing, volume of shipping traffic across the Baltic Sea. Narrow passes (bottlenecks) are found in the Kadet channel and in the coastal zones of the big ports such as Kiel, Lübeck, Rostock, Szczecin, Gdansk, Klaipeda, Riga, Tallinn, St. Petersburg, Helsinki, Stockholm and Malmö/Copenhagen.

Overall success in terms of pollution reduction can be gauged by the decrease in phosphate entry, both by a reduction of the use of manure and by building modern sewage-plants along the coast. In 1996, as part of the political reorganisation programme, HELCOM worked out a pollution chart of the Baltic Sea. Of the 132 hotspots identified in this chart, 98 are in Eastern Europe. In the course of redevelopment, the total number of hotspots was reduced to 119 by 2004.

The fate of substances entering the Baltic is determined by the specific water structure of the Baltic Sea. The presence of a permanent halocline at a depth of approximately 60 m makes water exchange between sediment and Pelagial considerably more difficult. This results in a mechanical purification of the Baltic due to sedimentation. Substances that are not deposited are transformed in the pelagic zone by biological and chemical processes and, at the same time, are gradually carried into the North Sea with the run-off of surface water. Due to the relatively long residence time of approximately 32 years, this exchange process is not very efficient. Below the halocline, predominantly anaerobic degradation takes place. Only occasionally, upon entry of oxygen via salt water input from the North Sea, is there an aerobic component.

One success story with regard to pollution reduction in the Baltic Sea is the partial reduction of heavy metals and pesticides. The proportion of heavy metals has been reduced in large regions of the western Baltic and the Baltic Proper. A decline

unfavourable lighting and temperature conditions in the northeast regions of the Baltic Sea (Bothnian Bay 16–29 g C m^{-2} a^{-1}, Lassig et al. 1978; Wikner 1996).

In accordance with the relatively small availability of primary producers, the production of meso-zooplankton is at a low level. Only in nearshore areas can higher turnover occur.

1.3 Benthal

The phytobenthos colonises the photic zone of the benthal. Traditionally this includes Cyanobacteria, macroscopic algae and higher plants that can tolerate brackish water. Salinity and lighting conditions as well as temperature and substrate conditions determine occurrence, growth and distribution. Due to the selecting and modifying effect of the brackish water, the Baltic Sea is considered as the remaining stock of an originally common fully marine North Sea flora (Schwenke 1995). Especially striking is the impoverishment of the brown seaweeds.

The composition of macrophytes in the benthal varies greatly, due to salinity differences, the composition of the substrate and the changing oxygen level in the sediment. Whereas in the western Baltic Sea, marine species predominate, these forms are disappearing in the central Baltic Sea and are largely being replaced by limnetic species in the Bothnian Sea. The distribution of the vegetation in different regions of the Baltic Sea is shown in Fig. 1.6.

In regions in which eutrophication is not yet conspicuous, a fairly extensive macrozoobenthos fauna exists. Zenkevitch (1963) listed the most important macrozoobenthos communities in the Baltic Sea:

- In the western Baltic the Abra-alba-coenosis with *Corbula gibba*, *Arctica islandica*, *Lagis coreni*, *Nephtys* spec., *Diastylis rathkei* and *Ophiura albida* predominates.
- Towards the east it is replaced by the Arctica-Astarte-coenosis.
- In the shallower areas of the Baltic Proper, the Macoma-balthica-coenosis is common.
- In the deeper regions the Macoma-calcarea-coenosis (with *Mya truncata* and Astarte-species) predominates.
- In the eastern soft bottom areas of the Baltic Sea, the Mesidothea-Pontoporeia-coenosis with *Mesidothea entomon*, *Pontoporeia affinis*, *P. femorata*, *Saduria entomon*, *Halicryptus spinulosum*, *Scolopos armiger* and *Terebellides stroemi* are prevalent. The two latter coenoses are postglacial communities, with representatives of the Pontoporeia-coenosis, which distinguish themselves by particularly euryhaline organisms.
- The Gotland Deep is colonised only by a species-poor hypolimnic community as a result of oxygen deficiency.

Table 1.3 shows the depth distribution of macrozoobenthos in a subarea of the Baltic Sea (Carman and Cederwall 2001).

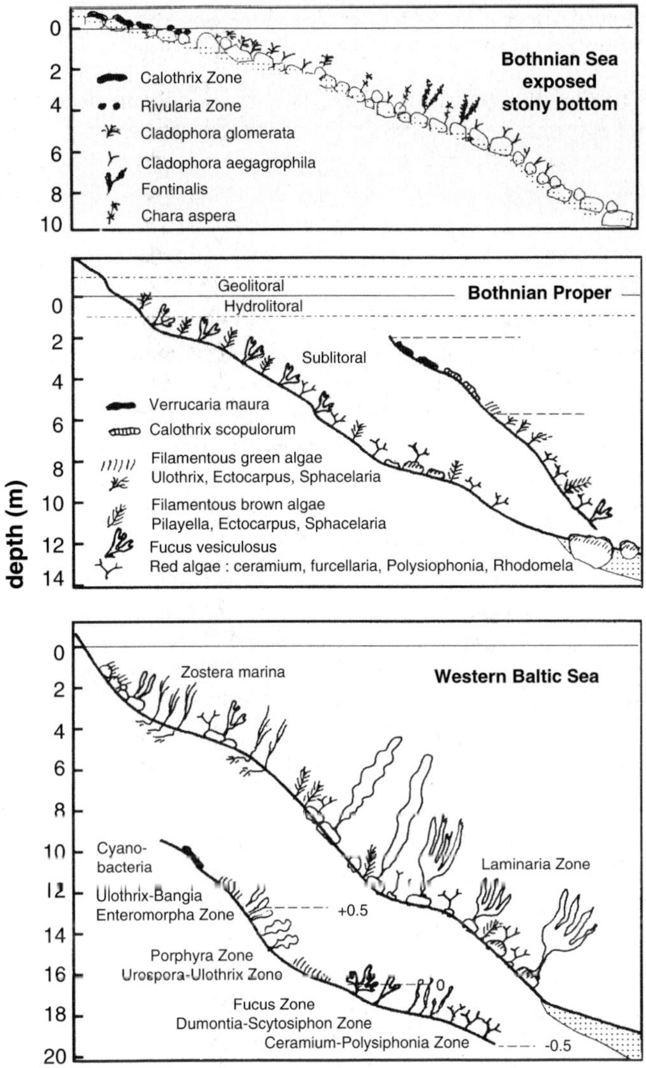

Fig. 1.6 Distribution of macrophytobenthos in different regions of the Baltic Sea (Schwenke 1995)

As a result of the unfavourable and, more importantly, the changing salinity and historically short development time of the Baltic Sea as a brackish sea, hardly any genuine brackish water species are found among the animals. Exceptions are the cosmopolitan *Fabricia sabella* and *Streblospio shrubsolii*, which also occur in the Mediterranean region, and *Alkmaria romijni*, which inhabits only the North Sea and the Baltic Sea.

Meiofauna, with a relatively large variety of species, is available in abundant amounts. Macrofauna elements are prevalent in sediment above 60 m. In deeper

Table 1.3 Depth distribution of macrozoobenthos biomass in different sub-areas of the Baltic Sea (measured as g wet weight, carbon and nitrogen) with standard errors of mean (after Carman and Cederwall 2001)

Depth zone (m)	Wet weight (g m^{-2})	Carbon (g m^{-2})	Nitrogen (g m^{-2})	No. of samples
Bothnian Sea, Åland Sea and Archipelago Sea				
0–30	96.9±13.1	5.7±0.8	0.8±0.1	82
30–70	57.8±8.1	6.6±0.9	0.8±0.1	71
70–120	17.0±2.2	3.3±0.4	0.4±0.1	37
120–250	14.0±3.0	2.3±0.5	0.3±0.1	26
Gulf of Finland				
0–30	93.7±19.4	6.0±1.2	0.8±0.2	11
30–70	82.1±42.6	6.0±3.1	0.8±0.4	7
70–100	4.0±1.2	0.8±0.3	0.1±0.0	15
Baltic Proper, Mecklenburg Bight and Kiel Bight				
0–30	117.2±8.0	5.5±0.4	0.8±0.1	528
30–70	57.8±3.3	4.2±0.2	0.6±0.0	352
70–120	1.9±0.4	0.2±0.0	0.0	156
120–350	0.1±0.1	0.0	0.0	8

zones are found mainly bacteria and fungi, which is due to the anoxic conditions and H_2S enrichment. Cyanobacteria and green algae occur only in the upper oxic sand layers. With increasing eutrophication, the anoxic regions are growing. The problem is getting worse because of the lack of salt invasions from the North Sea. A sequence of this negative development is depicted in Fig. 1.7:

- Stage 1: A stable "climax"-community, dominated by shells and echinodermates. Species colonising the well-mixed oxic sediment with deep redox layer are characterised by long-living mussels and echinodermates.
- Stage 2: As a result of increased nutrient loads (eutrophication), the community gains an increased biomass, but is also subject to strong fluctuations. Mussels and long-living polychaetes dominate.
- Stage 3 is characterised by the transition to a biomass-poor small polychaete community. There are considerable variations and occasional obliterations due to oxygen deficiency. The redox-discontinuity layer is located only a few millimetres below the sediment surface.
- Stage 4 already lacks macrozoobenthos. *Beggiotoa*-periphytons occur frequently, vagile epifauna (*Harmothoe*) occasionally. First laminations are to be found in the sediment, which can also be interpreted as "annual rings".
- Stage 5 is characterised by long-term azoic, laminated sediments (Landsort Deep).

Sediment metabolism in the less oxic regions is marked by bacterial, cyanobacterial and fungal biotic communities, which are arranged along a vertical gradient in accordance with the redox potential (Fig. 1.8). After oxygen, nitrate is the preferred electron acceptor, followed by manganese and iron oxidation, both of which are probably of lower significance for the conversion of substances. They are, however, important for

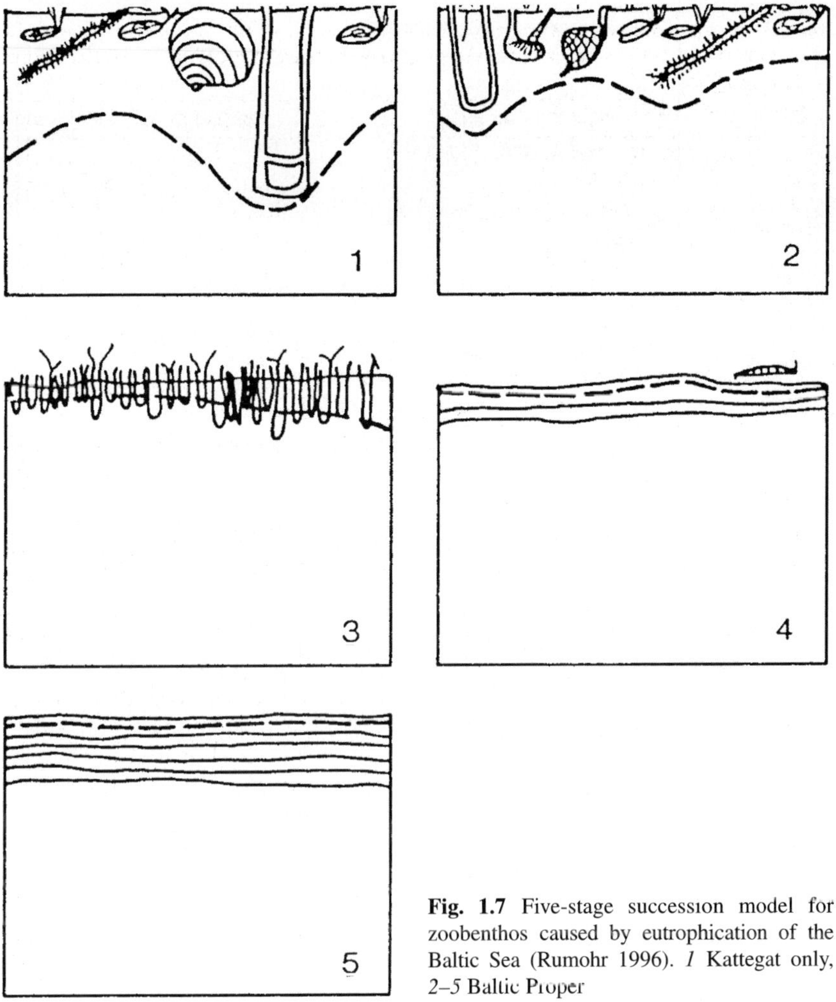

Fig. 1.7 Five-stage succession model for zoobenthos caused by eutrophication of the Baltic Sea (Rumohr 1996). *1* Kattegat only, *2–5* Baltic Proper

the mobilisation of phosphate as well as trace substances, which are able to diffuse massively from the sediment into the open water after this oxic layer falls away.

The most important electron acceptor is sulphate. About 50% of the degradation of organic material in anaerobic marine sediments is assumed to take place via sulphate reduction. The products of sulphate respiration are carbon dioxide and water as well as the toxic hydrogen sulphide. The final oxidation of organic material is carried out by methane-forming bacteria, which reduce CO_2.

More than 90% of microorganisms are attached to particles. Due to its high numbers, the biomass is comparable to that that of all other benthic organisms.

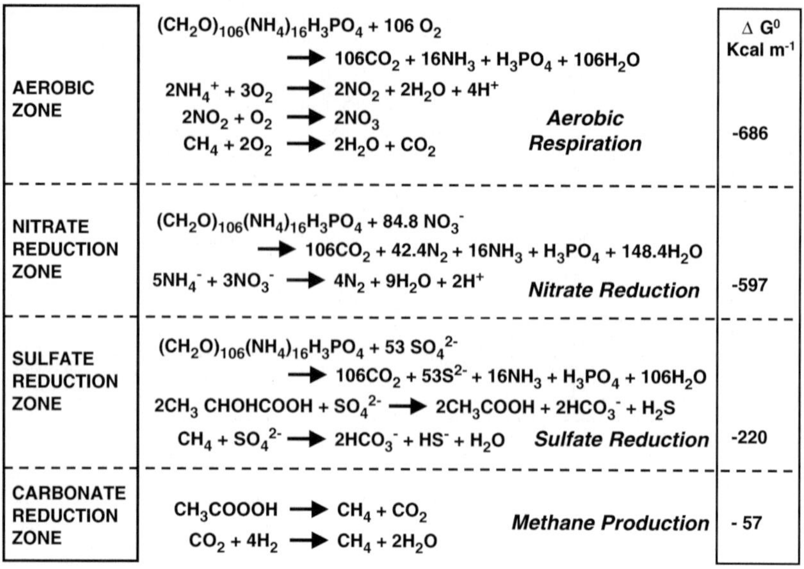

Fig. 1.8 Sequences of microbially mediated reactions in estuarine sediment, including stoichiometric decomposition equations (Martens 1978)

With regard to surface area and metabolic activity, however, microorganisms by far exceed other organisms. In sand–sludgy sediments, microbial production has been calculated to be 300 mg (spring), 140 mg (autumn) and 20 mg C m^{-2} day^{-1} (winter). For sludgy sediments the corresponding productivity amounts to 120, 370 and 10 mg C m^{-2} day^{-1} (Meyer-Reil 1995).

1.4 Conversion of Substances

The conversion of substances in the Baltic Sea is bound to relatively simple food webs. In the Baltic Proper, the microbial food web (MFW) is common. Figure 1.9 represents the conversion, according to investigations by Samuelsson and Andersson (2003), for the oligotrophic northern Baltic Sea.

Often, 10–15% of the total primary production is transferred into dissolved organic matter (DOM). This can be used only by bacteria and passed on to metazoans via protozoans. Throughout the year, however, clear differences occur in its composition and regulation:

- During spring, the MFW is of only minor importance, also the zooplankton is still missing. Due to the high availability of nutrients and radiation as well as the well-mixed water column, diatoms and dinoflagellates are prevalent. The food chains are short. A large part of the phytoplankton sediments and is consumed by benthic communities.

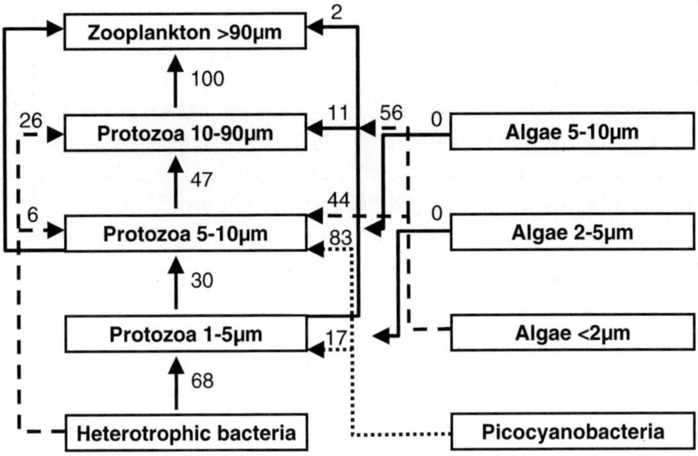

Fig. 1.9 Microbial food web (MFW) in the northern Baltic Sea (Samuelsson and Andersons 2003). Grazing rates (percent of production grazed day^{-1}) for the different groups within the MFW. *Dotted lines* denote grazing rates where the result is not supported by proportional changes in maximal biovolume concentration

- In summer the primary producers in the nutrient-poor, stratified pelagic zone, which is provided with a high energy input, are flagellates and small phytoplankton species with low sedimentation rates. Production is based mainly on regenerated nutrients, which are used by microzooplankton. On the other hand, "mosaics" of extensive blooms of N_2-fixing Cyanobacteria appear. These are degraded to a large extent in the pelagic via bacteria, fungi and protozoans. In addition, an intense production of heterotrophic bacteria supports predation by protozoans. With the summer advancing, the production of bacteria decreases as a result of nutrient deficiency and strong grazing of protozoans. Accumulation of dissolved organic carbon (DOC) in the pelagic is the result (Zweifel et al. 1995). On the whole, a clearly smaller amount of biomass reaches the higher trophic levels in the summer. Sedimentation, too, is considerably reduced.
- After completing the summer stratification, and with good mixing in the autumn, the nutrient content is high. The light is still sufficient, so that conditions are similar to those in spring. Over the whole Baltic Sea large plankton species, such as cold water diatoms and dinoflagellates, develop (Andersson et al. 1996). The mesozooplankton is well developed and controls, to a large extent, the development of the algal biomass. Bacterial production remains high, and a considerable fraction of the available carbon is transferred into higher trophic levels by the MFW.
- Winter is characterised by a high availability of nutrients, little light, a well-mixed water column and sinking temperatures. Primary production is low, which limits the growth of zooplankton. Bacterial productivity decreases sharply. The ice covering further decreases light levels. A low level of primary production, amounting to approximately 1% of the annual production, can take place in the ice pores.

The fate of the phytoplankton blooms that occur throughout the year is different. The diatoms of the spring bloom predominantly settle down and support the secondary production of the aphotic zone. In this process, sinking rates of 30–60% per day were measured (Heiskanen 1998). The dinoflagellates, in comparison, show sinking rates of only up to 10% per day (Heiskanen 1998). Thus, they support mainly the secondary production of the pelagic. However, the formation of cysts can lead to stronger sedimentation (Kremp and Heiskanen 1999). Blooms of filamentous Cyanobacteria in the summer are, as a rule, decomposed in the euphotic zone (Heiskanen and Kononen 1994). At ~10% (often even less), the faeces of secondary producers of the pelagic zone represent only an insignificant component in the downward flux of particulate material.

Important for vertical exchange is the presence of the permanent halocline and a chemocline. These determine the sedimentation speed of sinking particles and thus, the location and the speed of nutrient regeneration. The pool of particulate organic carbon (POC) is very small compared to the huge DOC-pool, and represents only about 3–7% of the total organic material in the Baltic Proper. Less than 50% of the POC is living organisms, the rest consists of detritus (see Chap. 17 by Schiewer, this volume).

1.5 Spatial and Temporal Variations in the Baltic Sea

Increasing algal blooms in the coastal zones, decreasing transparency in the water column and more frequent anoxic conditions in the deeper basins are leading to changes in the Baltic Sea. More detailed examinations confirm:

- A general increase in nitrate of 4–5% per year in all depth zones of the Bothnian Bay and below 20 m in the Bothnian Sea. An increase of 1–4% can also be seen in the Baltic Proper. Total N behaves similarly to nitrate N. Ammonium, on the contrary, has decreased in the entire surface of the Baltic Sea (Rahm and Danielsson 2001).
- Phosphate has increased with a significant rate of 2%, whereas dissolved silicate in the whole area has decreased.
- These changes are only weakly reflected in the behaviour of primary production in the Baltic Proper. Whereas from 1969 to 1991 a significant decrease in the Secchi-depth of 0.05 m a^{-1} was measured at five stations in the Baltic Proper, gross primary production increased by 0.4–1.0% annually
- On the whole, the "oceanification" of the Baltic Sea has stopped and has been replaced by a "brackification", as clearly shown by the decreasing salinity of the surface layers of the Baltic since 1976. The situation of the haloclines has not changed, their peculiarity, however, is becoming weaker.
- An increase in sea level caused by climatic changes will exert effects on the Baltic Sea. Assuming a doubling of the CO_2-concentration in the atmosphere within the next 60 years, an increase in the sea level of at least 70 cm is expected in the southern Baltic Sea within the coming 100 years. Together with a lowering of the coasts (see above), a considerable extension of the Baltic Sea to the south will result.

A particular problem is the entry of foreign species. To date, approximately 100 foreign species have been proved to exist in the Baltic Sea. Of these, about 70 species have adapted to the conditions and found a new living space in the Baltic. Of these 70 species, less than 20 species are nuisance species, causing damage to underwater constructions, fisheries, shores and embankments, or target species for hunting (Leppäkoski 2002). The best-known representatives of this group are *Teredo navalis*, *Cercopagis pengoi*, *Cordylophora caspia*, *Dreissena polymorpha*, *Anguillicola crassus*, *Alexandrium tamarense* and *Mustela vison*. The main reason for the invasion of new species is the increase in shipping traffic and, in particular, the growing transport of ballast water.

There have been no reports to date concerning the entry of protozoans, bacteria or viruses. Also, any knowledge of the costs or benefits connected with the entry of foreign organisms is fragmentary at best, or unavailable. Since the majority of organisms that have entered so far are indigenous to warmer regions, an increase in the number of foreign species establishing themselves in the Baltic Sea is expected in the course of global warming.

Table 1.4 Guidelines for the degree of changes in all parameters considered in relation to the scale of changes in the conceptual flow-models (Rönnberg and Bonsdorff 2004)

	Effects:		
Parameter	Small–moderate	Severe	Very serious
Transparency (m)	3–5	2–3	<2
Oxygen (ml l^{-1})	4–6	2–4	<2
Total-P (μg l^{-1})	15–19	19–24	>24
Total-N (μg l^{-1})	250–310	310–360	>360
Chlorophyll *a* (μg l^{-1})	1.5–2.2	2.2–3.2	>3.2
Harmful algal blooms (g m^{-3})	Few colonies (1–5)	Formation of floating algae (5–11)	Bloom areas and layer of Cyanobacteria (>11)
Macrovegetation	*Fucus* species, meadows of *Zostera* with associated charophytes, relatively sparse abundances of filamentous algae	Filamentous algae as epiphytes on *Fucus*, sporadic occurrence of *Zostera*, no charophytes. Filamentous algae dominate	No *Fucus* present. Filamentous algae dominate. Drifting algal mats and sulphur bacteria
Zoobenthos	Dominated by molluscs and long-lived polychaetes. Increased total biomass/production. Low species richness, high abundances, low biomass	Animals close to or at the sediment. Small worms (e.g. *Capitella capitata*) dominate	No macrofauna. Lack of bioturbation, lamination of the sediment
Fish	Decrease in flatfishes	Decrease in cod, herring, sprat and cyprinids	

1.6 Protection

With regard to the combat of eutrophication, it is first necessary to establish a set of clear and obligatory criteria for ascertaining the degree of eutrophication, a first draft of which has been submitted by Rönnberg and Bonsdorff (2004) (Table 1.4). The high degree of differentiation within the Baltic Sea also requires that region-specific approaches be taken (see above). These must start in any case in the respective catchment area and continue as far as the coastal zone, since only here can a direct influence be exerted. To what extent an influence of the sea area is possible depends on the extent of eutrophication and the steps which led to it. Any action taken in these fields has to be direct in nature and thus will be focussed on interactions.

Unquestionably, the Baltic Sea is currently one of the most interesting waters in the world. This is largely due its location in one of the most prosperous regions in the world. Situated between western industrial nations and newly orienting young eastern democracies, it is currently subject to widely varying influences. How these influences are controlled will determine whether the ecosystem of the Baltic Sea achieves a sustainable development or whether the eutrophication process continues. Developments in the past 10 years gives rise to a certain optimism, supported mainly by general declarations of intent from the countries bordering the Baltic Sea and the extensive research and education activities in the Baltic region. However, this must be weighed against the current decrease in economic development, the delayed reorganisation of the national economies in the eastern bordering countries as well as the considerable administrative problems that still exist among and within these states themselves.

References

Agrell C, Larsson P, Okla L, Agrell J (2002) PCB congeners in precipitation, wash out ratios and depositional fluxes within the Baltic Sea region, Europe. Atmos Environ 36:371–383

Andersson A, Hajdu S, Haecky P, Kuparinen J, Wikner J (1996) Succession and growth limitation of phytoplankton in the Gulf of Bothnian (Baltic Sea). Mar Biol 126:791–801

Carman R, Cederwall H (2001) Sediments and macrofauna in the Baltic Sea – characteristics, nutrient contents and distribution. In: Wulff F, Rahm LA, Larsson PA (eds) A systems analysis of the Baltic Sea. Springer, Berlin, pp 289–327

Fonselius S (1996) Västerhavets och Östersjöns oceanografi, Norrköping. SMHI Reports

Granéli E, Schulz S, Schiewer U, Gedziorwska D, Kaiser W, Plinski M (1988) Is the same nutrient limiting potential phytoplankton biomass formation in different coastal areas of the southern Baltic? Kiel Meeresforsch Sonderh 6:191–202

Grimvall A, Stalnacke P (2001) Riverine inputs to the Baltic Sea. In: Wulff F, Rahm LA, Larsson PA (eds) A systems analysis of the Baltic Sea. Springer, Berlin, pp 113–132

Hagström A, Azam F, Kuparinen J, Zweifel U-L (2001) Pelagic plankton growth and resource limitation in the Baltic sea. In: Wulff F, Rahm LA, Larsson PA (eds) A systems analysis of the Baltic Sea. Springer, Berlin, pp 177–210

Hakanson L, Jansson M (1983) Principles of lake sedimentology. Springer, Berlin

Heiskanen A-S (1998) Factors governing sedimentation and pelagic nutrient cycles in the northern Baltic Sea. Monogr Boreal Environ Res 8:1–80

Heiskanen A-S, Kononen K (1994) Sedimentation of vernal and late summer phytoplankton communities in the coastal Baltic Sea. Arch Hydrobiol 131:175–198

HELCOM (1998) The 3rd Baltic Sea pollution load compilation. Baltic Sea Environ Proc 70:1–133

HELCOM (2001) Environment of the Baltic Sea area 1994–1998. Baltic Sea Environ Proc 82A:1–24

Hübel H (1984) Die Bindung molekularen Stickstoffs durch autotrophe und heterotrophe Mikroorganismen in Küstengewässern der DDR und einigen Stationen der Ostsee. Habil-Arbeit, E-M Arndt Universität, Greifswald

Jonsson P, Carman R (1994) Changes in deposition of organic matter and nutrients in the Baltic Sea during the twentieth century. Mar Pollut Bull 19:417–426

Kononen K, Kuparinen J, Maekelae K, Laanemets J, Pavelson J, Nommann S (1996) Initiation of cyanobacterial blooms in a frontal region at the entrance of the Gulf Finland, Baltic Sea. Limnol Oceanogr 41:98–112

Kremling K (1995) Ionenanomalien. In: Rheinheimer G (ed) Meereskunde der Ostsee, 2nd edn. Springer, Berlin, pp 88–90

Kremp A, Heiskanen A-S (1999) Sexuality and cyst formation of the spring-bloom dinoflagellate *Scrippsiella hangoei* in the coastal northern Baltic Sea. Mar Biol 134:771–777

Kuparinen J (1984) Annual and seasonal fluctuation of primary productivity and overall respiration in a pelagic plankton community off Tvärminne, SW coast of Finland. Ophelia 3:111–122

Kuparinen J, Heinänen A (1993) Inorganic nutrient and carbon controlled bacterioplankton growth in the Baltic Sea. Estuarine Coast Shelf Sci 37:271–285

Laine A, Albertsson J, Katajisto J, Rissanen H, Luokkanen E, Kronholm M (2005) Integrated management system for the Bothnian Bay. Verh Int Verein Limnol 29:869–872

Lampe R (1995) Küstentypen. In: Rheinheimer G (ed) Meereskunde der Ostsee, 2nd edn. Springer, Berlin, pp 17–24

Lassig J, Leppaenen J-M, Niemi A, Tamelander G (1978) Phytoplankton primary production in the Gulf of Bothnia in 1972–1975 as compared with other parts of the Baltic Sea. Finn Mar Res 244:101–115

Leppäkoski E (2002) Harmful non-native species in the Baltic Sea – an ignored problem. In: Schernewski G, Schiewer U (eds) Baltic coastal ecosystems. Springer, Berlin

Lignell R (1990) Excretion of organic carbon by phytoplankton: its relation to algal biomass, primary productivity and bacterial secondary productivity in the Baltic Sea. Mar Ecol Prog Ser 68:85–99

Martens C-S (1978) Some of the chemical consequences of microbially mediated degradation of organic materials in estuarine sediments. Biogeochemistry of Estuarine Sediments. In: Proceedings of UNESCO/SCOR workshop, Melreux, Belgium, 29 November–3 December 1976, pp 266–278

Meyer-Reil L-A (1995) Mikrobiologie des Benthos. In: Rheinheimer G (ed) Meereskunde der Ostsee, 2nd edn. Springer, Berlin, pp 181–188

Nielsen R, Kristiansen A, Mathiesen L, Mathiesen H (1995) Distributional index of the benthic macroalgae of the Baltic Sea area. Acta Bot Fenn 155:1–51

Rahm L, Danielsson A (2001) Statistical analysis of spatial and temporal variations in the Baltic Sea. In: Wulff F, Rahm LA, Larsson PA (eds) A systems analysis of the Baltic Sea. Springer, Berlin, pp 329–351

Remane A (1955) Die Brackwasser-Submergenz und die Umkomposition der Coenosen in Belt- und Ostsee. Kiel Meeresforsch 11:59–73

Rheinheimer G (ed) (1995) Meereskunde der Ostsee, 2nd edn. Springer, Berlin

Rönnberg C, Bonsdorff E (2004) Baltic Sea eutrophication: area-specific ecological consequences. In: Kautsky H, Snoeijs P (eds) 17th BMB Symposium. Kluwer, Stockholm, pp 227–241

Rumohr H (1996) Veränderungen des Lebens am Meeresboden. In: Lozán JL, Lampe R, Matthäus W, Rachor E, Rumohr H, Westernhagen H von (eds) Warnsignale aus der Ostsee. Parey, Berlin, pp 162–168

Samuelsson K, Andersson A (2003) Predation limitation in the pelagic microbial food web in an oligotrophic aquatic system. Aquat Microb Ecol 30:239–250

Savchuk O, Wulff FV (2001) A model of the biogeochemical cycles of nitrogen and phosphorus in the Baltic. In: Wulff F, Rahm LA, Larsson PA (eds) A systems analysis of the Baltic Sea. Springer, Berlin, pp 373–415

Schwenke H (1995) Phytobenthos. In: Rheinheimer G (ed) Meereskunde der Ostsee, 2nd edn. Springer, Berlin, pp 163–172

Sommer U (1985) Comparison between steady state and non steady state competition: experiments with natural phytoplankton. Limnol Oceanogr 30:335–346

Tamminen T, Kaitala S, Kivi K, Kuparinen J (1985) Response of planktonic brackish water community to single and combined addition of ammonium and phosphate in a factorial mesocosm experiment. In: Gray JS, Christiansen ME (eds) Marine biology of polar regions and effects of stress on marine organisms. Wiley, New York, pp 363–378

Thiel R, Winkler H, Urho L (1996) Zur Veränderung der Fischfauna. In: Lozán JL, Lampe R, Matthäus W, Rachor E, Rumohr H, Westernhagen H von (eds) Warnsignale aus der Ostsee. Parey, Berlin, pp 181–188

Tiesel R (1995) Das Wetter. In: Rheinheimer G (ed) Meereskunde der Ostsee, 2nd edn. Springer, Berlin, pp 46–55

Wallström K, Johansson S, Larsson U (1992) Effects of nutrient enrichment on planktonic blue-green algae in the Baltic Sea. Acta Phytogeogr Suec 78:25–31

Wikner J (1996) State of marine environment of the Baltic Sea regions. Gulf of Bothnia. In: HELCOM 1996. Third periodic Assessment of the State of the Marine Environment of the Baltic Sea, 1989–1993; background document. Baltic Sea Environ Proc 64B

Winterthaler B, Flodén T, Ignatius H, Axberg A, Niemistö L (1981) Geology of the Baltic Sea. Elsevier Oceanography Series, Amsterdam

Wulff, F, Ærtebjerg G, Nicolaus G, Niemi Å, Ciszewski P, Schulz S, Kaiser W (1987) Pelagic biology. In: Lassig J (ed) First periodic assessment of the state of the marine environment of the Baltic Sea area 1980–1985; background document. Baltic Sea Environ Proc 17B. Baltic Marine Environment Protection Commission, Helsinki Commission, Helsinki, pp 216–255

Wulff FV, Rahm LA, Larsson PA (eds) (2001) A systems analysis of the Baltic Sea. Springer, Berlin

Zenkevitch L (1963) Biology of the seas of the U.S.S.R. Allen & Unwin, London

Zweifel UL, Norrman B, Hagström A (1993) Consumption of dissolved organic carbon by marine bacteria and demand for inorganic nutrients. Mar Ecol Prog Ser 101:23–32

Zweifel UL, Wikner J, Hagström A, Lundberg E, Norrman B (1995) Dynamics of dissolved organic carbon in a coastal ecosystem. Limnol Oceanogr 40:299–305

Chapter 2
The Baltic Coastal Zones

U. Schiewer

2.1 Coastal Structures and Types

Coastal waters surround, on the one hand, seas and oceans, while on the other they form transitional zones between inner coastal waters and offshore regions. With respect to hydrology, basic differences exist between the coastal waters of the North Sea and those of the Baltic Sea. Geographically, the two sets of coastal waters can be defined thus:

North Sea (with tidal effect)

- outer coastal waters seawards of the Friesian Islands
- inner coastal waters such as estuaries (e.g. classical estuaries such as mud flats or river mouths)

Baltic Sea (without significant tidal effect)

- outer coastal waters
- inner coastal waters, e.g. "Förden", inlets, boddens, haffs, lagoons and other river mouth areas (estuaries in the broadest sense)

In contrast to other waters, the Baltic Sea distinguishes itself by its high variety of coastal types. According to Klug (1985), 12 coastal types can be distinguished; Lampe (1996) grouped them into 10 basic types (Fig. 2.1).

In general, moraine material predominates at the south and southeast coastal regions, while hard bottom and rocky shores are typical of northern coasts. The underlying geology is usually Precambrian and Palaeozoic bedrock. Looking at each of the coastal regions of Denmark, Poland, Finland, Sweden, and even up to the Danish Islands, the following major areas can be classified:

- Narrow seabed coasts in north Jylland due to the uplift in this area.
- Starting from mid-Jylland, there is a predomination of moraine material, which forms "Förden" and a pattern of cliffs and shallow bights; the latter emerge from the southern part of Schleswig-Holstein onwards and are even more distinctive in Mecklenburg-Vorpommern.
- The coasts of Vorpommern have numerous lagoons and inner coastal waters, so-called Haffe and Bodden. The internal shores are filled up with phytogenic

U. Schiewer (ed.), *Ecology of Baltic Coastal Waters. Ecological Studies 197.*

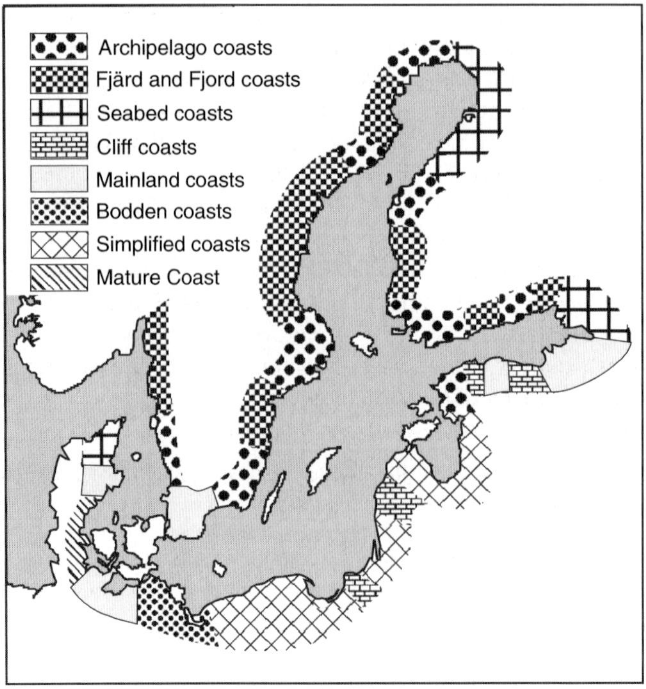

Fig. 2.1 Spatial distribution of coastal types of the Baltic Sea (Schubert and Blindow 2004)

material. Additionally, we find chalk or moraine cliffs on the islands of Rügen and Usedom.

- The Polish Pomeranian coast, from Wollin up to the Hel peninsula, is shaped by coastal levelling ("Ausgleichsküste"), which is also typical for the coasts between Klaipeda and Riga stretching as far as the Pärnu Bay. The coasts there are increasingly becoming a smoothed course. Beach lakes with dune embankments lying off them are typical (e.g. near Łeba).
- The mouth of the river Wisła represents a typical delta coast, whereas the sediment-rich river forms a wide transitional zone, which consists of alluvial islands, filled backwaters and residual lakes, between the mainland and the sea. The Neva delta also belongs to the delta coasts.
- The more eastern parts of the Polish, Russian and Lithuanian coastal regions are determined by accumulation of highly mobile sand material resulting in formation of large lagoons.
- Coastal formation that is shaped due to massive material transport stops at the coast of Estonia. This coastline is, at the same time, the border between uplift and sinking areas. Mainland-bay coasts are typical here. These are unfinished, deeply structured inner coasts that are sheltered with plant material due to the lack of bigger marine habitats.

- A pattern of cliff coasts is situated between Kaliningrad and St. Petersburg. Further east, different hard bottom coasts and rocky shores are characteristic.
- The latter applies also to the Finnish and the Swedish coasts to a great extent. Their further differentiation results from the formation of fjord and "fjärd" coasts, characterised by bays stretching deep into the mainland. Often accompanied by numerous islands in front of the coast, forming the "Schärenhof" (skerry) coasts and the Archipelago. The inner parts of the archipelago are sheltered and often consist of soft bottom shores.
- In the northeast area of the Gulf of Finland, and in the Bothnian Bay, seabed coasts dominate. These are relatively shallow marsh-like regions, the character of which is determined by uplift of the previous sea floor.
- The Swedish coast is shaped by hard bottom. The only exception is in Skåne, where coasts of the moraine type are typically found.
- The Danish islands mostly have coastal lagoons.

Different terms are assigned to coastal formations according to the diverse processes involved in their origin:

- "Förden" are glacial valley gaps stretching far into the areas of the mainland which were flooded when sea levels rose. Förden are separated from the Baltic Sea by sills.
- "Bodden" are wide and shallow coastal bays that emerged at the southern coastline of the Baltic where glacier tongue basins or deep-seated ground moraines had been flooded. Offshore islands are imbedded by strong accumulation of sandy material and coastal levelling.
- "Haffe" are former bights of the Baltic Sea that have been separated by the formation of sandy barrier beaches starting from the mainland. They have a seawards dune embankment coast.

A characteristic feature of Baltic coastal waters is the existence, within a relatively small area, of water types that are highly different in geographical, morphological, hydrological, hydrographical, chemical and thus also in biological terms. The missing levelling of conditions caused by tidal action is, among other factors, responsible for this diversity. Thus, the effects of anthropogenic impact are naturally highly modified in these waters.

2.2 Coastal Dynamics

The extent of marine forces is determined by topographical, hydrographical and meteorological conditions. The eight basic types illustrated in Fig. 2.2 will be discussed in more detail.

West of the Darß region, erosion areas are dominant. According to Sterr (1988), the retreat of the coastline amounts to 34–40 cm a^{-1} on average for Schleswig-Holstein and Mecklenburg-Vorpommern. The Darß-Zingst coastal zone – the first coastal area in the west–east direction – has developed through the impact of winds, waves, erosion and sedimentation during past centuries, and is continuing to develop in both inner and outer areas. In the external area, erosion- and accumulation-processes predominate, partly in narrowed spaces. At Darßer Ort, the accumulation of sand amounts to 85 cm a^{-1}. A phytogenic detritus sedimentation process delineates the inner parts of the separated water bodies. The characteristic sectional profile of cliff and low coastal zones in the southern Baltic Sea is given in Fig. 2.3. Comparable processes also formed the lagoons and the sand barrier beaches (Vistula and Curonian lagoon) that can be found further east.

In contrast to the German coast, the Pomeranian coast of Poland is morphologically almost uniform and represents a typical "Ausgleichsküste" (levelled coast). We can expect that, under the influence of the dominating south westerly winds, nutrient-loaded waters from the Oder river-mouth will be transported to the east ("coastal jet"), forming a gradient of decreasing loads from west to east. Results from the work of Furmanczyk and Musielak (2002) showed that the whole coastal

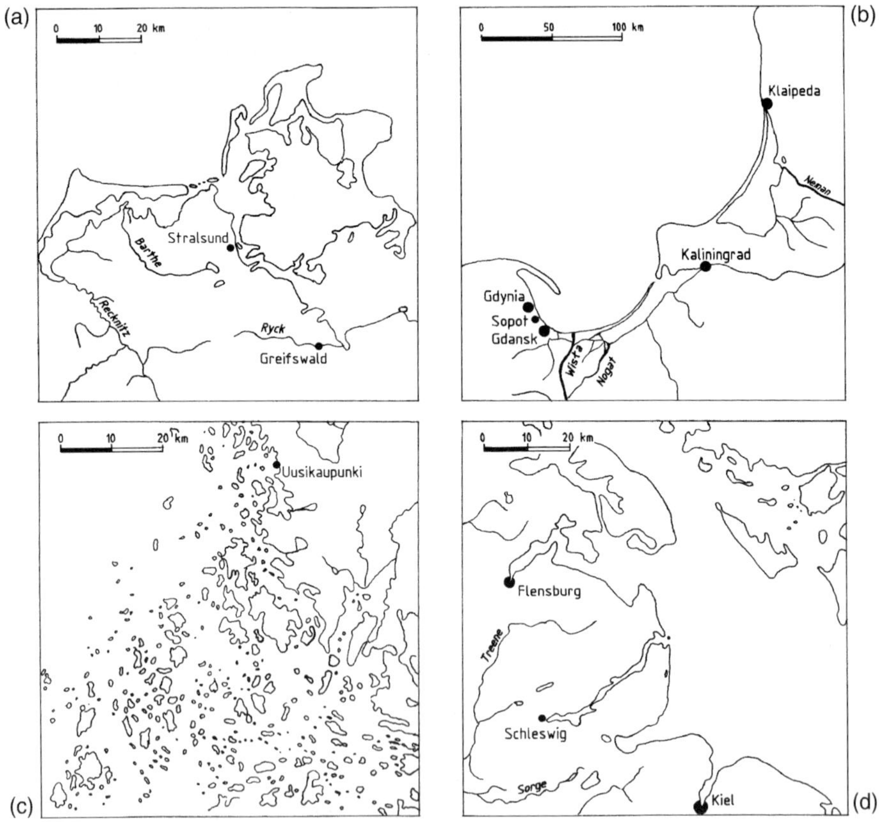

Fig. 2.2a–h Outlines of typical coastal types of the Baltic Sea (Lampe 1995). **a** Bodden coast in northern Germany. **b** Haff-sand-bar coast in the former east Prussia. **c** "Schärenhofküste" in south-western Finland. **d** Fjord coast in northern Germany.

(continued)

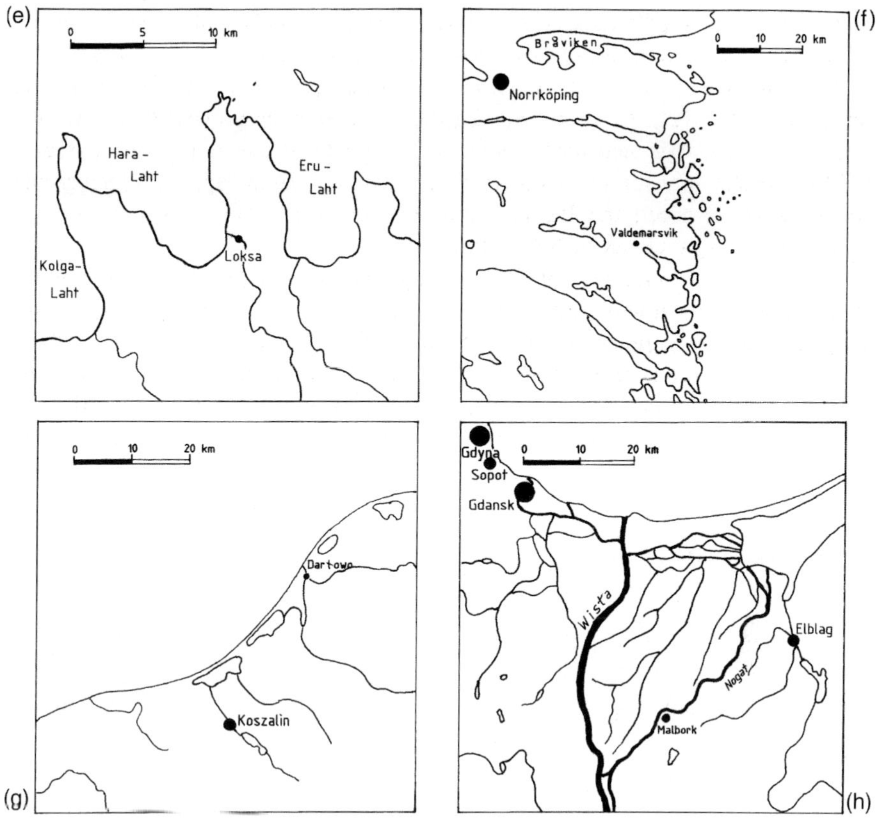

Fig. 2.2 (continued) **e** Core-land bight coasts in northern Estonia. **f** Fjärd-Schären coast in the middle of Sweden. **g** Equalisation coast in northern Poland. **h** Delta coast in northern Poland [mouth of the Vistula (Wisła) river]

zone is much more complex, in that it forms "gates" and "nodules" along this coast (Fig. 2.4). These "gates" are underwater cross-shore gates up to 3 km wide. Inside the gates, there are wide channels caused by cross-shore current flows. The water flow is directed towards the open sea. In this way, the gates dramatically increase the connection between coastal and open sea regions. The "gates" are stable for long time periods, and are also present on the eastern German coast. Such "gates" in the "coastal jet" will change the buffer and filter capacity of the coastal zone. However, results on the biological level are still lacking.

Completely different forming processes underlie the Delta coasts. These are preferably found in sinking areas (deltas of the rivers Wisła, Nogat, Neman) or in areas of minimal uplift (Neva Delta). The huge amount of moraine material carried along in rivers is deposited as the stream velocity decreases. These river mud sediments can be deformed by sand input caused by the impact of wind. Phytogenic detritus lead to the formation of peat areas.

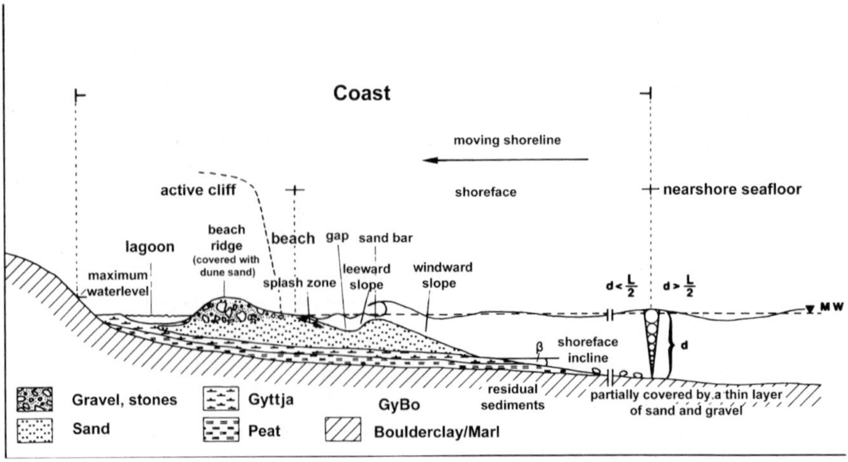

Fig. 2.3 Characteristic cross-profile of low coast and cliff coast areas in the southern Baltic Sea with typical morphometric elements (Schwarzer 1995). The coasts strive for an equilibrium between incoming streams (breakers and streams), morphological features (sand bars, gaps and beaches) and sediments present on the sea bottom (sand, boulder, clay/marl, peat, gyttja and gravel/stones)

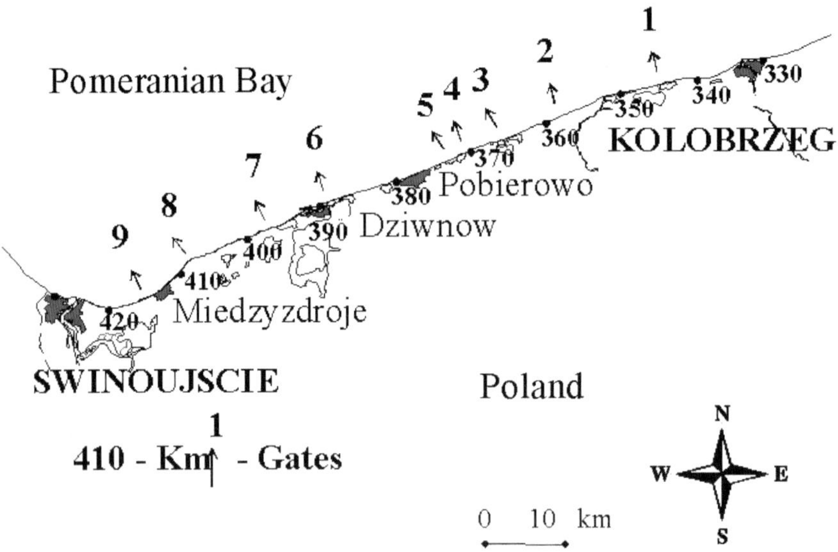

Fig. 2.4 Location of "gates" along the Polish coast of the Pommeranian Bight (Furmanczyk and Musielak 2002)

The sea approaches directly toward the cliff coasts. Characteristic of the southern Baltic coast from Denmark to Russia is the predominance of glacial moraine material, which is degraded by the influence of waves, winds and temperatures at different rates (see above).

The “Schärengürtel” (skerry belts) are a typical phenomenon of rocky uplifted coasts of the Baltic Shield, and mostly define the character of the coasts of Finland and central Sweden. The skerry belts comprise numerous offshore rocky isles of different size consisting of granite and gneisses, which have been eroded smooth and polished by glaciers of inland ice. Both the continental rocky coast and the nearby rocky islands close to the shore, which are mostly covered with forest, belong to the inner skerry belt. The offshore, mostly bare rocky islands are heavily influenced by the sea, and are classified as the external skerry belt. Thus, the “Schären” form an amphibious line between the land and the sea. The densest accumulation of skerries is located at the Åland islands in the Finnish–Swedish Archipelago Sea. Due to the progressing uplifting, landward islands are merging with larger islands and to the mainland, whereas seawards, new islands are emerging above the sea level.

“Schärenhofküsten” came into existence due to a rise in sea level. While the rocky areas are barely influenced by the sea, sections built of Pleistocene unconsolidated material will be subject to intensive form change and material decomposition.

Fjärd-Schären-coasts and Fjord-Schären-coasts are also altered by flooded downstream river valleys and changed river embankments, respectively. The uplift of land is the most distinctive process in the remodelling at present.

2.3 Ecology of Coastal Waters

The most striking feature of most coastal waters is their shallowness. The average depth is about 5 m or less. Consequently, longer thermal stratification can be excluded. On the other hand, continuous mixing and tight benthic–pelagic couplings are common [see Chaps. 7 (Kruk-Dowgiałło and Szaniawska), 8 (Chubarenko and Margoński), 9 (Gasiūnaitė et al.) and 16 (Riisgård et al.), this volume]).

Variations in the water level are moderate, usually about 0.5 m. The stochastic water exchange with the Baltic Sea causes a “washout-effect”, which raises the loading level of these waters.

The often shallow shores allow for extensive flooding. Floods pertaining to the Baltic coast produce greater levels of variation, which include sharp changes in salt content [see Chaps 3 (Schiewer), 4 (Schiewer), 14 (Hill and Wallström), and 16 (Riisgård et al.), this volume]; quick increases in salinity of 100% and more are not unusual. This results in strong hyperosmotic stress situations for both flora and fauna. High river water inputs can also result in a drastic decrease in salinity and thus cause hypoosmotic stress [see Chaps. 6 (Schiewer), 8 (Chubarenko and Margoński), 9 (Gasiūnaitė et al.) and 10 (Kotta et al.), this volume].

The shallowness of the waters allows high variations of light levels and temperature. In addition, both eutrophication and pollution are supported by loads and polymictic mixing, and can also cause drastic changes in the light climate (see Chaps. 3 and 6 by Schiewer, this volume). All these facts combine to yield a significantly lower diversity of marine species in contrast to the open Baltic Sea. On the other hand, the influence of freshwater hinders the development of the species minimum typical of the brackish Baltic Proper in the inner coastal waters. In coastal waters, only those organisms that can tolerate variability in environmental factors are able to survive, and those which can usually reach high abundances. The bottom of these shallow waters is normally covered by submerse macrophytes, which dominate primary production under normal conditions and make an essential contribution to the temporary immobilisation of nutrients. Furthermore, high ecosystem fluctuations, which are controlled mainly by physical factors (salinity, temperature, water exchange, light and other factors), are typical. But nutrient loads (usually nitrogen, also transiently phosphorus) play a bigger role, as does pollution in port and shipping areas.

2.4. Anthropogenic Loads and Protection Measures

The coastal zones belong to the territorial sea. Several, often conflicting, user demands are typical (Table 2.1). Coastal waters are important transportation, filtering and buffering systems of the Baltic Sea. They are the last barriers for loads before reaching the open sea.

The buffer and filter effect, and thus the self-purification of the fjords, boddens and haffs is based on:

- simple dilution effects by which arising gradients are softened, and material and organisms are transported to the Baltic Sea without further change.
- changes in transportation speed, whereby the substances entering are introduced into internal cycles, in which they remain for a certain time before being carried into the Baltic Sea.
- transformations, i.e. changes of the (bio-/geo-)chemical bond of chemical elements and compounds.

Changes in transportation speed determine whether a substance is stored in the coastal waters for a shorter or longer time, either in solution or bound to particles in the water body, in the sediment or in organisms.

The shallowness of the waters usually guarantees a steady mixing and thus high autotrophic and heterotrophic production and transformation. The good O_2-supply of the pelagic zone and the upper sediment layer ensures considerable degradation performances of heterotrophic organisms.

The discharge takes place:

- horizontally by episodic water exchange with the Baltic Sea (in the tide-estuaries periodically)
- vertically by accumulation in the sediment or transformation into a gaseous substance and transition to the atmosphere.

Anoxic zones in the deeper sediment parts are responsible for considerable denitrification rates, especially in the springtime, and thus for nitrogen balancing through N_2-production. In the shallow, and always well-mixed, coastal waters, the sediment, water body and atmosphere are in a constant state of intense mass transfer.

Table 2.1 User needs in the territorial sea area (modified from Obenaus and Köhn 2002)

Main user interests	Major and additional types of use
Business	Transportation
	Sea-borne transportation
	Roads
	Use of minerals (mining) and sea-water
	Sand/gravel
	Oil/natural gas
	Sea water use
	Maritime tourism
	Beach tourism
	Water sports / sailing
	Coastal passenger shipping
	Fisheries / aquaculture
	Coastal fishery
	Aquaculture
National defence	Uses for military purposes
	Exercise voyage above and below the sea surface
	Military shooting
Nature conservation, landscape and coastal protection	Nature and landscape protection
	Protection of ecosystems (fauna, flora)
	Protection in accordance to European Law and HELCOM Recommendations
	Coastal protection
Other user needs and restrictions	Other user needs
	Disposal of dredged material
	Conservation of monuments
	Funerals at sea
	Marine research/ monitoring
	Restrictions
	Wrecks and other impediments
	Ammunition

Depending on a large number of external and internal factors, as well as internal transformations of substances, they continuously or alternately act as sources and/ or sinks of these substances. Only a certain proportion of the entered material is accumulated as part of the sediments, or by organisms, and remains in the coastal waters. In the long run, there will always be a positive sinking, and thus filtering, function for all sedimentary substances. For the restoration of the waters considered here as well as the Baltic Sea itself, the positive source function of the atmosphere is of considerable importance.

Strong anthropogenic influences cross-react strongly with natural influences in coastal regions. This applies especially to eutrophication, which has caused serious structural and functional changes [see Chaps. 3 (Schiewer), 4 (Schiewer), 10 (Kotta et al.), 14 (Hill and Wallström), and 16 (Riisgård et al.), this volume]. Figure 2.5 shows a flow model of the eutrophication course in coastal Baltic waters

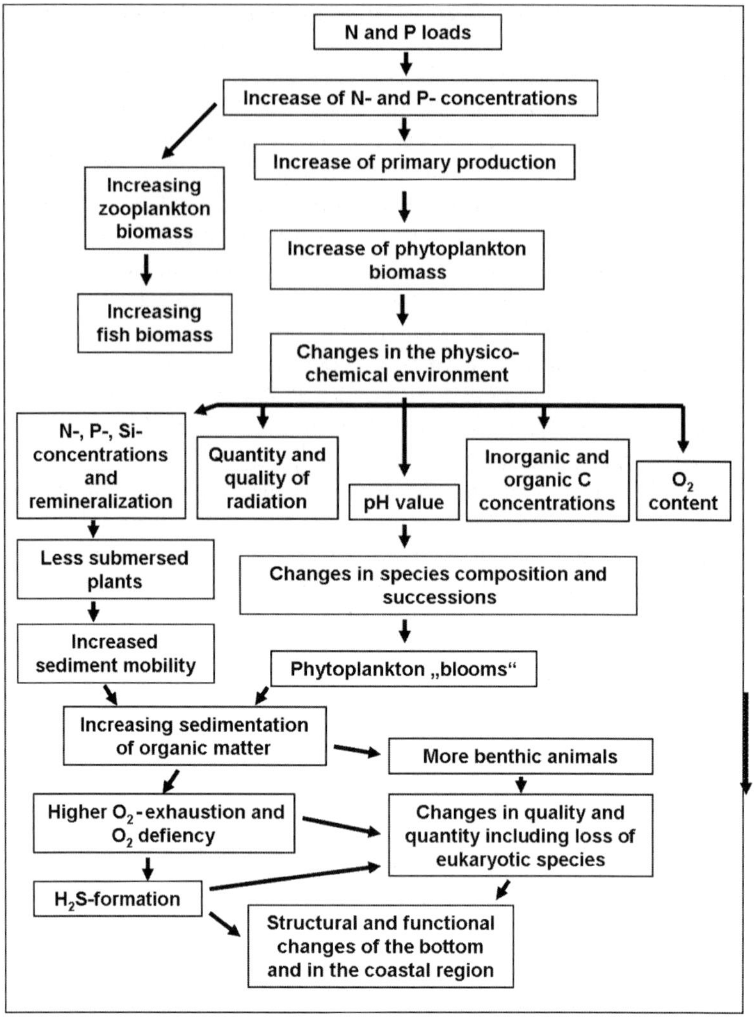

Fig. 2.5 Flow-model of coastal zone eutrophication. Benthic–pelagic coupling is included

from the point of view of benthic–pelagic coupling. Shallow inner coastal waters with hampered exchange with the Baltic Sea are especially sensitive.

In general, a one-sided and/or too intense use of an area reduces, at least partly, its multivalent capacity. Hence, over-exploitation of the self-purification potential has led to a loss of filtering and buffering capacity in many cases. Transformation of such waters into loading sources of the Baltic Sea is the consequence.

In contrast, a balanced, multivalent use of the self-purification, production, transportation and recovery potential of coastal waters, secures both ecological stability and optimal economic benefits.

A precondition for carrying out remedial measures is the regionally specific quantitative/qualitative determination of the achieved loading level by ascertaining the water quality by means of standardised parameters. An important step in this direction is the implementation of the European Water Framework Directive (1999a), application of which applies also to coastal waters of the Baltic Sea. A second requirement is precise quantitative and qualitative determination of loads and their sources. Rehabilitation and restoration objectives should then safeguard sustainable development and use. The closing off of point load sources is the simplest form of rehabilitation. The elimination of diffuse load sources from the catchment area and of input via the air is much more difficult to achieve. Since single measures have often only little success, integrated coastal zone management (ICZM) is considered the best approach at present.

The European Commission (1999b) defined ICZM thus: "ICZM is a dynamic, continuous and iterative process designed to promote sustainable management of coastal waters". ICZM attempts to integrate three basic objectives in a sustainable manner: coastal protection, nature and resource conservation as well as economic development.

The first preliminary work to reach this entire target has been carried out within the BERNET programme (Baltic Eutrophication Regional Network) of the European Union (BERNET Main Report 2001).

References

BERNET (2001) BERNET main report: Eutrophication management in the Baltic region – a regional approach. European Commission, Odense, pp 1–88

European Commission (1999a) European Water Framework Directive. In: European Commission, Brussels

European Commission (1999b) Toward a European Integrated Coastal zone Management (ICZM). European Commission, Brussels

Furmanczyk K, Musielak S (2002) Important features of coastline dynamics in Poland: "nodal points" and "gates". In: Schernewski G, Schiewer U (eds) Baltic coastal ecosystems. Springer, Berlin, pp 141–147

Klug H (1985) Küstenformen der Ostsee. In: Newig J, Theede H (eds) Die Ostsee: Natur und Kulturraum. Heide, Husum, pp 70–78

Lampe R (1995) Küstentypen. In: Rheinheimer G (ed) Meereskunde der Ostsee, 2 edn. Springer, Berlin, pp 17–25

Lampe R (1996) Die Küsten der Ostsee und ihre Dynamik. In: Lozán JL, Lampe R, Matthäus W, Rachor E, Rumohr H, Westernhagen H von (eds) Warnsignale aus der Ostsee. Parey, Berlin, pp 41–47

Obenaus H, Köhn J (2002) Important user needs in the coastal zone of Mecklenburg–Vorpommern and legal obligations in the German Baltic territorial sea. In: Schernewski G, Schiewer U (eds) Baltic coastal ecosystems. Springer, Berlin, pp 204–214

Schubert H, Blindow I (eds) (2004) Charophytes of the Baltic Sea. Gantner, Ruggell

Schwarzer K (1995) Dynamik der Küste. In: Rheinheimer G (ed) Meereskunde der Ostsee, 2nd edn. Springer, Berlin, pp 25–33

Sterr H (1988) Das Ostseelitoral von Flensburg bis Fehmarnsund: Formungs- und Entwicklungsdynamik einer Küstenlandschaft. Habilitationsarbeit, Universität Kiel, Kiel

Chapter 3
Darß-Zingst Boddens, Northern Rügener Boddens and Schlei

U. Schiewer

3.1 Introduction

The Darß-Zingst Boddens (DZB) represents the typical bodden coasts of northern Germany, which stretch from the Fischland-Darß-Zingst peninsula via the islands of Hiddensee, Rügen and Usedom to the Polish island of Wolin. The "double coasts" show the typical features of level coasts at the seashore while, at the back, the indented boddens and lagoons are inner seawaters that are subject to a progressive phytogenic silting-up.

The "jointed" structure of the DZB is also found in the Northern Rügener Bodden and at the mouth of the river Schlei (Fig. 3.1), and these areas are therefore considered together here. Investigations into the DZB were sporadic at first (Gessner 1937; Hupfer 1959, 1960; Fukarek 1961; Schwarz 1961a, 1961b, 1964; Brosin 1965a, 1965b; Hoppe and Pankow 1968; Hübel 1969). This changed fundamentally in the mid-1960s. The DZB has become the main area of investigation under the new focus on ecology and marine biology at Rostock University.

3.2 Darß-Zingst Boddens

3.2.1 Environmental Characteristics

The climate of the Baltic coast is determined by the "Ostseeklima". According to Kliewe (1951) and Hurtig (1957), a coastal area stretching 20–30 km landwards is basically involved. The "Ostseeklima" is identified by a transitional character with a relatively constant temperature, and by a delay in the start of the spring and summer seasons followed by mild autumns and relatively short winters. There is only little rain (405–790 mm per year), falling usually during autumn and winter. The mean annual air temperature is 9.1°C (1995–2004). The number of frosty days is around 75–80 (Niedermeyer et al. 1987). The main wind direction is southwest/west, with 80% of all winds reaching an average speed of 1.3–5 m s^{-1}. In shallow

U. Schiewer (ed.), *Ecology of Baltic Coastal Waters. Ecological Studies 197.*

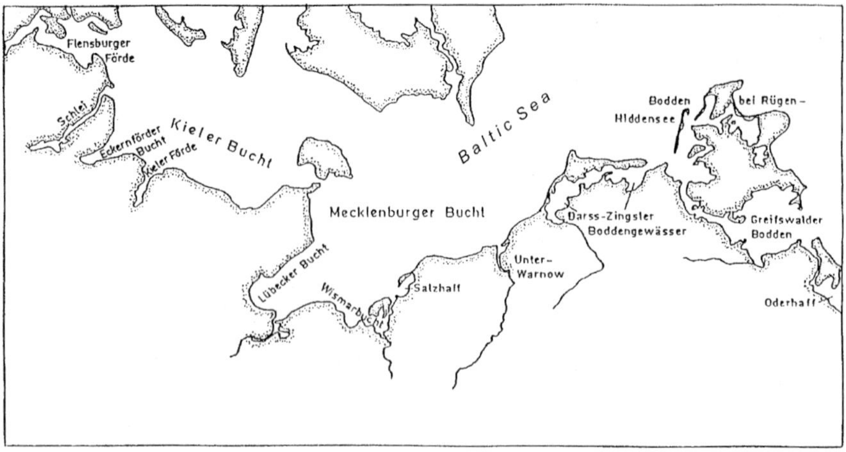

Fig. 3.1 The German Baltic coast. Fjords, boddens, haffs – typical inner coastal waters

waters wind speeds of 5 m s^{-1} are enough to mix the waters completely, stronger winds induce sediment resuspensions.

The DZB (Table 3.1) has a total area of 197 km^2 and is structured into four large basins (Fig. 3.2): the Saaler Bodden (81 km^2), Bodstedter Bodden (24 km^2), Barther Bodden (20 km^2) and Grabow (42 km^2). These basins are connected with each other through narrow transitional zones. As a result of anthropogenic interventions, the transitional zone between the Bodstedter and the Barther Bodden has become a serious bottleneck, obstructing water exchange between the eastern and western basins considerably.

The mean depth of the DZB is around <2 m; the Zingster Strom reaches a maximum depth of 12 m. The Bodstedter Bodden stands out for its maximum depth of 10 m. The water volume between Pramort and Ribnitz amounts to $397 \times 10^6 m^3$ at normal water level. The catchment area of approximately 1,600 km^2 is characterised mainly by agricultural production. Besides the small towns Ribnitz-Damgarten and Barth, the villages Wustrow, Ahrenshoop and Zingst play an important role as holiday resorts at the Bodden.

Because of the catchment area/water surface ratio of 8:1, the potential natural water quality is classified as mesotrophic/eutrophic. The very narrow outflow towards the Baltic as well as the subdivision into basins intensify the natural tendency to eutrophy in the western region.

The inflow of freshwater via the small rivers Recknitz and Barthe as well as via some streams is low ($290 \times 10^6 m^3 a^{-1}$). However, it may reach 60% in the Saaler Bodden, while it is around 10% on average for the entire bodden chain (Aurada 2000). The DZB is far more strongly influenced by the water of the Baltic Sea ($2{,}650 \times 10^6 m^3 a^{-1}$). This reinforcement of the exchange frequency compensates the unfavourable proportion of catchment area/boddens surface. The theoretical water residence times vary from 7 days for the Barther Bodden to up to 55 days for the

Table 3.1 Main characteristics of the Darß-Zingster Bodden (DZB), Northern Rügener Bodden (NRB) and Inner Schlei. *dw* Dry weight, *ww* wet weight, Chl *a* chlorophyll *a*, *POC* particulate organic carbon, *DOC* dissolved organic carbon, *POM* particulate organic matter, *DIN* dissolved inorganic nitrogen, *DIP* dissolved inorganic phosphorus

	Darß-Zingster Bodden[a]	Northern Rügener Bodden[b]	Inner Schlei[c]
a. Climate and hydrology			
Air temperature (°C)	9.1 (max 9.7; min 7.2; 1995/2004)		
Water temperature (°C)	+11 (max +22, min +1)		
Ice covering (days a^{-1})	43–90		
Annual global radiation	1,060 kW h m^{-2}	1,060 kW h m^{-2}	
Average precipitation (mm a^{-1})	556 (max: 790; min: 405; 1996/2004)		
Mean freshwater inflow ($10^6 m^3 a^{-1}$)	290	45	202
Mean inflow ($10^6 m^3 a^{-1}$)	2,750 (from Baltic)	3,360 (from Baltic)	
Mean total outflow ($10^6 m^3 a^{-1}$)	3,020	3,405	
Residence time (days)	8–61 (regionally differing)	51 (regionally differing)	90 (30–150)
Water exchange rate (a^{-1})	0.15	0.14	no data
b. Morphometry and sediments			
Area (km^2)	196.7	160	19.5 (total Schlei 54)
Volume ($10^6 m^3$)	397.1	554	49 (133)
Mean depth (m)	2.0	3.5	2.5
Maximum depth (m)	16.5	9	13
Catchment area (km^2)	1,594	352	440 (667)
Area/catchment relation	1:8	1:2	1:23 (1:12)
Organic matter content (% dw)	Mineral soil: 1.4–2.6 Muddy soil: 12.1–20.3	No data	No data
Sediment oxygen demand (mmol O_2 m^{-2} day^{-1})	1,630 mg	No data	No data
Phosphate release	17 mg m^{-2} day^{-1}	No data	No data
Ammonia release	90 mg m^{-2} day^{-1} (max: 230 mg m^{-2} day^{-1})	25–60 µmol $m^{-2}h^{-1}$	No data
Denitrification rate	Up to 73 mg N m^{-2} day^{-1} 1997: 40 µmol $m^{-2}h^{-1}$	37 g N $m^{-2}a^{-1}$ 1997: 0.2–8.6 µmol $m^{-2}h^{-1}$	No data

(continued)

Table 3.1 (continued)

	Darß-Zingster Bodden[a]	Northern Rügener Bodden[b]	Inner Schlei[c]
Sediment distribution	50–60 sandy	Mostly sandy	No data
c. Pelagic biological components			
Secchi depth (m)	0.3–0.4	6–8 / 0.14–0.25	0.4–1.5 (0.4–>4)
Bacteria (10^6 ml^{-1})	24	No data	9–28
Phytoplankton biomass (µg Chl *a* l^{-1})	8–227 (spring and autumn max 183–283)	1.3–4.5 (Rassower Strom, 1997)	29.4–162 (0.7–162)
Phytoplankton biomass (spring and autumn maxima (mm^3 l^{-1})	14.5–235	No data	No data
Dominating phytoplankton species	1975: *Gomphosphaeria pusilla, G. lacustris, Lyngbya contorta, Anabaena spiroides, Microcystis aeruginosa, Scenedesmus quadricauda, Monoraphidium contortum, Stephanodiscus hantzschii, Diatoma elongatum*; 1994: *Tetrastrum triangulare, Scenedesmus* spp., *Aphanothece clathrata*	1960/63 (Großer Jasmunder Bodden): *Microcystis* spp., *Gomphosphaeria* spp., *Coelosphaerium* spp, *Scenedesmus* spp., *Oocystis* spp., *Monoraphidium* spp., *Anabaena* spp.; 1998 (Libben): *Chaetoceros danicus, Ch. wighamii, Skeletonema costatum, Thalassiosira baltica, Achnanthes taeniata, Peridinium pellucidum*	*Cryptomonas* spp., *Rhodomonas* spp *Eutreptiella* cf. *gymnastica, Katodinium rotundatum, Chaetoceros* spp., *Skeletonema* spp., *Thalassiosira baltica, Dactylosphaerium jurisii, Monoraphidium contortum, Microcystis aeruginosa, Aphanocapsa, Chroococcus, Anabaenopsis arnoldii, A. elenkinii, Anabaena*
Flagellates			
Biomass (µg C l^{-1})	17–40 (ww)	No data	No data
Dominating species			
Ciliates			
Biomass (µg C l^{-1})	53–563 (ww)		ca. 20 (ww)

Dominating species	No data for dominating species	*Paramecium aurelia*, *Leprotintinnus bottnicus*, *Tintinnopsis* spp.	
Mesozooplankton			
Biomass (µg C l^{-1})	<12 mg l^{-1} (ww)		<2.5 mg l^{-1} (ww)
Dominating species	*Eurytemora affinis*, *Acartia tonsa*, *Keratella cochlearis f. tecta*. *Filinia*, *Brachionus*, *Synchaeta*, *Bosmina*, *Pleopsis*, *Chydorus*, *Alona*	*Keratella*-, *Filinia*-, *Brachionus*-, *Notholca*-, *Synchaeta*-, *Bosmina*-, *Asplachna*-, *Podon*-, *Eurytemora*-species, *Acartia bifilosa*, *Chydorus sphaericus*	*Acartia*, *Eurytemora*, *Bosmina*, *Synchaeta*, *Brachionus*, *Keratella*, *Filinia*, *Asplanchna*, *Hexarthra*, *Diaphanosoma*, *Cercopagis*, *Leptodora*, *Podon*, *Evadne*
Fish			
Number of species	47	ca. 47	
Dominating species	*Clupea harengus*, *Sander lucioperca*, *Abramis brama*, *Rutilus rutilus*, *Plathichthys flesus*, *Belone belone*, *Gadus morrhua*, *Psetta maxima*	*Clupea harengus*; see DZB	*Clupea harengus*, *Anguilla anguilla*, *Perca fluviatilis*, *Sander lucioperca*, *Rutilus rutilus*, *Abramis brama*, *Osmerus eperlanus*, *Esox lucius*, *Salmo trutta trutta*
Seston (mg l^{-1})	60–100 (ww)	No data	No data
POC	<16 mg C l^{-1} (421–1,162 µmol l^{-1})	No data	No data
DOC	<13 mg l^{-1} (824–1,018 µmol l^{-1})	No data	No data
C/N in POM	8.8–10.6	No data	No data
Primary production (g C m^{-2} a^{-1})	Outlet: 100–135; Saaler Bodden: >600	<180–>600	817 (223)
Primary production (mg C m^{-2} day^{-1})	200–>3,000	100–>6,000	Up to 2,430
Bacterial production (µg C l^{-1} h^{-1})	18	No data	0.8–8.8; (0.17–2.8)
Mesozooplankton production (g C m^{-2} a^{-1})	<60	No data	No data
Fish catches (t a^{-1})	305.2	No data	No data

(continued)

Table 3.1 (continued)

	Darß-Zingster Bodden[a]	Northern Rügener Bodden[b]	Inner Schlei[c]
d. Benthic biological components			
Macrophytes			
Biomass (g m^{-2})	Barther Bodden (1973): 4,990 (dw)	No data	No data
Number of species	No data	No data	No data
Dominating species	*Ruppia spiralis*, *Potamogeton pectinatus*, *Myriophyllum spicatum*, *Chara aspera*, *Ch. baltica*, *Najas marina*	*Ceramium rubrum*, *Pilayella litoralis*, *Ectocarpus siliculosus*, *Zostera marina*, *Potamogeton pectinatus*, *Ruppia marina*, *Zannichellia palustris*, *Chara canes-cens*, *Ch. baltica*, *Tolypella nidifica*	*Enteromorpha intestinalis*, *Ulva lac-tuca*, *Pilayella litoralis*, *Chorda filum*, *Zannichellia palustris*, *Potamoge-ton pectinatus*, *P. fluitans*, *Najas major*, *Scirpus maritimus*, *Phragmites*, *Typha*
Macrozoobenthos			
Biomass (g m^{-2})	>400 g ww m^{-2} (ww)		<100 g (ww)
Number of species	>53 (AFDW); >50 species		<20
Dominating species	*Marenzelleria viridis* (up to 85% of total biomass) Chironomidae, *Hydrobia*, *Nereis*, *Oligochaeta*	*Cardium edule*, *Nereis diversicolor*, *Corophium volutator*, *Cyathura carinata*, *Alkmaria romi-jini*, *Paranais litoralis*, *Prostoma obscurum*, *Manayunkia aestuarina*	*Marenzelleria (40–99 % of total biomass)*, *Nereis*, *Potamopyrgus*, *Hydrobia*, *Chironomus*, *Macoma*, Oligochaeta
e. Water chemistry, trophic status and pollution			
Salinity (psu)	5.6 (0.5–14.0)	10.1/3.4	6.8 (5.1–8.8); (5.1–19.5)
pH	7.5–11.5	No data	No data
Oxygen saturation (%)	90–180	90–160	No data
Total nitrogen (µmol l^{-1})	205 (max 253; min 138; 1996/2004)	No data	1–8

DIN (µmol l^{-1})	<1–190; winter max: 606	3.7–27.4; winter max: >10–>46.8	>250
Total phosphorus (µmol l^{-1})	1.2–4.1	No data	0.1–0.6
DIP (µmol l^{-1})	<0.5–1.0; winter max 4.1	0.8–3.0	<0.5–14; winter max: ca. 3; summer max: 14
Annual nitrogen input (t a^{-1})	3,370	100–230	1,638
Annual phosphorus input (t a^{-1})	99	13–35	68
Limitation of PP and period	light limitation vegetation period	N spring and autumn	nitrogen spring to autumn
Main pollutants	N, P	N, P	N, P
Trophic level	Eutrophic to hypertrophic	eutrophic to polytrophic (Kleiner Jasmunder Bodden: hypertrophic)	eutrophic to hypertrophic

[a]Data from Arlt 1984; Arndt 1985, 1991, 2001; Blümel et al. 2002; Heerkloss and Schnese 1994; Jost and Klinkenberg 1994; Kliewe 1951; Lindner 1978; Meyer-Reil 1999; Rieling 1999; Schiewer 1990, 1994a, 1994b, 1998a, 1998b, 2001; Schlungbaum and Baudler 2000; Schubert 1996; Schumann 1993; Schumann et al. 2001, 2003a, 2003b, 2005; Wasmund 1994; Wasmund and Schiewer 1992; Winkler 2001; Yap et al. 1987; Yousef et al. 1997; Zettler 1995

[b]Data from Aurada 2000; Dahlke and Hübel 1996; Dahlke et al. 1999; Schlungbaum 1994; Schmelzer 1994; Gessner 1957; Hübel 1984; Hübel et al. 1998

[c]Data from Feibicke 1994, 1995; Flenthe 1972; Gocke and Rheinheimer 1994; Gocke et al. 2003; Lupatsch and Nellen 1981; Nellen 1967; Ripl 1986; Stotz 1986

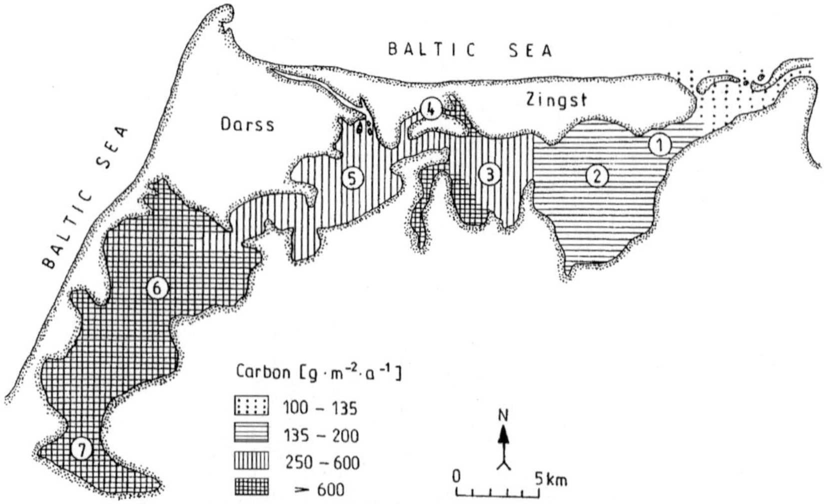

Fig. 3.2 The Darß-Zingst Boddens (DZB), showing structure and annual productivity in the different parts. Decreasing salinity and increasing nitrogen content from east to west; o-phosphate in all parts of the DZB is at a comparable and low level. *1* Pramort, *2* Grabow, *3* Barther Bodden, *4* Zingster Strom, *5* Bodstedter Bodden, *6* Saaler Bodden, *7* Ribnitzer Lake

Saaler Bodden, and are around 0.15 a^{-1} for the entire DZB (Schlungbaum 1994). The distinct horizontal salt gradient, with declining salinity from east to west, is the result of the eastern location of the outlet of the DZB and its small size. The mean salt content of the offshore Baltic amounts to 8–12 psu, that of the Barther Bodden is 5–6 psu, of the Bodstedter Bodden 3–4 psu and of the Saaler Bodden 1–3 psu. The existing measurement series on salinity for the Zingster Strom, which has been carried out for 40 years, verifies that this variability is relevant: mean 5.5 psu; minimum annual average 3.4 psu, maximum 8.3 psu. The average pH value of 8.0 also varies considerably; during the most productive months (May–July), pH values >9.0 often occur. Maximum values up to pH 11.5 have been measured repeatedly.

The oxygen content in the pelagic zone is 90–180% under normal conditions. Under highly productive conditions it may be higher. Normally, it does not fall below 10% near the sediment. Only when it is ice-covered for a longer time, and when there are high proportions of mud, does the sediment surface become anoxic. Despite maximum oxygen uptake rates of up to 1,145 mg O_2 m^{-2} day^{-1} at the sediment surface, the sediment usually remains oxic up to a depth of 2 mm. However, anoxic microhabitats may develop already in the fluffy sediment layer (FSL, see below). Phosphorus release also does not take place under anoxic conditions (see below). The mean BSB_5-values (the oxygen consumed by the bacteria in a waste water sample within 5 days) of the mostly sandy–muddy sediment with a mean carbon content of approximately 1% dry matter are between 3 and 7.5 mg O_2 l^{-1}.

The mean Secchi depth came to 30 cm (10–100 cm) in 1996/1998, which was very low. In the 1970s and up to the early 1980s, mean values of 60 cm were still measured. The solar radiation reaches its maximum in May (monthly mean ca.

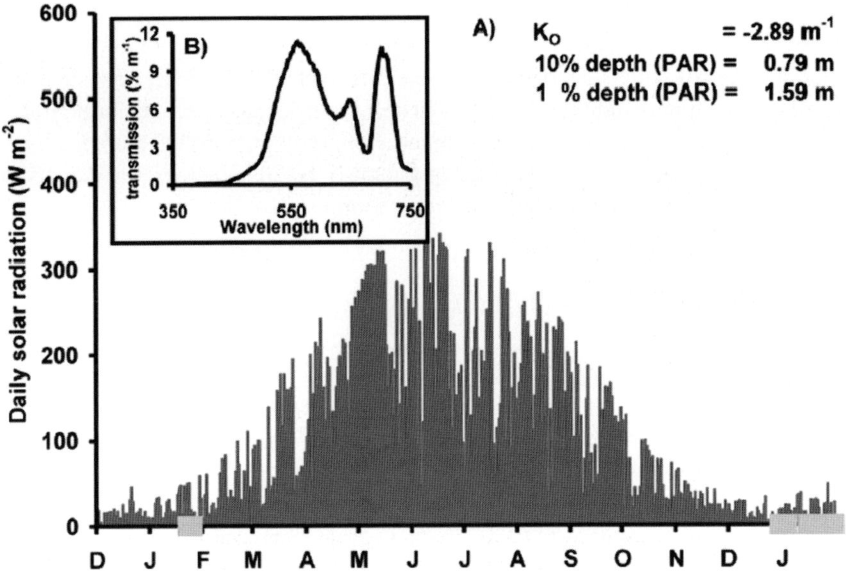

Fig. 3.3 Solar radiation (W m^{-2} day^{-1}) in Zingst near to the Zingster Strom from December 1999 to January 2001 (modified from Schumann et al. 2005). *Shaded rectangles* Ice cover. *Inset* Transmission (%) of the water body in Born, Bodstedter Bodden, 20 December 1999 (solar radiation 1.5 W m^{-2}). *PAR* Photosynthetically active radiation

12,000 W m^{-2}) and is lowest in December (monthly mean ca. <800 W m^{-2}). Figure 3.3 gives an overview of the conditions at the Born site at the Bodstedter Bodden from December 1999 to January 2001 (Schumann et al. 2005). The transmission of the water body is mostly 0% for UV light, usually lowest at 2.6% near the maximum of the chlorophyll absorption at 676 nm, and highest at 11.3% for green light. The 1% light penetration depth was 0.79 m and the K_0 value 2.89.

The water temperature follows a sine curve with mean values of +22°C in summer and +1°C in winter. The annual mean temperatures are around +11°C. The number of ice days fluctuates from year to year and may reach 30–60 days (Schmelzer 1994). Mainly shallow and low-salt sectors of the boddens are affected by longer ice coverage.

The annual average area-related nutrient inputs (g m^{-2} a^{-1}) of the DZB, according to Schlungbaum et al. (2001), are:

- phosphorus 0.068 (1983/1987); 0.056 (1993/1997)
- nitrogen 2.509 (1983/1987); 1.886 (1993/1997)

The proportions of diffuse impact (g m^{-2} a^{-1}) for the above-mentioned periods of time are:

- phosphorus 0.045 (1983/1987); 0.045 (1993/1997)
- nitrogen 2.452 (1983/1987); 1.848 (1993/1997)

The concentration of o-PO_4 in free water is very low (< 0.5–1.0 µmol l^{-1}), and subject to scarce fluctuation. The total phosphorus concentration in the water body is 1.2–4.1 µmol l^{-1}. In contrast, the NO_3 and NH_4 concentrations vary considerably over the year (Fig. 3.4a/b), both reach a maximum in winter. NO_3 declines already in spring and approaches zero in summer. NH_4 may still occur sporadically in spring and summer. The annual mean of total-N is 205 µmol l^{-1} for the period 1996–2004 (H. Baudler, personal communication).

3.2.2 *Planktonic Communities*

3.2.2.1 Structure, Dynamics and Productivity

The total number of phytoplankton species verified for DZB to date amounts to 250 species. This high number comes from the freshwater influence, which hinders the

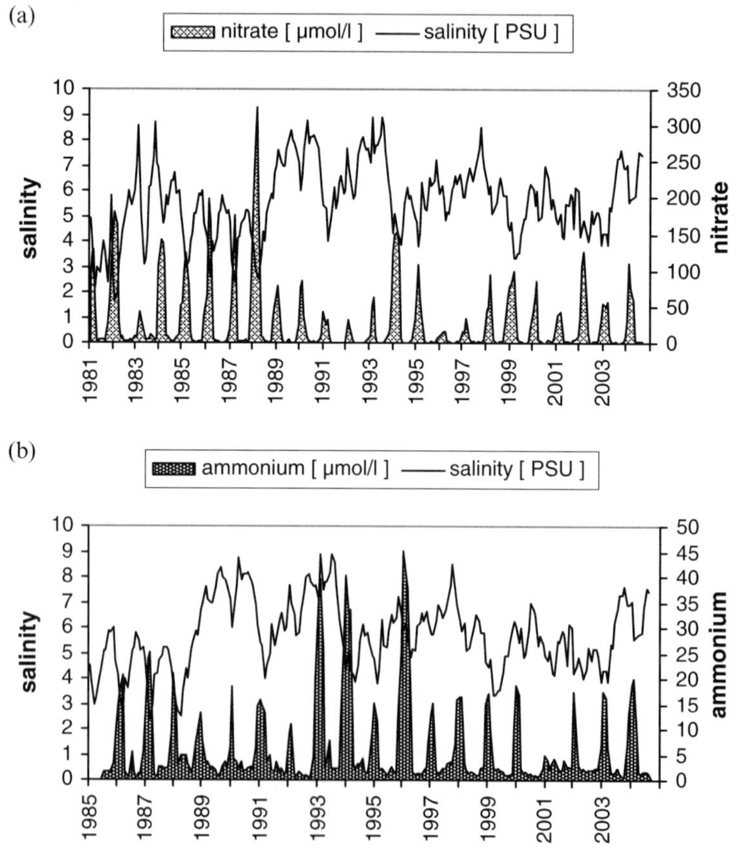

Fig. 3.4a,b Variability of nutrients (µmol l^{-1})and salinity (psu) in the Zingster Strom (H. Baudler and G. Schlungbaum, personal communication). **a** Nitrate: Monthly mean values 1981–2004. **b** Ammonium. Monthly mean values 1985–2004

development of a species minimum, which is typical for the Baltic Proper in the inner coastal waters (see Chap. 13 by Pitkänen, this volume). Species composition has changed markedly during the last few decades. According to Kell et al. (1975) *Gomphosphaeria pusilla, Oscillatoria limnetica, Scenedesmus quadricauda, Lyngbya contorta, Monoraphidium contortum, Gomphosphaeria lacustris, Microcystis aeruginosa, Lyngbya limnetica, Stephanodiscus hantzschii, Crucigenia quadrata* var. *secta, Diatoma elongatum, Microcystis wesenbergii, Chroococcus limneticus* and *Anabaena spiroides* were the most important species in the 1970s with respect to primary productivity (see Chap. 4 by Schiewer, this volume). As shown in Fig. 3.5, the annual periodicity of diatoms, green algae, Cyanobacteria and flagellates was characteristic up to the 1980s (Börner 1984). Cyanobacteria revealed summer peaks, while green algae were dominant during autumn and springtime. A higher occurrence of flagellates was observed during winter. In spring and autumn, the contribution of nitrogen was limiting.

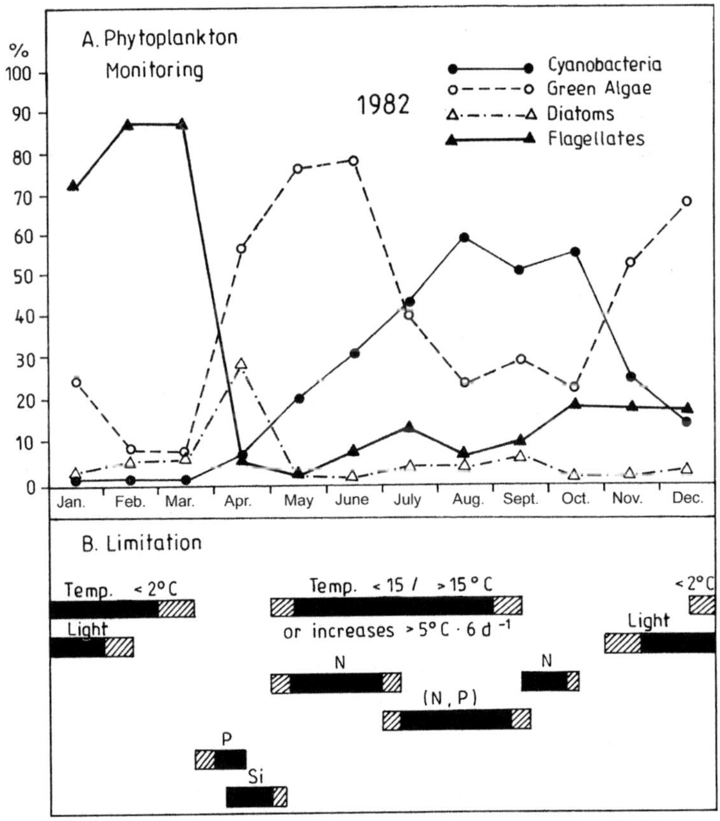

Fig. 3.5A,B Annual periodicity of phytoplankton in the Zingster Strom. **A** Relative frequency of phytoplankton groups in percent, calculated on the biomass level for 1982 (Börner 1984). **B** Generalised scheme of the limiting physico-chemical factors. Temperature >15°C promotes development of filamentous Cyanobacteria. A temperature increase of >5°C in 6 days induces a breakdown of the existing phytoplankton population. *N* Nitrogen-, *P* phosphorus-, (*N, P*) alternative N- or P-, *Si* silicate-limitation

Interactions between N- and P-limitations occurred in summer. The grazing pressure of rotifers and ciliates was low. A high grazing pressure of the copepod *Eurytemora affinis* was characteristic from May until the beginning of June (Heerkloss and Schnese 1994). During the winter months, firstly light and later temperature became limiting factors. In the second half of the 1980s, Wasmund (1994) designated *Tetrastrum triangulare*, *Scenedesmus* spp., *Monoraphidium contortum*, *Gomphosphaeria pusilla*, *Oocystis* spp., *Lyngbya contorta*, *Oscillatoria limnetica*, *Microcystis reinboldii* and *Stephanodiscus hantzschii* as the production-determining species. Thanks to higher resolving microscopes and UV-fluorescence, Schumann (1993) was able to prove that unicellular chroococcale Cyanobacteria and, mainly, *Aphanothece clathrata* must also have been essential parts of plankton at that time (see Chap. 6 by Schiewer, this volume).

Owing to progressive eutrophication (see below) and the associated water turbidity, light limitation has become a decisive regulating factor for much of the annual vegetation period since the end of the 1980s. As a result, the original annual periodicity has disappeared. At the end of the 1980s and beginning of the 1990s, the restructuring of the phytoplankton in the DZB was completed (Fig. 3.6). The permanent presence of high abundances of Cyanobacteria (slightly dominant in summer) and of green algae (slightly dominant in spring and autumn) is remarkable. According to Schubert (1996), the reasons for that are the interaction of short-time light inhibition (!) and recovery time. The wave movement of the water causes a focussing of light through a lens effect. Therefore, the light intensity can be up to eight times as high as it is at the water surface. Such radiation intensities inhibit the photosynthesis of the phytoplankton. A quick recovery is possible only via the temporary strong decline in light intensity that takes place through Langmuir spirals

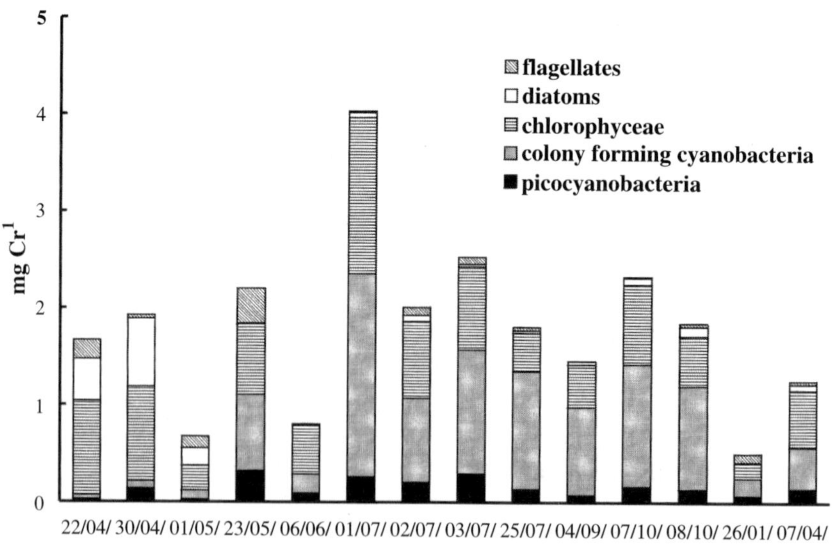

Fig. 3.6 Phytoplankton biomass (mg C l^{-1}) and taxonomical composition in the Kirr Bucht on distinct sampling dates between 22 April 1996 and 7 April 1997 (Schiewer et al. 2002)

in the water. Langmuir spirals come into existence at wind speeds of four or more in the DZB, and their turnover time has been calculated experimentally to be 10–20 min. The minimal light conditions thus caused are sufficient for the Cyanobacteria to recover from the light stress entirely. Green algae are, however, not capable of this as their adaptation reactions take effect only under a light stress of >90 min. This advantage of Cyanobacteria is sufficient to ensure successful competition with green algae.

The high proportion of nano- (2–20 µm) and pico-phytoplankton (Schumann 1993) is characteristic of the DZB. Pico-phytoplankton (0.2–2 µm) may reach a mean biomass proportion of 10% (in May and November up to 15%). Nano-phytoplankton representatives include species of green algae, such as *Tetrastrum*, and of Cyanobacteria, such as *Chroococcus*. The most important primary producers among the green algae are the genera *Tetrastrum*, *Scenedesmus*, *Monoraphidium*, *Crucigenia* and *Oocystis*. *Scenedesmus*-coenobia are often decomposed to single cells in waters rich in nutrients.

The pico-phytoplankton consists entirely of Cyanobacteria, with *Aphanothece clathrata* predominating. In addition, unicellular chroococcale Cyanobacteria are present but, like *Aphanothece*, they are capable of organising themselves into colonies [see Chaps. 6 (Schiewer) and 8 (Chubarenko and Margoński), this volume]. Recently, the prochlorophyte *Prochlorothrix hollandica* has been found (Geiß et al. 2003). Since the pico-phytoplankton is able to compete successfully with bacteria for inorganic nutrients, its nutrient limitation is excluded to a great extent. Originally, diatoms (Bacillariophyceae) were more important in the DZB than green algae and Cyanobacteria. Nowadays, their occurrence is only slightly increased in spring. *Stephanodiscus hantzschii* are present throughout the year. *Cyclotella* and *Skeletonema* species are also found, but they reach only 10% of the total phytoplankton biomass at most.

In summer, filamentous Cyanobacteria with heterocysts such as *Anabaena* and *Aphazinomenon* appear, whose N_2-fixation is active only under N-limitation.

According to Hammer (2003), phytoflagellates such as *Cryptomonas erosa* or *Rhodomonas* spp. may occur in winter time in large numbers (Fig. 3.7). When the water is covered with ice, they accumulate directly beneath the ice surface because of their mobility. In addition, they often have the capability of mixotrophy (Hammer 2003). Therefore, a positive net production can still be achieved in the upper 35 cm of the water column even in winter.

The shallowness of the DZB and its steady mixing winds cause algae that are normally found on the sediment or on plants as periphyton to exist in the plankton. The mean Chl *a*-values [see Chaps. 6 (Schiewer), 8 (Chubarenko and Margoński), 9 (Gasiūnaitė; et al.), and 12 (Telesh et al.), this volume], ascertained by Schumann (1993) for the years 1991 and 1992, were 131 µg Chl *a* l^{-1} (16–283) for the Saaler Bodden (station Dierhagen), and 71 µg Chl *a* l^{-1} (17–183) for the Zingster Strom.

An overview of the productivity of the phytoplankton in the DZB (see Fig. 3.2) reveals that the measured primary production rates are unusually high. The distinct gradient from east to west is striking; this is also reflected in the nitrogen gradient observed in the 1970s. Maximum primary production rates of >600 g C m^{-2} a^{-1} were

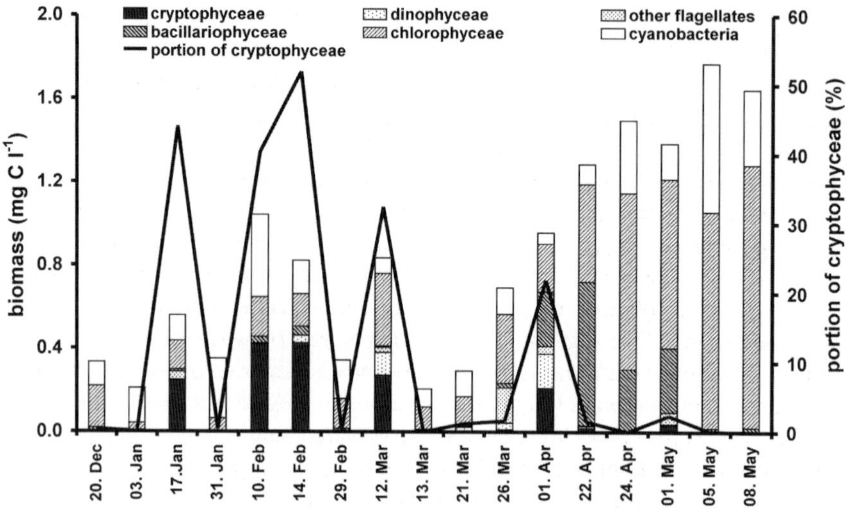

Fig. 3.7 Phytoplankton composition with high proportions of Cryptophyceae during the longest ice cover of the past decade in the Zingster Strom (Hammer 2003). 20 December 1995: before ice; 17 January and 14 February 1996: under ice; 1 April 1996: free of ice

reached in the Saaler Bodden. Heading to the mouth in an easterly direction, production rates fall to 100–135 g C m^{-2} a^{-1}, which corresponds to Baltic Sea conditions.

The mean abundance of bacterioplankton is 24 × 10^6 cells ml^{-1} (18 µg C ml^{-1}) at present, which is relatively high. Comparative investigations with cellular fluorescent markers for membrane integrity and metabolic activity revealed that 22–81% of all counted cells have impermeable membranes, and thus are morphologically intact (Schumann et al. 2003a). Furthermore, up to 48% showed a normal respiration activity. The proportion of cells with cellular esterase activity in the DZB was 12%, which is clearly higher than in the Baltic (5%) and in freshwater (9%), correlating positively to the concentration of dissolved organic carbon (DOC), particulate organic carbon (POC) and the Chl *a* content in the waters investigated. The development of the bacterial biomass over the year can be described by an approximate cosine function for the periods 1980/1982, 1988 and 1991/1992 (Jost and Klinkenberg 1994). The number of bacteria is in the range 10–40 × 10^6 cells ml^{-1}, with maxima occurring in summer.

Heterotrophic nanoflagellates (HNF) mainly control the bacterioplankton (Schumann and Schiewer 1994; Schiewer 1994a). HNF mean abundance is between 3,100 and 3,600 individuals (ind) ml^{-1} (17–40 µg C l^{-1}). The annual cycle of these flagellates reaches its peak in spring, with a plateau later in the year (Schiewer and Wünsch 1996). This is due to the grazing processes of the ciliates and of the mesozooplankton.

The pelagic–benthic protozoans belong to Prostomatida, Haptorida, Pleurostomatida, Scuticociliatida, Peritrichida, Oligotrichida and Hypotrichida (Schiewer 1994a). According to Arndt (1991), clearly higher biomass (< 56–563 µg C l^{-1}) is found in the coastal waters than in the open Baltic. Prena (1990) determined a mean biomass value of 53 µg C l^{-1} for the pelagic ciliates in the Zingster Strom for the year 1988. The peak value was 218 µg C l^{-1}. Subsequent investigations within the scope of the ÖKOBOD-project (Garstecki et al. 1999) verified that heterotrophic ciliates are the most important grazing component within the pelagic microbial food web (MFW) of the DZB (see Chap. 12 by Telesh et al., this volume). The large variety of heterotrophic ciliates (>100 taxa) was striking, with a considerable number of them being surface associated. The total biomass of protozoa was between 12 and 360 µg C l^{-1} in the open water of the Zingster Strom. Close trophic interactions within the ciliate community could be proved by means of complex mesocosm experiments in the pelagic zone (Schiewer 1994a, 1994b). The seasonal distribution of ciliates is controlled mainly by temperature. The increased grazing pressure of *Eurytemora affinis* in April and May has a negative effect on their occurrence. The weekly fluctuations of pelagic ciliates in the control enclosure of a mesocosm experiment in the Kirr Bucht are shown in Fig. 3.8.

The behaviour of flagellates and ciliates in the pelagic changed with increasing eutrophication and the onset of light limitation. An increased bond with the existing particles could be proved, with the proportion of particle-bound heterotrophic flagellates in free water being more than 90% and that of ciliates around 60% (Garstecki et al. 1999).

According to (Arndt 1985), the mesozooplankton includes 102 taxa (66 rotifers, 20 phyllopods and 14 copepods) and is joined by meroplanktonic larvae of various groups of the zoobenthos (polychaetes, mussels, snails and balanids). Since 1986, larvae of the invasive species *Marenzelleria neglecta* (formerly *M. viridis*) have occurred increasingly in the pelagic zone, having their major development in autumn, but their occurrence declined sharply at the end of the 1990s. A pronounced peak marks the annual cycles of the mesozooplankton in spring for *Eurytemora affinis*, while the development of rotifers begins only in early summer.

Under the aspect of primary productivity, only copepods *Eurytemora affinis* (Poppe) and *Acartia tonsa* Dana, rotifers *Keratella cochlearia* (O. F. Müller), especially *K. cochlearia* var. *tecta*, *K. quadrata* (O.F. Müller), *Filinia longiseta* (Ehrb.), *Brachionus calyciflorus* Palla, *B. quadridentatus* (Hermann), *Synchaeta cecilia* Rousselet, *S. vorax* Rousselet, *Trichocerca* spp. and *Asplanchna* ssp. have a determining influence. The horizontal distribution of the mesozooplankton and the biomass relations of phytoplankton to mesozooplankton in the DZB can be seen in Fig. 3.9. Figure 3.10 summarises the annual cycle of the most important heterotrophic plankton species.

Along with the genuine brackish-water species *Neomysis integer*, the nekton consists mainly of fish. From his own catches, Winkler (2001) has detected 47 autochthonous fish species in the boddens during the previous decades. Proof for the occurrence of six further species exist in earlier sources of indication. The

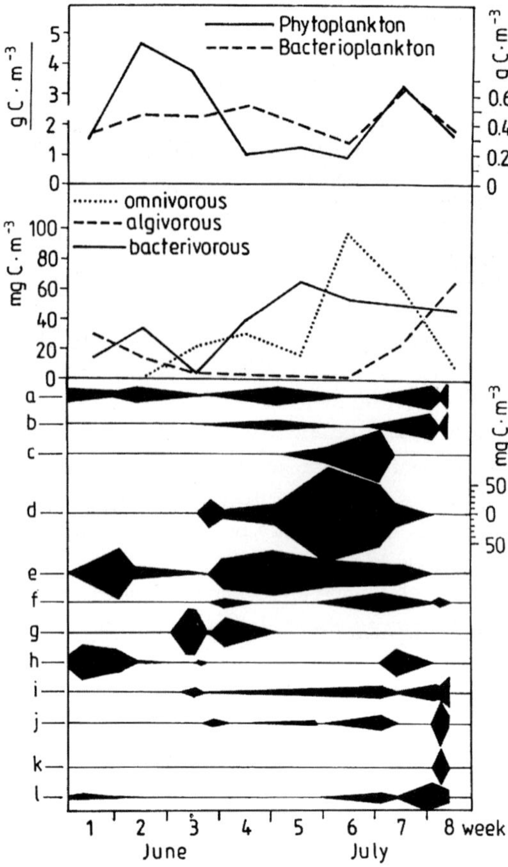

Fig. 3.8 Weekly fluctuation in biotic components in the pelagic zone of the control enclosure, shallow-water experiment 1986 (FLAK; Schiewer 1990). *Upper panel* Biomasses of phyto- and bacterioplankton, *middle panel* biomasses of planktonic ciliates separated in functional groups after Pratt and Cairns (1985), *lower panel* biomasses of single ciliate species. *a Mesodinium* sp. A; *b Mesodinium* sp. B; *c Didinium nasutum*; *d Monodinium balbianii*; *e* cf. *Enchelys*; *f Halteria grandinella*; *g Strombilidium velox*; *h Strombidium* sp. A; *i Strombidium* sp. B; *j Strombidium* sp. C; *k Condylostoma vorticella*; *l Euplotes* cf. *balteatus* (Arndt et al. 1990a)

potential spectrum comprises 25 freshwater fish species, 8 diadromous migratory fish species, as well as 21 marine fish species and 6 allochthonous species originating from implantation measures. The autochthonous species may occur permanently or may immigrate temporarily (seasonably or sporadically) to the boddens, both from fresh water and from the Baltic. Among freshwater species, the cyprinids with 12 species (such as roach *Rutilus rutilus*, bream *Abramis brama*) form the major share, but six of them occur only very rarely. The eight diadromous migratory species (among others eel *Anguilla anguilla*, *Coregonus laveratus balticus*, and river lamprey *Lampetra fluviatilis*) play only a minor role in the DZB. Among the 21 marine species, herring *Clupea harengus*, *Belone belone*, *Pomatoschistus*

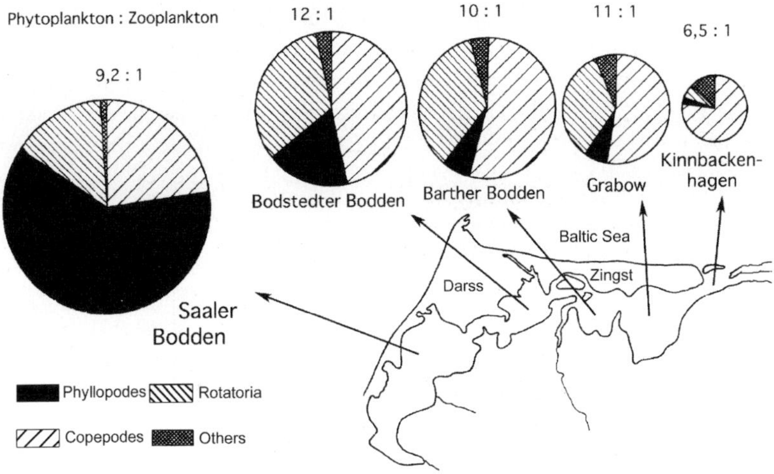

Fig. 3.9 Horizontal distribution of zooplankton and the relations of phytoplankton/zooplankton in the DZB (Heerkloß and Schnese 1994). The *circles* from left to right correspond to 9.58, 4.53, 3.90, 1.92 and 0.89 mg fresh weight (fw) l^{-1}

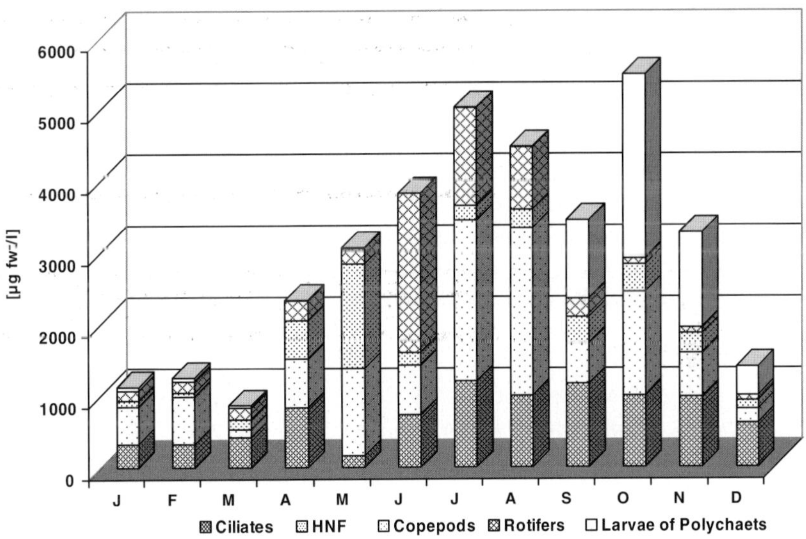

Fig. 3.10 Annual periodicity of all zooplankton groups (Heerkloss 2001). Mean values 1985/1993 for mesozooplankton, 1985/1987 for ciliates and 1991/1992 for heterotrophic nano-flagellates (*HNF*)

minutus, *P. microps*, *Zoarces viviparus* (eelpout), *Syngnathus typhle*, *Nerophis ophidion* and *Plathichthys flesus* should be mentioned.

The deterioration of the submersed macrophytes stands to induce a decline in the *Esox lucius* population, but the occurrence of *Sander lucioperca* was increased by the same process.

The total annual commercial fishing from the DZB amounts to 305.2 tons on average (Landesamt für Fischerei Mecklenburg-Vorpommern), consisting equally of freshwater and marine species. *Sander lucioperca* (pikeperch), *Abramis brama* and *Rutilus rutilus* predominate in the freshwater species. With the marine species, *Clupea harengus* accounts for 93.9% of the fish caught. *Plathichthys flesus*, *Belone belone*, *Gadus morhua* (cod) and *Psetta maxima* make up the rest.

3.2.2.2 Nutrient Cycles

The particulate phosphorus existing in the water body is available at 60–80%, incorporated mainly into the living biomass in the polythrophic DZB. Therefore, a high phosphorus release rate is observed, which is correlated with the high turnover performances of the food web. The first ideas regarding the structure and the function of the food web in the DZB using a mathematical model were developed by Vietinghoff (1982), and were based on the grazing and the detritivorous food web using records from the 1970s.

Today, the grazing food web is of only minor importance because of restructurings in the DZB caused by the strong eutrophication (Schiewer and Jost 1991; Schiewer 1990, 1994b, 1998). Investigations by Heerkloss et al. (1999) within the scope of the ÖKOBOD-project confirmed that only 0.2–6% of the gross primary production is consumed by the mesozooplankton. Egestion was 0.05–0.9% of the gross primary production. The decline in importance of the larger herbivores in the pelagic zone is caused by a diminution of *Eurytemora affinis* and a reduction in the size of rotifers. Furthermore, it should be taken into consideration that, under a wind velocity of 5 m s^{-1}, 25–30% of the consumed particulate material consists of the detritus of the resuspended sediment layer. The route from phytoplankton to fish via herbivorous and carnivorous zooplankton amounts only to 5–10% of the total turnover of substances. Most of the turnover (90–95%) goes through the MFW (Fig. 3.11), which comprises heterotrophic bacteria, heterotrophic flagellates and ciliates. Ciliates have internal carbon cycles at their disposal because of their species- and nutritional-diversity (Arndt et al. 1990b).

Table 3.2 provides an overview of the distribution of biomass, and the percentage of total microbial biomass in the POC and in the organic particle volume, at various sites of the DZB and of the outer coast at Ahrenshoop (AH) for the period April–October 1996.

3.2.3 *Benthic Communities*

3.2.3.1 Structure and Productivity

As a shallow water body, the DZB is predestined for submersed algae and macrophytes (Lindner 1972). Table 3.3 gives an overview of the potential occurrence of macroalgae and spermatophytes. This situation was present up to the 1960s.

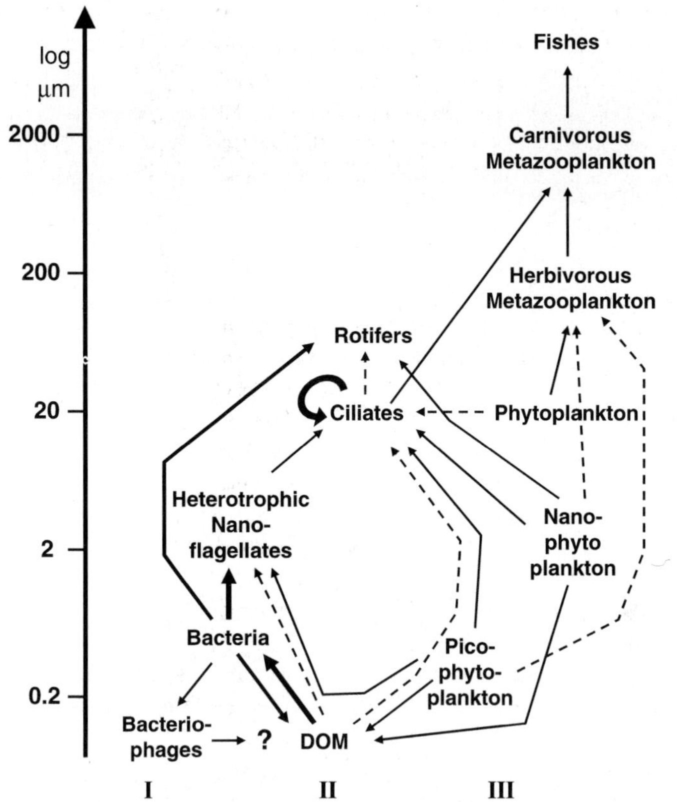

Fig. 3.11 Microbial food web (MFW) in the DZB – late spring/early summer situation. *I* Side chain through bacteriophages (proven abundances of virus-sized particles: $10^8\,ml^{-1}$). *II* MFW. Main pathway for carbon turnover mainly via nano- and picoplankton. Internal loop in the ciliate community (*arrow*) can involve up to three additional trophic levels. *III* "Classical" pelagic food web. The typical components are net plankton and fishes. Minor role only. *DOM* Dissolved organic matter

However, stand descriptions by Lindner from 1970/1971 verify already a clear decline in the dispersal of submersed macrophytes in the Saaler Bodden (Fig. 3.12). In contrast, stands in the Barther Bodden with extended charophyte meadows were still intact to a great extent. They have since been substituted by *Ruppia spiralis*, *Potamogeton pectinatus* and *Myriophyllum* stands in some areas. However, the occurrence of *Potamogeton* and *Myriophyllum* indicated the beginnings of eutrophication also in the eastern part of the Bodden chain already at that time. A sudden collapse of the submersed macrophytes occurred there in 1981 (see below). A "genuine" microphytobenthos in the DZB is lacking.

Benthic production is dominated by sedimentary planktonic primary producers. These form part of the complex FSL, which may have a thickness of some millimetres (see below). A comparison of the performance of the various primary producers in

Table 3.2 Average carbon converted biomasses (mg C l^{-1}), their contribution (%) to total microbial biomass and POC and biovolume percentage of organic particles volumes calculated from DTAF [5-(4,6-dichlorotriazinyl)aminofluorescein]-stainable objects (all averages are medians; range in parentheses) at the Baltic shore (*AH* Ahrenshoop) and in the five bodden stations (*BB* Barther Bodden, *ZS* Zingster Strom, *KB* Kirr Bucht, *WK* Wiek/Bodstedter Bodden, *DH* Dierhagen/Saaler Bodden) observed from April to October 1996 (Schumann et al. 2001)

	AH	BB	ZS	KB	WK	DH
Carbon biomass						
Phytoplankton	0.017 (0.002–0.053)	0.9 (0.8–2.1)	1.5 (0.7–2.0)	1.7 (0.7–2.0)	1.7 (0.9–2.6)	2.6 (1.5–5.7)
Bacteria	0.024 (0.013–0.062)	0.32 (0.20–0.50)	0.56 (0.41–0.62)	0.57 (0.41–0.62)	0.45 (0.23–0.80)	0.59 (0.28–1.23)
Protozoa	0.005 (0–0.080)	0.047 (0–0.153)	0.100 (0.005–0.197)	0.096 (0.031–0.212)	0.083 (0.011–0.247)	0.115 (0.004–0.542)
Contribution to total microbial biomass						
Phytoplankton	29 (6–78)	72 (61–83)	68 (47–80)	71(48–80)	75 (64–79)	78 (72–86)
Bacteria	44 (21–76)	24 (15–30)	26 (17–40)	24 (19–44)	23 (16–30)	19(5–26)
Protozoa	14 (0–71)	4 (0–8)	4 (8–12)	5 (0–8)	3 (8–14)	3 (1–13)
Contribution to POC						
Phytoplankton	3 (1–26)	25 (13–33)	19(11–36)	13 (6–23)	14 (8–40)	18 (11–49)
Total microbial biomass	28 (3–48)	34 (22–49)	26(21–54)	19 (10–32)	18 (12–53)	24 (15–63)
Contribution to organic particle volume						
Phytoplankton	2 (0–7)	18 (4–35)	16 (6–26)	23 (5–39)	18 (2–29)	21 (11–39)
Total biovolume	7 (1–31)	23 (5–46)	22 (11–36)	29 (7–46)	24 (2–35)	26(14–46)

Table 3.3 Submersed macroalgae and higher plants in the DZB in 1970/1971. Arranged according decreasing salinity amplitudes (modified from Lindner 1978)

Euryhaline-limnetic species	Growth in > 2 psu: *Chara aspera*, *Ch. tomentosa*, *Myriophyllum spicatum*, *Zannichellia palustris*, *Pedicellata* sp., *Najas marina*, *Ranunculus baudotii*
	Growth up to 1– 2 psu: *Potamogeton friesii*, *P. crispus*, *P. perfoliatus*, *P. lucens*, *Hydrocharis morsus-ranae*, *Stratiotes aloides*, *Elodea canadensis*, *Lemna trisulca*, *L. minor*, *L. gibba*, *Spirodela polyrhiza*, *Nymphea alba*, *Nuphar lutea*, *Ceratophyllum demersum*, *Fontinalis antipyretica*, *Nitellopsis abtusa*, *Chara globularis* var. *globularis f. globularis*, *Ch. hispida* var. *major f. intermedia*, *Ch. vulgaris* var. *vulgaris f. contraria*, *Hydrodictyon reticulatum*
Euryhaline-marine species	*Zostera marina*, *Ruppia spiralis*, *Chorda filum*, *Cladophora sericea*, *Ectocarpus confervoides*, *Polysiphonia nigrescens*, *P. violacea*, *Ceramium rubrum*, *C. diaphanum*, *Fucus vesiculosus*, *Chaetomorpha linum*
Brackwater species	*Chara canescens*, *Ch. hispida* var. *baltica f. baltica*, *Ch. hispida* var. *baltica f. fastigiata*, *Tolypella nidifica*
Euryhaline species	*Enteromorpha intestinalis*, *Cladophora glomerata*, *Potamogeton pectinatus*

the DZB (Pankow and Wasmund 1994) demonstrated the extraordinary role of the phytoplankton (4,570 t C a^{-1} = 95% of net production) versus the epipsammic microphytobenthos (160 t a^{-1} = 3.3%) and the macrophytobenthos and epiphytic microphytobenthos (82 t C a^{-1} = 1.7%).

The biomass of the sediment bacteria as reported by Yap (1987) amounts to 500–1,000 μg C cm^{-2}, with secondary production fluctuating between 70 and 100 μg C cm^{-2} a^{-1}. The bacterial biomass in the sediment was recorded again in 1996/1998 as part of the ÖKOBOD-project in the Kirr Bucht (Rieling 1999) and was found to be between 651 and 1,012 μg C cm^{-2}, measured up to a depth of 10 cm. The proportion of total carbon was between 0.7 and 1.1%. A count of bacterial cells in near-surface sediments (0–0.5 cm) yielded 5.9–10.0 × 10^9 cm^{-3}.

Several authors (Scharf 1979; Arndt et al. 1990a; Arndt 1991; Prena 1990) described the occurrence of a wide spectrum of benthic (Gymnostomatida, Hymenostomatida, Heterotrichida and Hyotrichida) or pelagic–benthic (Prostomatida, Haptorida, Pleurostomatida, Scuticocilliatida, Peritrichida, Oligitrichida and Hypotrichida) protozoa. Their mean biomass in the sediment was 12–105 mg C l^{-1}. A new discovery (Garstecki et al 1999) was that Rhizopoda, with a biomass of about 20%, form an essential part of the benthic protist fauna (sediment and FSL). In contrast, Actinopoda play only a minor part. As in the pelagic, the results in the sediment indicate that there are several trophic levels within the protist community that follow on from each other. Along with mainly bacterivorous (e.g. bodonids, scuticociliates) and herbivorous (e.g. big euglenids, hypotrich ciliates) forms, both omnivorous (e.g. chrysomonads, hypotrich ciliates) and carnivorous (e.g. *Metopus*, Karyrelictidea) organisms have relatively high biomass proportions.

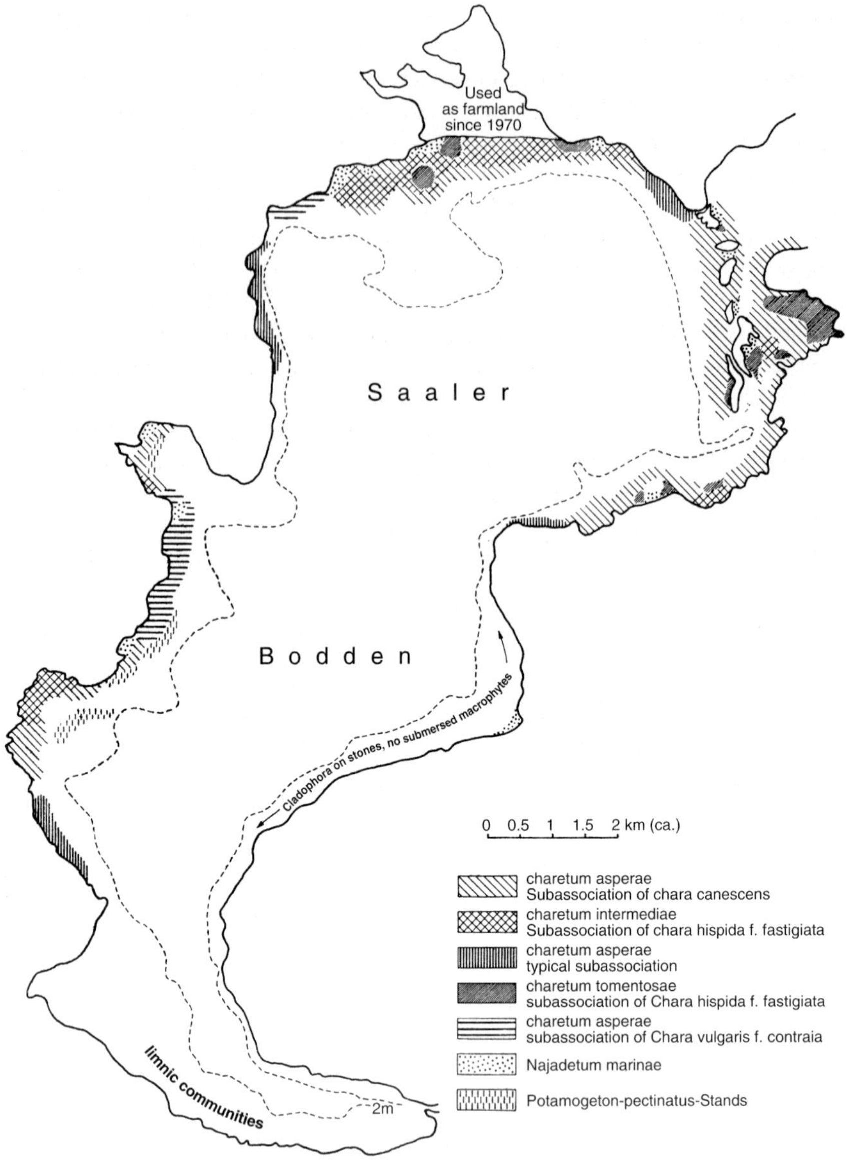

Fig. 3.12 Covering of the benthal region by submersed macrophyte associations in the Saaler Bodden 1970/1971 (Lindner 1978). The region most affected by eutrophication is the southern Saaler Bodden (Ribnitzer See). *Dotted line* 2 m depth border

The main groups of meiofauna (Arlt 1984) are Nematoda (28 species), Oligochaeta (23), Ostracoda (20), Harpacticoida (20) and Turbellaria. Nematoda are predominant with a relative abundance of 70–90% followed by Ostracoda, Harpacticoida, and Turbellaria with clearly lower numbers of individuals. Their occurrence is concentrated on the upper sediment area of 1–2 cm and they are frequently present in the sediment layer. Irregular annual fluctuations are typical for individual numbers. There is usually a minimum in winter (December–February).

Little is known about the function of the meiofauna, especially quantitative aspects. Although its biomass in the sediment is higher than that of the protozoa, the latter may exhibit higher turnover performances. ÖKOBOD-investigations by Arlt and Georgi (1999) proved the grazing pressure of meiofauna on protozoans. Laboratory experiments with *Cyprideis torosa* verified their role in the development of microbial mat structures. Their presence changed the capability of resuspension and permeability of the upper sediment layers within a short time, which has had an effect on the concentration of photosynthetic pigments and on the vertical oxygen distribution in the mats.

Decades of investigation into the macrofauna (1969–1977 for all regions; 1986–1990 for the Saaler Bodden and the Ribnitzer See) were summarised (Fig. 3.13)

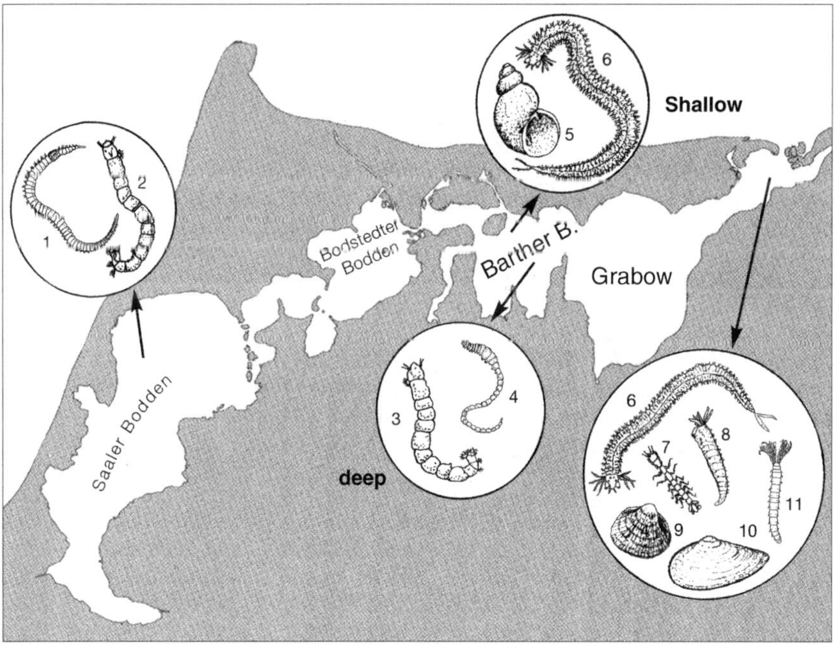

Fig. 3.13 Distribution of the main macrozoobenthos species in different parts of the DZB (Arndt 2001). *1 Potamothrix hammoniensis*, *2 Chironomus plumosus*, *3 Ch. halophilus*, *4 Monopylephorus rubroniveus*, *5 Hydrobia ventrosa*, *6 Nereis diversicolor*, *7 Cyathura carinata*, *8 Alkmaria romijni*, *9 Cerastoderma lamarcki*, *10 Arenomya arenaria*, *11 Fabricia sabella* (*flach* shallow, *tief* deeper parts of the Barther Bodden)

by Arndt (2001). The *Cyprideis-Manayunkia* community at the outflow of the boddens to the Baltic (Bock), the *Hydrobia-Nereis* community in the shallow area of the Barther Bodden (<2 m) and the *Monopylephorus-Chironomus halophilus* community in the deep area of the Barther Bodden as well as the *Ch. plumosus-Potamathrix* community in the Saaler Bodden can be differentiated. These are usually groupings with a lack of species, caused mainly by the dominance of soft soil and by the unfavourable salinity range of 3–8 psu (horohalinikum). A transitional zone to freshwater is proven in the area of the Bodstedter Bodden. As expected, genuine brackish-water species seldom occur (*Alkmaria romijni, Fabricia sabella, Streblospio shrubsoli*).

Chironomidae and *Nemertini* predominate within these groups. The biomass values (Arndt 1985, 2001) amount to 175 g fresh weight (fw) m^{-2} at the exit to the Baltic, 40 g fw m^{-2} in the shallow areas of the Barther Bodden, and 25 g fw m^{-2} in the deeper areas as well as 21 g fw m^{-2} in the Saaler Bodden and 54 g fw m^{-2} in the Ribnitzer See (most southerly, and the part of the Saaler Bodden with the lowest salinity, respectively).

The development of *Marenzelleria neglecta* (formerly *M. viridis*; Polychaeta), which was detected for the first time in the boddens in 1985, is remarkable. Clearly, it prefers eutrophic to hypertrophic brackish water with sandy or muddy–sandy substrates, respectively. In the Saaler Bodden, $21 \times 10^6 m^{-3}$ larvae were observed in the plankton in autumn 1992, which matured to a population density of 60,000 ind. m^{-2} (600 g fw m^{-2}) in winter 1992/1993 (Bochert et al. 1994). While the abundances in the oligohaline Saaler Bodden fell rapidly, the population in the mesohaline Barther Bodden remained at 1,000–10,000 ind. m^{-2} (up to 400 g fw m^{-2}). All the other fauna elements were barely influenced by this mass development (Zettler 1996). Apparently, the food basis in the polytrophic shallow waters is sufficient for invaders like these. It is not yet sure whether *M. neglecta* will manage to settle in the long term.

3.2.3.2 Nutrient Cycles

The anaerobic decomposition of organic material in the sediment by employing sulphate ("sulphate respiration") predominates, at 67% in winter/spring and 73% in summer. The aerobic decomposition via oxygen as an e-acceptor amounts to 22% in winter/spring or 26% in summer. In autumn, the proportions of oxygen and sulphate respiration come to 41% or 55%, respectively. Respiration of nitrate, manganese and Fe contribute only a few percent each to carbon oxidation, with manganese respiration considered to be the most important of the three (Rieling et al. 1999). The highest rates of sulphate respiration are measured in the upper 2–3 cm. Owing to low temperatures and the declining availability of organic substrates, sulphate reduction is lowest in winter and spring, although it then extends into deeper sediment horizons (Babenzien and Voigt 1999). H_2S – a toxic substance for vertebrates and invertebrates – usually develops as a final product.

The key processes of the nitrogen cycle are nitrification, denitrification and nitrogen fixation. Denitrification (annual mean about 40 µmol N $m^{-2} h^{-1}$; 5 g N $m^{-2} a^{-1}$) is dependent on the nitrate supply through nitrification to a great extent because of the low availability of nitrate during the growing season. This process depends on the availability of NH_4 and oxygen, and is also limited by competition for available organic carbon. Therefore, increased denitrification rates could be shown only during spring, parallel with high nitrate concentrations (Dahlke et al. 1999).

The relatively high benthic nitrogen fixation rates of 58 μmol N m^{-2} h^{-1} measured within the scope of the ÖKOBOD-project conducted from 1996 to 1998 in the Kirr Bucht were surprising (Dahlke et al. 1999). These data are comparable to the results of Hübel (1984) for the coastal waters of the Baltic. Thus, nitrogen fixation in the Kirr Bay exceeded denitrification in summer. A nitrogen net-input into the system might have occurred.

The organic material required for benthic respiration amounts to 95–106 g C m^{-2} a^{-1} (Schiewer 1994b). The total benthic respiration performance clearly rises from winter, over spring, to summer and autumn (Meyer-Reil 1999). About 30% of the gross production of carbon being respired at present by the benthic community probably involves mainly the oxygenated sediment layer.

3.2.4 Benthic–Pelagic Coupling and Eutrophication

The morphometry and hydrology of the DZB underlie its high sensitivity to nutrient loadings, especially in the western part of the Bodden chain. The first signs of eutrophication can already be found in investigations of the Saaler Bodden carried out at the beginning of the last century (Gessner 1957). Eutrophication has increased dramatically since the 1950s (Hübel 1973; Börner 1984; Wasmund and Schiewer 1992; Schumann 1993; Schiewer 1998a).

The phosphorus content in the sediment along the salinity gradient never goes beyond 1 mg g dw^{-1}. It is subject to neither horizontal nor seasonal large fluctuations. Only 0.1–0.3 mg P g dw^{-1} has been found in the sediment of the Kirr Bucht. The majority of the phosphorus is bound in carbonates both in the sand and in the mud and is unavailable in the DZB (Berghoff et al. 1999); only 10–50% is even potentially available. The redox process plays no role; rather there is a re-release of adsorptively and organically bound phosphorus. Phosphorus is supplied by resuspension, particularly back to the water body, where it is converted to o-PO_4 by chemical and microbiological processes. The release rates are high because resuspension occurs often in the DZB.

3.2.4.1 Biological Processes and Interactions

More than 30 years of continuous investigation have provided an excellent insight into the course and the background of eutrophication in the eastern DZB (Schiewer 1998a, 1998b). Especially striking was the loss of submersed macrophytes in the Barther Bodden that occurred suddenly in 1981. The coinciding of heavy rain in spring, increased nutrient input due to subsequent drainage, a rise in turbidity, and a decrease in salinity all contributed to the direct triggering of this process. The submersed macrophytes, already affected by eutrophication, collapsed almost simultaneously due to a lack of light supply. A fundamental recovery has not yet occurred. However, well-structured transitions from swamp forest zone via reed bed and floating leaf plants up to underwater lawns consisting of *Potamogeton* species and charophytes can still be found today in the still bays of the eastern parts of the boddens. From the mid-1990s onwards, denser *Potamogeton pectinatus* stands were sporadically found in the Kirr Bucht. Dense stock abundances of *Chara baltica* and *Chara tomentosa* reaching depths of up to 1.10m were observed in 2001 despite the ongoing light limitation, at least in the Kirr Bucht. No explanation for this has yet been proposed.

The lack of increase in phytoplankton biomass, and the fact that its productivity even declined temporarily, was surprising at first:

- 297 g C m^{-2} a^{-1} 1971/1975 Barther Bodden (H. Hübel, unpublished data)
- 205 g C m^{-2} a^{-1} 1976/1978 Barther Bodden (H. Hübel, unpublished data)
- 253 g C m^{-2} a^{-1} 1979/1983 Zingster Strom (Börner 1984)
- 194 g C m^{-2} a^{-1} 1984/1986 Zingster Strom (Wasmund 1987)
- >350 g C m–2 a–1 1991/1992 Zingster Strom (Schumann 1993)

This trend can be explained by the increased sediment mobility following the disappearance of the macrophytes. Thus, turbidity increases and the amount of available light is reduced. The phytoplankton had to adapt to the light limitation beginning at that time (see above). The dominance of the nano- and pico-phytoplankton thus revealed led to a rise in protozooplankton, especially ciliates. The biomasses of copepods were influenced negatively from the beginning of the 1980s, and rotifer biomasses also at the end of the 1980s. The in situ decline of copepod abundances, particularly *Eurytemora*, since 1980/1981 and 1989/1990, are remarkable (Heerkloß and Schnese 1994). Investigations with normalised abundances from 1983 to 2004 showed, at least for the Zingster Strom, an annual periodicity with reduced abundances (M. Feike, personal communication). The main species are *Synchaeta* spp., *Eurytemora affinis*, *Keratella cochlearia f. tecta*, *Filinia longiseta*, *Trichocerca* spp., *Brachionus quadridentatus*, *Acartia tonsa* and *Marencelleria viridis* larvae.

Recent results (Schumann et al. 1999; Schiewer 2001) show that turnover processes in the DZB take place mainly bound to particles. The FSL of 0.5–1.0mm plays a major role here. The FSL is a slightly movable conglomerate consisting of both dead and live material. While the DOC:POC ratio in the Baltic Sea is 6–10:1, the ratio in the inner part of the DZB is only 1.1:1. The absolute POC amounts to <16 mg C l^{-1} and DOC <13 mg C l^{-1}. The C/N-ratio in particles is between 9:1 and 10:1 in winter and spring samples. Bacterial abundance (24×10^6 ml^{-1}) and bacterial activity (18 μg C l^{-1} h^{-1}) are very high. These values are significant higher than the POC values of <0.8 mg C l^{-1}, DOC values of <5.5 mg C l^{-1}, bacterial abundance of $<3 \times 10^6$ ml^{-1} and bacterial production rate of 4.6 μg C l^{-1} h^{-1} of the Baltic shore (Schumann et al. 2003b).

The particles to be found are dominated by nanoparticles in the pelagic (Messal et al. 1999). There is no evidence of any hierarchical process of aggregation. Particles can be differentiated by employing fluorescence staining (Schumann and Rentsch 1998) and by using confocal laser-scanning microscopes. The results of such investigations have been available since 1999 (Schumann et al. 2001, 2003a, 2003b).

Figure 3.14 compares the composition of the POM along the eutrophic gradient starting at the Baltic shore and going on up to the Saaler Bodden (Schumann et al. 2001). It is striking that the detritus, mucous substances and the involvement of phytoplankton change. The abundances of autotrophic and heterotrophic organisms in all the organic particles are between 8 and 27%. The majority consists of mucous substances. Schumann et al. (2001) excluded a greater involvement of allochthonous material in the development of POM according to their results in investigations of aggregates.

Algae of the FSL are very similar to plankton algae with respect to their pigment composition. About 95% of all aggregates in the sediment layer are larger than 30 μm and smaller than 300 μm. They contain a relatively large fraction of mineral inclusions. Nanoparticles are not found in the FSL as free particles (Messal et al. 1999). The frequent resuspension obviously interrupts the further development to larger aggregates as found in the sea (Herndl 1992; Karner and Herndl 1992).

The FSL is very mobile and has a high metabolic activity. Concentrating at the sediment surface in windless times, it can remineralise or absorb directly nutrients that have been released from the sediment, and can immediately convert them effectively into organic biomass. This intensifies the productivity of the ecosystem.

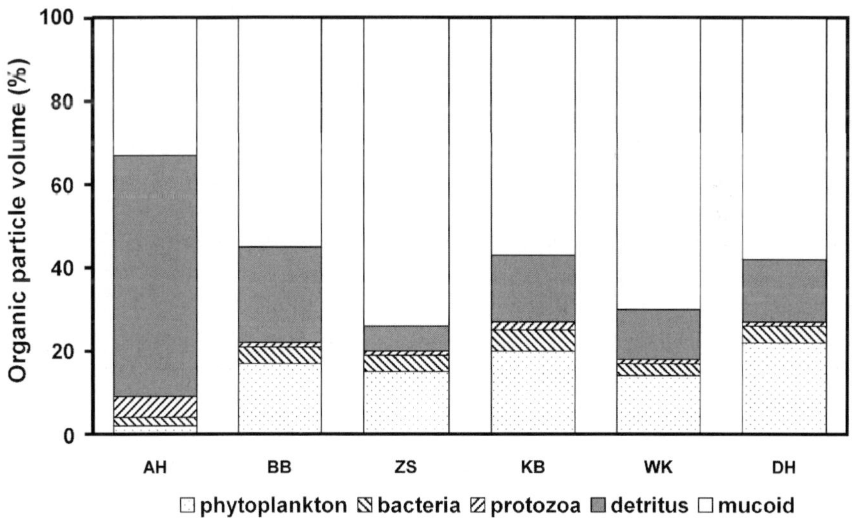

Fig. 3.14 Average composition of total organic particle volume determined by 5-(4,6-dichlorotri azinyl)aminofluorescein (DTAF) staining of bacteria, protozoa, phytoplankton, detritus (PI-biovolumes) and mucoid material (DTAF-PI) at the Baltic Sea coast near Ahrenshoop (*AH*) and at five bodden stations (Schumann et al. 2001). *DH* Dierhagen/Saaler Bodden, *WK* Wiek/ Bodstedter Bodden, *ZS* Zingster Strom, *KB* Kirr Bucht, *BB* Barther Bodden

However, the close coupling with heterotrophic organisms and aggregates at the same time also increases the turnover and decomposition rates. Thus, the entire system comes closer to conditions found in a highly active sewage plant.

A summarising overview of the gradual eutrophication process in the Barther Bodden is given in Fig. 3.15. The consequences of this restructuring of the food webs are diverse:

- The light-limited phytoplankton communities are not really controllable from outside. Long-term laboratory experiments involving microphyto-, bacterio- and microzoo- plankton has proved this impressively (Heerkloß and Klinkenberg 1993). Rather, coincidental and irregular patterns of sequences of events in the communities emerge, that can be interpreted causally only to a certain degree ("deterministic chaos").

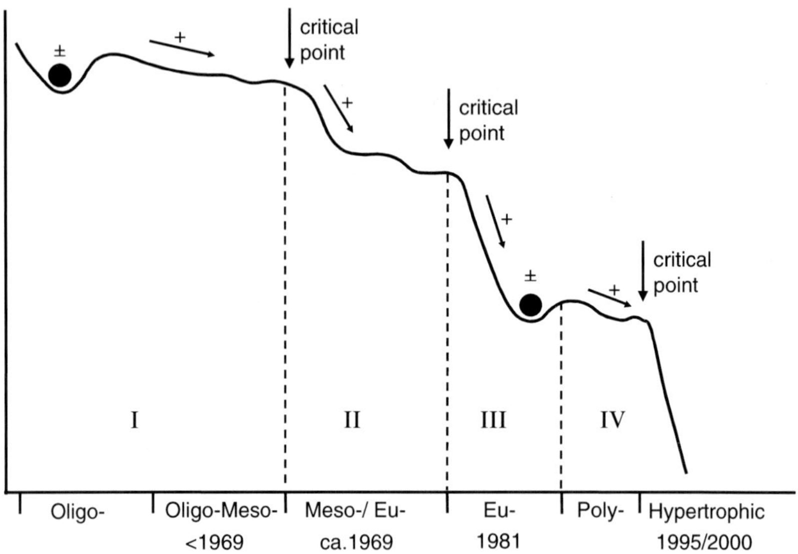

Fig. 3.15 Stepwise eutrophication of the Barther Bodden. *Step I* Oligo-mesotrophic before 1969. Nutrient limitation, low phytoplankton biomass and dominance of diatoms. Dominance of submersed macrophytes (charophyceae) in shallow parts. *Step II* Meso- to eutrophic 1969/1981. Nutrient limitation, mainly nitrogen; higher phytoplankton biomass; dominance of green algae and Cyanobacteria. Dominance of submersed macrophytes (charophyceae and potamogetoneceae) in shallow parts. *Step III* Eu- to polytrophic 1981/1986. Changes from nutrient- to light-limitation. Dominance of nano- and (pico)-phytoplankton (Cyanobacteria and green algae) and of the MFW. Dramatic loss of submersed macrophytes. *Step IV* Polytrophic starting after 1986. Light limitation of phytoplankton. The most important feature is the anthropogenically enhanced MFW concentrated mostly in the active fluffy sediment layer (FSL) (Schumann et al. 2001). Critical point for hypertrophy not reached, prevented by load reduction since 1991. Change from light- to nutrient-limitation expected between 2005 and 2015. First recoveries of submersed macrophytes (*Potamogeton pectinalis* and some charophytes) observed already in 2001 (Schubert et al. 2005)

- The turnover processes in the pelagic are highly accelerated as they are dominated by small organisms. For this reason, organic substances, which are easily decomposed, are respired more quickly, releasing inorganic nutrients simultaneously. These remineralisations are remarkably high, so that >30 mmol N m^{-2} day^{-1} and up to 3 mmol P m^{-2} day^{-1} are released during the vegetation period. Only the temporarily stronger development of copepods in May/June reduces the number of ciliates and flagellates so that a setback of degradation performances occurs during this time. Thus, remineralised nitrogen and phosphorus are almost always available, resulting in a high degree of self-eutrophication of the waters. This is the basis whereby a phytoplankton community with a Chl *a* concentration of about 1 μg l^{-1} is able to stabilise. Secchi depths of only 30 cm up to 50 cm maximum are the result. In a situation like this, only a few additional nutrients are necessary to increase the phytoplankton biomass up to the light limitation.
- Since the food web is microbial, it is capable of reacting very quickly to external and internal influences making it extremely stable. Thus, the DZB ecosystem has reached a stable maximum, which will be hard to move away from.

3.3 Northern Rügener Boddens

The first investigations into the Northern Rügener boddens (NRB) were realised by Gessner (1937, 1940), who focused on biological production. These were followed by taxonomic investigations into the plankton, benthos and nutrient recycling by Trahms (1937, 1939a, 1939b, 1941) as well as into questions regarding the occurrence of *Fucus*, the nano-plankton, and of the Fe/P-quotient in the sediment carried out by Overbeck (1956, 1962, 1964). These investigations were continued by Hübel (1968, 1969) and by Schnese and Hübel (1976) in the 1960s and were an integral part of the research into the boddens from 1970 to 1989, which was coordinated by Rostock University.

3.3.1 Environmental Characteristics

The NRB form a system of several shallow boddens (inlets) lying next to one another, which stretches from the Baltic Sea to the inner area of the Island Rügen in the easterly direction (Fig. 3.16). The climate corresponds to the "Ostseeklima" of the DZB (see above).

The morphometrical aspects of the NRB are comparable to those of the DZB (Table 3.1b). The Libben (23.3 km^2), which is situated close to the Baltic Sea, is followed by the Rassower Strom (22.1 km^2) to the east, the Breezer Bodden (11.6 km^2), the Breeger Bodden (9.7 km^2), the Großer Jasmunder Bodden (58.6 km^2), and the Kleiner Jasmunder Bodden (28.4 km^2). With a total area of 160 km^2, the water surface of the NRB is comparable to that of the DZB. The water volume,

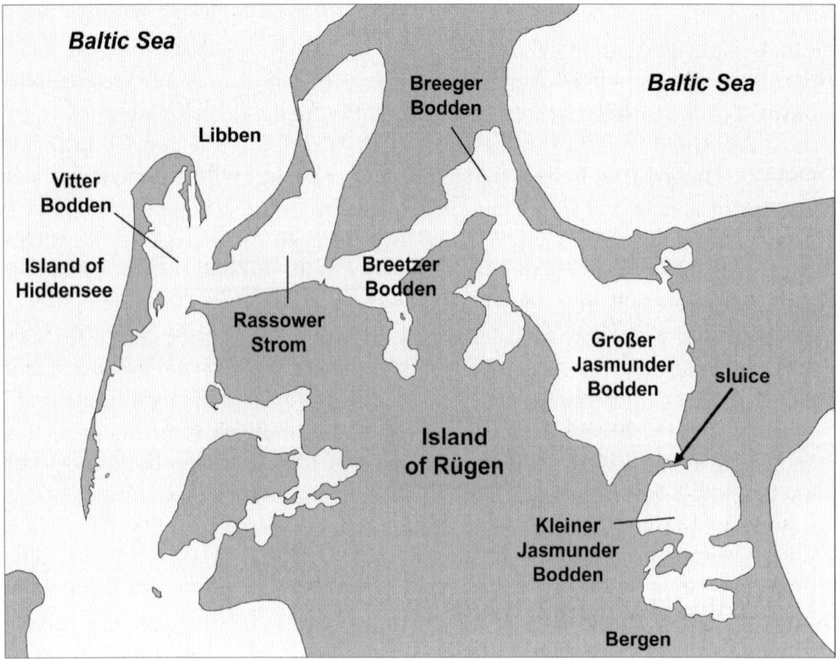

Fig. 3.16 The Northern Rügener Boddens (NRB) (modified from Hübel and Dahlke 1999). Sluice between Großen and Kleinen Jasmunder Bodden indicated

however, is significantly higher at ca. $554 \times 10^6\,m^3$. The single boddens are, as with the DZB, linked to each other by straits (Fig. 3.1). The Kleiner Jasmunder Bodden is considered to be a special case, as it has been connected to the Großer Jasmunder Bodden by a sluice only since the end of the nineteenth century. This sluice later suffered functional disorders, but its reconstruction in 1995/1996 enhanced water exchange. The mean depth of the NRB is about 3.5 m. The Großer Jasmunder Bodden has a mean depth of 5.3 m. Maximum depths of 9 m and 6 m are recorded for the Großer Jasmunder Bodden and the Kleiner Jasmunder Bodden, respectively. Vertical stratification occurs only rarely and does not last long.

The total catchment area of $352\,km^2$ is relatively small compared to that of the DZB. There is also the fact that $104\,km^2$ of this are apportioned to the Kleiner Jasmunder Bodden. The catchment/surface water area ratio is about 2:1, which is highly favourable. Agriculture in the catchment area is not well developed and has halved during the last 15 years. The main impact derives from municipal sewage waters. In this regard, the city of Bergen, with 85,000 inhabitants' equivalents, is the main source of impact, which concentrates on the Kleiner Jasmunder Bodden. During the summer season, higher impacts occur due to the relatively high number of holiday-makers. Input from rivers does not exist. Thus, the fresh-water input via rain, diffuse outlets and small creek courses is very small at $45 \times 10^6\,m^3\,a^{-1}$ (min/max $5–82 \times 10^6\,m^3\,a^{-1}$). The proportion of fresh-water input amounts to < 5% of the total input (average value for DZB ca. 10%). The water exchange with the Baltic Sea ($3,360 \times 10^6 m^3 a^{-1}$) is dominating. The theoretical residence time of water is 0.14 a^{-1} on average. In the Kleiner Jasmunder Bodden it is 1.5 years at present.

Salinity, Secchi depths, dissolved inorganic nitrogen (DIN) content, o-PO_4 content and the primary production of the single boddens are shown in Table 3.4. Since the outlet to the Baltic is situated in the west, gradients of salinity, nutrient, biomass and production stretch from west to east.

The mean salinity in the Libben is between 9 and 10 psu and in the Kleiner Jasmunder Bodden it is between 3.5 and 4.6 psu (before 1995 only 3.2–3.7 psu). The salinity increased significantly as a result of the deepening of the shipping channel in 1966/1967 (e.g. up to 8.8 psu in the Großer Jasmunder Bodden). However, after 1974 it fell back to its initial values, although it is still significantly higher than in the DZB.

Secchi depths of 6–8 m in the Libben (1960/1963) are correlated with the salinity. Secchi depths had declined to 1–4 m by 1980 but again reached 6–7 m after 1985. In the Großer Jasmunder Bodden, depths between 1.5 and 1.8 m were reached in the years 1960/1965. They levelled off at 2.3–2.8 m by 1981/1989, and are significantly higher than in the DZB. However, in the Kleiner Jasmunder Bodden, values between 0.14–0.25 m were measured until 1995, i.e. even lower than those in the DZB. Since 1995, the Secchi depth in the Kleiner Jasmunder Bodden has been just above 0.40 m (Dahlke and Hübel 1996).

The underwater light climate in the western boddens is more like that in the Baltic Sea, whereas that in the eastern boddens resembles more the light climate in the DZB.

Oxygen saturation varies between 90 and 160%, depending on the season and on the time of day. Even during periods of stagnation in summer, there is an oxygen saturation of >80% in the near-bottom region. Saturation may decline to 10% under long-lasting ice covering in winter. Only the sediment surface in the hypertrophic Kleiner Jasmunder Bodden is anoxic at times.

The nutrient load for 1970/1975 was 1.3 g P $m^{-2} a^{-1}$ (Schnese and Hübel 1976). For the isolated Kleiner Jasmunder Bodden, significantly higher values of 3.5 g P $m^{-2} a^{-1}$ were recorded in 1985/1989 (Hübel and Dahlke 1991). As far as N-input is concerned, values of 80 g N $m^{-2} a^{-1}$ were determined in 1985/1989 (Hübel 1984). In addition to this, there was an input via N_2-fixation of 0.71 g N $m^{-2} a^{-1}$. The N-input in all the other bodden areas was about 10 g N $m^{-2} a^{-1}$ at this time, with input through N_2-fixation was 0.47 g N $m^{-2} a^{-1}$. Schlungbaum and Baudler (1998) calculated the N-load as 23–21 g N $m^{-2} a^{-1}$ for 1990–1995. The Kleiner Jasmunder Bodden was released by a denitrification of 3.7 g N $m^{-2} a^{-1}$; the release of the other bodden areas was 37 g N $m^{-2} a^{-1}$.

3.3.2 *Planktonic Communities*

3.3.2.1 Structure, Dynamics and Productivity

Trahms (1939a) detected 69 phytoplankton species in the Großer Jasmunder Bodden. Among them are 28 Cyanobacteria (*Microcystis-*, *Coelosphaerium-*, *Gomphosphaeria-*, *Merismopedia*-spp., *Nodularia spumigena*, *Anabaena-*, *Oscillatoria-* and *Lyngbya*-spp.),

Table 3.4 Characteristics of the NRB at different observation periods (Hübel et al. 1998). Arithmetic mean values and rages of salinity, Secchi depth (vegetation period), nutrients (January–March), and primary production

	Observation period	Libben	Rassower Strom	Breetzer Bodden	Breeger Bodden	Grosse Jasmunder Bodden	Kleine Jasmunder Bodden
Salinity (psu)	1960/1963	9.3 (9.1–9.5)	8.4 (8.3–8.8)	8.0 (7.6–8.3)	7.8 (7.4–8.1)	6.9 (6.7–7.5)	3.4 (3.2–3.6)
	1964/1969	9.7 (8.8–11.0)	9.2 (8.2–10.1)	8.9 (8.3–9.7)	8.6 (8.1–9.6)	8.1 (7.8–8.6)	3.5 (3.2–3.7)
	1970/1980	10.1 (9.1–10.8)	9.8 (9.3–10.7)	9.1 (8.9–9.4)	9.2 (8.9–9.5)	8.8 (8.5–9.3)	4.6 (3.5–5.8)
	1981/1989	9.8 (9.3–10.3)	9.7 (8.8–10.1)	9.0 (8.7–9.4)	8.9 (8.6–9.1)	8.5 (8.3–8.7)	4.3 (3.9–5.3)
Secchi depth (m)	1960/1963	6–8	(2.8)[a]	2.4	2.2	1.5–1.8	0.15
	1964/1969	3–8	(2.8)[a]	2.8	2.5	1.7–2.4	0.20
	1970/1980	3–4	(2.8)[a]	2.8	2.5	2.1–2.5	0.25
	1981/1989	6–7	(2.8)[a]	2.8	2.6	2.3–2.8	0.20
DIN (NO_3 + NH_4 + NO_2) (μmol N l^{-1})	1960/1963	(2.5)	(3.5)	(6.0)	(8)	(10)	(15)
	1964/1969	3.7 (1.7–7.3)	4.7 (3.2–7.7)	5.9 (4.3–8.9)	5.9 (2.5–8.5)	7.0 (3.5–8.8)	14.0 (10.5–17.3)
	1970/1980	6.8 (3.7–9.4)	5.8 (3.4–7.9)	5.8 (2.7–8.1)	5.5 (2.7–7.9)	7.6 (4.2–10.0)	16.3 (9.8–33.5)
	1981/1989	7.9 (6.1–10.2)	7.1 (6.0–8.5)	7.2 (5.6–7.9)	6.9 (5.3–7.9)	7.7 (5.3–9.4)	27.4 (12.6–46.8)
PO_4-P (μmol P l^{-1})	1960/1963	0.3 (0.2–0.5)	0.5 (0.3–0.6)	0.6 (0.3–0.9)	1.4 (1.2–1.5)	1.6 (1.2–1.9)	2.9 (1.5–3.1)
	1964/1969	0.4 (0.3–0.5)	1.0 (0.5–1.6)	1.1 (0.5–1.9)	1.5 (0.9–2.8)	0.8 (0.5–1.1)	2.0 (1.7–2.8)
	1970/1980	0.1 (0.6–1.7)	1.4 (0.4–1.7)	1.5 (0.7–1.9)	1.4 (0.9–2.1)	1.4 (0.7–1.9)	2.8 (1.1–5.7)
	1981/1989	1.3 (0.8–2.1)	1.5 (0.8–1.7)	1.7 (1.0–1.8)	1.7 (1.1–2.1)	1.4 (0.9–2.6)	5.1 (2.1–6.6)
Primary production (g C m^{-2} a^{-1})	1960/1963	90 (65.9–112.3)	110 (105.2–118.4)	335 (305.2–362.5)	420 (310.4–530.0)	670 (605–785)	>800 (680–>800)
	1964/1969	115 (70.5–162.1)	140 (81.5–235.3)	278 (145.8–388.1)	310 (167.0–432.3)	375 (285–505)	>800 (650–>800)
	1970/1980	166 (107.1–203.3)	180 (141.7–248.6)	190 (128.2–275.9)	270 (193.6–339.4)	325 (240–365)	750 (610–>800)
	1981/1989	168 (127.4–187.7)	185 (144.6–250.5)	210 (157.4–263.5)	260 (243.5–342.6)	355 (270–445)	780 (610–>800)

[a]Visible down to the bottom (2.8 m)

22 diatoms (*Coscinodiscus-*, *Melosira-*, *Chaetoceros*-spp., *Actinocyclus ehrenbergii*, *Cerataulus turgidus*, *Thalassiothrix nitzschioides*; *Surirella-*, *Campilodiscus-*, *Amphiprora*-spp., *Bacillaria paradoxa*, *Fragilaria crotonensis* and *Achnanthes taeniata*), 13 chlorophyceans (*Botryococcus braunii*, *Oocystis*-spp., *Scenedesmus quadricauda*, *Pediastrum* spp.) and 6 dinoflagellates (*Prorocentrum micans*, *Glenodinium gymnodynium*, *Peridinium pellucidum*, *Dinophysis acuminata*, *D. ovum var. balticum* and *Ceratium tripos* var. *subsalsa*. Most of these are freshwater species.

According to Gessner (1957), the Cyanobacteria are predominant throughout the year, followed by diatoms in second place, but their spring and autumn peaks were often not detectable.

For the period 1960/1963, Hübel et al. (1998) detected Cyanobacteria (*Microcystis* spp., *Gomphosphaeria* spp., *Coelosphaerium* spp.) and green algae (*Scenedesmus* spp., *Oocystis* spp. and *Monoraphidium* spp.). N_2-fixing Cyanobacteria were *Nodularia spumigena*, *Anabaenopsis elenkinii* and *Anabaena* spp. during the summer.

Hübel et al. (1998) describes still clearly visible spring blooms for the Libben past 1970. The dominant species were *Chaetoceros danicus*, *Ch. wighamii*, *Skeletonema costatum*, *Thalassiosira baltica*, *Achnanthes taeniata* and *Peridinium pellucidum*.

Currently, phytoplankton is the main primary producer in the NRB. A summarising overview of primary production is given in Table 3.2 (see above). Significant relations to salinity and exchange rate of the boddens with the Baltic Sea have been proved. The primary production per year and the maximum production per day for the period of 1970/1989 characterise the bodden waters thus:

- Libben (< 180 g C m^{-2} a^{-1}) as mesotrophic
- Rassower Strom (142–251 g C m^{-2} a^{-1}) as mesotrophic to eutrophic
- Breetzer Bodden (128–276 g C m^{-2} a^{-1}) as mesotrophic to slightly eutrophic
- Breeger Bodden (194–343 g C m^{-2} a^{-1}) as eutrophic to hypertrophic
- Großer Jasmunder Bodden (240–445 g C m^{-2} a^{-1}) as polytrophic to hypertrophic
- Kleiner Jasmunder Bodden (>600 g C m^{-2} a^{-1}) as hypertrophic

Within the scope of the ÖKOBOD-project, analyses on the role of protozooplankton have been conducted by Garstecki et al. (1999). They verify, like in the DZB, the large diversity of heterotrophic flagellates with different nutrition behaviours. With regard to biomass, the protozoa dominate the zooplankton community. However, the total biomass of the protozooplankton in the Rassower Strom is less than that in the Kirr Bucht of the DZB. There is more significantly distinct seasonality in the Rassower Strom than in the DZB (see Chap. 3 by Schiewer, this volume).

According to Trahms (1939a), the mesozooplankton of the Großer Jasmunder Bodden is composed of 33 species, including *Paramecium aurelia*, *Leprotintinnus bottnicus*, *Tintinnopsis-*, *Keratella-*, *Filinia-*, *Brachionus-*, *Notholca-*, and *Eurytemora* species, *Synchaeta*, *Acartia bifilosa*, *Chydorus sphaericus*, *Bosmina*, *Asplachna* and *Podon*. Hübel et al. (1998) describe *Acartia bifilosa*, *Chydorus sphaericus* for the Libben. The relatively high proportion of marine species is striking. In the Kleiner Jasmunder Bodden freshwater species such as *Leptodora kindtii*,

Bosmina longirostris, *Alonella nana* and *Diaptomus* sp. are to be added. With regard to the nekton, there are only small differences to the DZB.

3.3.2.2 Nutrient Cycles

Long-term investigations into the nutrient development for the entire NRB are lacking. The following values derive from single years, and were gathered during scheduled field trips.

The Libben is characterised by a relatively uniform and low content of o-PO_4. Values rose from 0.2–0.5 μmol l^{-1} in the 1960s to 2.1 μmol l^{-1} in 1983. The content of DIN was about 2.3–3.7 μmol l^{-1} from 1964–1967, and continued to increase up to 10.2 μmol l^{-1} in 1968. A significant decline in the DIN concentration was measurable only after 1990.

In the Großer Jasmunder Bodden, the o-PO_4-values were 1.2–1.9 μmol l^{-1} in 1960/1963. They declined to 0.8–1.9 μmol l^{-1} after 1968, when the shipping channel was deepened. Definitely higher values up to 3 μmol l^{-1} were recorded only after 1984. The DIN content showed an average value of 10 μmol l^{-1} in 1960/1963 and fell to 3.5–7.5 μmol l^{-1} after dredging out. Following 1977, it again increased up to 10 μmol l^{-1}. The o-PO_4-content in the Kleiner Jasmunder Bodden fluctuated between 1.1 and 3.1 μmol l^{-1} over the period 1960/1978 and increased to 9.4 μmol l^{-1} after 1979. Values from 10 to 17.3 μmol l^{-1} DIN content were measured in 1960/1972. After 1973, there was a sharp rise to a maximum of 46.8 μmol l^{-1}.

Investigations into the nutrient cycle of plankton have rarely been carried out in the NRB. Heerkloß et al. (1999) verified a consumption of 0.5–65% of the gross primary production for the mesozooplankton in the Rassower Strom. The egestion values were 0.5–38.9%. Consequently, the MFW and the grazing food chain are thought to be almost equally active.

3.3.3 Benthic Communities

3.3.3.1 Structure and Productivity

Trahms (1939b) found five *Chara*-species, *Tolypella nidifica*, three species of green algae, one species of brown algae, three species of red algae as well as the spermatophytes *Potamogeton pectinatus*, *Ruppia maritima*, *Zannichellia palustris*, *Zostera marina* and *Myriophyllum spicatum*.

According to Hübel (1968, 1969) and Schnese and Hübel (1976), the benthic algae *Ceramium rubrum*, *Pilayella litoralis*, *Ectocarpus siliculosus*, *Chara canescens*, *Ch. baltica* and *Tolypella nidifica* as well as the higher plants *Zostera marina*, *Potamogeton pectinatus*, *Ruppia marina* and *Zannichellia palustris* are still significantly present in the outer boddens. Charophyceae have disappeared almost entirely. *Tolypella*, *Chorda filum*, *Fucus vesiculosus*, *Potamogeton pectinatus*,

Zannichellia, *Ruppia*, *Zostera* and *Ranunculus boudotii* have also been affected (Hübel and Dahlke 1999). However, in the coastal waters of the Baltic and in the outer boddens, filamentous algae (especially *Pilayella litoralis*, but also *Ceramium*, *Polysiphonia* and *Ectocarpus*) have increasingly appeared.

There are no exact determinations of the primary production of macrophytes.

In the investigations of Gerbersdorf and Meyercordt (1999), primary productions rates of 2.6–30.5 mg C m^{-2} h^{-1} have been reached for the microphytobenthos in the Rassower Strom under in situ incubation and depending on light conditions. The light saturation values fluctuated between 20 and 160 µE m^{-2} s^{-1}. The calculated gross primary production was 23 g C m^{-2} a^{-1} at a depth of 3.4 m. Thus, 16–22% of the gross primary production could originate from the microphytobenthos.

The abundances of bacteria in the sediments of the Rassower Strom (Rieling 1999) range between 1.1×10^9 and 11.3×10^9 cells cm^{-3}. The abundances in the 9–10 cm horizon ($1.1–2.2 \times 10^9$ cells cm^{-3}) are constantly low. The proportion of the bacterial biomass in the sediments was measured at between 0.5–0.6%.

With a total biomass of 12.5–105 mg C cm^{-3}, the protozoa in the sediment of the Rassower Strom are of the same order of significance as those in the DZB. Based on laboratory tests, both direct interactions with the meiofauna through grazing, and indirect interactions through oxygen supply caused by changes in the sediment structure, are assumed (Garstecki et al. 1999; Arlt and Georgi 1999). Field analyses are lacking.

Recent investigations (Arlt and Georgi 1999) emphasise the importance of the meiofauna. The highest abundances in the Rassower Strom and in the Kirr Bucht (DZB) are observed in springtime. The lowest abundance was found for the Rassower Strom in autumn, but for the Kirr Bucht in summer. At both sites, Nematoda are dominant (70–91%). Ostracoda, Harpacticoida and Turbellaria show distinctly lower proportions. The abundance in the Kirr Bucht is consistently higher by 1.8–6.8 times. This is also reflected in the higher biomass in the Kirr Bucht, which is 2.3 times as high.

According to Trahms (1939b), the macrofauna of the Großer Jasmunder Bodden consists of 36 species, 16 of which are marine, 11 are brackish water species and 11 are of limnetic origin.

In the sandy mud occur *Cardium edule*, *Nereis diversicolor*, *Corophium volutator*, *Cyathura carinata*, *Alkmaria romijini*, *Pygospio elegans*, *Paranais litoralis*, *Prostoma obscurum* and *Manayunkia aestuarina*. On sand, *Cardium edule*, *Nereis diversicolor*, *Corophium volutator*, *Cyathura carinata*, *Alkmaria romijini*, *Pygospio elegans*, *Paranais litoralis*, *Prostoma obscurum* and *Manayunkia aestuarina* have been found.

In the phytobenthos community, *Gonothyrea lovenii*, *Membranipora crustulenta*, *Jaera albifrons*, *Gammarus locusta* f. *zaddachi*, *Nais elinguis*, *Pisciola geometra*, *Neomysis vulgaris*, *Idothea viridis*, *Sphaeroma rugicaudum*, *Leptocheirus pilosus* and *Limnea ovata* f. *baltica* have been proved.

There are no recent investigations available.

3.3.3.2 Nutrient Cycles

The remineralisation of nutrients in the sediment occurs mainly through microbial substrate turnover. Rieling et al. (1999) measured inorganic nutrient flow rates of 25–60 µmol NH_4 m^{-2} h^{-1} and 2–9 µmol o-PO_4 m^{-2} h^{-1}. These flow rates range to above the level of the calculated releases for the Baltic (Koop et al. 1990; Stockenberg 1998). Consequently, a distinct contribution to the N and P need of phytoplankton comes from the sediment. The near-bottom water movements in the shallow bodden waters intensify the release of nutrients. The carbon decomposition in the sediment of the Rassower Strom is determined mainly by sulphate reduction and by aerobic respiration. The other e-acceptors (nitrate, manganese, iron) are of minor relevance.

The total respiration in the sediment ranged between 0.8 and 2.3 nmol C m^{-2} h^{-1}. These values correspond to 20–90% of the pelagic gross primary production measured in the field. A high concentration of iron in the pore water indicates a release of iron in the sediment. The high redox potential of the sediment apparently prevents its release into the pelagic zone. In contrast, a distinct release of manganese could be measured.

The investigations verify that bacteria in the sediment make a major contribution to the remineralisation of nutrients. Furthermore, Neuendörfer and Meyer-Reil (1997) succeeded in detecting a close connection between chlorophyll and available DOC.

3.3.4 Benthic–Pelagic Coupling and Eutrophication

3.3.4.1 Morphological, Hydrological Background

Essential differences from the DZB lie in the more favourable ratio of catchment area/water surface, in the lower freshwater and nutrient input, and (with the exception of the Kleiner Jasmunder Bodden) in the better exchange rates with the Baltic Sea. The higher "wash-out effect" reduces the degree of eutrophication.

During the last 30 years or so, the following incidents have been the major influences on the hydrographic conditions and biological productivity:

- The widening, deepening and straightening of the shipping channels from the Libben to the Großer Jasmunder Bodden in 1963/1965
- The long-term interruption and renewed improvement of the water exchange between the Großer and Kleiner Jasmunder Bodden
- The increase in nutrient content in the surface waters of the offshore Arkonasee
- The temporary intensification of agricultural production in the catchment area
- Intensification of sewage water discharge from the catchment area and the subsequent restoration of the point loads by building sewage plants

Due to the lower nutrient load, the increase in the primary production of the phytoplankton is not as high as in the DZB. The sediment keeps its normal function to a large extent (with the exception of the Kleiner Jasmunder Bodden).

3.3.4.2 Biological Processes and Interactions

As in the DZB, the submersed macrophytes were the first to be affected by eutrophication. Thus, the Charophyceae disappeared completely in the NRB in the 1980s. The rise in nutrients, the increase in primary production and the decline in Secchi-depths have all been the subject of measurements. However, the pelagic and benthic rates of primary production in the Rassower Strom were clearly lower than in the Kirr Bucht. However, there were no differences with respect to the Chl *a* content. In general, the Rassower Strom has to be classified as eutrophic, whereas the Kirr Bucht is hypertrophic. In the FSL, autotrophic and heterotrophic processes are closely connected with each other. The relationships, however, are more complicated than they are in the DZB. For example, the concentration of particles in the Rassower Strom was lower than that in the Kirr Bucht, and larger particles were dominant. This demonstrates a stronger succession of aggregates in the Rassower Strom. The share of organisms in organic particles is clearly lower than in the DZB. In contrast, "mini"-biofilms have been found in the NRB, which are able to merge to a genuine sediment layer. Additionally, small-scale differences in the shares of oxygen and sulphate respiration have been verified. Thus, oxygen respiration dominates in the muddy sand vs sulphate respiration (81% vs 19%). In the muddy sediments, the ratio of 6–23 % vs 74–92% shifted clearly in favour of sulphate respiration (Rieling et al. 1999).

In the Kirr Bucht, the sedimentary organic carbon has been re-mineralised with high turnover rates. Obviously, there was a balance between the oxygen supply and the carbon load of the sediments. In the Rassower Strom a limitation of microbial activities caused by the quality of organic carbon has been inferred from the increased C/N-values and enzyme activities (relation of hydrolysis of easily and less easily degradable carbohydrates as well as the relation of proteins to fatty acids). For this reason, the carbon content in the sediment was higher and its character more refractory. Additionally, an alternation of new and regenerated production over the year has been verified in the mesotrophic to eutrophic Rassower Strom. The meso-zooplankton was also of more importance as a consumer than in the Kirr Bucht.

Furthermore, total benthic respiration in the Rassower Strom was generally higher than in the Kirr Bucht. With regard to nutrient release, high concentrations of inorganic nutrients (ammonium, phosphorus) in the pore water correlate with high flows from the sediments into the open waters in the Rassower Strom. Thus, microbial remineralisation rates are an essential source of the nitrogen and phosphorus needed for benthic and planktonic primary production.

3.4 Schlei

3.4.1 Environmental Characteristics

The Schlei is one of the four large fjords of the Kieler Bucht that emerged during the last ice age. With its narrow, river-like shape (Fig. 3.17), it differs from the open Eckernförde Bay, the funnel-shaped Flensburger and the trumpet-shaped Kieler

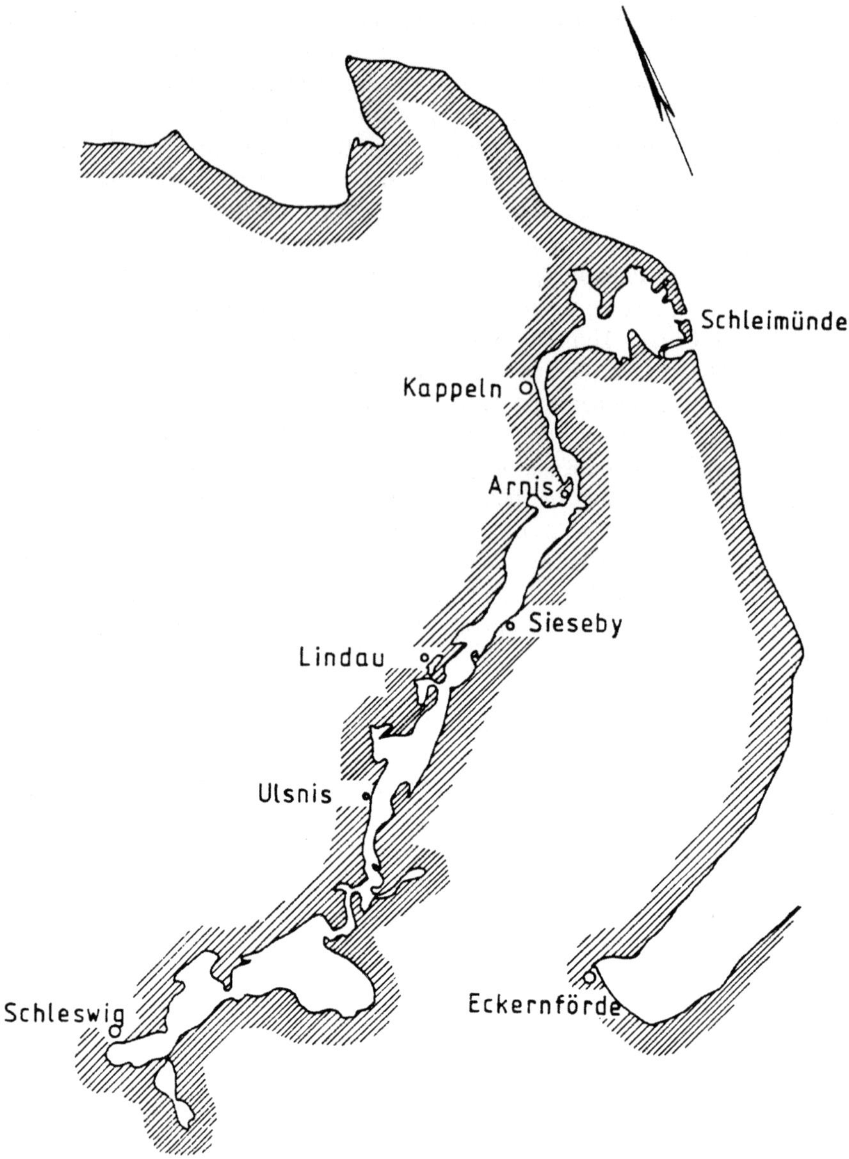

Fig. 3.17 The Schlei (Nellen 1967). Different parts and catchment area. Sample stations 1960/1961 indicated

Förde. Structurally, it resembles the eastern part of the DZB from the mouth to the Baltic Sea to the Barther Bodden. The individual data are summed up in Table 3.1.

The total length of the Schlei is 43 km, the average depth 2.5 m, and the maximum depth 13 m. With a water surface of 53 km^2 and a water volume of 133 km^3 it is clearly smaller than the DZB. The catchment area comprises 670 km^2. Due to the inflow of small freshwater rivers and the inflow of Baltic Sea water, a salinity gradient of 3–8 psu develops in the inner area and of 15–20 psu in the outer area. The freshwater inlet averages 307 × 10^6 m^3 a^{-1} and is considerably lower than the inflow of Baltic Sea water, which, at a range of tide of 10–15 cm, results primarily from local wind conditions and level changes of the Baltic Sea caused by the weather. The Secchi depth follows the salinity gradient. At the outlet to the Kieler Bucht, it can reach 1.4–4 m (average 2.8 m), in the Inner Schlei values range between 0.4 and 1.5 m (average 0.8 m). Longer phases without water movement, and thus without mixing of the water column, do not occur, explaining why there is neither a stable stratification nor anoxic conditions up to 30 cm above the bottom. Penetration of deep water containing H_2S due to longer-lasting offshore winds from the Kieler Bucht was not observed (Gocke et al. 2003).

With a ratio of catchment/surface water area of 12:1, there is a natural tendency towards a eutrophic water quality. The average residence time of only 30 days apparently compensates for this negative aspect, as in the summer the exchange time increases to 150 days. Particularly heavily loaded is the Inner Schlei, a lake-shaped extended area with a water surface of 20 km^2 and an average depth of 2.5 m. The city of Schleswig, with approximately 30,000 inhabitants, is located here, and the Füsinger Au, which drains a catchment area of 240 km^2, flows into it. Entry of phosphorus into the Inner Schlei amounts to approximately 3.8 g total-P m^{-2} a^{-1}, and that of nitrogen about 84 g total-N m^{-2} a^{-1}. The share from agriculture in the external entry amounts to approximately 85% for phosphorus and 90% for nitrogen. Whereas the annual average municipal entries are low, their share rises to >30% of the total entry in the summer months when the discharge is weaker (Ripl 1986).

Due to metabolic conversion, sedimentation and discharge, nutrient values decrease from inner towards outer parts of the Schlei. This results in gradients of phyto-, bacterio- and zooplankton (Feibicke 1995; Gocke et al. 2003).

3.4.2 Planktonic Communities

3.4.2.1 Structure, Dynamics and Productivity

In the spring, the phytoplankton of the Inner Schlei is, according to Feibicke (1994, 1995), characterised by the phytoflagellates *Cryptomonas* spp., *Rhodomonas* spp., *Eutreptiella* cf. *gymnastica*, *Katodinium rotundatum* or the diatoms *Chaetoceros* spp., *Skeletonema* spp., *Thalassiosira baltica*. In May, Chlorophyceae reproduce on a huge scale, above all *Dactylosphaerium jurisii*, *Monoraphidium contortum*, which form a first phytoplankton maximum. This goes down only after a strong development of rotifers. A mass reproduction of Cyanobacteria takes place as late as in July, with *Microcystis aeruginosa* reaching a share in the biomass of 70–90% in the summer. Apart from this, small-sized, chroococcale Cyanobacteria with a high diversity of species are available in subdominant populations (e.g. the genera *Aphanothece*, *Chroococcus*, *Gomphospheria*, *Lemmermaniella*, *Merismopedia* and *Romeria*). For a short time, small, centric diatoms (*Cyclotella* spp., *Thalassiosira proschkinae*) can also form subdominant populations. The highest phytoplankton densities are to be found from July to September. In August, N_2-fixing Cyanobacteria (*Anabaenopsis arnoldii, A. elenkinii, Anabaena* spp.) with biomasses of 12–28 mg fw l^{-1} appear. Their N_2-fixation performance, however, is only small. The generally declining biomass at the end of October is accompanied by a decline in Cyanobacteria, diatoms and phytoflagellates.

Measurements of the primary production by Schiemann (1974), Gocke and Rheinheimer (1994) for 1991/1993 as well as by Feibicke (1994) revealed approximately 850 g C m^{-2} a^{-1} for the Inner Schlei. The regional distribution of the phytoplankton biomass was determined on the basis of the Chl *a* content (Table 3.5). A distinct gradient has been proven. The maximum values of 151 μg Chl *a* l^{-1} are measured in the Kleine Breite of the Inner Schlei, where they are obviously typical (Lenz 1970; Gocke and Rheinheimer 1994). The lowest concentrations, amounting to 5.5 μg Chl *a* l^{-1} are to be found in the area of the estuary.

Gocke et al. (2003) have provided measurements (see Table 3.5) for the bacterioplankton, the bacterial production using thymidine incorporation, and the turnover rate for dissolved glucose in the longitudinal profile of the Schlei. A gradient could always be shown to be present, with the highest values in the Inner Schlei. This is also true for the number of saprophytes. Depending on the inflow of salt water, a plateau can occur also in the central Schlei. The maximum total number of bacteria amounted to 14×10^6 cells ml^{-1}, thymidine incorporation was up to 137 pmol l^{-1} h^{-1}, and the turnover rate for glucose 125% h^{-1}. Feibicke (1994) calculated the production of bacterioplankton for the Inner Schlei to be approximately 104 g C m^{-2} a^{-1}.

In the mesozooplankton, *Eurytemora affinis* is predominant towards the end of the winter half of the year. After a strong decline in their population, rotifers, above all *Brachionus* spp. and *Filinia longiseta*, develop in huge numbers in July. Copepods (*E. affinis, Acartia tonsa*) only occur in late summer. On the long-term average, protozoa account for approximately 41% of the zooplankton biomass. With 80%, they reach a disproportionately high share in the production of the zooplankton, and thus represent the most important group within the zooplankton. There is no obvious stable seasonal pattern; however, they often form large populations. According to Feibicke (1994), the mean total biomass of the zooplankton is approximately 2.5 g C m^{-2} (4 mg fw l^{-1}).

According to the statistics of catches in the Schlei there are ten important fish species in the nekton (*Clupea harengus, Anguilla anguilla, Perca fluviatilis, Sander lucioperca, Rutilus rutilus, Abramis brama, Osmerus eperlanus, Platichthys flesus, Esox lucius, Salmo trutta trutta*). The remaining fish species account for only about 1% of the total catches. With 95–90 t a^{-1}, *Clupea harengus* is the dominating commercial fish. From 1949/1961 to 1971/1980, above all catches of Anguilla declined strongly, due mainly to the eutrophication of the Schlei. The decline in landings of *Perca fluviatilis, Rutilus rutilus* and *Abramis brama* is caused by depreciation and reduced fishing activities (Lupatsch and Nellen 1981). The total catches of the Schlei range between 35 and 40 kg ha^{-1} a^{-1}.

Table 3.5 Annual means and variations of some parameters along the Schlei length axis (Gocke et al. 2003)

Station	Reference station (Kiel Bight)	Maasholm	Kappeln	Arnis	Lindaunis	Königsburg	Große Breite	Kleine Breite
Distance from river mouth (km)		2.9	6.8	10.2	19.5	26.9	32.6	39.6
Salinity (‰)	16.9 (13.1–19.8)	15.8 (12.1–19.5)	15.0 (11.7–18.5)	13.5 (10.9–16.4)	11.6 (9.2–16.6)	9.1 (7.9–11.3)	8.0 (7.2–9.6)	6.4 (5.1–8.8)
Secchi depth (m)	6.8 (4.0–11.0)	2.8 (1.4–>4.0)	2.6 (1.2–4.8)	2.0 (0.90–3.4)	1.4 (0.75–2.0)	0.95 (0.60–1.7)	0.90 (0.60–1.5)	0.80 (0.40–1.5)
Chl *a* concentration ($\mu g\ l^{-1}$)	2.6 (1.0–6.9)	5.5 (0.7–19.6)	14.3 (1.2–34.7)	23.3 (1.0–67.9)	41.8 (14.6–103)	56.1 (22.0–140)	66.8 (32.6–151)	87.8 (29.4–162)
Total number of bacteria ($10^6\ ml^{-1}$)	0.89 (0.44–2.28)	2.94 (0.47–6.38)	5.40 (2.10–11.8)	8.56 (1.30–16.2)	8.71 (3.05–16.0)	10.3 (4.96–17.5)	16.1 (4.03–53.8)	16.5 (4.12–53.2)
Average bacteria cell volume (μm^3)	0.134 (0.098–0.168)	0.122 (0.068–0.142)	0.115 (0.073–0.159)	0.122 (0.091–0.163)	0.119 (0.089–0.143)	0.120 (0.060–0.199)	0.131 (0.102–0.237)	0.114 (0.052–0.187)

3.4.2.2 Nutrient Cycles

The nitrogen content in the water column rises to about 6–8 mg total-N l^{-1}. With 70–80%, NO_3 has the main share. In the summer, the values decrease to about 1–2 mg l^{-1}. Losses due to denitrification amount to approximately 23 g N m^{-2} a^{-1}.

In contrast to this, total-P and o-PO_4 reach a minimum in the winter (0.1–0.2 mg l^{-1} or 0.02–0.05 mg l^{-1}, respectively). With approximately 0.5–0.6 mg total-P l^{-1} or 0.3–0.5 mg o-PO_4 l^{-1}, respectively, the maximum lies in the summer (Ripl 1986). During the period with stronger discharge, uptake and deposition prevail. During this period, about 8 mg total-P m^{-2} day^{-1} is deposited. In the summer, there is a strong release rate of 12 mg total-P m^{-2} day^{-1}.

According to Feibicke (1995), who balanced the turnover of carbon in the Inner Schlei (Fig. 3.18), the phytoplankton uses up about one-third of the gross production for respiration. The remaining two-thirds go into the bacteria-detritus pool of the pelagic, with the main part being recycled by protozoa.

The carnivorous use of the zooplankton amounts to 56 g C m^{-2} a^{-1}, which accounts for only 8% of total ingestion. Less than 12 g C m^{-2} a^{-1} of crustacean production is available as a food supply for fish.

3.4.3 *Benthic Communities*

3.4.3.1 Structure and Productivity

Being a typical shallow water body, the Schlei used to be characterised by submersed macrophytes, with their spread following the salinity gradient. Predominantly

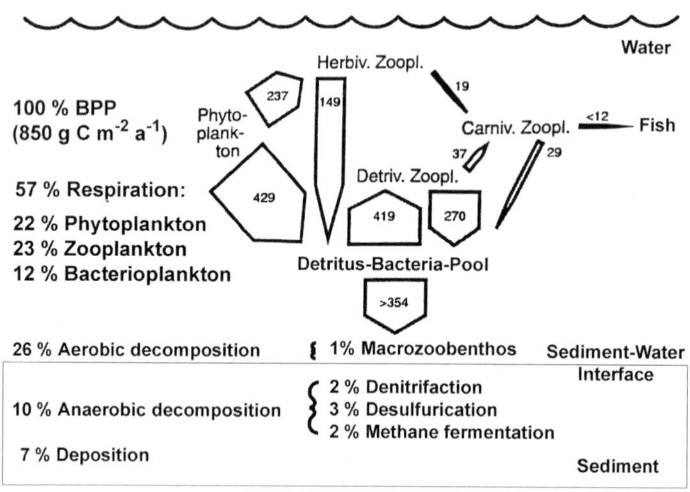

Fig. 3.18 Calculated carbon cycle of the Inner Schlei (Feibicke 1995)

marine or brackish water forms were found in the outer Kieler Bucht, whereas limnetic representatives prevailed in the Inner Schlei. At the beginning of the 1930s (Hoffmann 1937), various macroalgae (*Enteromorpha* spp., *Monostroma grevillei*, *Ceramium diaphanum*, *Fucus vesiculosus*, *Polysiphonia* spp., *Pilayella litoralis*, *Chorda filum*, *Ulva lactuca*) formed the population until far into the Inner Schlei.

Submersed spermatophytes used to grow throughout the entire area of the Schlei. With 8–9 psu, *Zostera marina* formed extensive meadows. *Zannichellia palustris* was available in large populations and advanced well into the Outer Schlei. *Potamogeton pectinatus*, *P. fluitans* and *Myriophyllum spicatum* were present from Lindaunis down the Schlei. For the Inner Schlei, *Najas major*, *Ranunculus baudotii*, *R. fluitans*, *Potamogeton perfoliatus*, and *Ceratophyllum demersum* were found to be frequently, often occurring over wide areas. In the lower littoral of the whole Schlei, charophytes that prefer brackish water (mostly *Chara baltica*, on a smaller scale *Tolypella* and *Nitella*) were shown to colonise extensive areas.

The coastal zone of the Schlei was almost completely covered by dense reeds (mainly *Phragmites australis*, less frequently *Scirpus maritimus*, *Sc. tabernaemontani*, and *Sc. parvulus*, and *Typha latifolia* in less saline areas). Hoffmann (1937) and Remane (1937), however, pointed to occasional dying of fish and accumulation of sewage already at that time. In the course of eutrophication, this state has changed completely (see below).

Investigations into the periphyton are not available. Estimations of the meiofauna are lacking.

For the macrozoobenthos, analyses by Remane (1937) and Nellen (1963) are available. The most important representatives are the Mollusca *Mytilus edulis*, *Cardium edule*, *Macoma balthica*, *Mya arenaria* and Hydrobia spp.; the polychaetes *Harmonthoe sarsi*, *Nereis* sp., *Pygospio elegans*, *Polydora ciliata*, *Streblospio shrubsoli*, *Alkmaria romijni* and *Capitella capitata*; Tubificidae; the Crustaceae *Gammarus* sp., *Microdeutopus gryllotalpa*, *Corophium volutator*, *Cyathura carcinata* and *Jaera albifrons* as well as chironomid larvae. Depending on the salinity and the subsoil, they can be divided into nine macroinvertebrate communities:

- 10–18 psu/dense layer of big mussel shells and living mussels, infiltrated by detritus: extensive communities with *Mytilus*, *Jaera albifrons* and *Harmonthoe sarsi*. In terms of quantity, Crustacea predominate along side Polychaeta.
- 10–18 psu/detritus-rich layer, infiltrated by broken mussel shells and cellulose particles: less varied communities with *Streblospio* and *Alkmaria*. Somewhat lower number of species, decrease of density of individuals. Crustacea and Polychaeta are on the decline.
- 5–10 psu/detritus-rich layer, infiltrated by broken mussel shells and cellulose particles: decrease of the number of species, *Capitella* and *Pygospio* are missing. Characteristic are *Alkmaria romijni* (salinity optimum) with dominating abundances.
- 10–18 psu/natural soil, gravel to sandy: very densely populated; apart from larvae of the Chironomidae, the same species occur as on the soft soil of the same salinity but in larger numbers and different composition. High proportion of mussels, and also *Corophium* and *Alkmaria*.

- 5–10 psu/natural, not sedimented soil, gravely to sandy: stronger emergence of Alkmaria and Hydrobia. Decline of mussels and Corophium.
- <5–8 psu/natural, not sedimented soil, gravely to sandy: decrease in species number, but increase in number of individuals. More Tubificida than *Alkmaria*, more *Hydrobia* than mussels.
- 10–18 psu/fine sewage, containing more or less H_2S: sewage is life-hostile in all psu ranges. Still 11 species present, predominantly *Alkmaria* and larvae of the Chironomida. Tubificida are absent.
- 5–10 psu/fine sewage, containing more or less H_2S: similar colonisation as at the previous site. *Alkmaria*, Chironomidae larvae and *Hydrobia* predominate.
- <5–8 psu/fine sewage, containing more or less H_2S: smallest colonisation of all. Low-species community; important are Tubificida and Chironomida. *Nereis* and *Alkmaria* can occur in addition.

This colonisation largely corresponds to the colonisation known for the Großer Jasmunder Bodden and the Greifswalder Bodden (see Chap. 4 by Schiewer, this volume). Increasing eutrophication and the subsequent extension of the formation of sewage in the years following led to a strong reduction in the macrozoobenthos (Stotz 1986).

3.4.3.2 Nutrient Cycles

Feibicke (1995) balanced the carbon turnover in the sediment of the Inner Schlei on the basis of a gross primary production of 850 g C m^{-2} a^{-1}. Proceeding from a production surplus of the pelagic of 365 g C m^{-2} a^{-1}, he calculated 10% for aerobic degradation, 6% for the desulphurisation, 2% each for denitrification and methane fermentation as well as 7% for sedimentary deposit. The latter results in an annual sediment formation of approximately 60 g C m^{-2} a^{-1} and thus in a growth rate of the sediment of around 3.5 mm a^{-1}. Since the macrozoobenthos is strongly reduced due to the current spread of sewage areas, its respiration rate is only 8 g C m^{-2} a^{-1} (approximately 1% of gross primary production; Stotz 1986). The majority of the degradation takes place via microorganisms. According to all results available for comparable coastal areas in Germany, the anaerobic share appears to be low. This is also true of desulphurisation. The ratio of denitrification and desulphurisation also clearly differs from other findings. Direct determinations would therefore be desirable.

3.4.4 Benthic–Pelagic Coupling and Eutrophication

3.4.4.1 Biological Processes and Interactions

Investigations into the sediment cores by Ripl (1986) showed that, as early as 120 years ago, there was a rapid increase in eutrophication due to the rapid increase in population density in the catchments area of the Schlei.

Based on initial data provided by Magnus (1875) and Hoffmann (1937), Gocke et al. (2003) developed a sequence chart of the eutrophication of the Schlei (Fig. 3.19). According to this, four stages of advancing eutrophication can be differentiated, based on the development of submersed macrophytes. The most advanced is eutrophication in the Inner Schlei, where submersed macrophytes communities have disappeared completely. This is also true of the formerly wide-spread charophytes in the entire Schlei. These have been replaced by phytoplankton, with massive blooms in the summer. Only in the shallow areas can spermatophytes occasionally be found. *Enteromorpha intestinalis* occurs on stones.

The disappearance of the submersed macrophytes also greatly affects the central Schlei. *Fucus vesiculosus* colonisation is restricted to the outer zones.

In the Outer Schlei, however, there are still coherent areas of macroalgae and spermatophytes. In deeper water, these are *Zostera marina*, in shallower parts *Ruppia maritima* and *Zannichellia palustris*.

Whereas the submersed macrophytes had already disappeared by the beginning of the 1970s (Flentge 1972), there has been a mass development of seasonal macroalgae, such as *Pilayella litoralis*, *Enteromorpha intestinalis* in shallow water as well as *Ceramium nodulosum*, and *Polysiphonia* spp. in somewhat deeper water. Algae that become detached from substrates float closely under the surface or on the bottom. Their degradation by bacteria can, in calm and warm weather, lead to anoxic conditions and the development of H_2S in layers close to the bottom. Death of soil fauna and of fish species can be the consequence.

The former wide-spread reed bed communities are strongly reduced. They seem, however, to be slowly recovering during the past few years.

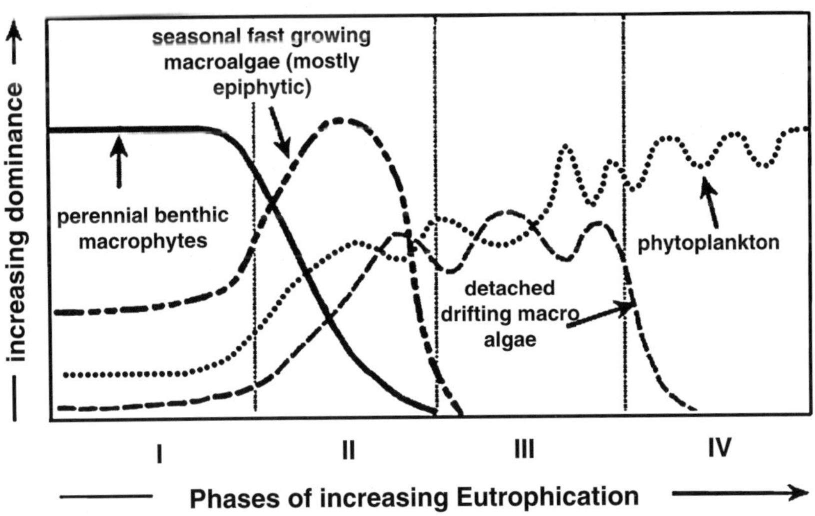

Fig. 3.19 Relative dominance of primary producers and eutrophication of the Inner Schlei in the last 30 years (Gocke et al. 2003)

3.5 Conclusions and Protection Measures

The specific features of the DZB, NRB and the Schlei resulting from their structure, are summarised in Table 3.6. A decisive factor is the high sensitivity of these inner coastal waters towards nutrient loads, which is based on the natural dominance of internal turnover rates over exchange processes with the Baltic Sea (restriction by subdivisions, narrow connection between the different parts and narrow outlets). On the other hand, this results in greater autonomy of the estuaries caused by their importance as bioreactors.

The dramatic decline in submersed littoral vegetation in the majority of the boddens leads to a marked reduction in filter and buffer capacity as well as a reduction in diversity. The long-lasting storage capacity for nutrients, which is closely correlated with the littoral vegetation, is lost. The uncovered areas promote sediment mobility and thus nutrient release. Together with the acceleration of turnover rates in the MFW, this increases the self-eutrophication of these waters.

The primary aim of redevelopment is a return to initial conditions. One possible route could be via re-colonisation with submersed macrophytes. Their dominance leaves phytoplankton with clearly fewer possibilities for development. If this condition is to be restored, it should be achieved more by changes in the catchment area rather than by steps taken in the boddens themselves. For this purpose, nutrient entry (nitrogen, phosphorus) has to be strongly reduced through elimination of point sources via the construction of sewage plants. In addition, diffuse nutrient sources, which exert their effect through direct discharge in the catchment area, must be reduced considerably. The re-mesotrophication process thus initiated would drastically reduce the load, and would be able to restore the original conditions in about 10–20 years. The building of efficient sewage plants and the decline in agriculture has definitely contributed to the restoration of the NRB in the past

Table 3.6 Common characteristics of the DZB, NRB and Schlei

1. Joint systems of several brackish shallow boddens, polymictic
   - Horizontal salinity and nutrient (nitrogen) gradients
   - High natural productivity
   - High detritus content
2. Tideless estuaries, irregular exchange processes with the Baltic Sea, dominance of outflow ("washout effects")
3. Large fluctuations in the ecosystems controlled mainly by physical–chemical factors
   - Salinity
   - Temperature
   - Wind
   - Water exchange
   - Ice covering
4. Dominance of species with high tolerance to environmental factors
5. Predisposition to eutrophication, which can induce the loss of submersed macrophytes dominance and light limitation of phytoplankton

10 years. First signs of improved water quality are the reoccurrence of *Tolypella nidifica* and *Chara canescens* in the Rassower Strom (Yousef et al. 1997). In the Kleiner Jasmunder Bodden, the Secchi-depth has increased. In addition, growth of the benthos vegetation (*Potamogeton pectinatus*) and the occurrence of marine species (*Ceramium diaphanum* and *Balanus improvisus*) have been shown (Hübel and Dahlke 1999). This process could be accelerated by reduced rainfall, sunny weather and an increased entry of Baltic Sea water in the spring. This would create better lighting conditions for the growth of submersed macrophytes. A further possibility would be to improve connections to the Baltic Sea. For example, the widening and deepening of the shipping channel into the NRB has brought, through the increased exchange rate, at least a temporary reduction in eutrophication. This restoration is, in the first stages, certainly at the expense of the Baltic Sea. It is therefore only useful if external restoration processes are carried out at the same time.

In the DZB, external measures have yielded only modest effects so far. The reasons are to be found in the strong storage of nutrients in the sediment and in the highly active MFW. The degree of self-eutrophication already reached, and the stability of the ecosystem are therefore very high. Under these conditions, redevelopment could perhaps be accelerated via an internal intervention, i.e. the massive use of *Dreissena* mussels.

References

Arlt G (1984) Taxonomische und ökologische Untersuchungen am Mesozoobenthos von Brackgewässern. Habilitationsarbeit, University of Rostock, Rostock

Arlt G, Georgi F (1999) Meiofauna in der Benthisch–Pelagischen Kopplung. Bodden 8:19–24

Arndt E-A (2001) Die Bodentierwelt der Darß-Zingster Boddenkette. Meer Museum 16:65–71

Arndt H (1985) Untersuchungen zur Populationsökologie der Zooplankter eines inneren Küstengewässers der Ostsee. Diss Biologie. University of Rostock, Rostock

Arndt H (1991) On the importance of planktonic protozoans in the eutrophication process of the Baltic Sea. Int Rev Hydrobiol 76:387–396

Arndt H, Jost G, Wasmund N (1990a) Dynamics of pelagic ciliates in eutrophic estuarine waters: importance of functional groups among ciliates and responses to bacterial and phytoplankton production. Arch Hydrobiol Beih Ergebn Limnol 34:239–245

Arndt H, Schiewer U, Jost G, Wasmund N, Walter T, Heerkloß R, Arlt G, Arndt E-A (1990b) The importance of pelagic and benthic microfauna in a shallow-water community of the Darß-Zingst estuary, Southern Baltic, during mesocosm experiments. Limnologica 20:101–106

Aurada K (2000) Hydrologische Aspekte der Bodden- und Haffgewässer. Bodden 10:23–30

Babenzien H-D, Voigt A (1999) Bedeutung der Sulfatreduktion. Bodden 8:67–72

Berghoff S, Schlungbaum G, Selig U (1999) Untersuchungen zum Phosphorkreislauf. Bodden 8:73–79

Bochert R, Zettler ML, Bick A (1994) Untersuchungen zur räumlichen Verteilung der Larven von *Marenzelleria viridis* (Polychaeta, Spionidae) in einem flachen Küstengewässer der Ostsee. Rostock Meeresbiol Beitr 2:227–240

Börner R (1984) Produktionsbiologisch–ökologische Untersuchungen am Phytoplankton des Zingster Stromes. Diss Biologie. University of Rostock, Rostock

Brosin H-J (1965a) Hydrographie und Wasserhaushalt der Boddenkette südlich des Darßes und des Zingstes. Veröff Geophys Inst Univ Leipzig 18:281–380

Brosin H-J (1965b) Die mittleren Salzgehaltsverhältnisse in der Boddenkette südlich des Darß und des Zingstes. Beitr Meerekunde 16:7–14

Dahlke S, Hübel H (1996) Der Kleine Jasmunder Bodden: Entwicklung eines hypertrophen Gewässers in Vergangenheit, Gegenwart und Zukunft. Bodden 3:83–98

Dahlke S, Louis A, Wolff C (1999) Kopplung zwischen Nitrifikation und Denitrifikation. Bodden 8:59–66

Feibicke M (1994) Strukturelle und funktionelle Studien zm Phytoplankton und Zooplankton in einer hypertrophen Brackwasserförde (Innere Schlei, Westliche Ostsee). Diss Botanik. Freie Univ Berlin, Berlin

Feibicke M (1995) Wege und Verbleib des Kohlenstoffs in der Inneren Schlei. Bodden 2:205–220

Flentge H (1972) Die Benthosvegetation der Schlei. Staatsexamensarbeit. Lehramt Schleswig-Holstein, Kiel

Fukarek F (1961) Die Vegetation des Darß und ihre Geschichte. Pflanzensoziologie 12:231–237

Garstecki T, Güber A, Premke K, Verhoeven R, Arndt H, Wickham SA (1999) Trophische Beziehungen innerhalb benthischer mikrobieller Nahrungsgewebe. Bodden 8:29–38

Geiß U, Bergmann I, Blank M, Schumann R, Hagemann M, Schoor A (2003) Detection of *Prochlorothrix* in brackish waters by specific amplification of pcb genes. Appl Environ Microbiol 69:6243–6249

Gerbersdorf S, Meyercordt J (1999) Primärproduzierende Prozesse in Pelagial und Benthal. Bodden 8:39–57

Gessner F (1937) Hydrographie und Hydrobiologie der Brackgewässer Rügens und des Darß. Kieler Meeresforsch II:1–80

Gessner F (1940) Produktionsbiologische Untersuchungen im Arkonabecken und in den Binnengewässern von Rügen. Kieler Meeresforsch 3:449–459

Gessner F (1957) Meer und Strand. VEB Deutscher Verlag der Wissenschaften, Berlin

Gocke K, Rheinheimer G (1994) Eutrophierung der Schlei – Vergangenheit und Gegenwart. Bodden 1:31–49

Gocke K, Rheinheimer G, Schramm W (2003) Hydrographische, chemische und mikrobiologische Untersuchungen im Längsprofil der Schlei. Schriften Naturwiss Vereins Schleswig-Holstein 68:31–62

Hammer A (2003) Ökophysiologie mixotropher Cryptophyceen. Diss Biologie, University of Rostock, Rostock

Heerkloss R (2001) Das Zooplankton der Darß-Zingster Boddenkette. Meer Museum 16:60–66

Heerkloß R, Klinkenberg G (1993) Chaotic dynamics of a plankton community in a species depleted mesocosm. Verh Int Verein Limnol 24:995–1000

Heerkloß R, Schnese W (1994) Jahreszeitliche Variabilität und Langzeitentwicklung des metazooplanktons im zentralen Teil der Darß-Zingster Boddenkette. Meeresbiolog Beitr 2:85–92

Heerkloß R, Burckardt R, Hammer A (1999) Funktionale Untersuchungen an dominanten Arten des Zooplanktons. Bodden 8:25–28

Herndl G (1992) Marine snow in the Northern Adriatic Sea: possible causes and consequences for a shallow ecosystem. Mar Microb Food Webs 6:149–172

Hoffmann (1937) Die Pflanzenwelt. In: Neubaur R, Jaeckel, S (eds) Die Schlei und ihre Fischereiwirtschaft. Teil III. Schriften Naturwiss Vereins Schleswig-Holstein 22:230–248

Hoppe E, Pankow H (1968) Ein Beitrag zur Kenntnis der Vegetation der Boddengewässer südlich der Halbinsel Zingst und der Insel Bock. Natur Naturschutz Mecklenburg 6:139–151

Hübel H (1968) Die Bestimmung der Primärproduktion des Phytoplanktons der Nord-Rügenschen Boddengewässer unter Verwendung der Radiokohlenstoffmethode. Int Rev Hydrobiol 53:601–633

Hübel H (1969) Die Primärproduktion des Phytoplanktons in Brackwässern unterschiedlichen Salz- und Nährstoffgehaltes. Limnologica 7:185–190

Hübel H (1973) Die Primärproduktion des Phytoplanktons in den Boddengewässern südlich des der Halbinseln Darß und Zingst im Jahr 1972 unter besonderer Berücksichtigung

der Ergebnisse einer synoptischen Aufnahme. Wiss Z Univ Rostock, Math nat R 31:1101–1104
Hübel H (1984) Die Bindung des molekularen Stickstoffs durch autotrophe und heterotrophe Mikroorganismen in Küstengewässern der DDR und an einigen Stationen der Ostsee. Habil. Univ Greifswald, Greifswald
Hübel H, Dahlke S (1991) Stoffhaushalt und Belastungsgrad des Kleinen Jasmunder Boddens. In: Landesamt für Umwelt und Natur Rostock, Rostock
Hübel H, Dahlke S (1999) Die Nordrügenschen Boddengewässer – Entwicklung in Vergangenheit, Gegenwart und Zukunft. Bodden 7:137–156
Hübel H, Wolff Ch, Meyer-Reil L-A (1998) Salinity, inorganic nutrients, and primary production in a shallow coastal inlet in the southern Baltic Sea (Nordrügenschen Bodden). Results from long-term observations (1960–1989). Int Revue Hydrobiol 83:479–499
Hupfer P (1959) Zur Hydrographie der Boddengewässer südlich des Darß. Wiss Z Univ Leipzig, Math-nat R 11:175–186
Hupfer P (1960) Beitrag zum Problem des Wasseraustausches in der Boddenkette südlich des Darß. Zur Fischerei und deren Hilfswiss. NF 9:447–456
Hurtig T (1957) Physische Geographie von Mecklenburg, Berlin
Jost G, Klinkenberg G (1994) Bakterien in der Darß-Zingster Boddenkette. Rostock Meeresbiol Beitr 2:69–75
Karner M, Herndl GJ (1992) Extracellular enzymatic activity and secondary production in free-living and marine-snow-associated bacteria. Mar Biol 113:341–347
Kell V, Martens B, Pankow H, Riesenweber S (1975) Die Mikroalgenbesiedlung der Darßer Boddengewässer (südliche Ostsee), Artenliste. Wiss Z Univ Rostock, Math-nat R 24:725–734
Kliewe H (1951) Die Klimaregionen Mecklenburgs. Eine geographische Untersuchung ihrer Ursächlichkeit nach mittel- und witterungsklimatischer Methode. Habil. Univ Greifswald, Greifswald
Koop K, Boynton WR, Wulff F, Carman R (1990) Sediment-water oxygen and nutrient exchanges along a depth gradient in the Baltic Sea. Mar Ecol Progr Ser 63:65–77
Lenz J (1970) Seston-, Chlorophyll- und Eiweißgehalt. Kieler Meeresforsch 26:343–346
Lindner A (1972) Soziologisch–ökologische Untersuchungen an der submersen Vegetation in der Boddenkette südlich des Darß und Zingst. Diss Biologie. Univ Rostock, Rostock
Lindner A (1978) Soziologisch-ökologische Untersuchungen an der submersen Vegetation in der Boddenkette südlich des Darß und Zingst (südliche Ostsee). Limnologica 11:229–305
Lupatsch I, Nellen W (1981) Der Zustand der Fischbestände in der Schlei und die Entwicklung der Fischerei im Zeitraum 1962–1981. Ber Inst Meereskd Univ Kiel 91:1–89
Magnus P (1875) Bericht über die botanischen Ergebnisse der Untersuchung der Schlei vom 7–10. Juni 1874. Verhdlg Bot Verein Provinz Brandenburg 17
Messal C, Göcke W, Holzhüter G, Gerber T (1999) Physik und Biologie der Aggregation. Bodden 8:13–18
Meyer-Reil L-A (1999) Verbundprojekt ÖKOBOD – Beitrag zum wissenschaftlichen Verständnis der Ökologie der Boddengewässer. Bodden 8:87–105
Nellen W (1963) Fischereibiologische und faunistische Brackwasseruntersuchungen in der Schlei, einer Ostseeförde Schleswig-Holsteins. Diss Math-Nat. Univ Kiel, Kiel
Nellen W (1967) Ökologie und Fauna (Makroevertebraten) der brackigen und hypertrophen Ostseeförde Schlei. Arch Hydrobiol 63:273–309
Neuendörfer F, Meyer-Reil L-A (1997) A microbial biosensor for the microscale measurement of bioavailable organic carbon in oxic sediments. Mar Ecol Progr Ser 147:295–300
Niedermeyer RO, Liewe H, Janke W (1987) Die Ostseeküste zwischen Boltenhagen und Ahlbeck, Gotha
Overbeck J (1956) Untersuchungen über die Standortverhältnisse an den Wuchsorten von *Fucus vesiculosus* in den Gewässern von Hiddensee und Rügen (*Fucus*-Untersuchungen II). Arch Protistenkde 101:401–442

Overbeck J (1962) Das Nanoplankton (μ-Algen) der Rügenschen Brackgewässer als Hauptproduzent in Abhängigkeit vom Salzgehalt. Kieler Meeresforsch 18:157–171
Overbeck J (1964) Der Fe/P-Quotient des Sediments als Merkmal des Stoffumsatzes in Brackgewässern. Helgol Wiss Meeresunters 10:430–447
Pankow H, Wasmund N (1994) Produktionsbiologie und Soziologie des Makro- und Mikrophytobenthos der Darß-Zingster Boddenkette. Rostock Meeresbiol Beitr 2:61–68
Pratt JR,Cairns J Jr (1985) Long-term patterns of protozoan colonization in Douglas Lake, Michigan. J Protozool 32:95–99
Prena A (1990) Zur Bedeutung heterotropher Protozoen im Nahrungsgefüge des Zingster Stromes (Darß-Zingster Boddenkette). Diss Biologie. Univ Rostock, Rostock
Remane A (1937) Die übrige Tierwelt. In: Neubaur R, Jaeckel S (eds) Die Schlei und ihre Fischereiwirtschaft. Teil III. Schriften Naturwiss Vereins Schleswig-Holsteins 22:209–224
Rieling T (1999) Remineralisierung organischen Materials in Boddengewässern Mecklernburg-Vorpommerns unter besonderer Berücksichtigung der Bedeutung von Partikeln und Aggregaten. Diss Biologie. Univ Greifswald, Greifswald
Rieling T, Stodian I, Blank J, Köster M, Meyer-Reil L-A (1999) Aspekte des mikrobiellen Kohlenstoffkreislaufes. Bodden 8:49–57
Ripl W (1986) Restaurierung der Schlei. In: Landesamt für Wasserhaushalt und Küsten Schleswig-Holstein, Kiel
Scharf E-M (1979) Die Bedeutung benthischer heterotropher Protozoen (Ciliaten) im Stoffkreislauf und im Energiefluss eines mesohalinen Brackgewässers (Barther Bodden). Diss Biologie. Univ Rostock, Rostock
Schiemann S (1974) Die Primärproduktion des Phytoplanktons in der Schlei und des Windebyer Noors im Jahre 1972. Diss Botanik. Univ Kiel, Kiel
Schiewer U (1990) Werner Schnese and the development of coastal waters ecology in Rostock, GDR. Int Rev Hydrobiol 75:1–13
Schiewer U (1994a) Vorkommen und Bedeutung der Protozoen in der Darß-Zingster Boddenkette. Rostock Meeresbiol Beitr 2:77–83
Schiewer U (1994b) Stoffkreisläufe in der Darß-Zingster Boddenkette: Kohlenstoffkreislauf. Rostock Meeresbiol Beitr 2:121–137
Schiewer U (1998a) 30 years eutrophication in shallow brackish waters – lessons to be learned. Hydrobiologia 363:73–79
Schiewer U (1998b) Hypertrophy of a Baltic estuary – changes in structure and function of the planktonic community. Verh Int Verein Limnol 26:1503–1507
Schiewer U (2001) Phytoplankton, Produktivität und Nahrungsnetze. Meer Museum 16:39–45
Schiewer U, Jost G (1991) Microbial food web in eutrophic shallow brackish estuaries of the Baltic Sea. Int Rev Hydrobiol 76:339–350
Schiewer U, Wünsch JU (1996) Heterotrophic nanoflagellates: Spring development in two different estuarine biotopes. In: Styczynska-Jurewicz E (ed) Proceedings of the 2nd International Estuarine Symposium. Crangon 1, Issues Mar Biol Centre Gdynia, Gdansk, pp 169–179
Schiewer U, Rentsch D, Görs S, Schumann R (2002) Aggregates and bacterial populations in fluffy sediment layers from shallow brackish waters. Verh Int Verein Limnol 28:1551–1557
Schlungbaum G (1994) Das Eutrophierungsproblem der Bodden und Haffe. Bodden 1:13–29
Schlungbaum G, Baudler H (1998) Erfordernisse, Möglichkeiten und Grenzen für eine nachhaltige Verbesserung der Gewässerbeschaffenheit in der Darß-Zingster Boddenkette. Studie im Auftrag des BAUN-MV, LAUN-MV und STAUN, Stralsund
Schlungbaum G, Baudler H, Krech M (2001) Das Eutrophierungsproblem der Darß-Zingster Bodden – Nährstoffeinträge und Nährstoffbilanzen. Meer Museum 16:25–34
Schmelzer N (1994) Die Eisverhältnisse in den Küstengebieten von Mecklenburg-Vorpommern. Die Küste, Arch Forsch Tech Nord Ostsee 56:51–56
Schnese W, Hübel H (1976) Über den Verlauf und die Größe der Phyto- und Zooplanktonproduktion in Brackgewässern mit abgestuftem Salzgehalt. Limnologica 10:459–465
Schubert H (1996) Ökophysiologie der Lichtanpassung des Phytoplanktons eutropher Flachgewässer. Habilitationsarbeit, University of Rostock, Rostock

Schubert H, Bahnwart M, Blümel C, Eggert A, Krause J, Rieling T, Sagert S, Schubert M, Selig U (2005) Ökologische Bewertung innerer Küstengewässer mittels Phytoplankton und Makroalgen und Angiospermen. In: Feld C, Rödiger S, Sommerhäuser M, Friedrich G (eds) Typologie, Bewertung und Management von Oberflächengewässern. Schweizerbart, Stuttgart, pp 151–164

Schumann R (1993) Zur Rolle des Pico- und Nanophytoplanktons im mikrobiellen Nahrungsgefüge der Darß-Zingster Boddenkette. Diss Biologie. Univ Rostock, Rostock

Schumann R, Schiewer U (1994) Influence of abiotic induced phytoplankton changes on protozoan communities from the Darss-Zingst Bodden chain. Mar Microb Food Webs 8:265–282

Schumann R, Rentsch D (1998) Staining particulate organic matter with DTAF – a fluorescence dye for carbohydrates and protein: a new approach and application of a 2D image analysis system. Mar Ecol Progr Ser 163:77–88

Schumann R, Rentsch D, Görs S, Schiewer U (1999) Charakterisierung und Klassifizierung von Aggregaten. Bodden 8:5–11

Schumann R, Rentsch D, Görs S, Schiewer U (2001) Seston particles along a eutrophication gradient in coastal waters of the southern Baltic Sea: significance of detritus and transparent mucoid material. Mar Ecol Progr Ser 218:17–31

Schumann R, Schiewer U, Karsten U, Rieling T (2003a) Viability of bacteria from different aquatic habitats. II. Cellular fluorescent markers for membrane integrity and metabolic activity. Aquat Microb Ecol 32:137–150

Schumann R, Rieling T, Görs S, Hammer A, Selig U, Schiewer U (2003b) Viability of bacteria from different aquatic habitats. I. Environmental conditions and productivity. Aquat Microb Ecol 32:121–135

Schumann R, Hammer A, Görs S, Schubert H (2005) Winter and spring phytoplankton composition and production in a shallow eutrophic Baltic lagoon. Estuarine Coastal Shelf Sci 62:169–181

Schwarz S (1961a) Produktionsbiologische Untersuchungen am Zooplankton der Rügenschen, Hiddenseer und Darßer Boddengewässer (1953 bis 1955). I. Ciliaten. Z Fischerei 10:401–428

Schwarz S (1961b) Produktionsbiologische Untersuchungen am Zooplankton der Rügenschen, Hiddenseer und Darßer Boddengewässer (1953 bis 1955). II. Rotatorien. Z Fischerei 11:641–672

Schwarz S (1964) Produktionsbiologische Untersuchungen am Zooplankton der Rügenschen, Hiddenseer und Darßer Boddengewässer (1953 bis 1955). III. Crustaceen. Z Fischerei 12:13–44

Stockenberg A (1998) The role of sediments in nitrogen cycling in a larger Baltic Sea. Diss Univ Stockholm, Stockholm

Stotz W (1986) Das Makrozoobenthos der Schlei. Produktion und Bedeutung als Fischnahrung. Diss Univ Kiel, Kiel

Trahms O (1937) Zur Kenntnis der Salzverhältnisse und des Phytoplanktons der Hiddenseer und Rügenschen Boddengewässer. Arch Hydrobiol 32:75–90

Trahms O (1939a) Beiträge zur Ökologie küstennaher Gewässer. 1. Das Plankton des Großen Jasmunder Boddens. Arch Hydrobiol 35:529–551

Trahms O (1939b) Beiträge zur Ökologie küstennaher Gewässer. 2. Die Bodenfauna und Bodenflora des Großen Jasmunder Boddens. Arch Hydrobiol 36:1–35

Trahms O (1941) Die Bedeutung des Untergrundes für die Nährstoffregeneration in stark eutrophen Küstengewässern. Z Fischerei 39:725–748

Vietinghoff U (1982) Die mathematische Modellierung des Ökosystems Barther Bodden als Beitrag zur Systemanalyse der Boddenkette West. Habil. Univ Rostock, Rostock

Wasmund N (1987) Die Primärproduktion des Phytoplanktons im Zingster Strom (südliche Ostsee) in den Jahren 1984–1986. Wiss Z Univ Rostock Math-Nat R 36:17–21

Wasmund N (1994) Phytoplankton periodicity in an eutrophic coastal water of the Baltic Sea. Int Rev Hydrobiol 79:259–285

Wasmund N, Schiewer U (1992) Überblick zur Ökologie und Produktionsbiologie des Phytoplanktons der Darß-Zingster Boddenkette (südliche Ostsee). Rostock Meeresbiol Beitr 2:41–60

Winkler H (2001) Fischgemeinschaften und Fischerei in den Darß-Zingster Bodden. Meer und Museum 16:76–84

Yap M (1987) Aspects of benthic energy dynamics in a shallow Baltic ecosystem. Diss Biologie. Univ Rostock, Rostock

Yousef M, Küster A, Schubert H, Nordheim H von (1997) Charakterisierung der Characeenbestände an der Küste Mecklenburg-Vorpommerns. Bodden 5:3–23

Zettler ML (1996) Successful establishment of spinoid polychaete, *Marenzellaria viridis* (Verrill, 1873), in the Darss-Zingst estuary (southern Baltic) and its influence on the indigenous macrozoobenthos. Arch Fish Mar Res 43:273–284

Chapter 4
Greifswalder Bodden, Wismar-Bucht and Salzhaff

U. Schiewer

4.1 Introduction

The Greifswalder Bodden, Wismar-Bucht and Salzhaff have an open connection to the Baltic Sea in common. Loads originating from the shore-side can thus be transported to the adjacent Baltic basins more quickly. The shorter residence times reduce the possibility of self-purification of these waters. Loads and disturbances in these ecosystems have an immediate effect on the Baltic Sea.

4.2 Greifswalder Bodden

Older studies by Abshagen (1908), Fraude (1906), Lemmermann (1901), Seifert (1936, 1938) and Stammer (1928) dealt with phytoplankton and/or the soil fauna. The first investigations into fishery were carried out by Henking (1904). In the past two decades, the Greifswalder Bodden (GB) has been repeatedly investigated within large projects. In the 1980s, the spawning and migration behaviour of herring was analysed, as well as the influence of the cooling water of a nuclear power station on the bodden ecosystem. In the 1990s, investigations concentrated mainly on exchange and conversion processes in the GB (GOAP 1998; TRUMP 1998).

4.2.1 Environmental Characteristics

Covering an area of 514 km^2, the GB (Table 4.1) is the largest bodden of the German Baltic coast. Its water volume amounts to 2,960 × 10^6 m^3. It has an average depth of 5.8 m and a maximum depth of 13.5 m. There is no pronounced stable stratification. The catchment area comprises only 665 km^2. The salinity ranges between 5.3 and 12.2 psu (average 7.5 psu). There is a broad connection to the Baltic Sea, which is limited in its effectiveness only in the region of the Ruden by a 2 m deep sill (Fig. 4.1). Exchange processes with the Baltic Sea dominate. At 34,419 × 10^6 m^3 a^{-1}, the water exchange with the Baltic Sea is intensive and the

U. Schiewer (ed.), *Ecology of Baltic Coastal Waters. Ecological Studies 197.*

Table 4.1 Main characteristics of the Greifswalder Bodden (GB), Wismar-Bucht (WB) and Salzhaff (SH).

dw dry weight, *ww* wet weight, Chl *a* chlorophyll *a*, *POC* particulate organic carbon, *DOC* dissolved organic carbon, *POM* particulate organic matter, *DIN* dissolved inorganic nitrogen, *DIP* dissolved inorganic phosphorus

	Greifswalder Bodden[a]	Wismar Bucht[b]	Salzhaff[c]
a. Climate and hydrology			
Air temperature (°C)	No data	No data	No data
Water temperature (°C)	+ 3.1 (max: 25.0)	No data	Min: –1.0; max: +24.8
Ice covering (days a^{-1})	50	No data	No data
Mean freshwater inflow ($10^6 m^3 a^{-1}$)	106	No data	No data
Mean inflow (from Baltic) ($10^6 m^3 a^{-1}$)	34,419	No data	No data
Mean total outflow ($10^6 m^3 a^{-1}$)	34,525	No data	No data
Residence time (days)	36	No data	No data
Water exchange rate (a^{-1})	0.10	No data	No data
b. Morphometry and sediments			
Area (km^2)	514	170	21
Volume ($10^6 m^3$)	2,960	No data	67
Mean depth (m)	5.8	5.5	2.3
Maximum depth (m)	13.5	12.0	10.0
Catchment area (km^2)	665	1,059	310
Area/catchment relation	1.3 : 1	6.3 : 1	15.0 : 1
Denitrification rate (µmol $m^{-2} h^{-1}$)	7	No data	No data
Sediment distribution	Mostly sandy; partly muddy	Mostly sandy; partly muddy	Mostly sandy

c. **Pelagic biological components**			
Secchi depth (m)	4–6.5; nearshore < 2.0	No data	No data
Bacteria (10^6 ml^{-1})	0.6–18.0	No data	No data
Phytoplankton biomass (μg Chl *a* l^{-1})	20–90	No data	No data
Phytoplankton biomass (mm^3 l^{-1})	< 1.0–8.0 (spring and autumn maxima)		No data
Dominating phytoplankton species	*Achnanthes taeniata*, *Cyclotella caspia*, *Detonula confervacea*, *Kirchneriella irregularis*, *Skeletonema costatum*, *Gomphosphaeria pusilla*, *Carteria cordiformis*, *Rhodomonas minuta*, *Stephanodiscus hantzschii*, *Merismopedia punctata*, *Eutreptia lanowii*, *Chaetoceros wighamii*, *Diatoma elongatum*	No data	56 diatom species: *Hantzschia spectabilis*, *Chaetoceros holsaticus*, *Ch. densus*, *Ch. decipiens*, *Thalassiothrix nitzschioides*, *Cocconeis scutellum*, *Skeletonema costatum*, *Coscinodiscus excentricus*, *C. radiatus*, *Bacillaria paradoxa* 12 dinoflagellate species: *Ceratium tripos*, *Prorocentrum micans*, *Nodularia spumigena*, *Aphanizomenon flos-aqua*, *Distephanus speculum*
Ciliates			
Dominating species	*Arsellus vulgaris*, *Cyphoderia ampulla*, *Difflugia lobostoma*, *Strombidium* sp., *Tintinnidium fluviatilis*, *Tintinnopsis beroides*, *T. tubulosa*, *T. lohmanni*, *Leprotintinnus bottnicus*, *Lohmaniella* sp.	No data	No data
Mesozooplankton			

(continued)

Table 4.1 (continued)

	Greifswalder Bodden[a]	Wismar Bucht[b]	Salzhaff[c]
Dominating species	*Asplanchna* spp., *Keratella cochlearis f. tecta*, *K. cochlearis* var. *f. typica*, *K. cochlearis*, *K. quadrata*, *K. cruciformis* var. *eichwaldi*, *Synchaeta* sp., *Brachionus* spp., *Nothalca* sp., *Bosmina coregoni maritima*, *Podon leuckarti*, *Evadne nordmanni*, *Acartia bifilosa*, *A. longiremis*, *Centropages* sp., *Cyclops* sp. *Eurytemora affinis*, *Temora longirostris*, *Pseudocalanus minutus elongatus*	No data	No data
Fish			
Number of species	47	51	51
Dominating species	*Clupea harengus*, *Gadus morhua*, *Platichthys flesus*, *Psetta maxima*, *Belone belone*, *Salmo trutta trutta*, *Coregonus lavaretus balticus*, *Anguilla anguilla*, *Esox lucius*, *Sander lucioperca*, *Perca fluviatilis*, *Rutilus rutilus*, *Abramis brama*, *Gymnocephalus cernuus*	*Clupea harengus*, *Anguilla anguilla*, *Gadus morhua*, *Platychthys flesus*, *Psetta maxima*, *Sprattus sprattus*, *Salmo salar*, *S. trutta trutta*, *Zoarces viviparus*, *Belone belone*, *Perca fluviatilis*, *Mugil cephalus*, *Pleuronectes platessa*, *Limanda limanda*. Other catches: *Palaemon adspersus*, *P. squilla*	*Clupea harengus*, *Anguilla anguilla*, *Gadus morhua*, *Platychthys flesus*, *Psetta maxima*, *Lampetra fluviatilis*, *Petromyzon marinus*
DOC (mg l^{-1})	46.6	No data	No data
Primary production (g C m^{-2} a^{-1})	279 (min: 140; max: 280)	No data	No data

Primary production (mg C m^{-2} day^{-1})	100–2,500	No data	No data
Fish catches (kg ha^{-1})	15.1	No data	343.7
d. Benthic biological components			
Macrophytes			
Number of species	56	> 60	> 50
Dominating species	*Cladophora glomerata, Enteromorpha intestinalis, Pilayella litoralis, Fucus vesiculosus, F. serratus, Polysiphonia nigrescens, Ceramium diaphanum, Zostera marina, Potamogeton pectinatus*	*Enteromorpha* spp., *Cladophora sericea, Cl. glomerata, Fucus vesiculosus, F. serratus, Chorda filum, Ceramium rubrum, Polysiphonia diaphanum, Zostera marina, Ch. baltica, Zannichellia palustris*	*Zostera marina, Potamogeton pectinatus, Ruppia Ruppia cirrhosa, R. maritima, Zannichellia palustris*; *Chara canescens, Ch. baltica, Tolypella nidifica*; *Enteromorpha* spp., *Chaetomorpha linum, Ulva lactuca, Fucus vesiculosus, Chorda filum*
Macrozoobenthos			
Biomass (g m^{-2})	ca. 18 (dw)		550–2,300 (ww)
Number of species	ca. 40		90

(continued)

Table 4.1 (continued)

	Greifswalder Bodden[a]	Wismar Bucht[b]	Salzhaff[c]
Dominating species	*Hydrobia ulvae, H. ventrosa, Potamopyrgus jenkinsi, Theodoxus fluviatilis, Mytilus edulis, Cerastoderma lamarcki, Macoma balthica, Mya arenaria, Hediste diversicolor, Manayunkia aestuarina, Fabricia sabella, Pygospio elegans, Polydora ciliata, Dreissena polymorpha, Marenzelleria viridis, Streblospio shrubsoli, Balanus improvisus, Cythomorpha fuscata, Jaera albifrons, Sphaeroma hookeri, Idothea chelipes, I. balthica, Cyathura carinata, Corophium voluntator, Gammarus salinus G. oceanicus* etc. *Leptocheiris pilosus*	see SH	*Mytilus edulis, Hydrobia ventrosa, H. ulva, Cerastoderma lamarcki, Hediste diversicolor, Microdeutopus gryllotalpa, Jaera albifrons, Pygospio elegans, Heteromastus filiformis, Sphaeroma hookeri, Cyathura carinata*
e. **Water chemistry, trophic status and pollution**			
Salinity (psu)	7.5 (min: < 5.3; max: 12.2)	10.5 (min: 7.8; max: 19.6)	10.5 (min: < 5.0; max: 15.0)
pH	7.8 (max. 9.0)	No data	No data
Oxygen saturation (%)	100 (never < 80)	60–160	No data
Total nitrogen (µmol l^{-1})	376	344	No data
DIN (µmol l^{-1}) (winter maximum)	< 5.0 (max: 20)	No data	No data
Total phosphorus (µmol l^{-1})	4.7	36.7	No data
DIP (µmol l^{-1})	2.0–5.0	No data	No data

Annual nitrogen input (t a^{-1})	> 626	No data	ca. 772
Annual phosphorus input (t a^{-1})	ca. 50	No data	ca. 12
	Temperature (early spring)		
	Phosphorus (April)	Temperature, P, N, grazing; see GB	Temperature, P, N, grazing; see GB
Limitation of PP and period	Nitrogen (May/August)		
	Grazing in late summer		
Main pollutants	N, P	N, P	N, P
Trophic level	Mesotrophic / eutrophic	Eutrophic / mesotrophic	Mesotrophic

[a]Data from Blümel et al. 2002; Brenning 1989; Geisel 1986; GOAP 1998; Hübel et al. 1995; Kell 1984, 1989; Schlungbaum et al. 1989; Schnese 1957; Steyer 1995; Winkler 1989; Yousef et al. 1997

[b,c]Data from Arndt et al. 1966; Bachor et al. 1995; Gosselck and von Weber 1997; Jaschhof 1990; Köhn et al. 1991; Walter 1996, 1997; Weber 1990

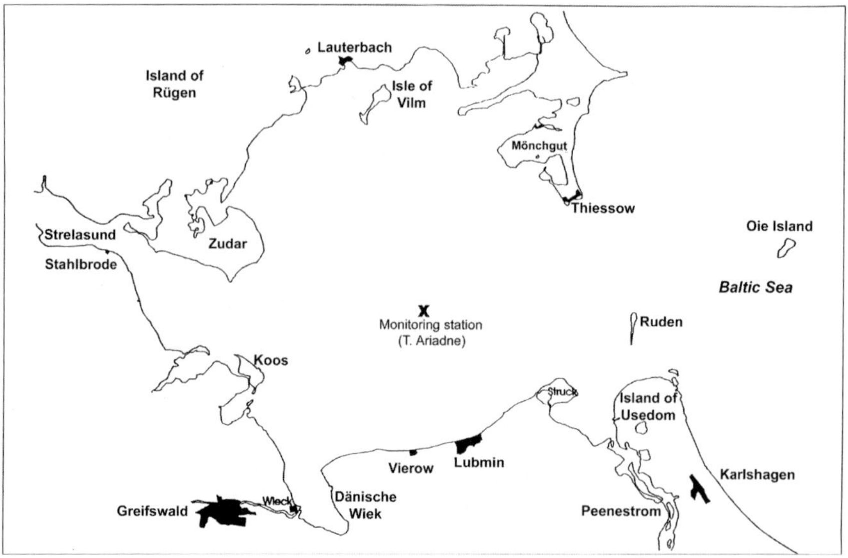

Fig. 4.1 The Greifswalder Bodden (GB) showing the position of the central monitoring station "Ariadne" (modified from Hübel et al. 1995)

residence time of $0.10\,a^{-1}$ very short. This means that the water volume of the GB is replaced almost 12 times per year. At $10^6 \times 10^6\,m^3\,a^{-1}$, the freshwater entry is very low and constitutes only 0.3% of the total inflow. Riverine water enters the Dänische Wiek via the brooks Ryck and Ziese.

The exchange with the Baltic Sea differs with the season (Jönsson et al. 1998). The inflow takes place from the beginning of the year until the end of July, preferably via the Strelasund – the outflow at the easterly edge of the GB. From August to October, the Strelasund and the eastern edge north of the Ruden are involved in the inflow, whereas the runoff takes place at the eastern edge south of the Ruden. From November until the end of the year the inflow comes from the eastern edge, and the outflow via the Strelasund. Larger water movements are characteristic of the winter, while small-scale, uniform exchange predominates in the summer. The decisive control factors are the wind conditions (frequency, direction and strength) as well as the air pressure over the central Baltic Sea. The main wind directions are south-west and west. Figure 4.2 presents the current conditions in the first half of the year.

Temporary water and nutrient loads from distant sources like the Peene and the Oder/Odra river enter the GB when south easterly or easterly winds are prevalent.

The ratio of catchment/water surface area is 1.3:1, which means that the amount of natural eutrophication is low. Furthermore, the share of agriculture is relatively small in the bodden catchment area. Tourism dominates in this region. In the 1970s and 1980s a nuclear power station was sited here, for which the GB served as a supplier of cooling water. The release of warm water showed no significant effects

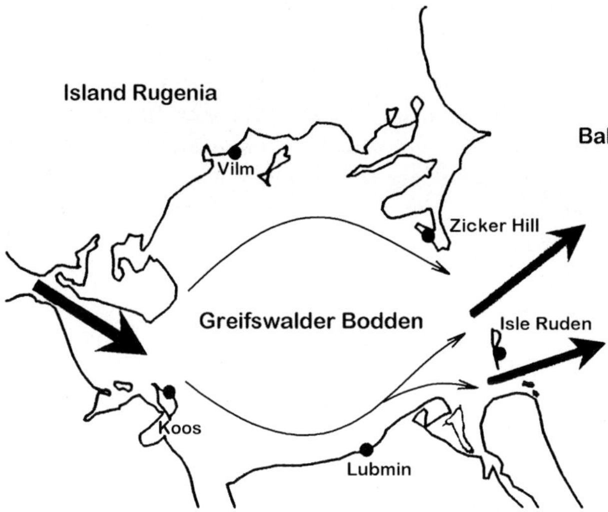

Fig. 4.2 Currents in the GB under the influence of the main wind direction – southwest/west (Jönsson et al. 1998)

on the aquatic environment. Currently, this region is being converted into an industrial area. There are no further industrial sites.

The GB is characterised by the "Baltic climate" (see Chap. 3 by Schiewer, this volume). The water temperature can reach 25°C in summer. Mean water temperatures are always close to mean air temperatures (9–10°C). On average, 50 days with ice-cover are observed annually in the GB (Blümel et al. 2002) but these are often separated into shorter ice cover periods during winter. Complete and long-lasting ice coverage, however, is shorter (max. 40 days).

The water transparency in the open parts was 4–6.5 m in 1993/1995, near the shore it was frequently <2 m. The pH-value can reach pH 9 during the vegetation period. The mean value is pH 7.8. In December, pH values seldom exceed pH 8.

Due to intensive mixing, the GB is usually vertically and horizontally homogeneously saturated with oxygen (ca. 8 mg O_2 l^{-1}) Thus, between 1966 and 1990 monthly average values never fell below 6 mg O_2 l^{-1}.

The biochemical oxygen demand (BOD_3) averaged 0.75 ml l^{-1} in 1993/1995. Maximum values up to 1.5 ml l^{-1} occurred in December/January, March/July and in the autumn. These are a measure of pollution of the water by aerobically degradable substances.

At values of 4.7 t a^{-1} total-P and 376 t a^{-1} total-N (HELCOM 1998), the total pollution of the GB via rivers is low compared to the water volume of the GB. Thus, a significant load is exerted mainly on the Dänische Wiek (see Fig. 4.1).

The concentrations of dissolved o-PO_4 inorganic nitrogen and silicate are very high at the start of the vegetation period. Particularly noticeable are the very variable NH_4 concentrations, which range from 0.2 to 6 µmol l^{-1} in the winter months (Hübel et al. 1995). Maximum values were measured in the Dänische Wiek with 8.5 μmol l^{-1}. The nitrite-concentrations, in comparison, are usually between 0.2–1 μmol l^{-1}. Values up to 2.5 μmol l^{-1} are seldom measured. The maximum NO_3 concentration measured between 1972 and 1981 was 100 μmol l^{-1} in winter (Schmidt 1998). A maximum of 13 μmol l^{-1} was measured during the period 1993/1994. The average value of the winter half-year was lower than 5 μmol l^{-1}. The o-PO_4 concentrations reached a maximum of 5 μmol l^{-1} (1972/1981) and remains relatively constant at 2 μmol l^{-1} (1993/1995). Conspicuous are concentrations of 5.0 or 3.5 μmol l^{-1} in August. A silicate minimum can occur for a short time in May. The average monthly values for nitrogen and phosphorus remain on a low level from May to August (nitrogen until October) and rise clearly after October.

By coupling the pelagic nutrient data to a 2D flow model, Meyer et al. (1998) were able to develop the first theories on the transport of substances. According to their hypothesis, the GB serves as a nitrate source for the Baltic Sea during strong water exchanges. If the water exchange is low, approximately 25% of the internal turnover is maintained by inflows into the GB, i.e. it works as a nitrate sink. Concerning phosphorus, the GB represents a permanent source, fed by internal release processes, irrespective of water exchange.

Breaking up these results into monthly values reveals that the particle transport into the Baltic Sea takes place mainly in the months of January to May, i.e. at a time of high wind speeds, the spring peak of the plankton production and of entry by rivers. The lowest transports took place between September and December.

4.2.2 Planktonic Communities

4.2.2.1 Structure, Dynamics and Productivity

Investigations carried out by Kell (1989) from 1976 to 1985 comprised 223 phytoplankton species. Of these, 33 species belong to the Cyanobacteria, 3 to the Chrysophyceae, 128 to the Bacillariophyceae, 3 to the Cryptophyceae, 10 to the Dinophyceae, 39 to the Chlorophyceae and 7 to the Euglenophyceae. The most frequent species were *Achnanthes taeniata*, with a maximum of 70×10^6 cells l^{-1}, *Cyclotella caspia* 16×10^6 cells l^{-1}, *Detonula confervacea* 11.2×10^6 cells l^{-1}, *Kirchneriella irregularis* 9.8×10^6 cells l^{-1}, *Skeletonema costatum* 7×10^6 cells l^{-1}, *Gomphosphaeria pusilla* 6×10^6 cells l^{-1}, *Carteria cordiformis* 5×10^6 cells l^{-1}, *Rhodomonas minuta* 5×10^6 cells l^{-1}, *Stephanodiscus hantzschii* 5×10^6 cells l^{-1}, *Merismopedia punctata* 2.5×10^6 cells l^{-1}, *Eutreptia lanowii* 2.3×10^6 cells l^{-1}, *Chaetoceros wighamii* 1.5×10^6 cells l^{-1} and *Diatoma elongatum* 1.1×10^6 cells l^{-1}.

In principle, this composition did not change in the 1990s (Hübel et al. 1995). However, long-term monitoring always reveals more species. Thus, the short-term analysis does not reflect a decrease in the number of species, but only a less comprehensive account of the species spectrum.

The phytoplankton of the GB in terms of annual average is still clearly dominated by diatoms: Bacillariophyceae 40%, Chlorophyceae 34.5%, Cyanobacteria 15% and others 0.5%. Of the diatoms, only 42% are meso- to polyhalobiont; 73% of the species can be classified as freshwater species.

A further characteristic feature of the phytoplankton is its pronounced annual periodicity, which is basically controlled by temperature, nutrients and grazing processes. In spring and autumn, stenothermal cold water species, mostly diatoms, prevail. With increasing temperatures, sedimentation- and grazing processes cause an aspect change via a "Klarwasserstadium" (clear water stage). For the Zicker Höft, the following sequence exists:

- spring with ice-covered ground: diatom blooms of *Achnanthes taeniata*, *Chaetoceros holsaticum*, *Ch. wighamii* and *Leosira arctia*
- spring without ice coverage: *Skeletonema costatum*, *Diatoma elongata*. The control factor is at first temperature, followed by silicate limitations. Later, N and P can become limiting.
- In April/May/June the diatoms are replaced by *Rhodomonas* and the green algae *Scenedesmus*, *Monoraphidium contorta* as well as *Pediastrum*. The development of the filamentous Cyanobacteria *Oscillatoria limnetica* and *Aphanizomenon flos-aquae* begins and is controlled by N-limitation.
- In the summer, Cyanobacteria blooms are predominant. Representatives are *Gomphospheria pusilla*, *Anabaena* sp., *Lyngbya* sp., *Merismopedia* sp., and *Microcystis* spp. High cell numbers can also be reached by small centric diatoms, such as *Thalassiosira oceanica* and the green alga genus *Scenedesmus*. The reasons for this remain unclear, but the overall effect should be exhaustion of nitrogen, sedimentation and grazing pressure.
- In the autumn, dinoflagellate blooms are characteristic (*Ceratium* spp., *Prorocentrum micans*). In addition, there are *Gomphosphaeria pusilla* until December. Furthermore, *Aphanizomenon flos-aquae*, *Merismopedia tenuissima*, *M. punctata*, *Scenedesmus quadricauda*, *Nitzschia* spp. and *Navicula* spp. are available in considerable amounts.

Close to the coast, the annual periodicity of phytoplankton is quite different. This applies mainly to the Dänische Wiek. In this clearly eutrophic area, *Diatoma elongata* now hardly occurs in the spring, whereas *Eutreptia lanowii* (Euglenophyceae) is becoming dominant. In the summer, the differences in the species spectrum disappear, but the biomass is nearly twice as high. The reason is to be found in the increase in Cyanobacteria.

Biomasses and primary production show a pronounced spring peak, which is followed by several summer peaks (Fig. 4.3). A real autumn peak is missing. The biomass varies between <1 mm^{-3} l^{-1} and approximately 8 mm^{-3} l^{-1} (0.6 g C m^{-2}). The chlorophyll values ranged between 20 and 90 µg Chl *a* l^{-1} in 1993/1995. The annual average production amounts to 140–280 g C m^{-2} a^{-1} (Kell 1984 for 1981/1982; Westphal and Lenk 1998). Production per hour varies between 20 and 250 mg C m^{-2} h^{-1} (Westphal and Lenk 1998), explaining why the GB is considered meso-/eutrophic with regard to its primary production.

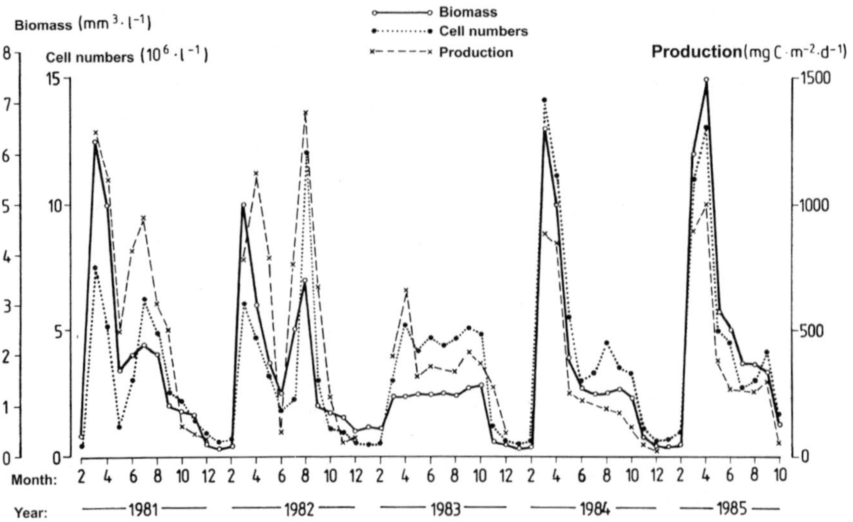

Fig. 4.3 Cell number, biomass and primary production 1981/1985 in the GB (Kell 1989). Data from station "Hagensche Wiek"

Analysis of saprophytic bacteria in the pelagial (1985/1996) revealed around 10^3 colony forming units (cfu) ml^{-1}. A considerable variability between space and time is typical (Westphal and Lenk 1998). In the central parts, which are less polluted, distinctly fewer cfu were found. A clear decline between 1993 and 1995 was noticeable in the whole investigation area. Annual variations are hidden by short-term hydrographic changes. Maximum values are to be found in the periods of November/February, March/June and August/October.

For the total number of bacteria, values are available only for the period from March 1994 to January 1995 from five standard sample stations (Steyer 1995). The total number of bacteria ranged between $0.6 \times 10^6 ml^{-1}$ and $18 \times 10^6 ml^{-1}$. The highest numbers were ascertained in May and August. A conversion into biomass produced a mean value of $2.8 \times 10^5 \mu m^3 ml^{-1}$ (0.3–8.1) and $0.085 \mu g$ C ml^{-1} (0.009–0.249). The cfu determined in parallel showed higher values in the western GB and thus indicated a higher load than in the eastern GB.

Extensive investigations of the zooplankton were undertaken by Schnese (1957). He proved the existence of 20 rotifer species, 10 copepod species, 3 phyllopode species, as well as larvae from polychaetes and balanides. In addition, he registered/found the following protozoans:

- the Rhipozoda *Arsellus vulgaris*, *Cyphoderia ampulla* and *Difflugis lobostoma*
- the ciliates *Strombidium* sp., *Tintinnidium fluviatilis*, *Tintinnopsis beroides*, *T. tubulosa*, *T. lohmanni*, *T. meunieri*, *T. parvula*, *T. nucula*, *T. campanula*, *T. karajacensis*, *Leprotintinnus bottnicus* and *Lohmaniella* sp.

Thus, the zooplankton of the GB is rather diverse. Table 4.2 gives an overview of species proved to be present in September 1993/1995. The majority of these species belong to the copepods. Within the rotifers, no further differentiation has been made. The coverage of the protozoans is restricted to the determination of the genus *Tintinnopsis* spp. In May, the tintinnides make up 70–80% of the counted individuals of the zooplankton. Due to their small size, however, their biomass rarely exceeds 10% of the zooplankton biomass. The species spectrum of the protozoans is probably distinctly more varied than one would think judging from the data available.

The mean biomass (100–1,300 mg fw m^{-3}) of the zooplankton follows a one-peak annual curve, with a maximum in June. Depending on the weather and the appearance of algae, strong variations can occur in the biomass of the major groups (Copepoda, Rotatoria, Tintinnida) throughout the year. Minor components are ciliates, Phyllopoda and larvae; in late autumn larvae of *Marenzelleria*.

What holds for the Darß-Zingst boddens (DZB) is also true of the nekton in the GB. Apart from *Neomysis integer* and the fish species (*Clupea harengus*, *Gadus morhua*, *Platichthys flesus*, *Psetta maxima*, *Belone belone*, *Salmo trutta trutta*, *Coregonus lavaretus balticus*, *Anguilla anguilla*, *Esox lucius*, *Sander lucioperca*, *Perca fluviatilis*, *Rutilus rutilus*, *Abramis brama*, *Gymnocephalus cernuus* and others), common jellyfish can occur in large numbers in the summer. Within the fish species, *Clupea harengus* dominates catches, at 93% of the marine species. The GB is one of the most important spawning grounds for herring in the Baltic coastal waters. Among freshwater fish species for human consumption, *Rutilus rutilus* is prevalent at 50%, followed by *Perca fluviatilis* (Winkler 1989). In the long-term development of catches we find an increase in the catches of *Esox lucius* and *Stizostedion lucioperca* from 1890 to 1985. However, herring is always dominant. The catches of herring in the GB reached a maximum of approximately 20,000 t a^{-1} in the 1980s. Compared to the situation 100 years ago, the spectrum of fish species caught has hardly changed. Only *Acipenser sturio* (sturgeon) and *Alosa fallax* (twaite shad) are nowadays completely lacking. Changes are also to be found in the catches of other species. From 1891 to 1900 the catches of freshwater fish amounted to 5.52 kg ha^{-1} and from 1976–1985, 15.12 kg ha^{-1}. This is rather low compared to the other internal boddens, where catches exceeded those of the GB by 20–40 kg ha^{-1} as a result of a better food supply due to eutrophication.

Table 4.2 Mesozooplankton species of the GB (Hübel et al. 1995). Species detected from September 1993 to March 1995

Rotatoria	*Asplanchna* spec., *Keratella* c. var. *tecta* f. *typica*, *K. cochlearis*, *K. quadrata*, *K. cruciformis* var. *eichwaldi*, *Synchaeta* spec., *Brachionus* spec., *Notholca* spec.
Phyllopoda	*Bosmina coregonie maritima*, *Podon leukarti*, *Evadne nordmanni*
Copepoda	*Acartia bifilosa*, *A. longiremis*, *Centropages* spec., *Cyclops* spec., *Eurytemora affinis*, *Temora longicornis*, *Pseudocalanus minutus elongatus*
Larvae	*Balanus*, *Nereis*, *Maranzelleria*, *Lamellibranchiata*, Gastropods

4.2.3 Benthic Communities

4.2.3.1 Structure and Productivity

Comprehensive studies by Seifert (1938) and Geisel (1986) on the coverage of submersed macrophytes are available, as well as additional investigations by Barthels and Klüber (1998) and Conrad et al. (1998). Investigations by Geisel (1986) at 31 stations revealed a species diversity of 9 Cyanobacteria, 15 Chlorophyceae, 7 Phaeophyceae, 9 Rhodophyceae, 3 Charophyceae and 13 Spermatophyta. An overview of the current colonisation areas at the single coastal sections is shown in Fig. 4.4.

Due to its relatively low inorganic and organic load, the GB is still rich in macrophytes. The main differences exist between the north-west and the south-west area. Whereas the former is definitely influenced by marine water, the latter shows, to a large extent, the influences of freshwater and eutrophication. Charophyceae have seldom been found in the GB since the 1980s. The three species proved by Geisel (1986) also occur in Lake Neuensien (*Chara aspera*, *Ch. canescens*), at the entrance of the Schoritzer Wiek (*Ch. aspera*, *Ch. canescens*) and on the east coast of Lake Zicker (*Ch. aspera*, *Ch. baltica*) (Yousef et al. 1997).

The following seasonal aspect change is characteristic of the coastal zone:

- In March/April, *Ulothrix* and *Urospora* develop. In April, *Pilayella litoralis* (according to later findings, *Ectocarpus siliculosus*) becomes the dominant species.

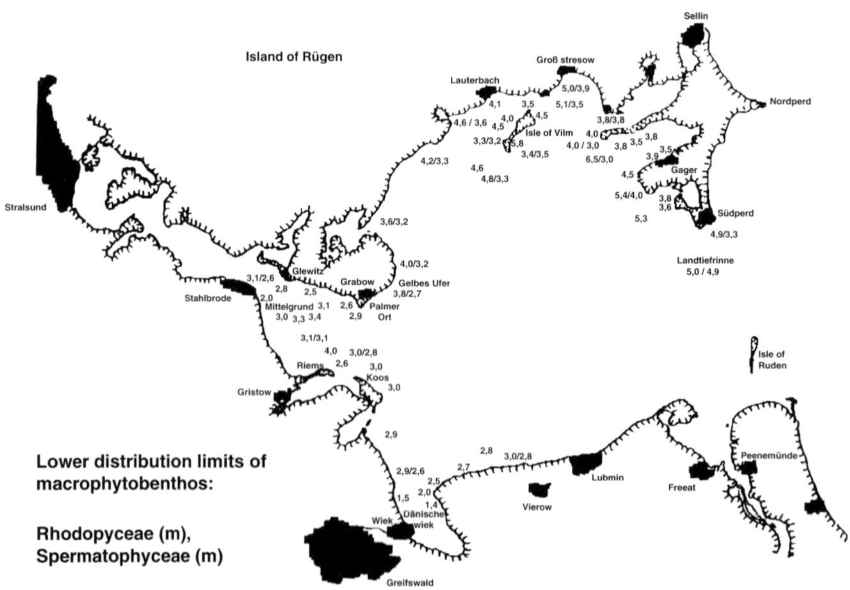

Fig. 4.4 Depth limits (growth depth in metres) of submersed macrophytobenthos (Hübel et al. 1995)

- Depending on the depth, the community restructures itself into an Enteromorpha community in May/June (0–0.5 m) or Cladophora glomerata community (0.5–2 m), or into a Potamogeton pectinatus community on sandy-to-muddy bottoms (0.5–3.5 m).
- In July/August, *Polysiphonia nigrescens* and *Ceramium diaphanum* (1.5–5 m) or *Zostera marina* (permanently on sandy bottom at 2.5–4 m depth) or *Fucus vesiculosus* and *Fucus serratus* (only Zickersches Höft at 0.2–3.5 m) become dominant.

The submersed macrophytes are estimated to fix a maximum of 2.2 g C m^{-2} a^{-1} (Conrad et al. 1998), which, according to Asmus et al. (1997), can be considered as net primary production. About two-thirds of submersed macrophytes are released again in the winter half-year. Estimates by Messner and Oertzen (1991) still assume a deposition in the sediment of 140 g C m^{-2} in the 1930s. The decline is due to eutrophication (see below).

The species number of the macrozoobenthos (Table 4.3) is clearly higher than in the DZB, but also clearly lower than in the Salzhaff (see below). The abundances and species composition depend largely on the salinity, flow, sediment structure and on primary production (Günther 1999).

The species *Dreissena polymorpha*, which has occurred to an increasing extent in recent years, is becoming important for the conversion of substances within the bodden. It currently covers an area of approximately 5.5 km^2 and its high filtration rate exerts a considerable influence on phytoplankton and suspended material.

The quantification of the macrozoobenthos is uncertain, mainly because investigations have been carried out only over two annual cycles (Günther 1999). However, macrozoobenthos is subject to an annual rhythm. The dominant groups are Mollusca (variation range of dominance over 2 years: 73–21%), Polychaeta (38–23%) and Crustacea (41–7%). Within the dominant representatives, it is above all the instability in the spring that is striking. For the southern GB, an average biomass of 12.95 g C m^{-2}, a production rate of 30.61 g C m^{-1} a^{-1} and a respiration rate of 38.67 g C m^{-1} a^{-1} have been reported (Günther 1999).

Table 4.3 Macrozoobenthos species of the GB (Hübel et al. 1995). Species detected from September 1993 to March 1995

Bryozoa	*Electra crustulenta*
Gastropoda	*Hydrobia ulvae, H. ventrosa, Potamopyrgus jenkinsi, Theodoxus fluviatilis*
Bivalvia	*Mytilus edulis, Cerastoderma lamarcki, Macoma balthica, Mya arenaria*
Polychaeta	*Hediste diversicolor, Neanthes succinea, Manayunkia aestuarina, Fabricia sabella, Pygospio elegans, Polydora ciliata, Marenzelleria viridis, Heteromastus filiformis, Bylgides sarsi, Streblospio shrubsoli*
Oligochaeta	*Tubifex costatus*
Crustacea	*Balanus improvisus, Cythomorpha fuscata, Diastylis rathkei, Jaera albifrons, Sphaeroma hookeri, Idotea chelipes, I. balthica, Cyathura carinata, Bathyporeia* spec., *Corophium voluntator, C. multistosum, Gammarus oceanicus, G. salinus, G. zaddachi, G. locustra, Leptocheirus pilosus, Crangon crangon, Neomysis integer*, Larvae of chironomides, *Gobiidae*

Jönsson et al. (1998), in comparison, ascertained biomasses of only 18 mg dw m^{-2} (approximately 9 g C m^{-2}). Since the majority of the benthos-organisms are filter feeders, a large part of the carbon required should stem from the phytoplankton.

4.2.3.2 Nutrient Cycles

More than 70% of the C-conversion takes place in the pelagic zone. Data on the role of the microbial food web (MFW) is lacking.

The conversion of nitrogen is controlled mainly by biological processes in the pelagic and benthic zones. Thus, the autotrophic nitrification rate in the pelagic averages 370 mmol N l^{-1} day^{-1}, which constitutes about 20–70% of the total turnover. The conversion process is positively correlated with the water temperature. Due to denitrification, averages of 7 µmol N m^{-2} h^{-1} are eliminated, which make up approximately 50% of the entry of riverine water.

In contrast to carbon and nitrogen, the conversion of phosphorus is basically controlled by sorption and fixation processes, which are closely correlated to the oxygen regime of the sediments and the near-bottom water as well as to resuspension processes. Release rates in the MFW have not been determined. For the GB, a permanent P-release is calculated (Meyer et al. 1998), the reason for which remains unclear. Showing a molar C_{org}/P-ratio of 70–100, the sediments of the GB are very rich in phosphorus compared to other bodden waters (Schlungbaum et al. 1989; Lampe and Meyer 1995). Therefore, above all a release due to balance-conditioned desorption is worthy of consideration.

The share of C-degradation in the benthic zone and the sediment amounts to about 30%. A precise analysis is missing, especially for the microbial processes in the sediment. However, a calculation of the long-term accumulation of nutrients in the sediments of the GB is available (Lampe 1998, 1999), amounting to 0.66 g dw ml^{-1}. This results in 0.5 mm sediment a^{-1} in a sewage area of 260 km^2 and a total area of 510 km^2, which corresponds to 4.4% of the carbon (7.4 g C m^{-2} a^{-1}; 3,780 t C a^{-1}), 0.54% of the nitrogen (0.90 g C a^{-1}; 460 t N a^{-1}) and 0.21% of the phosphorus (0.35 g P m^{-2} a^{-1}; 180 t P a^{-1}). This means that, as a result of the high water exchange with the Baltic Sea, only those substances that can be deposited in the benthic zone remain permanently in the GB. Statements about the transport of the sediment are not possible to date.

4.2.4 Benthic–Pelagic Coupling and Eutrophication

4.2.4.1 Biological Processes and Interactions

Seifert (1938) describes a 90% colonisation of the bottom area by submersed macrophytes.

Reconstructing macrophyte colonisation in the GB for the period 1851–1951 based on historical data, Blümel et al. (2002) attain similar results. Analysis of ecophysiological studies reveals that, under the photoclimatic conditions at that time, red algae were able to advance well to a depth of 10–1 m. The eutrophication that started in the 1960s led to a decline of the colonisation areas. On the basis of aerial photographs and scuba-diving evaluations, Barthels and Klüber (1999) assume only a 4.5% colonisation of the whole bottom area of the GB. For the shallow regions, percentages of coverage between 19.5 and 22.5% were ascertained (Hübel et al. 1995). This was also correlated with a decrease in colonisation depth. Whereas it was clearly below 6 m before 1960, it is currently between 3 and 4 m.

The reasons for the decline of the submersed macrophytes are complex. When classifying eelgrass, macrophytes and phytoplankton with regard to the biomass per square metre, the eelgrass ranks first and the phytoplankton takes third place. As far as the specific growth rate and the turnover rate are concerned, the order is reversed. Regarding efficiency and capacity of photosynthesis, phytoplankton is first, followed by eelgrass. Furthermore, grazing pressure, which is highest at the phytoplankton and lowest at the eelgrass, must be considered. A decline in these populations can generate a considerable increase in grazing pressure. Under the influence of eutrophication, slowly growing macrophytes, such as *Fucus* and *Furcellaria* are disappearing and being replaced by fast-growing green algae (above all *Cladophora* species) and fast-growing brown algae (mainly *Pilayella* and *Ectocarpus*). Table 4.4 summarises the eutrophication process for the specific conditions of the GB.

Due to the disappearance of submersed macrophytes, sediment mobility and water turbidity increase. The increased production of phytoplankton results in a muddy sediment. This leads to changes in soil fauna, and thus to changes in sediment structure. These follow-up processes can also be proved in the GB. The eastern part

Table 4.4 The eutrophication process in the GB (Schiewer and Schernewski 2004)

1. Increase in nutrient concentrations
2. Increased growth of phytoplankton
3. Reduced light for submersed macrophytes
   - Reduction of growth depth
   - Loss of slow-growing red and brown algae
   - Loss of eelgrass communities
   - Increase of fast-growing green and brown algae
4. Increased sediment mobility and turbidity
5. Expansion of mud covered bottoms

Bottoms and zoobenthos

Eastern bodden

- Sandy bottoms, dominated by high abundances and diversity of molluscs and ostracods
- Dominant ostracod species *Cytheromorpha fuscata*

Western bodden

- Muddy bottom, dominated by low abundances and low diversity of molluscs and ostracods
- Dominant ostracod species *Cyredeis torosa*

of the GB is characterised by sandy sediments. High abundances and a considerable diversity of Mollusca and Ostracoda are typical. The prevalent ostracode species is *Cytheromorpha fuscata*. In the western part of the GB, sewage muddy bottom, which has developed due to the increasing eutrophication within the past 40 years, predominates. Low abundances and a small diversity of molluscs and ostracods are typical of this area. The predominant ostracode species is *Cypredeis torosa.*

Summarising ecosystem analyses on the balance of materials of the GB stem from Hubert et al. (1995), who also modelled specific aspects. Within GOAP, Lampe (1998, 1999) published a preliminary balance of materials. It is calculated for the whole area and comprises the southern GB, the northern and southern Peenestrom, the "Achterwasser" and the Kleines Haff. Only the GB is considered here (Table 4.5). The secondary production of the zooplankton affects the phytoplankton mainly in the spring and in the summer. In individual measurements, 70% of the pelagic primary production were used up by respiration, which means that about 30% of the gross production would be available for benthic turnover. Lampe (1998, 1999) calculated a C-balance based on net primary production, the assimilation of consumers and bacterial turnover, (see Table 4.5).

The turnover of fish, submersed macrophytes and fungi makes hardly any difference. The same is true of the large number of migratory birds during a resting period (Leipe 1985).

4.3 Wismar-Bucht and Salzhaff

The Wismar-Bucht (WB), being a southern part of the Mecklenburg Bucht, reaches far into the hilly ground moraine landscape of North-West Mecklenburg. We differentiate between the outer Wismar-Bucht, with its very shallow Hannibal and

Table 4.5 Mean pools and mean annual process rates in the southern GB (Lampe 1999). Macrozoobenthos (Günther 1999; Jönsson et al. 1998), calculated after Heerkloss (1996) into carbon by Jönsson et al. (1998)

DOC (g C m^{-2})	46.6
PC = POC + PIC (g C m^{-2})	5.8
Brutto-primary production (g C m^{-2} a^{-1})	279.0
Respiration (g C m^{-2} a^{-1})	195.0
Netto-primary production (g C m^{-2} a^{-1})	84.0
Macrophytobenthos production (g C m^{-2} a^{-1})	2.2
Biomass macrozoobenthos (g C m^{-2})	12.0
Production (g C m^{-2} a^{-1})	24.0
Respiration (g C m^{-2} a^{-1})	35.0
Assimilation (g C m^{-2} a^{-1})	59.0
Microbial CO_2 and CH_4 emission from the sediment (g C m^{-2} a^{-1})	28.0
Fish biomass (g C m^{-2})	0.003
Consumption rate by fishes (g C m^{-2} a^{-1})	0.3
Carbon balance rate (g C m^{-2} a^{-1})	−1.0

Schweinskötel reefs as well as the Kraken Deep and the Wohlenberger Wiek, and the inner Wismar-Bucht (iWB), which comprises the area south of the Hohen Wischendorfer Huk and the Island of Poel including the Salzhaff (Fig. 4.5).

The Salzhaff (SH) is located east of the Isle of Poel and south of the Wustrow peninsula. It is connected with the outer WB by a 4 m deep channel, the Kielung. It is linked to the iWB by the shallow water area of the Breitling south of Poel.

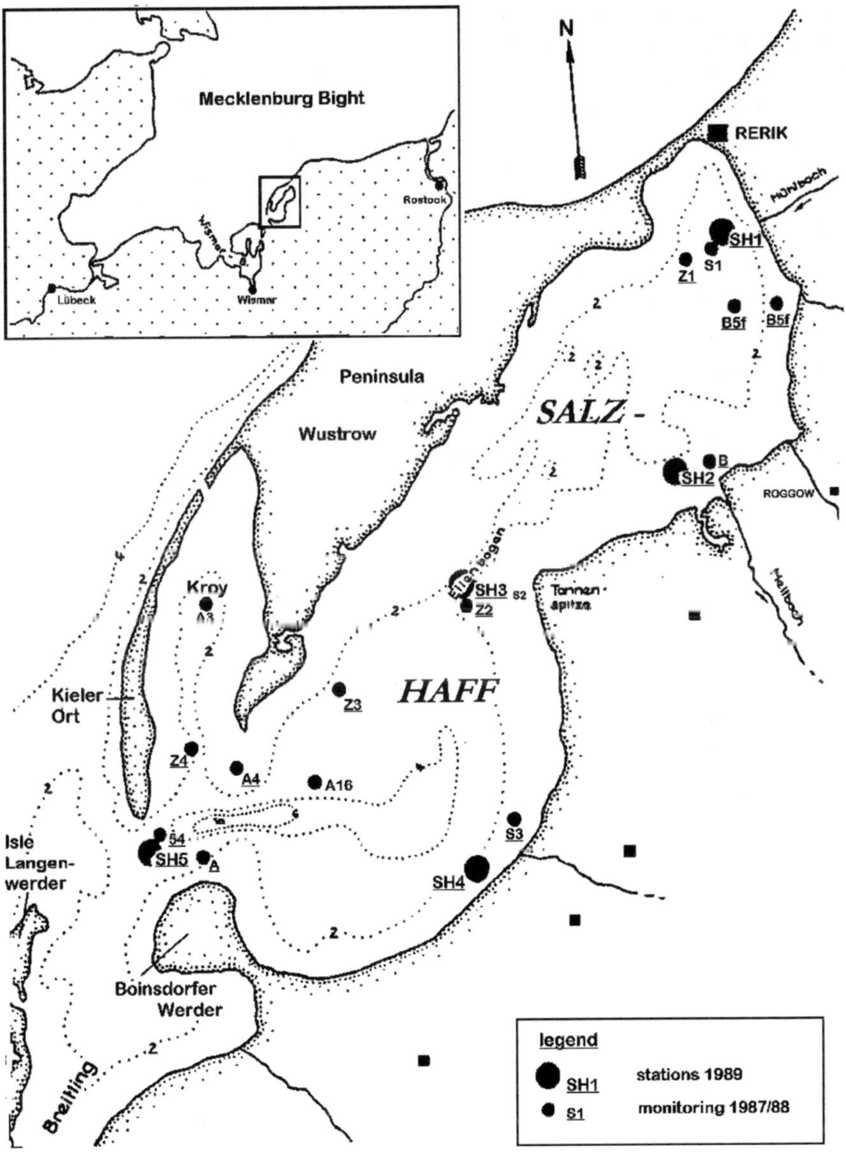

Fig. 4.5 The Wismar-Bucht (WB) and Salzhaff (SH)

In terms of flora and fauna, the WB and SH have not been much investigated. Important early studies were carried out by Krüger and Meyer (1937). Only in the 1980s were such investigations again taken up. The observation of coastal birds, however, has had a long tradition since 1859. Since 1924, the Isle of Langenwerder has continuously been a nature protection area with its own observation station.

Only the iWB and the SH will be dealt with here.

4.3.1 Environmental Characteristics

The WB has a surface area of 170 km^2 and an average depth of 5–6 m. The maximum depth is 12 m. The catchment area covers 1,059 km^2. This results in a ratio of catchment/water surface area of 6.3:1. The region is basically characterised by the town of Wismar with its shipyard and port, and by tourism. The agricultural influence can be seen above all by diffuse entries.

The iWB (see Table 4.1) is a shallow water area of a depth of 3–4 m with broad littoral zones between 0 and 1 m in depth. It is cut by a 9.5 m broad shipping channel. The port of Wismar is permanently dredged out to a depth of 9.5 m. The catchment area covers approximately 40 km^2. Extensive mudflats, which are formed depending on the wind ("wind mud flats" or pseudo-mud flats), are characteristic of the iWB. The largest coherent shallow water zones of the Wismar-Bucht are found between iWB and SH. The fresh water entry is low and takes place predominantly via the Wallensteingraben, which has a catchment area of 156 km^2. The water balance of the Wismar-Bucht is determined mainly by the inflow and outflow of Baltic Sea water. Normally, the iWB and the SH are well mixed, both horizontally and vertically. Owing to the shallow depths, the water bodies are well supplied with oxygen. An O_2-saturation of 62–160% was measured in the SH. Off the Isle of Walfisch, mean values of 7–8 mg O_2 l^{-1} have been observed in the deep water in summer and 10–13 mg l^{-1} in winter. In the surface water, the over-saturation can reach as much as 140% due to oxygen production. In the area of the port, values of over 200% have even been measured.

From summer to autumn, when there are continuous south-west winds, upwelling, low-oxygen deep water from the offshore Mecklenburg Bucht can enter the iWB, causing an oxygen deficiency near the bottom. Due to the high respiration activity of the macrophytes, and the decomposing processes at high temperatures, short-term oxygen fluctuations of >200% and an oxygen deficiency near the bottom of 1.5 mg O_2 l^{-1}, lasting up to 2 weeks, occur in the SH (Walter 1996). Colonies of *Beggiatoa*, which were proven, suggest H_2S-formation.

The medium salinity amounts to 10–11 psu, the maximum fluctuation limits are 7.8–19.6 psu. The iWB has two main sources of pollution: the brook Wallensteingraben, which drains a catchment area of 156 km^2, and the sewage plant of Wismar, with 91,000 inhabitant equivalents. Prior to 1990, Wismar had only a mechanical waste treatment stage, and this was not redeveloped until 1996. Since 1996 there has been a temporary solution in the form of the chemical simultaneous precipitation of phosphorus. The current entries of materials by loads from both sources (1991/1995) when considering an average sewage water discharge of 57 × 10^6 m^3 a^{-1} amount to around 342 t a^{-1} organic material (BSB_5), 13.2 t a^{-1} o-PO_4, 36.7 t a^{-1} total-P, 12.7 t a^{-1} NH_4, 246 t a^{-1} NO_3 and 344 t a^{-1} total-N (Gosselck and von Weber 1997). The iWB is highly polluted, mainly in the port area (LAUN M-V). In 1996, the average nitrate contents in the port area and near the Isle of Walfisch were 6.1 µmol N l^{-1} and 2.0 µmol N l^{-1}, respectively. The two offshore outer stations showed average values of 1.3 or 1.5 µmol l^{-1}. With 1.15 µmol l^{-1} in the port area, the average o-PO_4 concentration was also the highest (long-term average value 1977/1995: 3.0 µmol l^{-1}). At the two offshore outer stations, 0.36 or 0.26 µmol l^{-1} was measured (long-term average 1977/1995: 1.0 µmol l^{-1}). Since 1990, a clear decrease in the o-PO_4-concentration has taken place in the water body of the iWB. This is due to the use of phosphate-free detergents and the introduction of phosphate-simultaneous precipitation in the yet-to-be modernised sewage plant of Wismar. Total nitrogen is also on the decline (Bachor et al. 1996), whereas the entries of nitrate and ammonium caused by agriculture are as high as before.

The SH is a relatively closed smaller water body covering a total area of approximately 21 km^2 (see Table 4.1). The average depth is 2.3 m, the maximum depth 10 m. The water volume amounts to approximately 67 × 10^6 m^3 a^{-1}. The catchment area comprises about 310 km^2, which results in a ratio of catchment/water surface area of 15:1. With an average salinity of 10.5 psu, the variation range is <5–15.0 psu.

The "Ostseeklima" (see Chap. 3 by Schiewer, this volume) affects the annual cycle of the temperature. The average summer temperature of the water is around +20°C, the maximum being +27°C (Walter 1996). In the surface water, temperatures between –1 and +24.8°C were measured from 1977 to 1996 (data of LAUN M V). The long-term average value on the Isle of Walfisch amounts to +10.6°C, the number of days with ice-cover averaging 30 (Blümel et al. 2002).

The SH is divided into a south-west (outer) and a north-east (inner) bay by the Tesmannsdorfer Haken, which continues underwater. Sufficient water exchange is ensured by a small channel ('Ellenbogen') with an average depth of 2.3 m. Current speeds on the bottom can reach up to 40 cm s^{-1}. The outer SH is 3–5 m deep, the inner area only 2–3 m. Around 50% of the SH is occupied by shallow edge areas with depths of 0–2 m.

An opening 1.5 km wide and 4 m deep at the western edge between Boiensdorfer Werder and Kieler Ort connects the SH with the Baltic Sea. Since 1987 there has also been an opening to the Kroy in the northern part of the Wustrow peninsula, which is about 20 m wide. With a range of tide of approximately 20 cm, the sea level variations of the Baltic Sea, depending on wind and air pressure, are decisive for water exchange with the Baltic Sea. Water exchange with the iWB via the Breitling is low.

The fresh water inflow comes primarily from the Hellbach brook. At 53 × 10^6 m^3 a^{-1}, it is rather high for such a small water body, but is considerably lower than the water inflow from the Baltic Sea. Its catchment area covers 210 km^2. The local anthropogenic loads of the SH, via the Hellbach (Table 4.6), from the local sewage of Rerik and a poultry farm near Roggow, have not had a widespread effect. The intensive water exchange with the Baltic Sea leads to a lasting reduction of the

Table 4.6 Anthropogenic loads from the Hellbach into the SH (Gosselck and von Weber 1997). *MQ* Mean discharge

	Mean Discharge	Organic load ($t\ ha^{-1}$)				Nutrient load ($t\ ha^{-1}$)		
Year	MQ $10^6 m^3$	BOD_5	COD	o-Phosphate	Total P	NH_4	NO_3	Total N
1990	41	136	265	9.46		15.40	252	
1991	37	162	322	11.10		18.40	295	
1992	24	101	205	4.04		7.94	209	
1993	65	331	512	5.20	10.3	24.60	716	957
1994	74	249	701	7.49	13.6	27.50	695	835
1995	61	161	554	4.56	9.6	24.90	456	523

effects of this anthropogenic load. Thus, the water transparency rarely drops below 4 m, with > 8 m often being reached.

4.3.2 *Planktonic Communities*

4.3.2.1 Structure, Dynamics and Productivity

Current studies on the phytoplankton are rare. Only Arndt et al. (1966) published data on this subject: 56 Bacillariophyceae, 1 Dictyochophyceae (Silicoflagellates), 12 Dinophyceae and 2 Cyanobacteria were shown to be present, Chlorophyta were missing. The composition is characterised by marine species and is similar to that of Kiel Bay.

No investigations of the bacterio-, protozoo- and mesozooplankton are available.

In the nekton of the WB, fish species predominate. In addition, the genera *Praunus* and *Neomysis* are present, as well as prawns (*Palaemon adspersus, Palaemon squilla*). In the summer, larger populations of *Aurelia aurita* and *Cyanea capillata* may occur. Providing habitat for 49 bony fish species and 2 Cyclostomata (*Lampetra fluviatilis* and *Petromyzon marinus*), the WB is a water body rich in fish species (Walter 1997). The reasons for this are to be found in the relatively high salinity of the brackish water and the good mixing of the water body. The well developed submersed macrophytes with extensive eelgrass meadows offer protection and spawning places (Weber 1990; Jaschhof 1990). A varied fauna and a rich zoobenthos in the iWB provide food for many fish species. About 30% of the fish species in the WB are used for commercial purposes (Walter and Jansen 1994). The most important commercial fish species are eel (*Anguilla anguilla)*, herring (*Clupea harengus*), cod (*Gadus morhua*), flounder (*Platychthys flesus*) and turbot (*Psetta maxima*). Apart from these there are sprat (*Sprattus sprattus*), salmon (*Salmo salar*), sea trout (*Salmo trutta trutta*), eelpout (*Zoarces viviparus*), garfish (*Belone belone*), river perch (*Perca fluviatilis*), flathead mullet (*Mugil cephalus*), plaice (*Pleuronectes platessa*) and *Limanda limanda*. Catches between 1972 and 1996 averaged 343.7 t

a^{-1}. Within the catches of this period, marine species predominated, accounting for more than 90% (312.4 t a^{-1}). The amount of freshwater- and migratory fish caught amounted to only 22.3 t a^{-1}. The catch statistics clearly show the strong decline from 1986 to 1990. This is due to the absence of cod and flounder, the reasons for which are to be found in the over-fishing of the Baltic region and eutrophication in the Baltic Proper. A special characteristic is the Baltic prawn (*Palaemon squilla*) in the WB, of which 2–10 t are landed annually (Walter 1997).

4.3.3 Benthic Communities

4.3.3.1 Structure and Productivity

Krüger and Meyer (1937) observed *Zostera marina* cover up to a water depth of 7 m. By the 1990s, this depth maximum had decreased to only 4 m (Gosselck and von Weber 1997).

At the same time, the proportion of fast-growing algae, such as *Ulva lactuca, Enteromorpha* spp., *Cladophora* spp., *Ectocarpus* spp. and *Ceramium* spp. rose. The depth-dependent distribution of the iWB, including the port, shows *Ruppia cirrhosa* and *Zostera noltii* for 0.2–0.5 m, for 0.5–1.5 m mainly *Zannichellia*, rarely *Potamogeton*, at 1.5 m *Zostera marina* and small amounts of *Zannichellia*. Below approximately 2 m water depth, *Ulva lactuca* spreads area-wide. In deeper zones, areas with *Chaetomorpha linum* occur.

The dominating sediments of the SH are medium and fine sands. In slack water areas and in the channels, sewage mud is deposited. In the inner SH of Rerik, sewage mud deposits up to 25 cm thick have been measured. Stones are occasionally present. Owing to its depth, the SH is even today characterised by submersed macrophytes. Its stable sand bottom vegetation consists basically of spermatophytes (*Zostera marina*, *Potamogeton pectinatus*, *Ruppia cirrhosa*, *R. maritima* and *Zannichellia palustris*) and includes five benthic algae (*Enteromorpha* spp., *Chaetomorpha linum*, *Ulva lactuca*, *Fucus vesiculosus* and *Chorda filum*).

Around 50–70% of the bay's surface is covered by submersed macrophytes. A shift of the depth limit of submersed macrophytes to lower depths, as described for the GB and as is typical of the iWB, did partially take place in the SH for *Zostera marina*. Thus, the depth-dependent distribution of plants known from earlier times (Krüger and Meyer 1937) still exists: *Ruppia* spp. in 0.2 m to 0.5 m depth, *Potamogeton/Zannichellia* mixed stocks in 0.5–2 m depth and *Zostera-Zannichellia* mixed stocks, infiltrated by *Chaetomorpha linum* in 1.5–3.5 m depth. The first symptom of eutrophication is a luxuriant "Aufwuchs" on the surface area, predominantly by *Ceramium*- and *Ectocarpus* species. In the inner part of the SH, near Rerik, we find a pattern of green algae in the 4 m zone, with *Ulva lactuca*, *Chaetomorpha linum* and the cyanobacterium *Spirulina major*. In the mid-1980s, there was a denser colonisation of submersed macrophytes with Cyanobacteria also during the summer months, which disappeared in the 1990s. Reliable quantitative statements on biomass density and productivity of submersed macrophytes are not available. Investigations of sediment bacteria, protozoobenthos and meiofauna are also lacking.

The investigated macrofauna of the area consists of eurybiotic marine and brackish water species. To date, 90 macrofauna taxa have been found. Table 4.7 presents an overview of species occurring among the macrophytes and on the sandy bottom, and species occurring as periphyton ("Aufwuchs"). Sandy bottom species are prevalent. The productivity of these species is comparatively high; however, quantitative data are rare. Weber (1990) identifies 1,000–5,000 ind. m^{-2}, with biomasses between 500–23,000 mg fw. As far as weight is concerned, mussels dominate the biomass with 80%, with *Mytilus edulis* constituting the main proportion (190 g fw m^{-2}). Characteristic species of the SH are *Hydrobia ventrosa*, Oligochaeta, *Cerastoderma lamarcki*, *Hediste diversicolor*, *Microdeutopus gryllotalpa*, *Hydrobia ulvae*, Chironomidae, *Jaera albifrons*, *Mytilus edulis*, *Pygospio elegans*, *Heteromastus filiformis*, *Sphaeroma hookeri* and *Cyathura carinata*. In 1981, the percentage of the systematic groups in the inventory of species was 32.5% for Malacostracae and Polychaeta, 22.5% for Mollusca and 12.5% for Turbellaria, Nemertini, Oligochaeta and Chironomidae together. The nutritional types are divided into 52.5% deposit feeders, 20% herbivorous, 15% carnivorous and 12.5% suspension feeders.

Table 4.7 Distribution of macrozoobenthos at different habitats (modified from Köhn et al. 1991). Habitat preferences given in brackets: *P* phytal, *H* hard substrates, *S* sandy and muddy bottom

Porifera	*Haliclona limbata* (P,H)
Cnidaria	*Laomedea lovenii*, *L. flexuosa*, *Clava multicomis* (P, H)
Turbellaria	*Planaria torva*, *Procerodes ulva* (S)
Nemertini	*Lineus* spec. (S)
Gastropoda	*Theodoxus fluviatilis* (P,H); *Radix ovata*, *Lacuna pallidula*, *Littorina saxatilis*, *Turboella neglecta*, *T. inconspicua*, *Zippora membranacea*, *Elysia viridis*, *Limapontia capitata* (P), *Potamogpygus jenkinsi*, *Hydrobia ulvae*, *H. ventrosa*, *H. neglecta* (S); *Littorina litorea*, *Embletonia pallida* (H)
Mollusca	*Cerastoderma hauniense* (P); *Mytilus edulis* (P,S): *C. lamarcki*, *Scrobicularia plana*, *Macoma balthica*, *Arenomya arenaria* (S)
Polychaeta	*Antionella sarsi*, *Harmothoe imbricate*, *H. impar*, *Eteone longa*, *Hediste diversicolor*, *Scoloplos armiger*, *Polydora ligni*, *Pygospio elegans*, *Streblospio shrubsoli*, *Heteromastus filiformis*, *Capitella capitata*, *Arenicola marina*, *Manayunkia aestuarina* (S); *Platynereis dumerili* (P,S): *Fabricia sabella* (P,H,S); *Spirorbis spirorbis* (P)
Oligochaeta	*Tubificoides benedeni*, *Nais elinguis* (S)
Crustacea	*Idothea balthica*, *I. chelipes*, *Gammarus salinus*, *G. oceanicus*, *G. locusta*, *G. inaequicauda*, *Melita palmate*, *Calliopius laeviusculus*, *Amphithoe rubricata*, *Palaemon adspersus* (P); *Heterotanais oerstedti*, *Jaera albifrons*, *Corophium insidiosum* (P,H); *C. volutator*, *Cyathura carinata*, *Crangon crangon* (S); *Neomysis integer*, *Sphaeroma hookeri*, *Microdeutopus gryllotalpa*, *Carcinus maenas* (P,S)
Insecta	*Chironomus* sp. (P,S)
Bryozoa	*Membranipora lineate*, *Electra crustulenta*, *Alcyonidium polyoum* (P,H)
Tunicata	*Ciona intestinalis* (P)

4.3.4 Benthic–Pelagic Coupling and Eutrophication

4.3.4.1 Biological Processes and Interactions

Being ecologically impaired by long-lasting sewage entries, mainly in the port area, the WB is highly eutrophic. Precise analyses are not available.

Low freshwater inflows, very good conditions for exchange with the Baltic Sea and low nutrient loads, however, have made the SH a mesotrophic water reference zone in the southern Baltic. Above all in the SH, but also in large parts of the iWB, submersed macrophytes are still prevalent. Thus, the constellation is the same as 60 years ago, although eutrophication in the SH has becomes apparent locally by the massive occurrence of Cyanobacteria (*Spirulina*, *Oscillatoria*), associating with Conjugatophyceae and filamentous fast-growing brown- and red algae. The coverage of brown algae and other submersed macrophytes with filamentous brown- and red algae is also typical. In addition, there are the green algae *Ulva lactuca*, *Ulvaria obscura* and *Chaetomorpha linum*, predominantly in the inner SH. Already in 1937, Krüger and Meyer described the occurrence of the charophytes as being rare. In the outer SH, *Chara canescens*, *Ch. baltica* and *Tolypella nidifica* are present as before (Yousef et al. 1997). Large stocks are reported to have been available in the nineteenth century and at the beginning of the twentieth century. They were used by farmers for fertilising their fields with lime.

4.3.5 Protection Measures

The area around the WB and the SH plays an important economic role as a local recreation area. Thanks to the construction of sewage plants in Rerik, Neubukow and several villages in the catchment area, and a reduction in pollution from agriculture, the SH is no longer endangered by eutrophication. However, a threat for the submersed macrophytes, mainly in the southern part, is the rapid increase in windsurfing activities. The final consequences of this load cannot yet be estimated. A prerequisite for the redevelopment of the WB is the construction of a modern sewage plant and a more environmentally orientated cultivation of the arable land.

4.4 Conclusions and Summary

GB, WB and SH represent a second group of the inner coastal waters of the German Baltic coast, in addition to those of the DZB, Northern Rügener boddens (NRB) and Schlei (see Chap. 3 by Schiewer, this volume). GB, WB and SH are characterised by the dominance of efficient horizontal exchange processes with the Baltic Sea via internal turnover rates in the boddens. The absence of a pronounced horizontal salinity gradient reflects this intensive water exchange.

Table 4.8 Control factors of the self-purification potential of German coastal waters (Schiewer/Schernewski 2004)

a. Morphology and hydrology
   - Mean depth
   - Surface/catchment area
   - Exchange with the Baltic Sea
   - River inflow
   - Water residence time
b. Physical–chemical processes
   - Salinity
   - Nutrient loads
   - Sedimentation/resuspension
   - Accumulation
c. Biological processes and regulation
   - Changes in phytoplankton, e.g. decline of diatoms
   - Phytoplankton versus submersed macrophytes dominance
   - Grazing versus microbial food web dominance
   - Formation of fluffy sediment layer
   - Decline of diversity
   - Deterministic versus stochastic regulation
   - Stepwise changes of trophic levels

The efficiency of these waters to act as bioreactors is clearly reduced due to short residence times. The dominance of submersed macrophytes results (at least partly) in a longer settling of nutrients and more particles entered. Relatively intense biological activity is normally secured by grazing food web activities. If these waters are anthropogenically overloaded and their structures destroyed, they will act direct as sources of pollution for the adjacent Baltic Sea. Table 4.8 gives a final summary of the most important characteristics on which the self-purification potential of the coastal waters of the German Baltic coast can be evaluated.

References

Abshagen G (1908) Das Phytoplankton des Greifswalder Boddens. Diss Biologie. Univ Greifswald, Greifswald

Arndt E, Pankow H, Kell V (1966) Über das Phytoplankton der Wismar-Bucht. Int Rev Hydrobiol 51:127–146

Asmus R, Jensen MH, Murphy D, Doerffer R (1997) Energiefluß und trophischer Transfer im Sylt-Römö-Wattenmeer. In: Gätje C, Reise K (eds) Ökosystem Wattenmeer. Springer, Berlin, pp 367–391

Bachor A, Weber M von, Wiemer R (1996) Die Entwicklung der Wasserbeschaffenheit der Küstengewässer Mecklenburg-Vorpommerns. Wasser Boden 48:26–32

Barthels S, Klüber U (1998) Die räumliche Verbreitung des Makrophytobenthos und seine Akkumulation von Nährstoffen und Schwermetallen. Teil 1: Erfassung des Bedeckungsgrades des Greifswalder Boddens mit submersen Makrophyten. Greifswalder Geographische Arbeiten 16:316–325

Barthels S, Klüber U (1999) Determination of submerged macrophyte coverage of Greifswalder Bodden. Limnologica 29:311–313

Blümel C, Domin A, Krause J, Schubert M, Schiewer U, Schubert H (2002) Der historische Makrophytenbewuchs der inneren Gewässer der deutschen Ostseeküste. Rostock Meeresbiol Beitr 10:5–111

Brenning U (1989) Das Zooplankton des Greifswalder Boddens. Meer Museum 5:36–43

Conrad K, Bernhardt R, Miske V, Schröder K (1998) Die räumliche Verbreitung des Makrophytobenthos und seine Akkumulation von Nährstoffen und Schwermetallen. Teil 2: Akkumulation von Stickstoff, Phosphor und anderen Elementen im Makrophytobenthos des Greifswalder Boddens sowie sein Gehalt an kompatiblen Substanzen und Photosynthesepigmenten. Greifswalder Geographische Arbeiten 16:326–340

Fraude H (1906) Grund- und Planktonalgen der Ostsee. Diss Biologie. Univ Greifswald, Greifswald

Geisel T (1986) Pflanzensoziologische Untersuchungen am Makrophytobenthos des Greifswalder Boddens. Diplomarbeit Biologie. Univ Rostock, Rostock

GOAP (1998) Greifswalder Bodden und Oder-Ästuar – Austauschprozesse. Greifswalder Geographische Arbeiten 16:1–400

Gosselck F, Weber M von (1997) Pflanzen und Tiere des Meeresbodens der Wismar-Bucht und des Salzhaffs. Meer Museum 13:41–52

Günther B (1999) Die Bedeutung des Makrozoobenthos für die Kohlenstofftransformation im Greifswalder Bodden und im Oderästuar. Bodden 7:63–78

HELCOM (1998) The third Baltic Sea pollution load compilation. Baltic Sea Environment Proceedings 70:1–133

Henking H (1904) Orientierungsfahrten im Greifswalder Bodden. Mitt Dtsch Seefisch-Verein

Hubert M-L, Kob A, Vietinghoff U (1995) Ergebnisse der Modellierung des Ökosystems Greifswalder Bodden. Rostock Meeresbiol Beitr 3:123–145

Hübel H-J, Vietinghoff U, Hubert M-L (1995) Ergebnisse des ökologischen Monitorings Greifswalder Bodden September 1993 bis März 1995. Rostock Meeresbiol Beitr 3:5–67

Jaschhof M (1990) Zur Sedimentbesiedlung des Salzhaffs durch die Makrofauna. Diplomarbeit Biologie. Univ Rostock, Rostock

Jönsson N, Busch A, Lorentz Th, Korth B (1998) Struktur und Funktion von Boddenlebensgemeinschaften im Ergebnis von Austauch- und Vermischungsprozessen. Greifswalder Geographische Arbeiten 16:250–285

Kell V (1984) Primärproduktionsmessungen im Greifswalder Bodden in den Jahren 1981 und 1982. Wiss Z Univ Rostock, Math-nat R 33:53–55

Kell V (1989) Das Phytoplankton des Greifswalder Boddens. Meer Museum 5:25–35

Köhn J, Jaschhof M, Weber M (1991) Das Salzhaff. Meer Museum 7:8–21

Krüger K, Meyer P-F (1937) Biologische Untersuchungen in der Wismarschen Bucht. Z Fischerei Hilfswiss 5:665–703

Lampe R (1998) Das Oder-Ästuar als Filter und Transformator. Greifswalder Geographische Arbeiten 16:451–483

Lampe R (1999) Das Oder-Ästuar – Filter oder Bypass fluvialer Einträge? Bodden 7:45–62

Lampe R, Meyer H (1995) Sedimentqualität und hydrographische Bedingungen – ein Vergleich verschiedener Boddengewässer. Bodden 2:7–26

Leipe T (1985) Zur Nahrungsökologie der Eisente (*Clangula hyemalis*) im Greifswalder Bodden. Beitr Vogelkd 31:121–140

Lemmermann E (1901) Beiträge zur Kenntnis der Planktonalgen. XII: Das Phytoplankton des Ryck und des Greifswalder Boddens. Ber Dtsch Bot Ges 29:92–95

Messner U, Oertzen J-A von (1991) Long-term changes in the vertical distribution of macrophytobenthic communities in the Greifswalder Bodden. Acta Ichth Piscat 21 [Suppl]:135–143

Meyer H, Lampe R, Jonas P, Buckmann K (1998) Nährstoffe im Oder-Ästuar – Transporte und Inventare. Greifswalder Geographische Arbeiten 16:99–129
Schiewer U, Schernewski G (2004) Self-purification capacity and management of Baltic coastal ecosystems. J Coastal Conserv 10:25–32
Schlungbaum G, Stolle S, Westphal H, Nausch G (1989) Sedimentchemische Untersuchungen in den Küstengewässern der DDR. Zur Beschaffenheit der Sedimentoberflächenschicht des Greifswalder Boddens. Wiss Z Univ Rostock, N Reihe 38:18–22
Schmidt I (1998) Langzeitdatenreihen aus dem westlichen Teil des Oder-Ästuars – Ergebnisse der Bearbeitung von Monitoringdaten der WWD Küste/Stralsund aus dem Zeitraum 1972-81. Greifswalder Geographische Arbeiten 16:194–227
Schnese W (1957) Faunistisch-ökologische Untersuchungen über das Zooplankton des Greifswalder Boddens. Diss Biologie. Univ Greifswald, Greifswald
Seifert R (1936) Die tiergeographische Stellung des Greifswalder Boddens. Mitt Naturwiss Verein Neu-Vorpommern 63/64:1–15
Seifert R (1938) Die Bodenfauna des Greifswalder Boddens. Z Morphol Ökol Tiere 34:221–271
Stammer H-J (1928) Die Fauna der Ryckmündung, eine Brackwasserstudie. Z Morphol Ökol Tiere 11:36–101
Steyer D (1995) Untersuchungen des Jahresganges der Bakterienpopulation an ausgewählten Stationen im Greifswalder Bodden. Diplomarbeit Landeskultur und Umweltschutz. Univ Rostock, Rostock
TRUMP (1998) Transport- und Umsatzprozesse in der Pommerschen Bucht. Forschungs-bericht. Institut Ostseefrschg Warnemünde, Rostock-Warnemünde
Walter U (1996) Zur Entwicklung wichtiger abiotischer Daten an verschiedenen Pegeln des Salzhaffes sowie weiteren Messpunkten der Wismar-Bucht. Bericht. Pädagogische Hochschule, Güstrow
Walter U (1997) Fische, Fischerei und Garnelenfang in der Wismar-Bucht. Meer Museum 13:53–61
Walter U, Jansen W (1994) Untersuchungen zur Fischfauna in der Wismar-Bucht. Mitt LFA Landwirt u Fisch 6:111–115
Weber M v (1990) Untersuchungen an der Makrofauna des Phytals des Salzhaffs (Wismarer Bucht, westl. Ostsee). Diplomarbeit Biologie. Univ Rostock, Rostock
Westphal H, Lenk B (1998) Die räumlich-zeitliche Verteilung von Primärproduktion und Bakterienkeimzahlen. Greifswalder Geographische Arbeiten 16:228–249
Winkler H (1989) Fische und Fangerträge im Greifswalder Bodden. Meer Museum 5:52–58
Yousef M, Küster A, Schubert H, Nordheim H v (1997) Charakterisierung der Characeenbestände an der Küste Mecklenburg-Vorpommerns. Bodden 5:3–23

Chapter 5
The Szczecin (Oder-) Lagoon

T. Radziejewska and G. Schernewski

5.1 Introduction

The Odra (German: Oder) estuary, located in the southern Baltic Sea, consists of the Szczecin (Oder-) Lagoon and the Pomeranian Bay (Fig. 5.1). The Szczecin Lagoon (687 km^2) is shared by Poland and Germany and is naturally subdivided into the "Large Lagoon" (Polish: Wielki Zalew) on the Polish territory and the "Small Lagoon" (German: Kleines Haff) on the German side. The Lagoon connects with the Pomeranian Bay via three outlets. With an average depth of 3.8 m, the Szczecin Lagoon is a shallow water body: its maximum natural depth is 8.5 m, but dredging in the shipping channel produces depths exceeding 10.5 m. The channel intersects the Lagoon to link the River Odra mouth and the harbour of Szczecin with the Baltic Sea. Data of the Maritime Office of Szczecin show ca 1.5 million m^3 sediment per year to have been dredged during the last 50 years to maintain the average channel depth of 10.5 m (Minning 2004).

The Lagoon is of great ecological and economic importance for the two countries that share it. The economic importance of the Lagoon is a result of its having served for centuries as a fishing ground and since the eighteenth century as a major transportation pathway for both goods and people (Piesik 1993; Robakiewicz 1993). Since the second half of the twentieth century, another dimension to the Lagoon's economic importance has been provided by tourism (Andrulewicz and Lamp 1994; Schernewski and Sterr 2002), the tourist attraction of the Lagoon stemming from its appeal as a major water body amenable to supporting several types of recreational activity. However, the different uses of the Lagoon and its surroundings have resulted in considerable pressure on its environment and biota. Nevertheless, the Lagoon remains of great ecological importance to Poland and Germany. The Lagoon and its surroundings support valuable species, habitats, and landscapes and offer great conservation potential. In recognition of those qualities, the Szczecin Lagoon has been intensively studied in a number of large interdisciplinary and international (Polish–German) projects (see e.g. Lampe 1998; Humborg et al. 2000b; Schernewski et al. 2004) and features prominently in European-scale conservation-oriented programmes (NATURA 2000, European Commission's Bird and Habitat Directives).

U. Schiewer (ed.), *Ecology of Baltic Coastal Waters. Ecological Studies 197.*

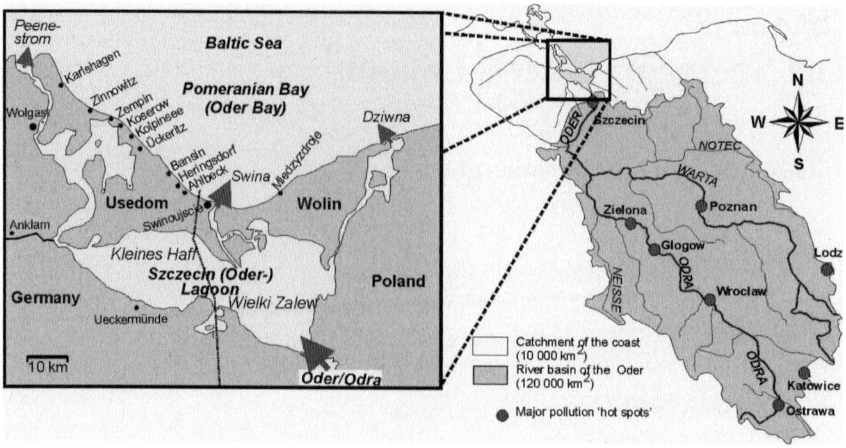

Fig. 5.1 The Szczecin (Oder-) Lagoon and the Oder (Odra) river basin (http://www.ikzm-oder.de)

5.2 Environmental Characteristics

The entire estuary is dominated by the discharge of the River Odra (Oder) into the Lagoon. With its length of 854 km and basin area of 120,000 km^2, the Odra is one of the most important rivers in the Baltic region. Between 1980 and 1999, the annual Odra discharge (as measured at Gozdowice, Poland) varied between about 9.5 km^3 (1990) and 25 km^3 (1980), with an average of 17 km^3 (530 m^3 s^{-1}) (IMWM 1980–1983, 1987–1999). The combined discharge of all other rivers into the Lagoon amounts to ca. 1 km^3 a^{-1} (Wielgat 2002). Therefore, the Odra contributes at least 94% to the Lagoon's water budget. The major features of the Lagoon's climate and hydrology are summarised in Table 5.1a.

The Lagoon is brackish and its salinity in the central part ranges between 0.5 and 2 psu. Periodic intrusions of Baltic water of ca. 6 psu salinity into the Lagoon occur mainly via the Świna Channel (Bangel et al. 2004). These salt water influxes usually reach no farther than to an area around the Świna Channel, but particularly dense saline water can penetrate, along the bottom, several kilometres into the Lagoon.

Sediments in the Lagoon (Table 5.1b) have been intensively investigated (e.g. Breitenbach et al. 1999; Leipe et al. 1998; Majewski 1980; and see recent reviews by Osadczuk 2004 and Tejakusuma 2004). In general, the shallow areas are dominated by eroded sand. The sandy areas are temporarily covered by a thin and mobile organic-rich fluffy material, and contain numerous subfossil mollusc shell fragments and abundant ostracod valves (Borówka et al. 2005; T. Radziejewska unpublished observations). Sediments on the bottom areas at depths of 4 m and more are dominated by silt. According to Osadczuk (2004), sediments in the southern and western parts of the Wielki Zalew are under a strong influence of the Odra, whereby they feature high contents of organic matter (>12%) and heavy metals

Table 5.1 Main characteristics of the Odra Lagoon. *dw* Dry weight, *ww* wet weight, *Chl a* chlorophyll *a*, *POC* particulate organic carbon, *DOC* dissolved organic carbon, *POM* particulate organic matter, *DIN* dissolved inorganic nitrogen, *DIP* dissolved inorganic phosphorus

a. Climate and hydrology	
Average annual air temperature (°C)	0.8 (min), 8.7 (average), 17.9 (max)[a]
Water temperature (°C)	~11 (average), 26.0 (max)[b]
Ice coverage (days a^{-1})	59[c]
Yearly global radiation (kW h m^{-2})	No data
Average precipitation (mm a^{-1})	550
Average freshwater inflow (km^3 a^{-1})	18
Average Baltic inflow (10^6 m^3 a^{-1})	No reliable data
Average total outflow (10^6 m^3 a^{-1})	18
Residence time (days)	55
b. Morphometry and sediments	
Surface area (km^2)	687
Volume (km^3)	2.58
Average depth (m)	3.8
Maximum depth (m)	8.5 (10.5)[d]
Catchment area (km^2)	129,000
Catchment/surface relation	188
Organic matter content (% dw)	13.8–35.2 (average ~23)[e]
Sediment fraction < 63 μm (range) (% dw)	<10– >90[e]
C/N ratio (0–10 cm depth)	8.01–30 (average ~11)[b]
Sediment oxygen consumption (mol O_2 m^{-2} a^{-1})	1.6–4.7[f]
Phosphate release (μmol P m^{-2} day^{-1})	up to 2–10[g]
Denitrification rate (mmol N m^{-2} day^{-1})	2.0[h]–4.0[i]
c. Pelagic biological components	
Secchi depth (m)	0.2–3.0 (average 0.9)[a]
Bacteria (10^6 ml^{-1})	No data
Phytoplankton (μg Chl*a* dm^{-3})	70[j]
Phytoplankton (mm^3 dm^{-3})	15[k]
Number of species	231[l]
Dominant species	*Microcystis aeruginosa*
Flagellates (mg ww dm^{-3})	No data
Ciliates (mg ww dm^{-3})	No data
Metazoa (mg ww dm^{-3})	0.80[l]
Dominant species	*Keratella cochlearis*, *Daphnia cucullata*[m,n]
Fish	
Number of species	56[o]
Dominant species	*Rutilus rutilus*, *Abramis brama*, *Perca fluviatilis*[o,p,q]
Seston (mg dw dm^{-3})	9–60[r]
POC (mg C dm^{-3})	15.4[s]
DOC (mg C dm^{-3})	32.5[t]
C/N in POM	No data
Primary production (g C m^{-2} a^{-1})	591–648[u]
Bacterial production (μg C dm^{-3} d^{-1})	No data
Metazoan production (g C m^{-2} a^{-1})	No data
Fish catches (t a^{-1})	2,528[q]

(continued)

Table 5.1 (continued)

d. Benthic components	
Macrophytes	
Biomass (g dw^{-1} m^{-2})	2,664.16[v]
Number of species	22[w]
Dominant species	*Phragmites australis*, *Schoenoplectus lacustris*[v]
Macrozoobenthos	
Biomass (g ww m^{-2})	0.45–1,640.26[x]
Number of species	97[x]
Dominant species	*Chironomus f. l. plumosus*, tubificid oligochaetes
e. Water chemistry, trophic status and pollution	
Salinity (psu)	0.3–4.5 (average 1.4)[y]
pH	6.8–9.5[aa]
Surface oxygen saturation (%)	40–220[y]
DIN (µmol l^{-1})	170 (winter)[y]
DIP (µmol l^{-1})	3.5 (winter)[y]
Dissolved silicate (SiO_4-Si) (µmol l^{-1})	150 (winter)[y]
Nitrogen load (1995–1999) (t a–1)	66,000–82,000[z]
Phosphorus load (1995–1999) (t a–1)	5,000–6,400[z]
Suspended particulate matter (t a^{-1})	425,000[c]
Heavy metal loads (t a^{-1})	Pb: 85, Zn 700[c]
Limitation	Light
Trophic level	Hypertrophic[a, d] Eutrophic[e, f]

[a]Data for Ueckermünde weather station for the period 1991–2000 (Deutscher Wetterdienst); [b]LUNG data for the period 1989–2000 in the central part of the Lagoon (Schernewski et al. 2006); [c]Girjatowicz (2001): average number of days with ice for 1950/1951–1989/1990 recorded at Karnin (Kleines Haff); recent unpublished observations show prolonged periods of ice-free water in both parts of the Lagoon; [d]Average depth in the dredged shipping channel; [e]Osadczuk (2004); [f]Model calculations of Humborg et al. (2000a); [g]Release during short anoxic periods in summer; Schernewski and Wielgat (2001); [h]Annual average in the central Kleines Haff; Dahlke et al. (1998); [i]Annual average based on model simulations; Wielgat and Witek (2004); [j]Averaged over April-August; Bangel et al. (2004); [k]Averaged for Kleines Haff; Lampe (1999); [l]Luścińska (2005); [m]Wolska and Piasecki (2004); [n]Chojnacki (1991); [o]Wysokiński (1998); [p]Dunin-Kwinta (2000); [q]2003 total in Wielki Zalew; Szostak et al. (2004); [r]T. Radziejewska and H. Dworczak, unpublished data; [s]Total particulate carbon (POC+PIC), averaged for Kleines Haff; Lampe (1999); [t]Günther (1998); [u]Westphal and Lenk (1998); [v]Rycak and Kowalski (2007); [w]Aquatic and rush vegetation; unpublished data of Department of Botany, Agricultural University of Szczecin, Poland; [x]Masłowski (1992); [y]Bangel et al. (2004), based on WIOS and LUNG data for the period 1990–1999 covering the entire lagoon; [z]Wielgat (2002): total loads including Odra, Zarow, Ina, Peene and local sources; [aa]Leipe et al. (1998)

(>150 µg Pb g^{-1}, >80 µg Cu g^{-1} and >1,250 µg Zn g^{-1}). Baltic water intrusions into the north-eastern part of the Polish Wielki Zalew are reflected in the sediments as low contents of organic matter (c. 2%), calcium carbonate (c. 5%), and heavy metals. The high proportion of calcium carbonate of up to over 50% in the central part of the German Kleines Haff (Tejakusuma 2004) results mainly from calcite precipitation due to a high primary production, and also partly from the high proportion of shells and ostracod valves in the sediment. The shipping channel sediment features the highest phosphorus and manganese contents, probably as a result of frequent changes in the redox potential due to ship traffic and Baltic water intrusions.

Although complex, the hydrography and transport processes in the Lagoon have been studied in detail (Majewski 1980; Robakiewicz 1993; Bruckmann et al. 1998; Wolff et al. 1998; Schernewski et al. 2000, 2001; Tejakusuma 2004). The flow field in this shallow, polymictic system (Fig. 5.2) is predominantly wind-driven. Generally, southerly to westerly winds prevail for 58% of the time and low wind velocities between 3 and 5 m s^{-1} are predominant (50% of the time). During strong wind periods (velocity higher than 10 m s^{-1}), wind directions between south and west prevail more than 90% of the time.

The three outlets connecting the Lagoon with the Pomeranian Bay (Baltic Sea) are the Świna, Peenestrom, and Dziwna (Fig. 5.1). Considering wind stress, water level changes, and salinity gradients, the proportions of the total outflow from the Lagoon accounted for by the Świna, Peenestrom, and Dziwna have been estimated at 69, 17, and 14%, respectively (Mohrholz and Lass 1998). Due to sea level changes, in- and outflow events usually occur with a periodicity of a few days.

To visualise the spatial transport of water entering with the Odra, Fig. 5.2 shows the trajectories of passive particles drifting with the current. The trajectories are based on calculated depth-averaged flow fields in the Lagoon. The depth-averaged flow field and the wind show similar directions in the shallow and coastal parts of the Lagoon. The flow velocity is about 0.5 to 1% of the wind speed. In central parts, especially in the Kleines Haff, the average flow is directed against the wind, but the surface layer in those areas also moves with the wind. A strong compensation flow directed against the wind is found in deeper water layers

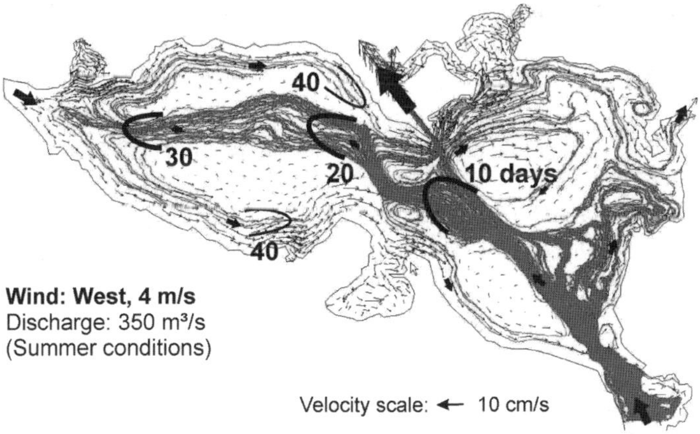

Fig. 5.2 Simplified depth-averaged flow fields in the Szczecin Lagoon under typical summer west-wind conditions

(Schernewski et al. 2002). Under average conditions, the two opposite processes produce a weak current directed against the wind.

Northerly and south-westerly winds enhance the rapid transport of Odra water through the shipping channel into the Baltic Sea (Podsetchine et al. 2004). During easterly and westerly winds situations, which happen with high likelihood, larger amounts of polluted Oder river water enter the Kleines Haff before it finally reaches the Baltic Sea. At a wind velocity of 4 m s^{-1}, transport from the Odra mouth to the Kleines Haff takes about 12 days.

The available data indicate the presence of a large number of variable and relatively small eddies within the first kilometres offshore. The eddies are very likely to increase the efficiency of water and nutrient exchange between the reed belts and the open water. The Kleines Haff shows the presence of large eddies, several kilometres in diameter, which trap drifting suspended particulates and phytoplankton. These large eddies are relatively stable and persist even when the wind direction changes.

5.3 Pelagic Communities

In her recent paper, Luścińska (2005) reported the Szczecin Lagoon phytoplankton to consist of a total of 231 species (Table 5.1c). The phytoplankton standing stocks in the Lagoon are very high: water chlorophyll *a* (Chl *a*) contents (Table 5.1c) – a measure of phytoplankton abundance – may be at times as high as 220 µg dm^{-3} in the Wielki Zalew (Mutko et al. 1994). Consequently, annual phytoplankton blooms, particularly conspicuous in the Wielki Zalew (Mutko et al. 1994), are a constant feature of the Lagoon. During the year, two phytoplankton biomass peaks are usually observed, their timing shifting somewhat from one year to the next. The first, smaller peak, formed mostly by diatoms, occurs in spring; the contributing taxa include *Cyclotella* sp., *Diatoma elongatum*, and *Fragillaria crotonensis* (M. Bąk, personal communication). The major peak, produced primarily by mass development of the cyanobacterium *Microcystis aeruginosa*, is formed later in the spring or in early summer and usually extends until autumn (cf. Chap. 3 by Schiewer, this volume).

Judging by the few actual measurements, the Lagoon's primary production is very high (Table 5.1c): Westphal and Lenk (1998) reported the gross primary production to oscillate around ca. 600 g C m^{-2} a^{-1} (cf. Chap. 3 by Schiewer, this volume).

It is difficult to tell whether this high primary production, and the cyanobacterial blooms in particular, are accommodated in the pelagic food chain, as postulated by Paerl (1988). That this could be the case is suggested by the results reported by Bogusławska-Wąs and Dą browski (2001), who found microbial heterotrophs (yeasts and yeast-like cells) to have been enhanced in the Lagoon's water column and sediment during the periods of increased phytoplankton biomass.

Recent data on the Lagoon's zooplankton are very scant. The existing information shows rotifers and veligers of the bivalve *Dreissena polymorpha* to dominate zooplankton abundance, rotifers and cladocerans (cf. Chap. 3 by Schiewer, this volume) supplying the bulk of the biomass (Wolska and Piasecki 2004) (Table 5.1c).

The Lagoon – although heavily anthropogenically impacted – still supports a numerous and diverse fish fauna (Table 5.1c) composed of fresh- and brackish-water (e.g. roach, perch, pikeperch, bream, ruffe), migratory (e.g. eel, sea trout), and marine (e.g. herring) species (Wysokiński 1998). Most species of commercial importance (e.g. pikeperch, perch, roach, whitefish) form populations migrating between the Lagoon and the adjacent inshore waters of the Baltic, while eel and herring perform much more extensive migrations outside the Lagoon.

Since time immemorial, fishing in the Lagoon has been a traditional occupation of the coastal zone inhabitants (Piesik 1993; Kłyszejko et al. 2004). At present, the fish resources of the Polish part of the Lagoon are exploited by approximately 260 professional (subsistence) fishermen from about 20 small fishing harbours around the Lagoon who operate based on individually issued governmental fishing permits (licenses) and are required to report their catches to state fisheries inspectors (Wysokiński 1998). Those fishermen use mainly stationary gear (fyke nets and gillnets) (Wysokiński 1998; Dunin-Kwinta 2000). In addition, the Lagoon's fish are a valuable target for a multitude of anglers, operating based on angling permits, whose catches – thought to be "enormous" (Dunin-Kwinta 2000) – go unreported and unregulated (Wysokiński 1998; Dunin-Kwinta 2000). Consequently, the data describing the dynamics of catches (and fish resources) in the Lagoon are based solely on the reports of subsistence fishermen. The dynamics of annual catches of those fishermen (Wysokiński 1998; Dunin-Kwinta 2000) provides evidence of the sufficiently ample overall abundance of the resources. After a major decrease to well below 2,000 tonnes in the 1950s, the 1960s and the early 1970s witnessed a general increase in the total annual catch to about 4,000 tonnes reported in 1974 (Wiktor and Garbacik-Wesołowska 1993), following which the annual landings generally decreased and fluctuated rather widely within about 2,000–3,500 tonnes (Wiktor and Garbacik-Wesołowska 1993; Dunin-Kwinta 2000). In 2001–2003, the total annual landings reported oscillated around 2,500 tonnes (Ministry of Agriculture and Rural Development 2003; Szostak et al. 2004). However, notwithstanding the high fish stock abundance, the present catch species structure has changed compared to the period between late 1960s and early 1970s. In contrast to the earlier period, those fish species highly valued on the market (e.g. eel, pike, and pikeperch) do not feature prominently in present landings (Wysokiński 1998; Dunin-Kwinta 2000; Szostak et al. 2004). This is a reflection of both the eutrophication-induced changes in the biotic system of the Lagoon and the dynamics of the fishing effort.

5.4 Benthic Communities

The overview of the Lagoon's meiobenthos presented in this chapter focuses on the Wielki Zalew, where the benthic meiofauna has been found to be dominated by nematodes and ostracods. It displays a very distinct estuarine gradient (sensu Heip et al. 1995): the meiobenthic abundance and diversity decrease with increasing distance from the Lagoon's connection with the Baltic Sea (Radziejewska and

Drzycimski 1988, 1990; T. Radziejewska unpublished data). In addition, when observed along a temporal axis over the year, the dynamics of meiobenthic communities in the Lagoon showed a time-lagged response to the phytoplankton biomass sedimentation to the bottom (T. Radziejewska unpublished data). However, as meiobenthic research in the Lagoon began only in the 1980s (Radziejewska and Drzycimski 1988), no conclusions as to the generalised community responses in the form of, e.g. a shift in community structure, abundance, or diversity, can be drawn. There are grounds to infer that, within a year, changes in the meiobenthic communities are a manifestation of benthic–pelagic coupling in the Lagoon, responses to differences in the hydrodynamic regime being very strong on the spatial scale. In contrast to the macrobenthos, no explanation of long-term changes in the meiobenthic communities could be provided by climatic and hydrological variables (Radziejewska and Chabior 2004).

The Lagoon's benthic macrofauna is at present typical of fresh, eutrophic temperate water bodies, and thus shows a pronounced domination, in terms of both abundance and biomass, of oligochaetes and chironomids (Table 5.1d) (Masłowski 1992; Wolnomiejski 1994), known for their ability to build up high biomass on organic-rich muddy bottoms (cf. Chaps. 3 and 4 by Schiewer, this volume). Consequently, the overall benthic diversity is rather low: although the published faunal lists report up to 97 species and higher taxa (e.g. Masłowski 1992), most of these occur at very low abundances and/or frequencies (Masłowski 1992; Hensel 1994; Wolnomiejski 1994). This contrasts sharply with the earlier (pre-1960s) situation when the most striking characteristics of the Lagoon's bottom (except for the muddy areas in the central part) were dense aggregations of the bivalve *Dreissena polymorpha* (Wiktor 1969). Like other bank-forming suspension feeding bivalves (e.g. Tsuchiya and Nishihira 1985), the presence of *D. polymorpha* enhanced both benthic habitat complexity and macrofaunal diversity. However, after a sharp decline in *D. polymorpha* coverage and abundance in the 1970s–1980s (Stańczykowska et al. 1997), the most recent information (Wolnomiejski and Woźniczka 2007) shows the zebra mussel to have re-established itself in the Lagoon.

Despite qualitative impoverishment of the macrobenthos, its abundance and biomass remain high (Table 5.1d) (Masłowski 1992) and were reported to have increased since the 1970s (Masłowski 1993): the huge organic enrichment of the sediment is beneficial for non-selective detritus feeders such as chironomid larvae and oligochaetes (cf. Chaps. 3 and 4 by Schiewer, this volume). Notwithstanding the generally high level of macrobenthic abundance and biomass, these parameters do show temporal variation. The interannual component of that variation has at certain periods been quite substantial, e.g. 1985–1989 (Masłowski 1992). Explanations of the changes, particularly the sharp reduction observed in 1986–1987, invoke an increased load of unspecified pollutants (Wolnomiejski and Grygiel 1992) or responses to climatic and hydrological changes (Radziejewska and Chabior 2004).

As elsewhere in the Baltic Sea coastal waters, introduction of non-native species is a factor that may be important for the Lagoon's ecosystem structure, biodiversity and functioning (Gruszka 1999). Some of those species (e.g. the polychaete *Marenzelleria neglecta*, the amphipod *Gammarus tigrinus*, and the Chinese mitten crab *Eriocheir sinensis*) have successfully established thriving populations (Hensel 1994; Gruszka 1999; Czerniejewski et al. 2003). The impact those populations will have on the native fauna and flora is not yet known.

5.5 Eutrophication and Benthic–Pelagic Coupling

Over the past 50 years, increased anthropogenic influence has accelerated the eutrophication process in the Lagoon. Intensive algal blooms, low water transparency, oxygen depletion in some parts, and fish kills have become common. Toxic blooms of cyanobacteria, mainly *M. aeruginosa*, have also been observed. Depending on the trophic system adopted, the present trophic state of the Lagoon is described as polytrophic (Schernewski et al. 2006), hypertrophic (OECD 1982), and eutrophic (Wasmund 1990; Nixon 1995) (Table 5.1e)

Primary production in the Lagoon is largely light-limited, nutrient limitation playing a minor role (cf. Chap. 3 by Schiewer, this volume). As already mentioned, the eutrophication impact on the Lagoon's fish fauna is manifested in its altered structure, whereby, compared to the pre-1970s situation, the abundance of plankti- and benthivorous species (roach and perch) has greatly increased at the expense of fish predators (most notably pikeperch and pike) (Wysokiński 1998).

The Odra discharge is the main source of nutrients and major pollutants, e.g. heavy metals or human pathogenic viruses (Schernewski and Jülich 2001; Schernewski et al. 2002). The Odra discharge controls, to a high extent, the nutrient dynamics in the Szczecin Lagoon. Nutrient loads in the Odra increased until the late 1980s, declined steeply during the early 1990s, and increased slightly thereafter. This pattern was reflected in nitrogen and phosphorus concentrations in the Lagoon. The average total phosphorus content showed a decline between the late 1980s (11 μmol l^{-1}) and late 1990s (6 μmol l^{-1}), while the average annual total nitrogen contents declined from 160 to 130 μmol l^{-1} during that decade (Bangel et al. 2004). In wet years, the P and N load discharged by the Odra can be up to twice as high as in dry years. The reduction in nutrient contents observed in the early 1990s was largely an effect of the warm, dry years at that time and cannot be attributed primarily to anthropogenic nutrient load reductions (Schernewski and Wielgat 2001).

However, long-term data series on nutrient contents in the Lagoon have always shown a high interannual variability. Internal nutrient cycling processes in the Lagoon are driven largely by short-term weather conditions. A thermal stratification in the water column and oxygen depletion above the sediment can develop during rare and short calm summer periods. Model-based estimates indicate an anoxic P-release from sediments of up to 10 μmol P m^{-3} day^{-1} or up to 400–600 t P for the entire Lagoon (Schernewski and Wielgat 2001). Such situations last no longer than several days and happen only during warm summers. The water column is already fully mixed by a wind of an average daily speed above 2–3 m s^{-1}, and mixing puts an end to the anoxic P release. Compared to a monthly summer P load of 100–150 t supplied by the Odra, internal eutrophication in the Lagoon can be a very important source of P, and can counteract long-term measures to improve water quality.

Sedimentation and denitrification are important nutrient sinks the Lagoon (cf. Chap. 3 by Schiewer, this volume). Denitrification, nitrification, and the release of N_2O and CH_4 were investigated by Dahlke et al. (1998) in the central Kleines Haff. They found annual average denitrification rates of 2 g N m^{-2} a^{-1} and calculated the annual nitrogen loss due to denitrification in the Kleines Haff to be 440–500 t N a^{-1}. Model-based calculations estimated a long-term annual average denitrification of 4 g N m^{-2} a^{-1} (Table 5.2; Wielgat and Witek 2004) for the entire Lagoon. Based on the model, the annual loss of N to the atmosphere is in average 15,800 t N a^{-1} for the entire Lagoon (Table 5.2).

Sedimentation of phytoplankton bloom-derived organic material to the bottom is the major pathway of nutrient burial and sediment organic enrichment in the Lagoon (Kowalewska et al. 1997; Grelowski et al. 2000; T. Radziejewska unpublished data). The resulting organic enrichment of the sediment is very substantial: as measured by the weight loss on ignition, the sediment organic matter content may exceed 30% (Table 5.1b), and the sediment chloroplastic pigment contents may exceed 500 μg g^{-1} dry weight (dw) in the uppermost 1 cm layer (T. Radziejewska and H. Dworczak unpublished data).

According to Leipe et al. (1998), the uppermost 15 cm sediment surface layer accumulated an average of 1,000 tons nitrogen and 300 tons phosphorus per year during the last century. This calculation takes into account only the soft, silty sediments of the central part (about 56% of the total area). Assuming N and P sedimentation rates of 0.74 and 0.11 m^{-2} a^{-1}, respectively, Meyer and Lampe (1999) arrived at much lower values (510 t N and 76 t P). for the silty areas of the Lagoon. Intensive turbulence by methane bubbles, wind- and ship-propeller-induced sediment resuspension, and biotic activity result in thorough mixing of the sediment surface and prevent stratification.

Coarse and very simplified model simulations by Humborg et al. (2000a) showed a long-term annual net sedimentation of 2,650 t N and 716 t P. This does not take into account the littoral zones, and neglects the removal of sediment due to regular dredging of the shipping channel. As already mentioned, the amount of sediment removed annually during the last 50 years as a result of shipping channel dredging is about 1.5 million m^3 (Maritime Office Szczecin data; Minning et al. 2003). Taking this into account, Wielgat and Witek (2004) calculated a much higher annual N and

Table 5.2 Average annual budgets of nitrogen and phosphorus based on model simulations for the period 1980–1999 (Wielgat and Witek 2004)

	Wielki Zalew		Kleines Haff	
	Nitrogen (t N a^{-1})	Phosphorus (t P a^{-1})	Nitrogen (t N a^{-1})	Phosphorus (t P a^{-1})
Input	80,900	7,100	19,000	1,330
Output	64,200	5,900	10,500	1,100
Denitrification	8,800	–	7,000	–
Burial	7,900	1,200	1,500	230

P burial of 9,400 and 1,430 t, respectively (Table 5.2). The data suggests that the Lagoon is still an efficient nutrient sink (cf. Chap. 3 by Schiewer, this volume).

Due to its eutrophication and pollution status, the Szczecin Lagoon has been placed on the HELCOM hot spot list as an area in need of a management programme (http://www.helcom.fi/stc/files/Projects/JCP/LISTperJUNE06.pdf). As the Lagoon is affected by riverine nutrient loads, efficient management requires measures to be undertaken in the Odra basin. Against the background of the European Water Framework Directive (WFD), Wielgat and Schernewski (manuscript submitted) simulated the impact of nutrient load reductions on the Lagoon water quality. Literature data were used to calculate historical riverine loads for the pre-industrial state (100 years ago). As indicated by the model, the pre-industrial spring phytoplankton biomass (as Chl *a* content) was between 17 $\mu g\ l^{-1}$ in the Wielki Zalew and 12 $\mu g\ l^{-1}$ in the Kleines Haff, the corresponding summer values being 24 and 17 $\mu g\ l^{-1}$. On average, Chl *a* concentrations throughout the growing season were only one-third of those recorded at present. At that time, nutrient-limitation played a more important role in primary production. However, the Lagoon was already eutrophic. The pre-industrial level cannot be regained and possible nutrient load reductions (Behrendt et al. 2001) will barely be sufficient for the Lagoon to reach a good water quality according to the WFD.

5.6 Summary and Conclusions

As described in this chapter, the Szczecin Lagoon is an area that has, for the last 50 years, experienced heavy anthropogenic pressure. The pressure has been exerted by agriculture-, forestry-, and industry-related activities both in the surrounding land areas and in the entire basin of the Odra, a major supplier of nutrient and pollution load into the Lagoon. The Lagoon has always acted as a nutrient sink and nowadays internal nutrient regeneration has a certain potential to undermine basin-wide efforts aimed at improved coastal water quality. The pelagic system reflects the Lagoon's eutrophication by featuring persistent phytoplankton blooms and by supporting high biomass of planktivorous fish. The pelagic–benthic couplings emphasise eutrophication effects by enhancing nutrient storage and exchange, and by increasing organic enrichment of bottom sediments. The consequences for benthic communities involve mainly the abundant presence of deposit feeders typical of organic-rich environments. Shipping is an important vector of non-native species introduction.

Eutrophication of the Lagoon cannot be managed directly and requires efficient basin-wide programs. Parallel to, and contingent on, efforts aimed at containing the progress of eutrophication and pollution are measures taken towards nature conservation. The state of the Lagoon's aquatic environment is monitored by the state agencies of Poland and Germany. A number of particularly valuable and/or attractive coastal habitats and wider systems are subject to legal protection of various stringency, from nature reserves to landscape parks to the National Park level. The establishment of NATURA 2000 sites will add a new dimension in this respect.

Because the Szczecin Lagoon and its resources are shared by two countries, various processes affecting the Lagoon are transboundary in character. To list just a few, the effects of eutrophication and pollution are perceived on both sides of the border, as are, e.g., migrations of fish stocks or dispersal of alien species. In addition, in view of the fact that the Lagoon's coastal environments on both sides of the border are located in areas attractive for visitors, it is acknowledged that tourism will soon represent an additional source of external pressure on the Lagoon's coasts and on the Lagoon itself (Andrulewicz and Lamp 1994). Such issues call for joint and cooperative activities and management programs, which are, in fact, underway (ECOBALTIC 2001; IOŚ, 1999, 2001). However, the issues to be dealt with are very complex and will require a jointly implemented integrated coastal management system (ECOBALTIC 2001; Schernewski and Schiewer 2002).

Acknowledgements The authors are very grateful to the many friends and colleagues who willingly shared the data and information necessary for this chapter to take form. G. Schernewski additionally wishes to thank the German Federal Ministry of Education and Research for supporting the project "IKZM-Oder" (BMBF 03F0403A; http://www.ikzm-oder.de) and the ongoing research in the Odra estuary.

References

Andrulewicz E, Lamp J (1994) Odra/Szczecin Lagoon – spoiled but hopeful green link to the Baltic. WWF Baltic Bull 1/94:29–34

Bangel H, Schernewski G, Bachor A, Landsberg-Uczciwek M (2004) Spatial pattern and long-term development of water quality in the Oder estuary. In: Schernewski G, Dolch T (eds) The Oder Lagoon – against the background of the European Water Framework Directive. Mar Sci Rep 57:17–65

Behrendt H, Dannowski R, Deumlich D, Dolezal F, Kajewski I, Kornmilch M, Korol R, Mioduszewski W, Opitz D, Steidl J, Stronska M (2001) Point and diffuse emissions of pollutants, their retention in the river system of the Odra and scenario calculations on possible changes. UBA Research Report, Project 298–28–299

Bogusławska-Wąs E, Dąbrowski W (2001) The seasonal variability of yeasts and yeast-like organisms in water and bottom sediment of the Szczecin Lagoon. Int J Hyg Environ Health 203:451–458

Borówka RK, Osadczuk A, Witkowski A, Wawrzyniak-Wydrowska B, Duda T (2005) Late Glacial and Holocene depositional history in the eastern part of the Szczecin Lagoon (Great Lagoon) basin – NW Poland. Quatern Int 130:87–96

Breitenbach E, Lampe R, Leipe T (1999) Investigations on mineralogical and chemical composition of suspended particulate matter (SPM) in the Odra estuary. Acta Hydrochim Hydrobiol 27:298–302

Bruckmann K, Duwe K, Frey A, Gebhardt U, Hellmann B, Post J, Weidauer A (1998) Simulation und Messung von Zirkulations- und Transportprozessen im Greifswalder Bodden, Oderästuar und angrenzenden Küstengewässern. In: Lampe R (ed) Greifswalder Bodden und Oder-Ästuar-Austauschprozesse (GOAP), Synthesebericht des Verbundprojektes. Greifs Geogr Arb 16:1241

Chojnacki J (1991) Zooplankton succession in the River Odra estuary. Acta Ichthyol Piscat 21[Suppl]:41–45

Czerniejewski P, Filipiak J, Radziejewska T (2003) Body weight and morphometry of the Chinese mitten crab (*Eriocheir sinensis* H. Milne-Edwards, 1853) in the River Odra estuary (North-Western Poland). Acta Sci Pol Piscaria 2:29–40

Dahlke S, Wolf C, Bange HW (1998) Mikrobieller N-, S- und C- Umsatz und die Emission von N_2O, und CH_4 im Greifswalder Bodden und im westlichen Teil des Oderästuars. In: Lampe R (ed) Greifswalder Bodden und Oder-Ästuar-Austauschprozesse (GOAP), Synthesebericht des Verbundprojektes. Greifs Geogr Arb 16:370–402

Dunin-Kwinta I (2000) The state of fisheries in the Szczecin Lagoon (in Polish). Western Pomeranian Congress of Science at the Turn of the 20th and 21st Centuries, Szczecin, pp 261–266

ECOBALTIC (2001) Integrated management of the Szczecin Lagoon coastal areas (in Polish). Ecobaltic Foundation, Gdansk

Girjatowicz J (2001) Effects of atmospheric circulation on ice conditions in the Southern Baltic coastal lagoons. Int J Climatol 21:1593–1605

Grelowski A, Pastuszak M, Sitek S, Witek Z (2000) Budget calculations of nitrogen, phosphorus and BOD_5 passing through the Oder estuary. J Mar Syst 25:221–237

Gruszka P (1999) The River Odra estuary as a gateway for alien species immigration to the Baltic Sea Basin. Acta Hydrochim Hydrobiol 27:374–382

Günther B (1998) Die Bedeutung des Makrozoobenthos für den Kohlenstoffumsatz im Sediment. Greifs Geogr Arb 16:286–315

Heip CHR, Goosen NK, Herman PMJ, Kromkamp MJ, Middelburg JJ, Soetaert K (1995) Production and consumption of biological particles in temperate tidal estuaries. Oceanogr Mar Biol Annu Rev 33:1–149

Hensel S (1994) Status-quo-Erfassung des Makrozoobenthos im Kleinen Haff unter Berücksichtigung ökologischer Parameter. Diploma Thesis, University Greifswald

Humborg GC, Fennel K, Pastuszak M, Fennel W (2000a) A box model approach for a long-term assessment of estuarine eutrophication, Szczecin Lagoon, Southern Baltic. J Mar Syst 25:387–403

Humborg GC, Schernewski G, Bodungen B v, Dannowski R, Steidl J, Quast JV, Wallbaum V Rudolph K-U, Müller C, Mahlburg S, Erbguth W (2000b) Oder Basin – Baltic Sea Interactions (OBBSI), Final Report. Mar Sci Rep 41:1–84

IMWM (1980–1983) Institute of Meteorology and Water Management. Hydrological Annuals of the Oder River 1980–1983, IMGW, Warsaw

IMWM (1987–1999) Institute of Meteorology and Water Management. Water Resources – Monthly Bulletins 1987–1999, IMGW, Warsaw

IOŚ (1999) Environmental Report for the Province of Western Pomerania 1997–1998 (in Polish). Bibl Monit Środ, Szczecin

IOŚ (2001) Environmental Report for the Province of Western Pomerania 2000 (in Polish). Bibl Monit Środ, Szczecin

Kłyszejko B, Chełkowski Z, Chełkowska B, Sobociński A (2004) Identification of fish remains from early medieval layers of the Vegetable Market excavation site in Szczecin, Poland. Acta Ichthyol Piscat 34:85–102

Kowalewska G, Konat J, Wawrzyniak-Wydrowska B (1997) Polynuclear aromatic hydrocarbons (PAHs) in the sediments of the Oder Estuary (Szczecin Lagoon) before the flood of 1997. Oceanologia 39 255–266

Lampe R (ed) (1998) Greifswalder Bodden und Oder-Ästuar-Austauschprozesse (GOAP), Synthesebericht des Verbundprojektes. Greifs Geogr Arb 16:1–490

Lampe R (1999) The Odra estuary as a filter and transformation area. Acta Hydrochim Hydrobiol 27:292–297

Leipe T, Eidam J, Reinhard L, Hinrich M, Neumann T, Osadczuk A, Janke EPT, Blanz T, Gingele FX, Dannenberger D, Witt G (1998) Das Oderhaff, Beiträge zur Rekonstruktion der holozänen geologischen Entwicklung und anthropogenen Beeinflussung des Oder-Ästuars. Mar Sci Rep 28:1–61

Luścińska M (2005) Phytoplankton of the Szczecin Lagoon (The Great Lagoon). Oceanol Hydrobiol Stud 34:3–29

Majewski A (1980) Szczecin Lagoon (in Polish). WKiL, Warsaw

Masłowski J (1992) Bottom macrofauna of the Szczecin Lagoon (north-western Poland). Acta Hydrobiol 34:253–274

Masłowski J (1993) Long-term changes in the bottom macrofauna of the Szczecin Lagoon (northwestern Poland). Acta Hydrobiol 35:341–355
Meyer H, Lampe R (1999) The restricted buffer capacity of a South Baltic estuary – the Oder Estuary. Limnologica 29:242–248
Ministry of Agriculture and Rural Development (2003) Sector Operational Programme Fisheries and Fish Processing 2004–2006 (in Polish). Ministry of Agriculture and Rural Development, Warsaw
Minning M (2004) Der Einfluss des Schifffahrtkanals auf den Stoffhaushalt im Oderhaff. In: Schernewski G, Dolch T (eds) The Oder Lagoon – against the background of the European Water Framework Directive. Mar Sci Rep 57:177–222
Minning M, Leipe T, Schernewski G (2003) Auswirkungen der regelmässigen Kanal-Ausbaggerungen auf die Trophie des Stettines (Oder) Haffs. Dtsch Ges Limnol Tagungsber 2002, Braunschweig, Tutzing
Mohrholz V, Lass HU (1998) Transports between Oderhaff and Pomeranian Bight – a simple barotropic box model. Ger J Hydrogr 1998:371–383
Mutko T, Stechman J, Landsberg-Uczciwek M, Garbacik-Wesołowska A, Protasowicki M, Nied źwiecki E, Ciereszko W, Meller E (1994) The Szczecin Lagoon – Wielki Zalew – Long-term qualitative changes (in Polish). Bibl Monit Środ, Warsaw
Nixon SW (1995) Coastal and marine eutrophication: a definition, social causes, and future concerns. Ophelia 41:199–219
OECD (1982) Eutrophication of waters. Monitoring, assessment and control. Organisation for Economic Cooperation and Development (OECD), Paris
Osadczuk A (2004) The Szczecin Lagoon: environmental conditions of recent lagoonal sedimentation (in Polish). Uniwersytet Szczeciński, Rozpr Stud 549
Paerl HW (1988) Nuisance phytoplankton blooms in coastal, estuarine, and inland waters. Limnol Oceanogr 33:823–847
Piesik J (1993) The Odra estuary architecture (in Polish). Pr Nauk Polit Szczec 502
Podsetchine V, Schernewski G, Tejakusuma I (2004) Simulations of water flow, transport and temperature in the Oder (Szczecin) Lagoon. In: Schernewski G, Dolch T (eds) The Oder Lagoon – against the background of the European Water Framework Directive. Mar Sci Rep 57:85–97
Radziejewska T, Chabior M (2004) Climatic and hydrological controls over the zoobenthos in a southern Baltic coastal lagoon. Hydrobiologia 514:171–181
Radziejewska T, Drzycimski I (1988) Meiobenthic communities of the Szczecin Lagoon. Kieler Meeresforsch Sonderh 6:162–172
Radziejewska T, Drzycimski I (1990) Dynamics of meiobenthic communities in a eutrophic and polluted estuary. Limnologica 2:83–88
Robakiewicz W (1993) Hydrodynamic regime of the Szczecin Lagoon and the straits connecting the Lagoon with the Pomeranian Bay (in Polish). Bibl Nauk Hydrotechnol 16
Rycak SA, Kowalski WWA (2007) Growth dynamics of rush vegetation in the coastal zone of the Szczecin Lagoon. Acta Sci Pol (in press)
Schernewski G, Jülich W-D (2001) Risk assessment of virus infections in the Oder estuary (southern Baltic) on the basis of spatial transport and virus decay simulations. Int J Hyg Environ Health 203:317–325
Schernewski G, Schiewer U (2002) Status, problems and integrated management of Baltic coastal ecosystems. In: Schernewski G, Schiewer U (eds) Baltic coastal ecosystems. structure, function and management. Springer, Berlin, pp 1–16
Schernewski G, Sterr H (2002) Tourism and environmental quality of the German Baltic coast: conflict or chance? In: Schernewski G, Schiewer U (eds) Baltic coastal ecosystems. structure, function and management. Springer, Berlin, pp 215–229
Schernewski G, Wielgat M (2001) Eutrophication of the shallow Szczecin Lagoon (Baltic Sea): modelling, management and the impact of weather. In: Brebbia CA (ed) Coastal engineering: computer modelling of seas and coastal regions. Witpress, Southampton, pp 87–98
Schernewski G, Podsetchine V, Siegel H, Huttula T (2000) Instruments for water quality management and research in coastal zones: flow and transport simulations across spatial scales. Period Biol 102[Suppl 1]:65–75

Schernewski G, Neumann T, Podsetchine V, Siegel H (2001) Spatial impact of the river Oder plume on water quality and seaside summer tourism at the south-western Baltic coast. Int J Hyg Environ Health 204:143–155

Schernewski G, Huttula T, Jülich W-D, Podsetchine V, Tejakusuma I (2002) Water quality problems in Baltic coastal waters: the Odra river as a source of human pathogenic viruses. In: Proceedings of the International Conference 'Sustainable Management of Transboundary Waters in Europe', UNECE, 21–24 April 2002, Międzyzdroje, Poland, pp 341–345

Schernewski G, Behrens H, Bock S, Dehne P, Erbguth W, Glaeser B, Janssen G, Kazakos W, Neumann T, Permien T, Schuldt B, Sordyl H, Steingrube W, Vetter L, Wirtz K (2004) Integriertes Küstenzonenmanagement in der Odermündungsregion (IKZM Oder). In: Schernewski G, Dolch T (eds) Geographie der Meere und Küsten, Coastline Reports 1, pp 183–193

Schernewski G, Hoffmann J, Löser N, Dreisewerd M, Stavenhagen P, Grunow B (2006) Measuring the progress and outcomes of Integrated Coastal and Ocean Management: the German Oder Estuary case study. Report to the UNESCO Intergovernmental Oceanographic Commission (IOC). Baltic Sea Research Institute, Warnemünde

Stańczykowska A, Lewandowski K, Świerczyński M (1997) Summary of studies on *Dreissena polymorpha* (Pall.) conducted in the period 1993–95 in the Masurian and Pomeranian lakelands. Pol Arch Hydrobiol 54:505–515

Szostak S, Kuzebski E, Budny T (2004) Marine fisheries in 2003 (in Polish). MIR, Gdynia

Tejakusuma I (2004) Investigations into the hydrography and dynamics of suspended particulate matter and sediments in the Oder Lagoon, southern Baltic Sea. IKZM-Oder Berichte 1. http://www.ikzm-oder.de/download.php?fileid=298

Tsuchiya M, Nishihira M (1985) Islands of *Mytilus* as a habitat for small intertidal animals: effect of island size on community structure. Mar Ecol Prog Ser 25:71–81

Wasmund N (1990) Characteristics of phytoplankton in brackish waters of different trophic levels. Limnologica 20:47–51

Westphal H, Lenk B (1998) Die räumlich-zeitliche Verteilung von Primärproduktion und Bakterienkeimzahlen. Greifs Geogr Arb 16:228–249

Wielgat M (2002) Compilation of the nutrient loads for the Szczecin Lagoon (Southern Baltic). In: Schernewski G, Schiewer U (eds) Baltic coastal ecosystems: structure, function and coastal zone management. Springer, Berlin, pp 75–92

Wielgat M, Witek Z (2004) A dynamic box model of the Szczecin Lagoon nutrient cycling and its first application to the calculation of the nutrient budget. In: Schernewski G, Dolch T (eds) The Oder estuary, against the background of the Water Framework Directive. Mar Sci Rep 57:99–125

Wiktor J (1969) The biology of *Dreissena polymorpha* (Pall.) and its ecological importance in the Firth of Szczecin (in Polish). Stud Mat MIR, Ser A 5:1–88

Wiktor J, Garbacik-Wesołowska A (1993) Fish resource management in the Szczecin Lagoon over 1949–1992 (in Polish). In: Draganik B, Szlauer L, Garbacik-Wesołowska A (eds) The status and perspectives of hydrobiological and fisheries research in the Odra estuary. Stud Mat MIR, Świnoujście, pp 7–12

Wolff T, Lou J, Rosenthal W (1998) Anwendung eines gekoppelten Modells im Oderhaff. In: Lampe R (ed) Greifswalder Bodden und Oder-Ästuar-Austauschprozesse (GOAP), Synthesebericht des Verbundprojektes. Greifs Geogr Arb 16:42–77

Wolnomiejski N (1994) Ecological study on muddy bottom macrofauna of the Szczecin Lagoon (1982–1992) (in Polish). Stud Mat MIR Ser A 31:1–126

Wolnomiejski N, Grygiel I (1992) Muddy-bottom macrofauna of the Szczecin Lagoon in 1987–1990 (in Polish). Stud Mat MIR Ser B 60:5–23

Wolnomiejski N, Woźniczka A (2007) A drastic reduction in abundance of *Dreissena polymorpha* Pall. in the Skoszewska Cove (Szczecin Lagoon, River Odra estuary): effects in the population and habitat. Ecol Quest 8 (in press)

Wolska M, Piasecki W (2004) Planktonic organisms as indicators of trophic conditions in a channel of the Odra River mouth (in Polish). Zesz Probl Post Nauk Roln PAN, 501:485–490

Wysokiński A (1998) Fishery management in the Szczecin Lagoon. Bull Sea Fish Inst 3:65–81

Chapter 6
Near-Shore Zones: Koserow and Tromper Wiek

U. Schiewer

6.1 Introduction

Because of their special geological and hydrographical properties, near-shore zones are thought of as highly productive ecosystems serving as a potential source for the organic carbon deposition of the deeper Baltic Sea areas (Bodungen and Zeitschel 1996; Emeis et al. 1999; Witt et al. 1999). However, since the hydrodynamic, chemical and biological features of near-shore coastal zones are highly diverse, this general statement was tested by carrying out small-scale investigations at two distinct sites taking into consideration the specific properties of the investigated areas. The two distinct areas of the German coast in the southern Baltic Sea selected were "Tromper Wiek" (TW), at the northwest part of the island of Rügen, and "Koserow" (KW), at the middle part of the island of Usedom. The test sites are located at the south-western end of the Baltic Proper, near to the Darß sill. TW is located near to the Arkona Basin.

6.2 Environmental Characteristics

The two sampling sites lie in the Pomeranian Bay, Mecklenburg-Vorpommern (Fig. 6.1), and differ in eutrophication at 5 m water depth. Station KW (54° 3.9′N; 14° 0.9′E), located at the coast of the island Usedom, is influenced by the supply of freshwater from the Odra River (Schiewer et al. 2005). The Odra is the main supplier of freshwater and the largest source of nutrients for the Pomeranian Bight (Pastuszak et al. 1996). Its tributaries make up to 86.3% of the entire drainage area, with an annual water outflow of about 18 km^3. Differences in the average monthly outflows in a given year are rather large, varying from 359 $m^3\ s^{-1}$ in September to nearly 900 $m^3\ s^{-1}$ in March during the spring snowmelt. Between December and May, as much as two-thirds of the entire annual volume of freshwater enters the sea, and outflow into the Bay peaks between January and April. Being rather shallow, the Pomeranian Bay has a small volume (73 km^3), fed by a relatively large volume of riverine water (15–20 km^3) (Pastuszak et al. 1996).

U. Schiewer (ed.), *Ecology of Baltic Coastal Waters. Ecological Studies 197.*

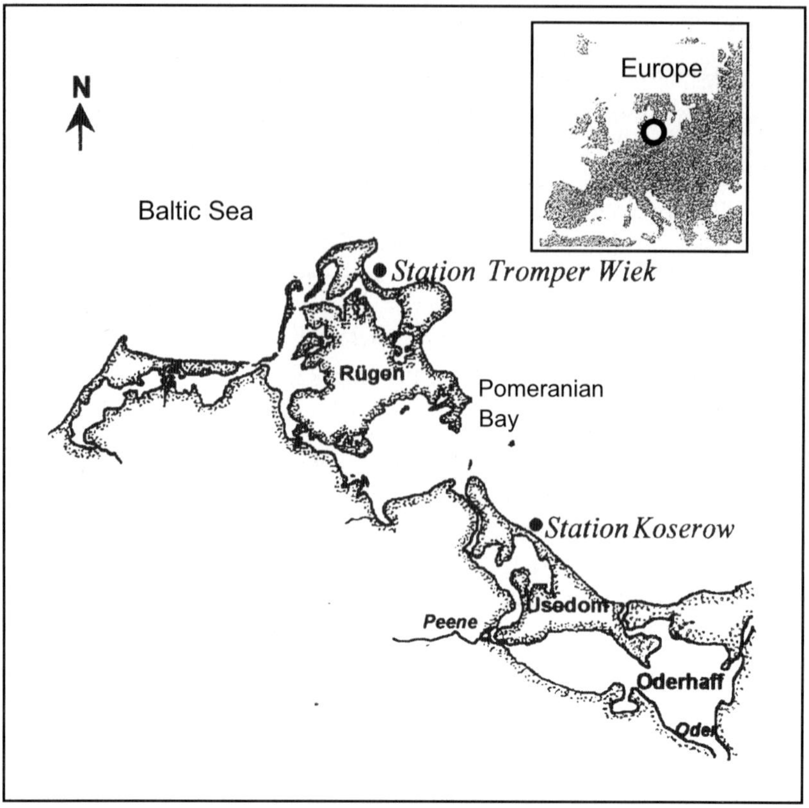

Fig. 6.1 The two near-shore sampling sites "Koserow" (KW) and "Tromper Wiek" (TW) selected in Pomeranian Bay, Mecklenburg-Vorpommern, north-eastern Germany (Schiewer et al. 2005). KW and TW differ in eutrophication at 5 m water depth

The second station, TW (54° 37.3′N; 13° 23.4′E), is located in the north-west region of the Isle of Rügen. A considerable direct supply of river freshwater into this area is rare. Thus, a predominant influence of salt water from the Baltic Sea is more likely.

In 1997, pelagic samples from both stations in the Pomeranian Bay were taken monthly (February–November) at noon with a Ruttner-type sampler. Integrative samples for the whole water column taken from 2, 5 and 10 m depth were placed into 30 l polyethylene containers for immediate transport to the laboratory and subsequently analysed for chemical and biological parameters.

6.3 Results

The coastal area in KW is dominated by erosion processes, while TW is characterised by equilibrium between erosion and accumulation. Both selected stations have sandy bottoms, low and patchy distributed benthic biomasses (Dehmlow et al. 2004), and are dominated by the pelagic food web. Production of macrophytes and macrozoobenthos are of lesser importance, but are significantly more expressed and dominated by a soft-bottom fauna at KW. At this station the net-production of the bottom-fauna was four times higher than in TW. The two ecosystems are controlled mainly by physical factors, e.g. temperature, light and wind forces. During the vegetation period, phytoplankton growth is controlled by nutrient, mainly nitrogen, limitation. This limitation is more expressed in TW, which is influenced mainly by the Baltic Sea. For KW, a greater riverine-based nutrient load from the Odra River is documented by nutrient analysis (Tables 6.1, 6.2) and differences in the biota.

In general, benthic–pelagic coupling is well expressed in both areas. The biodiversity at TW is at a natural level, at KW a near to natural level.

The annual periodicity of phytoplankton development starts in January–February with light and temperature limitation. Sometimes ice covering of the surface waters occurs. During such conditions, nutrient limitation changes to Si-limitation for a short time, followed by nitrogen limitation. In summer, there is a mixture of nitrogen or phosphorus limitation and, at TW, a greater influence of zooplankton grazing pressure. At KW this general behaviour is disturbed by nutrient inputs from the Odra River. Towards the end of the year, light limitation and, later, temperature limitation again become evident.

TW can be classified as a meso-/oligotrophic and KW as a eutrophic/mesotrophic water body (Tables 6.1, 6.2; Fig. 6.2). These differences are also expressed in differences in phytoplankton annual periodicity (Feuerpfeil et al. 2004b), the grazer-controlled food chain in TW, and the dominance of the microbial food web (MFW) and more fluffy sediments at KW (Estrum-Yousef et al. 2004).

High bacterial activity is correlated with aggregates (Estrum-Yousef et al. 2004). In general, aggregates seem to play an important role in carbon turnover, nutrient remineralisation and transport of matter.

Calculation of the carbon turnover rates demonstrate a balanced ecosystem at TW while at KW an important part of the organic carbon net-production is exported into the Baltic Proper and its basins (Fig. 6.3; for details see Feuerpfeil et al. 2004a, 2004b). This characterises KW as a typical transition zone.

6.4 Discussion and Conclusions

Despite the generally high degree of eutrophication of the Baltic Sea coastal zone, coastal waters such as TW, with pristine character and natural biodiversities, still exist. From the nutrient input, primary production rates and biomasses, TW could be classified as a meso-/oligotrophic area. Moderate nutrient input from the catchment area led to gross production rates that could be respired by the heterotrophic community over the year, leading to a balanced carbon budget. Net primary production rates of organic carbon (Feuerpfeil et al. 2004a) in March 1997 (Bacillariophyceae), July and August 1997 (Dinophyceae and Cyanobacteria), respectively, were compensated by high respiration rates in April and during the

Table. 6.1 Abiotic parameters of the near-shore stations “Koserow” (KW) and “Tromper Wiek” (TW) during 1997. Salinity values ranged from 5.9 to 8.2 psu and 6.9 to 11.6 psu at KW and TW, respectively, showing significant differences between both areas (Wilcoxon signed rank test, $P > 0.02$, $n = 10$) throughout the year, reflecting the influence of freshwater from the Odra River on salinity regime at KW throughout the year. DIN Dissolved organic nitrogen, DIP dissolved organic phosphorus

Station	Date (1997)	Temperature (°C)	Salinity (psu)	NO_3^- (μM l^{-1})	NO_2^- (μM l^{-1})	NH_4^+ (μM l^{-1})	DIN (μM l^{-1})	DIP (μM l^{-1})	DIN/DIP
TW	25 February	2.4	10.2	5.85	1.29	2.78	9.92	0.86	11.5
	18 March	3.1	8.3	0.24	0.64	0.5	1.38	0.64	2.1
	21 April	5	7.9	0.67	0.18	0.8	1.65	0.76	2.1
	20 May	9.3	8.3	0.48	0.28	4.08	4.84	0.58	8.3
	17 June	1.1	7.6	0.73	0.54	3.18	4.45	0.54	8.2
	25 July	20.3	10.1	0.08	0.56	0.52	1.16	0.64	1.8
	19 August	22.7	6.9	0.1	0.38	0.46	0.94	0.42	2.2
	24 September	15.6	11.6	0.4	0.37	1.43	2.2	1.56	1.4
	22 October	9.8	7.9	0.09	0.41	1.06	1.56	1.08	1.4
	19 November	9.5	8.4	1.01	0.54	3.98	5.53	1.17	4.7
KW	26 February	2.8	7.2	16.19	1.98	5.6	23.77	0.91	26.1
	19 March	2.9	6.2	16.23	1.4	3.98	21.61	1.22	17.7
	22 April	6	7	0.86	0.14	0.8	1.8	0.6	3.0
	21 May	10.9	6.9	4.75	0.54	4.5	9.79	0.41	23.9
	18 June	15.8	5.9	0.49	0.55	3.61	4.65	0.49	9.5
	15 July	19.6	6.7	1.71	0.87	1.3	3.88	1.81	2.1
	20 August	20.8	6.7	2.05	0.86	5.41	8.32	1.14	7.3
	23 September	15	7.6	4.85	1.68	7.22	13.75	6.96	1.9
	21 October	10	8.2	0.77	0.74	1.09	2.6	2.81	0.9
	18 November	6.5	8.1	3.99	0.61	3.87	8.47	1.11	7.6

Table 6.2 Chemical parameters of the near-shore stations KW and TW during 1997. Significant differences between the sampling sites are shown with significant limits (*P*) and number of pairs (*n*). *DIN* Dissolved inorganic nitrogen, *TP* total phosphorus, *PC* particulate carbon, *PN* particulate nitrogen, *DOC* dissolved organic carbon

Parameter	Units	KW		*P*	*n*	TW
Salinity	psu	5.9–8.2	<	<0.02	10	6.9–11.6
DIN	µmol l^{-1}	1.8–23.8 (NH_4^+ > NO_3^- > urea > NO_2^-)[a]	>	<0.02	10	0.9–9.9 (NH_4^+ > urea > NO_3^- > NO_2^-)[a]
TP	µmol l^{-1}	1.62–4.05	>	<0.05	10	1.39–2.11
PN	µmol l^{-1}	0.08–0.48	>	<0.05	10	0.03–0.17
PC	µmol l^{-1}	0.35–4.35	>	<0.02	10	0.25–1.51
DOC	µmol l^{-1}	3.18–6.77	~			3.79–6.05
C/N		4.89–10.62	~			6.03–10.76

[a]Relative abundance

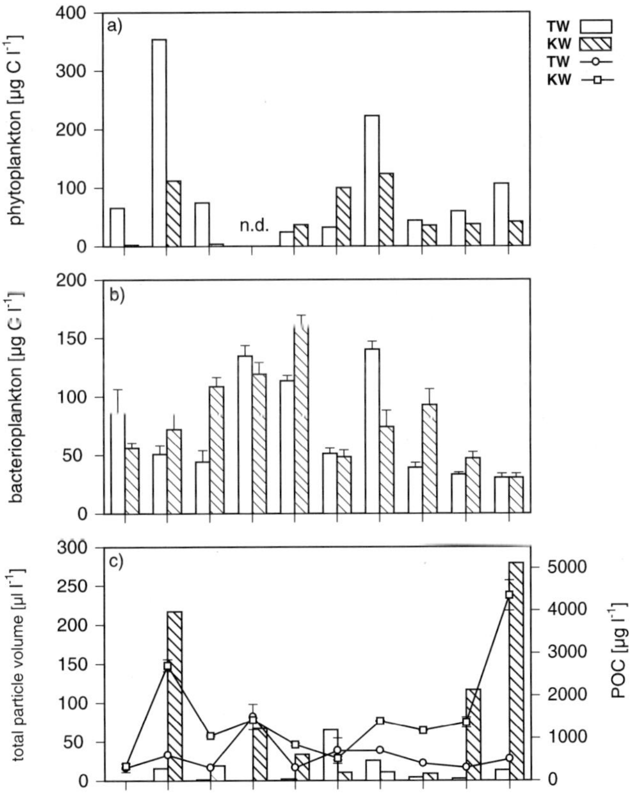

Fig. 6.2 Seasonal course of phytoplankton biomass (**a**), bacterial biomass (**b**) and total particle per volume (**c**) (*bars*) as well as bacterial production rates and particulate organic carbon (*POC*) concentrations (*lines*) at the two sampling sites KW and TW

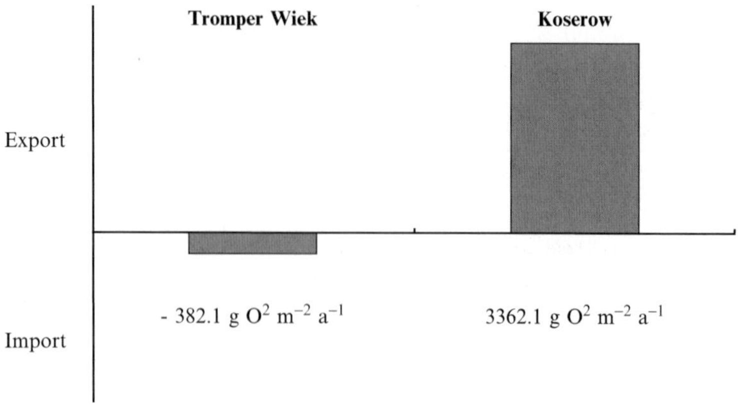

Fig. 6.3 Annual balance of primary production 1997 of the near-shore stations KW and TW

period from September to November 1997 (Fig. 6.4b). The net production of organic matter in spring and autumn at the TW study site might also contribute to organic carbon deposition at deeper areas of the Arkona basin, but relative high zooplankton biomasses and a clear seasonality of the copepod biomass indicated a major part of the organic matter produced in spring was channelled to the classical food web, transferring the organic carbon to higher trophic levels. The Cyanobacteria bloom in summer should be more resistant against metazoan grazing, and potentially more suitable for forming aggregates. These are degraded to a large extent in the pelagic via bacteria, fungi and protozoans (see Chap. 1 by Schiewer, this volume). Comparing photoautotrophic biomasses at TW with biomasses built up at study site KW in summer, TW was less important as source of organic matter deposition for the Arkona basin.

KW was influenced directly by the high nutrient load of the Odra River. In contrast to TW, primary production at KW (Feuerpfeil et al. 2004a) could not be remineralised by the pelagic heterotrophic community over an annual cycle, leading to a net production (Fig. 6.4a) of organic carbon of 3,362 g C m^{-2} a^{-1} (assuming a photosynthetic quotient of 1.2). Since the general transport direction of fine-grained matter from the Odra River, as revealed by three-dimensional (3D) modelling, including time scales of several months, is towards the Arkona basin in the NNW direction (Edelvang et al. 1999), this material acts as a potential carbon source for the deeper basins. Underwater camera observations at KW showed a high total particle volume >40 µm (4–280 µl l^{-1}), indicating a high potential for aggregation and sedimentation at this site, while at TW particle volume was much lower (1–5 µl l^{-1}). High particle volumes >40 m (220 µl l^{-1}) at KW were found in spring in combination with a high percentage of Bacillariophyceae taxa. A second peak (280 µl l^{-1}) was observed in November 1997 following the autumn peak of the Chloropyhceae-dominated phytoplankton. However, the particle volume did not follow primary production or biomass parameters significantly. Therefore, lateral transport processes or resuspension of organic matter from the Odra River

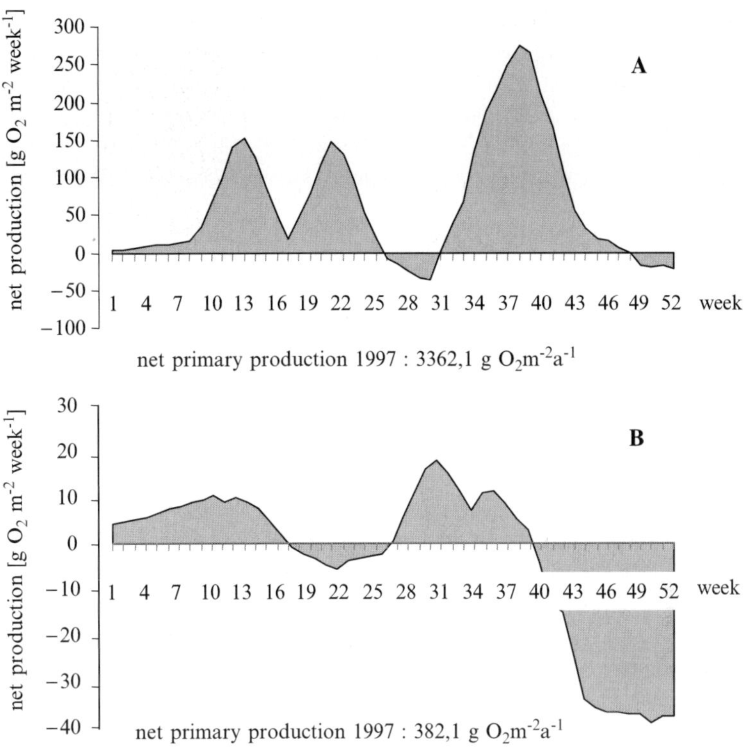

Fig. 6.4A,B Monthly calculated net primary production of phytoplankton and annual rate for 1997 (Feuerpfeil et al. 2004a). **A** Station KW, **B** Station TW

seemed to be the main factor controlling the abundance of particles >40 μm at study side KW, which are not included in the calculation. Periodic investigations on the benthic meiofauna indicated that sedimentation should play an important role at KW since the sediment was inhabited by a high biomass of deposit feeders dependent on the energy input from the water column. Estrum-Yousef and Schoor (2000) found potential sedimentation rates of 1 g m^2 day^{-1} at KW, indicating that at least part of the net production should be used by benthic organisms or buried in the sediments at KW. Furthermore, the accumulation of particulate organic matter (POM) at the sediment surface can lead to the formation of a fluffy sediment layer inhabited by an active microbial community (Schumann and Rentsch 1998; Rieling 1999) that is also mineralising a part of the organic matter originated in the water column. Because of the unknown contribution of benthic decomposition it was not possible to quantify the amount of POM that is transported to the deeper areas. However, since the threshold velocity for the fluffy layer is about 5 cm s^{-1}, which is average near-bottom velocity for shallow coastal waters (Edelvang et al. 1999), we hypothesise that KW acts as a source of organic matter by sediment resuspension and vertical near-bottom transport processes. The fate of the organic matter transported along the coast to the deeper basins is, as yet, unclear but since the material underlies strong biological degradation processes it should be altered significantly during transport processes to the Arkona basin. This biological alteration and the associated shifts in microbial processes and communities are still unknown, and further investigation into the successive decay of organic matter originating from the Baltic coastal areas is required to obtain a more detailed insight into its role in biogeochemical cycling in the deeper basins of the Baltic Sea.

References

Bodungen B von, Zeitschel B (1996) Die Ostsee als Ökosystem. In: Rheinheimer G (ed) Meereskunde der Ostsee. Springer, Heidelberg, pp 230–244

Dehmlow J, Heerkloß R, Schiewer U (2004) Characterisation of zooplankton and macrozoobenthon parameters of inshore waters in the Pomeranian Bay – a comparison of two sampling sites. Final Report, http://www.io-warnemuende.de/Projects/Basys/reports/final/en_home.htm

Edelvang K, Lund-Hansen LC, Christiansen C, Petersen OS, Laima M (1999) Vertical and horizontal fluxes of suspended matter in a coast to basin depth profile of the SW Baltic Sea. In: Third BASYS (Baltic Sea System Study) Annual Science Conference, proceedings. Institut für Ostseeforschung, Warnemuende

Emeis K-C, Christiansen C, Jähmlich S, Leipe T, Lund-Hansen L, Löffler A, Miltner A, Pazdro K, Pempkowiak J, Pollehne F, Shimmield T, Voß M, Witt G (1999) Origin and character of material transported from the nearshore to the basinal environment in the southern Baltic Sea. In: Third BASYS (Baltic Sea System Study) Annual Science Conference, proceedings. Institut für Ostseeforschung, Warnemuende

Estrum-Yousef S, Schoor A (2000) Quality of particulate matter and its potential sinking rates in pelagic samples of inshore and Bodden waters of the southern Baltic Sea. Int Rev Hydrobiol 85:341–357

Estrum-Yousef S, Schiewer U, Schoor A (2004) Chemical, bacteriological an sedimentological characterisation of nearshore waters in the Pomeranian Bay – comparison of two sampling sites over the annual period. Final Report, http://www.io-warnemuende.de/Projects/Basys/reports/final/en_home.htm

Feuerpfeil P, Rieling R, Estrum-Yousef S, Dehmlow J, Papenfuß T, Schoor A, Schiewer U, Schubert H (2004a) Carbon budget and pelagic community compositions at two coastal areas that differ in their degree of eutrophication, in the southern Baltic Sea. Estuarine Coastal Shelf Sci 61:89–100

Feuerpfeil P, Schiewer U, Schubert H (2004b) Characterisation of the phytoplankton component – comparison of two coastal environments of the southern Baltic over an annual cycle. Final Report, http://www.io-warnemuende.de/Projects/Basys/reports/final/en_home.htm

Pastuszak M, Nagel K, Nausch G (1996) Variability in nutrient distribution in the Pomeranian Bay in September 1993. Oceanologia 38:195–225

Rieling T (1999) Remineralisation organischen Materials in Boddengewässern Mecklenburg-Vorpommerns unter besonderer Berücksichtigung der Bedeutung von Partikeln und Aggregaten. Dissertation, Universität Greifswald

Schiewer U, Rieling T, Feuerpfeil P, Estrum-Yousef S, Dehmlow J, Papenfuß T, Schoor A, Schubert H (2005) Coastal zones of the southern Baltic Sea – sinks or sources of organic carbon fort he Arkona Basin? Verh Int Verein Limnol 29:517–520

Schumann R, Rentsch D (1998) Staining particulate organic matter with DTAF – a fluorescence dye for carbohydrates and proteins: a new approach and application of a 2D image analysis system. Mar Ecol Prog Ser 163:77–88

Witt G, Emeis K-C, Leipe T, Löffler A (1999) Seasonal and regional variability of Polycyclic Aromatic Hydrocarbons (PAHs) in fluffy layer material from the Odra river estuary. In: Third BASYS (Baltic Sea System Study) Annual Science Conference, proceedings. Institut für Ostseeforschung, Warnemuende

Chapter 7
Gulf of Gdańsk and Puck Bay

L. Kruk-Dowgiałło and A. Szaniawska

7.1 Introduction

Located on the shores of the Gulf of Gdańsk, the Tri-Cities agglomeration, which, in addition to other small towns, has a population of 760,000, is comprised of the port cities Gdańsk and Gdynia and the spa town of Sopot. The gulf is subjected to different rates of anthropogenic pressure and is extraordinarily interesting from a scientific point of view. Poland has recommended that two regions in this basin be declared Baltic Sea Protected Areas HELCOM BSPA: the Nadmorski Landscape Park and the Mierzeja Wiślana Landscape Park (Kruk-Dowgiałło et al. 2000a, 2000b). The western region of this basin was included in the NATURA 2000 network in 2004 (Fig. 7.1).

The Gulf of Gdańsk is a system of estuaries in which there is a mix of brackish and marine waters that is typical of this type of basin. Freshwater comes from terrestrial sources, mainly the Vistula River, which is the second largest river flowing into the Baltic Sea. This causes spatial differentiation in the hydrological and hydrochemical characteristics of this basin. This differentiation is the basis upon which the surface waters (0–5 m) of the gulf are divided into seven surface water divisions (Nowacki and Jarosz 1998; Fig. 7.1a). Taking into consideration that hydrological differentiation shapes biocenotic diversity, Andrulewicz et al. (2004) identified natural sub-systems in the Gulf of Gdańsk (Fig. 7.1b) based on hydrographic and hydrochemical conditions, anthropogenic pressure, and the states of phytoplankton, macroalgae and angiospermae, macrozoobenthos, and ichthyofauna. The quality status of the different environmental units, which were assessed according to criteria proposed by the Water Framework Directive (EU WFD 2000) in comparison with the selected reference period of the 1950s, was assessed and then categorised into one of four quality ranges (bad, poor, moderate, and good).

In accordance with systems A and B of the EU WFD (2000), in 2004 the Polish Baltic zone was divided into two categories: transitional waters (TW) and coastal waters (CW). While a portion of Gulf of Gdańsk waters was designated as internal marine waters, those of Puck Bay and the Vistula Lagoon were categorised as TW, and those along the shores as CW (Fig. 7.1c). Within these two categories, five types were identified based on coastal morphology and bottom sediment type. All transitional

U. Schiewer (ed.), *Ecology of Baltic Coastal Waters. Ecological Studies 197.*

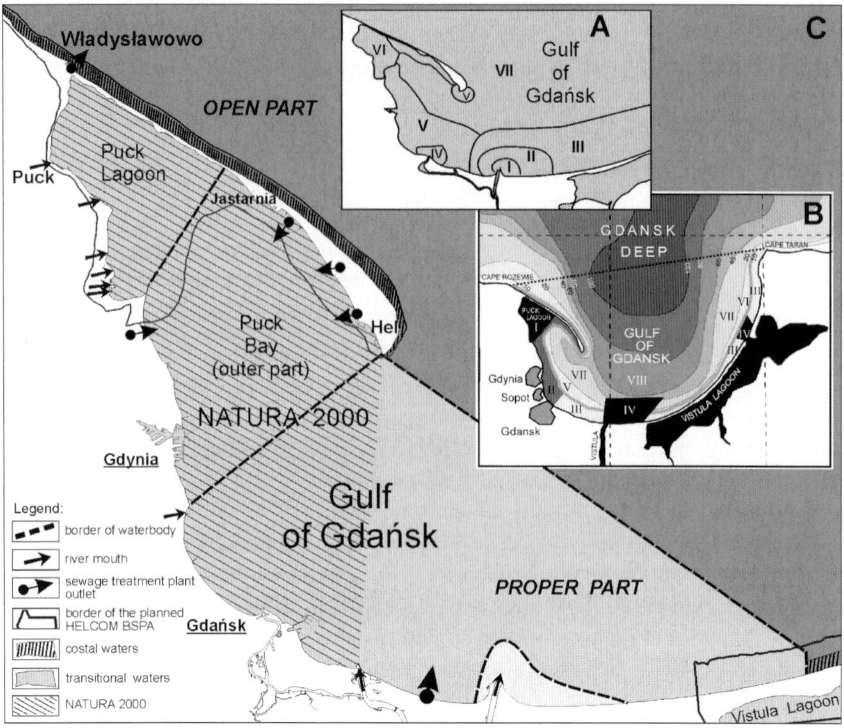

Fig. 7.1A–C Gulf of Gdańsk and Puck Bay. A Divisions based on hydrological and hydrochemical characteristics according to Nowacki and Jarosz (1998). B Divisions into natural subsystems according to Andrulewicz et al. (2004). C Polish typology of marine waters according to Krzymiński et al. (2004)

water bodies (WB) are represented by different types. Coastal waters were classified as the spit type to which one water body was assigned. According to this classification, the basins described below were given the following codes: Gulf of Gdańsk proper – PL TW IV WB4; Puck Lagoon – PL TW II WB2; Puck Bay outer – PL TW III WB3; and Vistula mouth – PL TW V WB5. The southern border of the Gdańsk Deep was classified as open Baltic waters (Krzymiński et al. 2004; Fig. 7.1c).

7.2 Gulf of Gdańsk

7.2.1 Environmental characteristics

The Gulf of Gdańsk (Table 7.1) is part of the Gdańsk Basin, which is comprised of the southeastern part of the Baltic Proper as designated for the purposes of HELCOM (Majewski 1994). This basin is differentiated as regards hydromorphological regime, and its borders lie at the Gulf of Gdańsk proper, the open sea that forms the southern border of the Gdańsk Deep, and Puck Bay (Fig. 7.1b,c). The recognised border between the gulf and the Baltic Proper is a line 58 nautical miles (Nm) in length that connects Rozewie Cape (54° 50′N, 18° 20′E) with Taran Cape (54° 58′N, 19° 69′E) (Majewski 1990; Fig. 7.1b).

The borders of the gulf include the coastal cliffs, sand dunes (western part of the Vistula spit), and the lowlands and the alluvial cone of the Vistula River. The bottom area is comprised of relicts shaped by glaciers and glaciofluvial and relict forms from delta and marine accumulation and erosion. The bottom sediments are composed primarily of coarse-grained sand, sandy marine mud, and delta mud and sand (Krzymiński et al. 2004).

One of the characteristic features of the climate is the great changeability of the weather. Due to its spatial differentiation, the gulf is divided into several regions: open sea (southern part of the Gdańsk Deep), coastal region up to the 50 m isobath, and the seaside region (Table 7.1a). The winds blowing in the gulf region are from the western sector, i.e. SW, W, and NW winds comprise 40–50% of the total winds. The average annual wind speed ranges from 4 to 9 m s^{-1}. Sea level rises of up to +1.5 m are observed during storms (Majewski 1990). Since 1950 there has been an increase in the temperature of gulf surface waters, especially in the first 3 months of the year (Miętus et al. 2004). This is also confirmed by the decreasing number of days with ice; a total of 76 were noted in the 1940–1969 period, while there were only 47 in the 1970–1999 period (Sztobryn et al. 2004). The number of ice days has continued to decrease in recent years (Table 7.1a).

The surface area of the gulf is approximately 1.3% of the surface area of the entire Baltic Sea. Some environmental characteristics, including the surface area, volume, average depth, catchment area, and the area-to-catchment relationship of the Gulf of Gdańsk, are presented in Table 7.1b.

The waters of the gulf are divided into surface waters of equal salinity that undergo significant seasonal temperature variation, intermediary waters of equal salinity and low temperature (so-called winter waters), and the saline near-bottom waters with little temperature variation that remain primarily under the influence of saline near-bottom waters from the North Sea. The salinity of waters (Table 7.1b) in the Gdańsk Deep region oscillates from 6.9 psu in the surface layer to more than 12 psu near the bottom, while near the Vistula mouth it drops periodically to as low as 1 psu. The entire volume of the Puck Bay waters is in the isohaline surface water layer of the Gulf of Gdańsk (Nowacki 1993b). The relative mean value of water transparency measured with Secchi discs changed significantly in the gulf from 7.6 m in the 1957–1969 period to 4.7 m in the 1980–1990 period (Trzosińska 1992).

Waters from 20 streams and rivers flow into the Gulf of Gdańsk proper. The Vistula River shapes the hydrological conditions of the gulf and introduces the largest load of pollutants.

The Vistula drains an average of 34.48 km^3 a^{-1} of waters into the gulf, which include the following loads (the percentages of the load introduced by the Vistula in comparison with the loads from 11 other rivers, including the Oder, that drain into the Polish coastal zone of the Baltic are noted in parentheses): biochemical

Table 7.1 Main characteristics of the Gulf of Gdansk and Puck Bay. *dw* Dry weight, *ww* wet weight, *Chl a* chlorophyll *a*, *DIN* dissolved inorganic nitrogen, *DIP* dissolved inorganic phosphorus

	Gulf of Gdansk	Puck Bay
a. **Climate and hydrology**		
Air temperature (°C)	min: –10.1; max: +33.2[a]	Min: –23.8; max: +35.5[u]
Water temperature (°C)	min: –0.43; max: +24.7[a]	1–21[t]
Ice covering (days a^{-1})	22[c]	60–80[u]
Annual global radiation (J cm^2)	4,304–65,320[c]	4,304–65,320[w]
Average precipitation (mm a^{-1})	42.4 (min: 15.6; max: 128.1)[c]	16.0–104.8[w]
Mean freshwater inflow (km^3 a^{-1})	35.14[b]	8,091[t]
Mean inflow (km^3 a^{-1})	1,200 (from Baltic)[b]	No data
Mean total outflow (km^3 a^{-1})	1.240[b]	No data
Residence time (days)	< 7[l]	Puck Lagoon: 38[v] Outer Puck Bay: 7–30[v]
b. **Morphometry and sediments**		
Area (km^2)	4,940[b]	Total: 359.20[s] Puck Lagoon: 102.69[s] Outer Puck Bay: 256.51[s]
Volume (10^6 m^3)	291,460[b]	Total: 5.58[s] Puck Lagoon: 0.32[s] Outer Puck Bay: 5.26[s]
Mean depth (m)	59[b]	Total: 15.53[s] Puck Lagoon: 3.13[s] Outer Puck Bay: 20.50[s]
Maximum depth (m)	111[b]	Total: 54.00[s] Puck Lagoon: 9.40[s] Outer Puck Bay: 54.00[s]
Catchment area (km^2)	323,200[b]	Total: 908.8

		Puck Lagoon: 693.5 Outer Puck Bay: 215.3
Area/catchment relation	0.015	Total area: 0.395 Puck Lagoon: 0.148 Outer Puck Bay: 1.191
Organic matter content (% dw)	0.2–25[f]	0.1–25[aa]
Sediment fraction <63 μm (% dw)	0– > 50[f]	0.1–65[aa]
C/N ratio (0–10 cm depth)	5–25[f]	8.50 ± 1.30[z]
Sediment oxygen demand (mmol O_2 m^{-2} day^{-1})	< 500[c]	No data
Phosphate release (μg P m^{-2} day^{-1})	1–612[f]	0.021[y]
Ammonia release (μg N m^{-2} day^{-1})	<400[f]	No data
Denitrification rate (μmol N m^{-2} h^{-1})	No data	11.0 (min: 0; max: 40.2)[z]
Sediment composition	Variously grained sands; silty sands; sandy silts; clayey silts[e]	Clays, sands, gravel, silty sandy sediments, silts, peat, mud
c. Pelagic biological components		
Secchi depth (m)	3.3 (max: 6.0)[l]	3.7 (max. 14.5)[ee]
Bacteria (10^6 ml^{-1})	3.8–13.45[m]	No data
Phytoplankton biomass (μg Chl *a* l^{-1})	0.62–51.93 (spring max: 51.93; autumn max: 10.00)[a]	Spring and fall max: 52
Dominating phytoplankton species	*Actinocyclus octonarius*, *Chaetoceros wighamii*, *Ch. Holsaticus*, *Ch. ceratosporus*, *Dinophysis acuminata*, *Heterocapsa rotundata*, *H. triquetra*, *Periainiella catenata*, *Plagioselmis prolonga*, *Pyramimonas* spp.	*Aphanizomenon flos-aquae*, *Nodularia spumigena*, *Melosira granulata*, *M. nummuloides*, *Sceletonema costatum*, *Coscinodiscus granii*, *Diatoma elongatum*
Ciliates		
Biomass (mg l^{-1})	0.01–0.48 (wet weight)[m]	No data
Dominating species	No data	No data
Fish		

(continued)

Table 7.1 (continued)

	Gulf of Gdansk	Puck Bay
Number of species	74	57
Dominating species	Herring, sprat, flatfish	Garfish, black goby, three-spined stickle back, *Neogobius melanostomus*[ff]
Primary production (g C m^{-2} a^{-1})	156	198[gg]
Bacterial production (mg C m^3 day^{-1})	0.29–28.1[m]	No data
Fish catches (t a^{-1})	5,963.8[o]	728.9[ff]
d. Benthic biological components		
Macrophytes		
Mean biomass (g m^{-2} dw)	Stony and sandy bottom: 35.5	Sandy bottom: Puck Lagoon 38.0; outer Puck Bay 50.1
Number of species	35	24
Dominating species	*Pilayella littoralis*, *Zostera marina*, *Potamogeton* spp., *Zannichellia palustris*, *Polysiphonia fucoides*, *Ceramium* spp., *Cladophora glomerata*[q]	*Pilayella littoralis*, *Ceramium* spp., *Zostera marina*, *Potamogeton* spp., *Zannichellia palustris*[q]
Macrozoobenthos		
Mean biomass (g m^{-2} ww)	Sandy bottom: 57.5	Sandy bottom: 590.3
Number of species	37	55
Dominating species	*Hydrobia* sp., *Macoma balthica*, *Mytilus trossulus*[r]	*Hediste diversicolor*, *Hydrobia* sp., *Oligochaeta*[hh]
e. Water chemistry, trophic status and pollution		
Salinity (psu)	0.65–12.2[d]	Puck Lagoon: 7.31[x] Outer Puck Bay: 7.65[x]
pH	7.52–8.70[d]	
Oxygen saturation (%)	20 (min: 40; max: 80) (near-bottom); 80–170 (surface)[h]	No data
Total nitrogen (µmol l^{-1})	26.96[d] (winter max: 15.90)[e]	6.96[cc]
DIN (µmol l^{-1})	14.38[d]	0–3.36[z]

Total phosphorus ($\mu mol\ l^{-1}$)	1.29[d] (winter max: 0.79)[e]	No data
DIP ($\mu mol\ l^{-1}$)	0.50[e]	Puck Lagoon: 0.09–19.20[z] Outer Puck Bay: 0.02–36.90[z]
Annual nitrogen input ($t\ a^{-1}$)	118,000[j]	2,275[bb]
Annual phosphorus input ($t\ a^{-1}$)	7,000[j]	No data
Heavy metal loads ($mg\ kg^{-1}$) (0–10 cm, dw)	Zn: 52–74; Cu: 12–16; Cd: 0.4–0.7; Pb:23–16[h]	No data
Limitation of PP and period	Winter: temperature Summer: present – phosphorus[k,l]	No data
Main pollutants	Vistula River, Vistula Lagoon	Detergents, phenols, fats, oils, pesticides, nutrients
Trophic level	High trophic	Eutrophic

[a]Łysiak-Pastuszak (2004); [b]Majewski (1994); [c]Krzymiński et al. (2001); [d]Łysiak-Pastuszak (2004); [e]Majewski (1990); [f]Kruk-Dowgiałło and Dubrawski (1998); [g]Krzymiński et al. (2000); [h]Krzymiński et al. (2004); [i]Andrulewicz and Witek (2002); [j]Renk (2000); [k]Łysiak-Pastuszak and Piątkowska (2004); [l]Łysiak-Pastuszak et al. (2004); [m]unpublished data; [n]Atlas (2003); [o]Jackowski (2000); [p]Renk (2000); [q]Kruk-Dowgiałło (1991, 1994a, 1998, 2000a, 200b); Pliński and Florczyk (1993a); [r]Osowiecki (1998); [s]Nowacki (1993), [t]Cyberski (1993), [u]Cyberski and Szefler (1993), [v]Łysiak-Pastuszak et al. (2004), [w]Krzymiński et al. (2001); [x]Nowacki (1993b); [y]personal communication; [z]Graca (2004); [aa]Jankowska and Łęczyński (1993); [bb]Pempkowiak (1994); [cc]Łysiak-Pastuszak. (2004); [dd]Bolałek and Graca (1996); [ee]**Krężel** (1993); [ff]Jackowski (1998); [gg]Renk (1993)

oxygen demand over a 5 day period (BOD_5) 145,211.6t O_2 a^{-1} (66%); chemical oxygen demand (COD) 1,060,602.6t O_2 a^{-1} (58%), 74,662.3t To_xN a^{-1} (62%); 120,514.3t N_{tot} a^{-1} (63%); 3,659.6t P-PO_4^{3-} (70%); 7,672.2t P_{tot} a^{-1} (61%); 560,7t Zn a^{-1} (82%); 1.64t Cd a^{-1} (71%); 15.8t Pb a^{-1} (60%); 33.0t Hg a^{-1} (97%) (Krzymiński et al. 2001). In the 1987–2001 period, the average load of inorganic compounds of phosphorous (phosphates) and nitrogen (nitrites, nitrates and ammonium salts, N_{in}) introduced into the gulf from the waters of the Vistula alone were 70,772.2t N_{in} a^{-1} and 10,213.0t P-PO_4 a^{-1} (Łysiak-Pastuszak and Piątkowska 2004). The combined total of sewage and river waters released into the Gulf of Gdańsk and Puck Bay, excluding Vistula waters, is 0.655 km^3 a^{-1}. The combined pollution load released from municipal sewage treatment facilities, industrial sources, and the remaining rivers is as follows: BOD_5 3,983.01t O_2 a^{-1}; COD 2,189.97t O_2 a^{-1}; 2,990.96t N_{tot} a^{-1}; 188.14t P_{tot} a^{-1} (Pilecki and Roszman 2005). Of these sources, sewage treatment facilities introduce the greatest pollution load into the gulf (Fig. 7.1c). The Vistula Lagoon, which is presently on the list of Baltic Hot Spots, plays a significant role in the pollution of the gulf.

The sanitary state of the coastal waters of the gulf has improved significantly in comparison with conditions in 1992, when 13 of 34 beaches were closed to swimming. In 2002, only five beaches were closed to swimming (WIOŚ 1993, 2003). The mean values of log MPN faecal coliform bacteria on the beaches of the Gdańsk Administrative District ranged from 3.09 to 4.91 in 1993, and from 1.88 to 2.33 in 2002 (Szumilas et al. 2004). There has been considerable improvement in recent years in the quality of river and sewage waters flowing into the gulf. This is indicated by the mean values of log MPN faecal coliform bacteria near outflows of municipal sewage and rivers of Puck Bay, which ranged from 0.7 to 2.3 at the end of 2003, and from 0.7 to 1.56 in spring 2004 (Kruk-Dowgiałło et al. 2004).

In the early 1990s, pathogenic bacteria from the genera *Escherichia coli*, *Staphylococcus* spp., *Salmonella* sp. and *Shigella* sp. were confirmed in 46–48% of tested cod, herring, sprat, flounder, and eelpout samples caught in Puck Bay and in the region from Gdynia to Sopot. Of conditionally pathogenic organisms, *Streptococcus faecalis* was isolated most frequently in sprat and flounder (in 36.9% and 32.2% of the tested fish, respectively) caught in the Gulf of Gdańsk proper and in the Gdańsk Deep region (Grawiński et al. 1998). This type of research has not been conducted in Poland in recent years.

In the early 1990s, the levels of Pb, Cd, and Hg in the muscle tissues of herring, a characteristic species in the Gulf of Gdańsk, were lower than in fish caught in the Słupsk Furrow (Polak-Juszczak 1996). Concentrations of these heavy metals did not exceed norms and were as follows: Pb 0.029±0.015mg kg^{-1} wet weight (ww), Cd 0.017±0.005mg kg^{-1} ww, and Hg 0.057±0.017mg kg^{-1} ww (Polak-Juszczak and Domagała 1994). Lower concentrations of Hg (0.024±0.016mg kg^{-1} ww) were detected in herring muscle tissue in 2000 than in the 1990s (Krzymiński et al. 2001), and, at 0.517±0.129mg kg^{-1} ww, the concentration of Cd in herring liver was within the limits of 0.16–0.91mg kg^{-1} ww reported for other Baltic regions (Environment 2002).

The characteristics of the Gulf of Gdańsk as regards hydrochemical conditions, bottom sediments, and trophic status are presented in Table 7.1b. Within the Polish Baltic zone, the gulf is the second most eutrophic basin after Pomeranian Bay

(Trzosińska and Łysiak-Pastuszak 1996). Temporal analyses of variations in winter concentrations of phosphates and total nitrites and nitrates (TO_xN) in the 0–10 m depth layer in the 1959–2001 period indicate that in the 1960s and 1970s there were rapid increases in the concentrations of phosphates to levels of 0.82 µmol $P\text{-}PO_4$ l^{-1} and, until the 1980s, of TO_xN to levels of 14 µmol l^{-1}. In subsequent years until 2001, successive declines were noted in these concentrations to levels of 0.65 µmol $P\text{-}PO_4$ l^{-1} and 11 µmol TO_xN l^{-1} (Łysiak-Pastuszak and Piąkowska 2004). The N:P mole ratio also decreased from 25.7 noted in the 1979–1998 period to 14.5 in 1999 (Krzymiński et al. 2000). The nutrient concentrations observed in recent years might indicate that the eutrophication process in the Gulf of Gdańsk is decelerating. This is not, however, confirmed by the results of summer water transparency measurements conducted over the 1959–2000 period. Such measurements indicate that, since the 1990s, the waters of the gulf have exhibited the strongest decline in transparency of any of the waters of the Polish part of the Baltic Sea, caused, in part, by abundant phytoplankton blooms in the summer season. One consequence of this is the low oxygen concentration in the near-bottom layers of the gulf that is observed at the end of the vegetation season. This is an indication of a significantly negative trend.

Despite the inflow of marine waters that occurred in the southern part of the Gdańsk Deep in 2000, the oxygen saturation of the near-bottom waters was 20% throughout the year and in December this figure fell to practically zero (Łysiak-Pastuszak and Piątkowska 2004). Hydrogen sulfide occurred throughout the year, with the highest recorded concentration being 61.6 µmol m^{-3}. On the other hand, the oxygen concentration in the near-bottom water layer in the Polish coastal zone in September was 34% (Krzymiński et al. 2001).

Evidence of eutrophication in the gulf can also be drawn from conditions prevailing in the bottom sediments. The content of organic matter in the autumn of 1994 in the coastal zone to a depth of 20 m ranged from 0.25 to 20.14% dry weight (dw), while in the summer of the following year it had decreased to 0.24–7.0% dw. The concentration of N_{tot} and P_{tot} rose proportionally to the content of organic matter. In autumn, their ratio in the 3–10 m depth zone was 2.7 and at 20 m 7.4. In the summer, the ratios were 4 and 6, respectively, in these depth zones (Kruk-Dowgiałło and Dubrawski 1998). The results presented indicate that organic matter decomposes intensively in winter reducing the oxygenation of sediments.

This is also why the gulf sediments undergo a reduction process in summer that is stronger in the western part of the basin, where in the 1–5 m depth zone the range of E_h varied from −170 mV to 100 mV, than in the eastern part where the range was from 10 mV to 400 mV. In the majority of the gulf bottom zone from depths of 10 m to 60 m, E_h ranged from 0 mV to −180 mV.

7.2.2 Planktonic Communities

During the 1995–1997 period, the maximum bacterial abundance (4.53×10^6 ml^{-1}), biomass (60.5 µg C l^{-1}), and production (18.1 µg C l^{-1} day^{-1}) were noted in the Gulf of Gdańsk in summer (see Chap. 3 by Schiewer, this volume). The highest production

(10.7 μg C l^{-1} day^{-1}) was noted in the region of the Vistula mouth (Table 7.1c). Bacterial production in winter decreased by 60-fold in comparison to that in summer (0.31 μg C l^{-1} day^{-1}) (Mudryk 1999). The summer maxima probably resulted from numerous phytoplankton blooms in the gulf as well as from the quantity of allochthonous matter introduced by rivers; the creation of a food web independent of primary production was based on this latter introduction. Witek (1995) also emphasised the dominant role of microorganisms in the energy flow of the Gulf of Gdańsk.

That the gulf is eutrophic is also indicated by the annual mean value of the N:P mole ratio of 19.8 in the surface waters. This ratio is shaped largely by the waters of the Vistula River. Concentrations of chlorophyll *a* (Chl *a*) are higher closer to the shore, but the thickness of the euphotic layer decreases. The coefficients of diffusive downward light attenuation (K) are 0.3, 0.4, and 0.5 in Gdańsk Deep, Puck Bay, and the Gulf of Gdańsk, respectively (Renk 2000). Primary production in the Gulf of Gdańsk increased from levels of 149 g C m^{-2} a^{-1} in the 1970s to 304 g C m^{-2} a^{-1} in 1987 [see Chaps. 3, 4 (Schiewer), 5 (Radziejewska and Schernewski), 8 (Chubarenko and Margoński), 9 (Gasiūnaitė et al.), 10 (Kotta et al), and 12 (Telesh et al.), this volume]. This increasing tendency clearly began in the region of the Gdańsk Deep, where the levels of production rose from 107 g C m^{-2} a^{-1} in the 1970s to 190 g C m^{-2} a^{-1} in the 1992–1996 period (Renk 2000).

The eutrophication of the gulf is also indicated by the approximate 100-fold increase in the overall phytoplankton abundance and the 10-fold diatom increase that have been noted since the 1970s (cf. Chap. 3 by Schiewer, this volume). Additionally, since the 1980s the structure of the phytoplankton biomass has shifted towards the domination of dinophyceae, Cyanobacteria, and flagellata. The differentiation of Chl *a* and biomass is presented in Figs. 7.2 and 7.3. In the summers of the mid-1990s, the mean phytoplankton biomass of the coastal zone, which was dominated by dinophyceae, ranged from 213 to 474.8 μg C l^{-1}, and the average values of chlorophyll ranged from 1.24 to 6.69 μg l^{-1}. The highest values of 17.4 μg l^{-1} were detected in the region of the Vistula mouth, where the biomass, dominated by dinophyceae and Cyanobacteria, reached 355.2 μg C l^{-1} (Niemkiewicz and Wrzołek 1998). In this region in the summer of 2003, although the biomass reached only 150 μg C l^{-1}, Cyanobacteria dominated and comprised 50% of total biomass (Niemkiewicz 2004).

In the mid-1990s in the region of the Gdańsk Deep, the phytoplankton biomass, dominated by flagellata and Cyanobacteria (see Chap. 5 by Radziejewska and Schernewski, this volume), exhibited the highest values of the 1984–1994 period at 205 μg C l^{-1}, while the concentration of Chl *a* was 4.3 μg l^{-1} (Cyberska et al. 1996). In 2000, the biomass was markedly lower, ranging from 50 to 90 μg C l^{-1}, although the decided dominant was flagellata. The concentrations of Chl *a* did not exceed 1 μg l^{-1}.

Since the 1990s, phytoplankton blooms have lengthened and intensified in the summer in the Gulf of Gdańsk proper (see Chap. 5 by Radziejewska and Schernewski, this volume). Species from harmful algae groups such as *Aphanizomenon flos-aquae*,

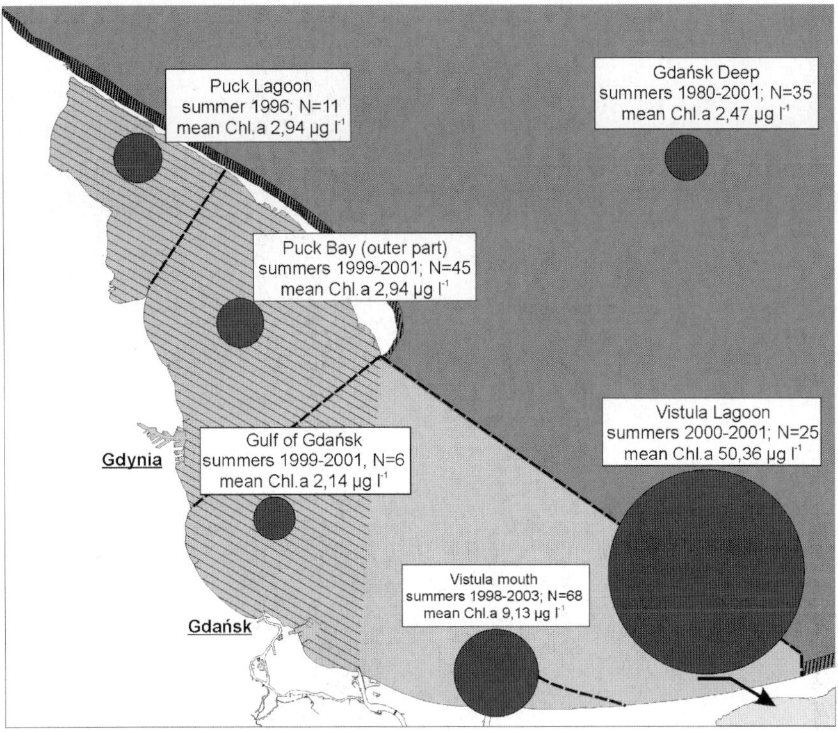

Fig. 7.2 Mean concentration of chlorophyll *a* (Chl *a*) in the Gulf of Gdańsk and Puck Bay in summer seasons from 1980 to 2003 (according to Niemkiewicz and Wrzołek 1998; Kruk-Dowgiałło et al. 2000b; Niemkiewicz 2004; Rybicka 2004)

Nodularia spumigena, and *Dinophysis norvegica* have begun to appear in these blooms, and their share of the summer phytoplankton biomass in 1995 was 6%, 0.1%, and 13%, respectively (Niemkiewicz and Wrzołek 1998). The share of cyanobacterium from the genera *Aphanizomenon* was 41% in the vicinity of the Vistula mouth in 2003 (Niemkiewicz 2004).

The taxonomic composition of the zooplankton in the Gulf of Gdańsk proper did not change substantially during the 1977–1995 period (cf. Chap. 5 by Radziejewska and Schernewski, this volume). In the warm period it is dominated heavily by two copepod species – *Acartia bifilosa* and *A. tonsa* (Wolska-Pyś 1998) – and the share of the rotifers *Synchaeta* sp. and *Keratella quadrata* as well as cladocerans increases (Opioła and Błachowiak-Samołyk 2001). The high abundance of larval pelagic epifauna, especially veligers of Lammellibranchiata (mollusc), is characteristic of the zooplankton of coastal zones (Fig. 7.4). The taxonomic composition of mezozooplankton is typical of the open waters of the southern Baltic (Opioła and Błachowiak-Samołyk 2001). Its primary component is comprised of the copepods *Acartia* sp., *Temora longicornis*, and *Pseudocalanus min. elongatus*. During the summer, other

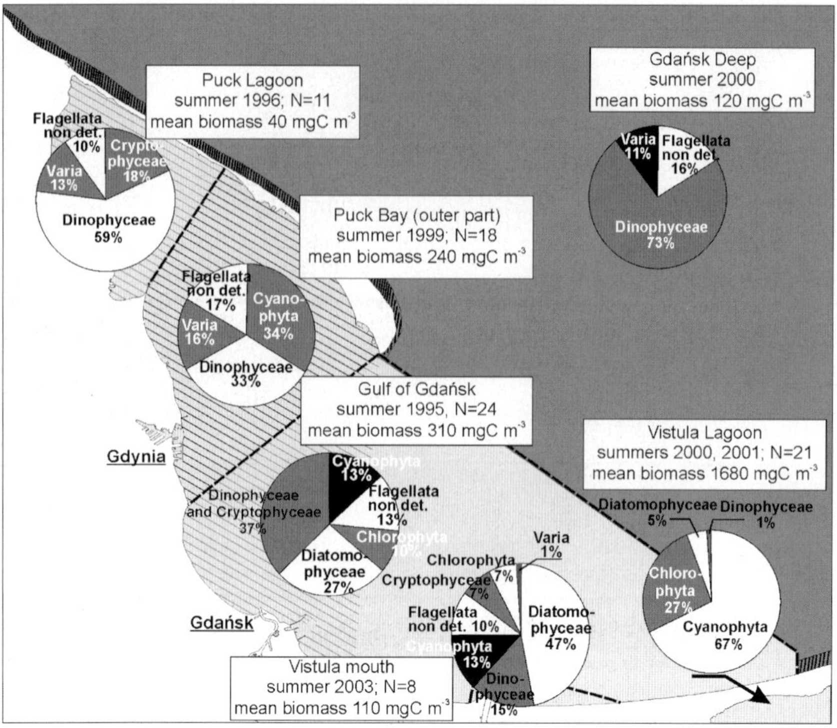

Fig. 7.3 Structure and mean values of phytoplankton biomass in the Gulf of Gdańsk and Puck Bay in the summer seasons from 1996 to 2003 (according to Niemkiewicz and Wrzołek 1998; Kruk-Dowgiałło et al. 2000b; Niemkiewicz 2004; Rybicka 2004)

significant components of the pelagic fauna are the cladocerans *Bosmina coregoni maritima* and *Evadne nordmanni*. Over the course of more than 20 years (1979–2001), the taxonomic composition of the mezozooplankton in the deep water zone did not exhibit fundamental change; however, in the 1994–1998 period an increasingly abundant population of the halophilous copepod *Oithona similis* was noted in the near-bottom water of the Gdańsk Deep (Korshenko et al. 2002).

7.2.3 Benthic Communities

The most significant changes resulting from the eutrophication of Gulf of Gdańsk waters can be confirmed in the macrophytes. Currently, 32 taxa occur in the gulf, including 8 Chlorophyceae taxa, 2 Charophyceae, 6 Fucophyceae, 7 Bangiophyceae, and 9 Angiospermae (Pliński and Florczyk 1993; Kruk-Dowgiałło 1991, 1994a, 1998). The eutrophication of gulf waters has influenced the current vertical and horizontal range of occurrence of macroalgae in comparison with that of the

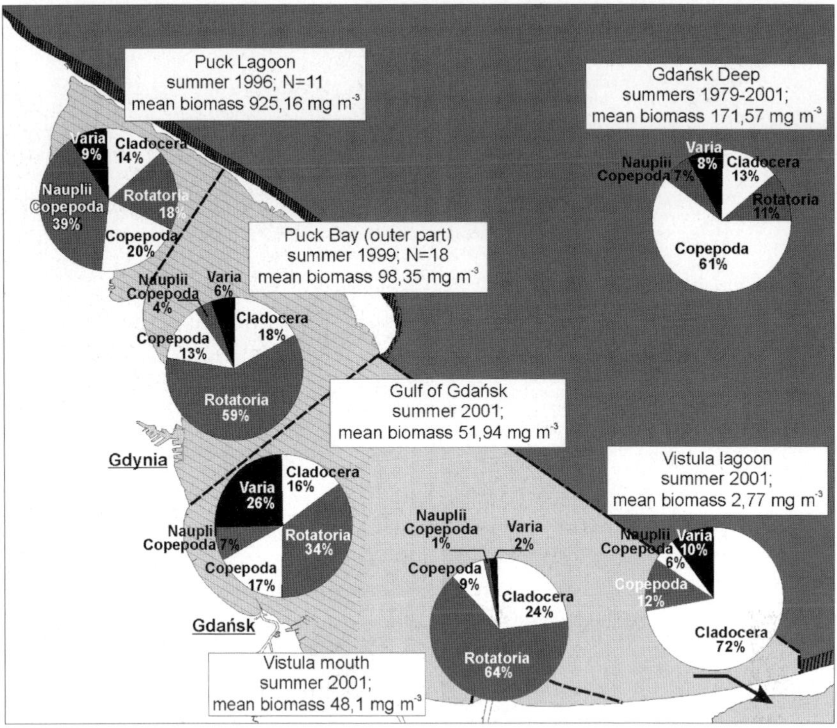

Fig. 7.4 Mean biomass of mezozooplankton and their quantity composition in Gulf of Gdańsk in the period 1979–2001 (Opioła and Błachowiak-Samołyk 2001), and in the 1990s according to Kruk-Dowgiałło (2000b)

1885–1905 period [cf. Chaps. 4 (Schiewer), 5 (Radziejewska and Schernewski), and 10 (Kotta et al.), this volume]. Macroalgae, mainly in the form of aegagropila, then occurred throughout the western part of the Gulf of Gdańsk up to Sopot at depths of up to 25 m (Fig. 7.1c) (Lakowitz 1907). Currently, they occur on the sandy bottom of the Puck Lagoon and in the coastal zone of the outer Puck Bay, and on the rocky bottom at the base of the Orłowo cliff (Fig. 7.5). Macroalgae with a maximum vertical depth range of up to 8–9 m (filamentous brown algae of the genera *Pilayella* and *Ectocarpus*) were noted in the rocky bottom area at the base of the Orłowo cliff. However, *Zostera marina* (Angiospermae) grows at a maximum depth of 3–4 m, while this species grew on bottoms as deep as 10 m as recently as the 1950s. The gulf regions mentioned above also differ as regards the qualitative and quantitative structure of macrophytes (Fig. 7.5, Table 7.1d). In the 1990s, a mean biomass of 35.5 g dw^{-1} m^{-2} was found in the Gulf of Gdańsk, with the highest mean being found in the coastal zone of the outer Puck Bay, where *Potamogeton* spp., *Zostera marina*, and *Pilayella littoralis* dominated. In addition to changes in ranges of occurrence caused by the eutrophication of gulf waters, there is the marked domination of filamentous brown algae of the genera *Pilayella* and *Ectocarpus*

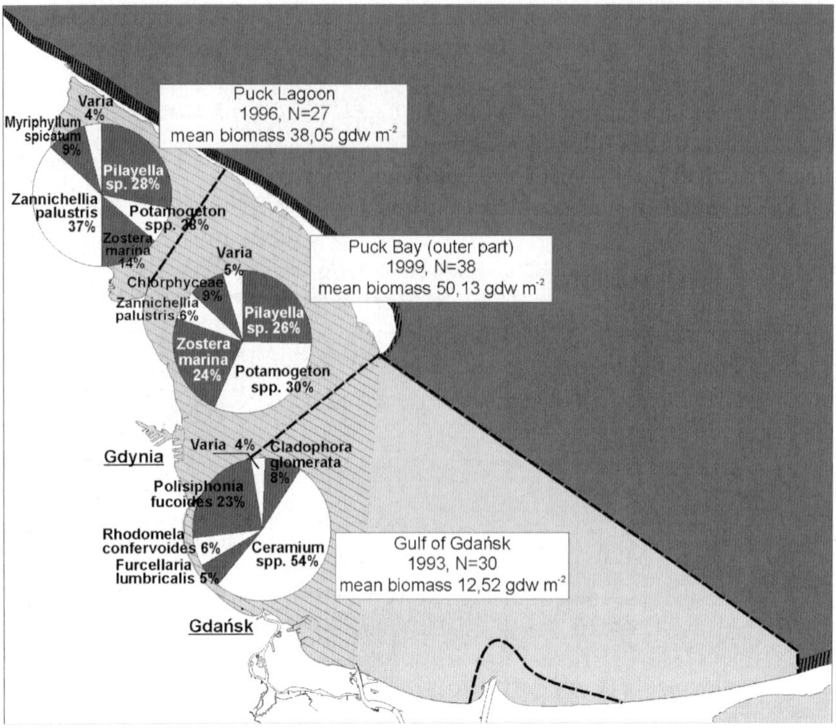

Fig. 7.5 Percentage of the biomass (g dw m^{-2}) of the dominating components of the macrophytobenthos of the Gulf of Gdańsk and Puck Bay in the 1990s according to Kruk-Dowgiałło (2000b) and Kowalczuk and Żmudziński (1993)

(Kruk-Dowgiałło 1994b, 1996). In comparison with the 1950s, the most significant changes in macrophyte structure were confirmed in the Puck Lagoon, where this parameter has been most frequently studied.

The eutrophic character of the Gulf of Gdańsk is also indicated by the macrozoobenthos structure, which is dominated in biomass by suspension and deposit feeders [cf. Chaps. 4 (Schiewer), 5 (Radziejewska and Schernewski), 10 (Kotta et al.), and 12 (Telesh et al.), this volume]. The dominating number of species belongs to *Crustacea* (18), while fewer species represent *Bivalvia* (4) and *Gastropoda* (3) (Table 7.1d). In comparison to the late 1970s and early 1980s, there has been a decided increase in the mean abundance of fauna in the 3–20 m depth zone. A similar, although slightly smaller, increase was noted regarding the mean biomass of benthic fauna, especially in the 10 m and 20 m depth ranges.

In the 1990s, Bivalvia dominated in terms of biomass (58–89%) in the five regions analyzed (Fig. 7.6). The lowest biomass (57.5 g ww m^{-2}; 44 sampling events) was noted in the area of the Gulf of Gdansk proper, where, as in other regions, Bivalvia (*Mytilus trossulus* and *Macoma balthica*) dominated (Osowiecki 1998). In the region of the Vistula mouth foreland, the biomass was only slightly

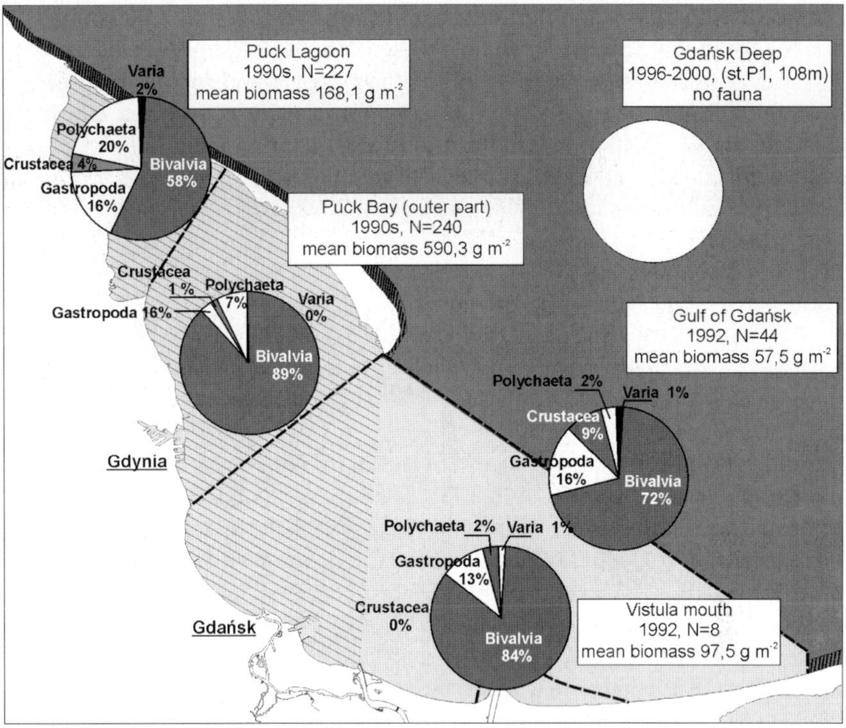

Fig. 7.6 Percentage of the biomass (g ww m^{-2}) of the dominating components of the macrozoobenthos of the Gulf of Gdańsk and Puck Bay in the 1990s according to Osowiecki (2000)

higher than in the remaining parts of the gulf. The state of the fauna on the southern slopes of the Gdańsk Deep, a station that has been monitored since 1979, is decidedly poor.

No benthic fauna was observed there in the 1996–2000 period. The last time benthic fauna was noted there was in 1995, when 80% of the biomass was comprised of *Harmothoe sarsi*; *Scoloplos armiger* and *Pontoporeia femorata* were also noted there. The lack of fauna in this region is due primarily to the presence of hydrogen sulfide in the sediments (Krzymiński et al. 2001).

Gdańsk Bay is populated by 74 species of fish. The following species are caught commercially: saltwater fish (flounder, turbot, herring, sprat, cod); diadromous fish (salmon, bull trout, eel, Baltic whitefish, trout); and freshwater fish (zander, perch, bream, burbot, roach and, very rarely, pike). During the 1950s, as well as at present, cod, herring and sprat were the basic catches in ICES sub-division 26. Between 1999 and 2003, fish catches in this sub-division increased from 54,600 tonnes to 76,100 tonnes. Sprat constitutes over 50% of the catch, and its largest contribution (60%; 57,000 tonnes) was recorded in 2003. The herring catch contribution also increased from 15% to 24% by 2002; whereas in 2003, a 3% (10,000 tonnes) drop in herring catch was recorded (Szostak et al. 2000–2003). The catch contribution of cod diminished successively from 21% in 1999, to 10% in 2002; 3,500 tonnes of cod were caught in 2003. In 2002, 11 species of fish were caught in fishery no. 101, with a predominance of sprat (3,700 tonnes), and herring (1,900 tonnes). In the Gdańsk Depth area, in fishery no. 102, eight species of fish were caught in a total amount of 17,400 tonnes, with a predominance of sprat (16,000 tonnes) (Linkowski 2003). Catches of fish of the salmon family have been systematically increasing thanks to fish restocking (Jackowski 2002).

7.3 Puck Bay

7.3.1 Environmental Characteristics

Puck Bay (Table 7.1) is the southwestern part of the Gdańsk Basin that is separated from the deep-water regions by the Hel Peninsula. The widely accepted eastern border of the bay is the line that connects the Cypel Helski with Kamienna Góra (Słomianko 1974). Puck Bay is divided into the inner Puck Lagoon and the outer Puck Bay (Fig. 7.1c). The border between them stretches 8.6 km from the Rybitwia Sandbank to the Cypel Rewski, and is divided by two straits through which there is intense water exchange between the outer Puck Bay and the Puck Lagoon. The area of Puck Bay comprises a mere 1.4% of that of the Gdansk Basin, and only 0.38% of its volume. The outer Puck Bay comprises approximately 70% of the area and 94% of the volume of Puck Bay (Table 7.1a). The depths and bottom morphology of these two parts of the bay differ significantly. The outer Puck Bay is notably deeper (average depth 20.5 m, maximum 54 m) and its bottom has a steep slope. The Puck Lagoon is considerably shallower (average depth 3.1 m, maximum 9.4 m) and its bottom profile is more varied with numerous furrows, shoals, and depressions (Nowacki 1993a).

The coastline of the Puck Lagoon is comprised of cliffs, sand dunes, and low-lying coasts in river beds. The bottom zone is comprised of lacustrine and riparian accumulation relicts, shoals, hollows, lamellar flats ranging in depth from 0.1 m to 5 m as well as dredging depressions of up to 14 m deep. The bottom sediments are primarily sandy and muddy, but there are also gravely and rocky areas. The coasts of the outer Puck Bay are either cliffs or sand dunes, and the bottom area is comprised of glacial and glaciofluvial relicts and delta and marine accumulation-erosion relicts; the sediments are primarily sand and mud (Krzymiński et al. 2004, Table 7.1a). The Puck Lagoon comprises 76.3% of the Puck Bay catchment area. In effect, there are 6.5 km^2 of catchment area for each square kilometre of the Puck Lagoon and only 0.84 km^2 for each square kilometre of the outer Puck Bay (Cyberski 1993).

The climate of Puck Bay is shaped by the land, and especially by the coastal zone, but it is also under the influence of the Gulf of Gdańsk and the waters of the Gdansk Deep. The prevailing winds blow from the west, particularly from the west and northwest. Although they are very strong and often exceed 10 ms^{-1}, their combined contribution is small at 20% (Cyberski and Szefler 1993).

Puck Bay is a coherent system of two strongly linked basins that exert an impact on each other. The hydrological conditions are shaped by inflows of marine and terrestrial waters as well as by water exchange between the different parts of the bay. Marine waters with a higher salinity than that of the bay flow in from the Gdańsk Deep through the eastern border. These waters are also transported to the Puck Lagoon. Terrestrial waters come from variously-sized streams and communal wastewater collectors. Estimates indicate that approximately 0.67 km^3 water flows annually from the outer Puck Bay into the Puck Lagoon, while the estimated portion of waters derived from rivers is only about 0.21 km^3 (Nowacki 1993a).

As regards hydrological and hydrochemical characteristics, Puck Bay is divided into three basins (Fig. 7.1a, Nowacki and Jarosz 1998) – two in the outer Puck Bay, and the third in the Puck Lagoon. Puck Bay is classified as a semi-enclosed gulf. In the Puck Lagoon, vertical water exchange is more than three-fold higher than horizontal water exchange, but in the outer Puck Bay there are only trace inflows of terrestrial water, and the vertical water exchange is several times greater.

The impact of marine and terrestrial factors on hydrological conditions in the Puck Lagoon mean that this basin can be classified as an estuary that exhibits characteristics typical of a lagoon, and its shallow depths influence water mixing throughout the volume of the Puck Lagoon. The far greater depths in the outer Puck Bay, and its opening for water exchange with the Gdańsk Deep, impact the shaping of seasonal vertical water stratification and periodic increases in salinity. In the Puck Lagoon, salinity depends primarily on water exchange with the outer Puck Bay (Table 7.1b).

Spatial and temporal variability in water transparency in Puck Bay is strongly dependent on the seasonal cycle of the ecosystem's biological activity. Based on research results from the 1980–1997 period, statistically significant, long-term changes in water transparency have occurred in the outer Puck Bay at –0.04 m a^{-1}. This represents a decrease in Secchi disc depth of approximately 0.6 m over the course of 16 years (Matciak 1998).

Puck Bay is polluted with wastewaters from three sewage treatment facilities and seven rivers. Approximately 60% of the BOD_5 was 7 mg O_2 l^{-1}, while the COD loads carried into the bay come from the Reda River. The sewage treatment facilities contribute a mere 10%, of which 90% is the load from the Dębogórze wastewater treatment facility. The estimated loads contributed by Reda River in 2002 were BOD_5 454.68 t O_2 a^{-1} and COD 3,772.21 t O_2 a^{-1}. The Dębogórze facility contributed BOD_5 136.51 t O_2 a^{-1} and COD 854.25 t O_2 a^{-1}. The estimated average load of nitrogen contributed by the Reda River in 2002 was 277.86 t N_{tot} a^{-1}, and of phosphorous 21.89 t N_{tot} a^{-1}. In comparison, the largest share of sewage (95%) was discharged from the Dębogórze facility, which contributed 548.13 t N_{tot} a^{-1} and 22.5 t P_{tot} a^{-1} in 2003. The largest decline in the magnitude of suspension loads contributed by the Reda River from 8,946 t a^{-1} in the 1980s to 2,021 t a^{-1} in 2002 (Pilecki and Roszman 2005) were recently recorded.

The Puck Lagoon is classified as moderately contaminated with heavy metals. As in other Baltic regions, this is indicated by the coefficients of anthropogenic enrichment, with values as follows: Cd 2.0, Pb 2.4; Zn 3.1, Cu 2.1, Cr 2.2 (Pempkowiak 1994).

The seasonal variation observed in the occurrence of nutrients stems from the biochemical transformation cycle and their inflow from the land. The average annual concentration of phosphates in the water of the Puck Lagoon in 1975–1976 was 0.97 μmol l^{-1} (Falkowska 1979), in 1981–1985 it remained at the level of 0.57 μmol l^{-1}, while in 1986–1991 it increased to 1.35 μmol l^{-1} (Bolałek et al. 1993). In recent years, it is believed that there has been an increase in the concentrations of nitrogen and phosphorous compounds in aerosols and rainfall (Bolałek et al. 1993). The average annual inflow of wet nitrogen from the atmosphere was 306,040 t, and 20,833 t phosphorous. In the outer Puck Bay the average value of phosphates released from sediments was 0.021 μmol m^{-2} day^{-1}. In the Puck Lagoon, this value was 0.025 μmol m^{-2} day^{-1}, which means that approximately 97 t phosphates are released annually from the sediments to the water. In the deeper regions of the outer Puck Bay the average annual stream was 6.28 t NH_4-N km^{-2} a^{-1}, while in the central part it was approximately 3.47 t NH_4-N km^{-2} a^{-1}. The sediments release approximately 825 t ammonium nitrogen into the near-bottom waters annually (Bolałek et al. 1993).

Annually, approximately 60.5×10^3 t phosphates and 205.4×10^3 t inorganic nitrogen compounds flow from the Gulf of Gdansk into Puck Bay, while the outflow of these substances, which occurs primarily in winter, is 58.7×10^3 t and 197.6×10^3 t, respectively. The annual flow into Puck Bay from terrestrial sources, the atmosphere, and the bottom is close to the phosphate content stored in the water column.

Short-term oxygen supersaturation or deficit is observed in gulf waters. Local sewage discharges cause significant oxygen deficits. The waters of Puck Bay are well oxygenated. The dominating concentration range in the 1986–1991 period in the Puck Lagoon was 7–8 ml l^{-1}, while in the outer Puck Bay it was 8–9 ml l^{-1}. Concentrations of less than 2 ml l^{-1} have not been noted (Bolałek et al. 1993). During the vegetation season, which begins in April, the strong increase in oxygen concentration is accompanied by a decrease in the concentration of nutrient substances. In the summer months, the oxygen concentrations decrease; increases are noted again starting from November.

The character of the bottom sediments and their distribution is reflected in the genetic and morphological variation observed between the two basins of Puck Bay: the Puck Lagoon and outer Puck Bay. The bottom sediments in the Puck Lagoon are typical of those found in lagoons and attest to the low lithodynamics of the basin. The organic matter content oscillates from 0.1% to 25% and depends on the lithographic sediment type. Organic matter occurs in the outer Puck Bay in quantities from 1% to 5%. In the vertical sediment profile, the content of organic matter generally decreases with increasing depth. Clear decreases are observed at depths of 15–20 m (Jankowska and Łęczyński 1993).

7.3.2 Planktonic Communities

Puck Bay can be divided into regions in terms of phytoplankton taxonomic composition. There is a significantly larger share of diatoms, mainly from *Pennatae*, which are often epiphytic and benthic, while the amount of green algae is small

(Fig. 7.3). In the coastal zone of the outer Puck Bay there is a paucity of phytoplankton species, primarily with regard to blue-green and green algae (Pliński 1993). Diatoms dominate in spring and Cyanobacteria in summer, but in autumn the diatoms are again dominant (cf. Chap. 4 by Schiewer, this volume). Green algae contribute only an insignificant share to the phytoplankton throughout the year, with a slight increase in summer and autumn. Blue-green algae and diatoms occur in the phytoplankton throughout the year. The intense development of Cyanobacteria begins in late spring and their maximum occurrence is noted in July and August. The maximum occurrence of diatoms is seen in spring and autumn (Pliński 1995). Table 7.1c gives some further information regarding Secchi depth, maximum Chl *a* content and species composition.

Relatively, there is little zooplankton variation, with zooplankton being comprised of both euryhaline species common to the Baltic Sea and those originating from fresh and brackish waters (see Chap. 4 by Schiewer, this volume). The zooplankton of the Puck Lagoon is dominated by *Copepoda*, and in outer Puck Bay, Rotatoria comprise as much as 59% of the zooplankton biomass (Fig. 7.4). Freshwater *Cyclopidae* are noted most frequently in shallow waters (Wiktor 1993). *Cladocera* occur in the zooplankton from October to May. *Rotatoria* also achieve maximum abundance in the warmer months when they comprise, on average, from 16% to 40% of all the zooplankton components. The larvae of snails and molluscs are also an important component. The zooplankton of Puck Bay experiences a seasonal variation in both abundance and structure that is typical for the entire Baltic Sea.

The zooplankton composition includes representatives of macrozooplankton such as jellyfish or *Mysidacea*. Although the eutrophication of gulf waters has caused the least relative variation in the structure of zooplankton, a decline in species diversity has been observed in the Puck Lagoon. This has resulted from a decrease in the abundance of taxa such as *Eurytemora* sp., *Bosmina coregoni maritima* and the larvae of some benthic animals. Another symptom of eutrophication is the increase in the overall abundance of zooplankton that results from the intense development of omnivores. Additionally, increases in the abundances of free-living protozoans have also been observed. Tintinnidae and epibiontic and parasitic protozoans have been observed on copepods. The spread of copepods infected with epibiontic and parasitic protozoans worsens the condition of this dominating component of the zooplankton, and this can lead to declines in population abundance (Wolska-Pyś 1994).

7.3.3 Benthic Communities

Environmental conditions in Puck Bay have been deteriorating for the past several decades; this has caused a transformation in the quantitative and qualitative structure of the macrophytes, which has been particularly evident in the Puck Lagoon since the mid 1970s. Despite this, the lagoon still has the richest variety of benthic fauna, both quantitatively and qualitatively, of any area in the Polish coastal zone of the Baltic (Table 7.1d, Fig. 7.1c). The effects of changes in environmental conditions are numerous. They included a decline in the number of species – particularly Fucophyceae and Bangiophyceae, the clear dominance in biomass of filamentous brown algae of the genera *Pilayella* and *Ectocarpus*, the decline of the overall phytobenthos biomass in comparison to that of the 1950s, and limitation of the distribution of Angiospermae *Zostera marina* in the Puck Lagoon from depths of 5 m to 2 m as well as a decline in its area of occurrence [see Chaps. 4 (Schiewer) and 10 (Kotta et al.), this volume]. The disappearance from the Puck Lagoon of two species that dominated in the 1960s – *Fucus vesiculosus* and *Furcellaria lumbricalis* – has been observed since the mid 1970s (Ciszewski et al. 1992; Pliński and Florczyk 1993; Kruk-Dowgiałło 1991, 1994a, 1994b). These species had occurred in the Puck Lagoon as aegagropila (rounded tufts) that formed layers as thick as 50 cm in some areas near the sandy bottom. Along with Angiospermae, mainly *Zostara marina*, the spatial structure of these underwater meadows was advantageous for the development of invertebrate fauna and fish fry. The mean biomass of *Fucus vesiculosus* and *Furcellaria lumbricalis* in the 1950s was 15 g dw m^{-2} and 23.7 g dw m^{-2}, respectively, while by the mid-1970s it had fallen to 4.3 g dw m^{-2} and 1.5 g dw m^{-2}. By the 1980s these species were no longer observed in this location. The biomass of *Zostera marina*, another dominant species, was 37.0 g dw m^{-2} in the 1950s; it fell to 8.1 g dw m^{-2} in the 1970s, and by the 1980s it did not exceed 4.5 g dw m^{-2}. In the 1990s, an increase in its biomass to 20.8 g dw m^{-2} was noted. This indicates that there was a slight improvement in the environmental conditions in the Puck Lagoon in comparison to those in the 1980s, when filamentous brown algae of the genera *Pilayella* and *Ectocarpus* dominated at 68%, with a mean biomass of 16.6 g dw m^{-2}. Their 32% share of the overall phytobenthos biomass in the 1990s also attests to the improvement of the state of the environment, even though the mean biomass of these algae was 16.5 g dw m^{-2} (Fig. 7.5).

Crustacea were the most taxonomically diverse phylum identified in Puck Bay, accounting for almost one-half of the benthic invertebrate taxa. Several species have most probably become extinct in Puck Bay (*Laomedea flexuosa, Dendrocoelum lacteum, Procerodes ulvae, Alkmaria romijni, Asellus aquaticus, Gammarus inaequicauda* and *Corophium multisetosum*); however, some new species (*Marenzelleria neglecta*) were noted for the first time in the 1990s. A non-indigenous amphipod species, *Gammarus tigrinus*, has successfully competed with other Puck Bay amphipods since 2001.

From the inception of research on Puck Bay zoobenthos in the mid 1930s until the end of the 1990s, 77 taxa of macrozoobenthos were found in 785 sampling events; 69 taxa were identified in the inner part (the Puck Lagoon), and 68 in the outer Puck Bay. The number of macrozoobenthos taxa found in the whole Puck Bay increased over the investigated period, from 34 in the 1930s to 61 in the 1990s (see Chap. 4 by Schiewer, this volume). Increases of taxonomic diversity were observed in both sub-regions of the Puck Bay, and the highest number of taxa were observed in the Puck Lagoon (50) and the outer Puck Bay (55) in the 1990s (Osowiecki 2000).

The macrozoobenthos biomass in Puck Bay has increased constantly, and in the 1990s it was over 2.5-fold higher than that noted in the 1930s [see Chaps. 4

(Schiewer), 5 (Radziejewska and Schernewski), 10 and 11 (both Kotta et al.), this volume]. The spatial distribution of macrozoobenthos biomass has also changed significantly since the 1930s. Areas in the Puck Lagoon which had had high biomass (>600 g m^{-2}) became virtually bereft of it in the 1990s. Some new areas with biomass ranges of 300–600 g m^{-2} have developed in the outer Puck Bay since the 1970s; this has been due mainly to increases in Bivalvia biomass, particularly of the blue mussel *Mytilus trossulus* (Osowiecki 2000). Polychaeta and Bivalvia had the highest increases in biomass, especially *Hediste diversicolor* and the soft clam *Mya arenaria*. The share of polychaetes in the total mean biomass increased from 0.9% in the 1930s to 9.9% in the 1990s. The contribution of *H. diversicolor* increased in the same period from 0.1% to 9.8% (Osowiecki 2000). *H. diversicolor* was the most common species throughout the area and occurred at 78% of the sampling sites, whereas Hydrobiidae snails and *Macoma balthica* occurred at 77% and 58% of the sites (Fig. 7.6), respectively (Osowiecki 2000).

7.3.4 Fish Communities

Puck Bay is inhabited by saltwater, freshwater and migratory species of fish. The existence of 57 species of fish was recorded between 1924 and 1998 (Jackowski 2002; Kruk-Dowgiałło 2000b). Since the 1980s, the structure of ichtiofauna has changed towards a predominance of Gasterosteiformes, and since the 1990s towards a predominance of Gobiidae (Jackowski 2002). The changes in fish catches are proof of adverse changes in the condition of the Puck Bay environment. Saltwater fish, with a predominance of *Pleuronectes flesus, Gadus morhua callarias* and *Clupea harengus*, constitute 90% of catches. Towards the end of the 1990s, the catch contribution of *Pleuronectes flesus* dropped to 50%. Since the beginning of the 1990s the catch contribution of *Gadus morhua callarias* has increased from 18.9 tonnes in 1993 to 79 tonnes in 1996 (Kruk-Dowgiałło 2000b). Since 1995, the catch of *Clupea harengus* has been continuously decreasing from 30–74 tonnes to 7–14 tonnes (ibidem). As far as freshwater species are concerned, *Perca fluviatilis* and *Sander lucioperca* constitute 90% of the catch; whereas during previous decades *Rutilus rutilus* and *Esox lucius* constituted a 90% contribution. At present, there are no intentional catches of *Rutilus rutilus*. Also, *Anguilla anguilla*, which was a predominant species in catches between 1993 and 1996, lost its significance towards the end of the 1990s (Jackowski 2002). The population of *Esox lucius* has undergone the most serious deterioration: during the 1990s this species appeared very rarely in catches, whereas during the 1960s and 1970s it used to be caught in amounts of 40–50 tonnes a^{-1} (Jackowski 1998).

Fisheries in Puck Bay are threatened by the mass occurrence of filamentous brown algae, primarily *Pilayella littoralis*, and the decreasing amount of bottom area overgrown with macrophytes. The disappearance of populations of fish such as pike and roach, and the decline in the perch population have caused the massive development of Gasterosteiformes fish, which have become the primary food competitor for the fry of other fish species. Since *Neogobius melanostomus* was first sighted in Puck Bay in 1990, this species has spread throughout the waters of this basin and beyond. Its increasing role has been documented since 1996, when the share of *N. melanostomus* in research catches in the outer Puck Bay was 0.16% of the catch weight. In subsequent years its share increased to 2.55% in 1997, 4.48% in 1998 and to as much as 15.16% in 1999 (Wandzel 2000). Changes occurring in the structure of the ichthyofauna since the 1980s were leading towards the domination of Gasterosteiformes fish but since the 1990s this has shifted towards Gobiidae fish (Jackowski 2002). The differences in catches recorded over the past several decades are also evidence of the disadvantageous changes that have occurred in the environment of Puck Bay. In the 1960s, garfish was caught at a rate of 4–28 t a^{-1}, but dramatic declines were observed in the 1980s. In the late 1990s, it was caught at a rate of 13 t a^{-1}. In the 1960s, roach was caught in quantities of 200 t a^{-1}, but by the 1990s this figure had fallen radically to a mere 0.3 t a^{-1}. Catches of perch also decreased from 40 t a^{-1} to 2 t a^{-1} in the 1990s. The most significant deterioration was observed in pike stocks; although this species was caught at rates of 40–50 t a^{-1} in the 1960s and 1970s, by the 1990s it was caught only sporadically (Jackowski 1998).

7.4 Eutrophication and Benthic–Pelagic Coupling

The Gulf of Gdańsk is a system of mutually interacting estuaries. Mixtures of marine and terrestrial waters typical of this type of basin occur here. Inflows of marine water are the result of exchange with the Gdańsk Deep, and the primary source of terrestrial water is the Vistula River, along with smaller streams and wastewater from communal sewage collectors. The nutrients, suspensions, and pollution that have flowed into the waters have caused changes in the gulf environment.

These changes began to become apparent in the mid-1970s, mainly in the macrophyte structure of the Puck Lagoon. The biomass of the seasonally occurring filamentous brown algae, mainly *Pilayella littoralis*, began to increase successively each year in early spring at water temperatures of approximately 8°C. The robust development of these algae limited the development of perennial algae (F*ucus vesiculosus* and *Furcellaria lumbricalis*) until they disappeared, thus spurring the further development of filamentous brown algae. Occupying a free ecological niche, these species formed algal mats and, by the 1980s, comprised over 70% of the phytobenthos biomass. On the one hand, the massive development of filamentous brown algae in the early spring limited the development of Angiospermae *Zostera marina* and the benthic fauna and fish fry associated with meadows. On the other hand, it did not permit the robust development of phytoplankton, as the filamentous algae were first to deplete the stores of nutrients collected over the winter. This is why phytoplankton blooms in the Puck Lagoon are not as intense as those observed in the Gulf of Gdańsk. Among other factors, this is caused by the lack of competition for nutrients, which results from the fact that only a small portion of the Gulf of Gdańsk bottom is overgrown with macrophytes, and the filamentous algae, which occurs as aegagropila (rounded tufts), is carried out of the gulf by currents or washed up on the shore.

Large amounts of allochthonous and autochthonous organic matter, originating from the Puck Lagoon in autumn in the form of dead filamentous brown algae, and from the Gulf of Gdańsk from phytoplankton blooms and large bacterioplankton production, impact trophic change by altering the biomass structure of the organisms occurring in the gulf. In effect, a decrease in the biomass of the meso fraction (0.27 g C m^{-2}) occurred, especially of mesozooplankton, which is unable to utilise the larger amounts of food. The largest increase was noted in the biomass of the macro fraction (7.77 g C m^{-2}). This resulted from the disproportionately strong development of macrozoobenthos, which took advantage of unused organic matter that fell to the bottom from the pelagic zone. Such a large spectrum of biomass was accompanied by a distribution of respiration size. In the gulf, this decreases most sharply as the size of the organisms increase (Witek 1995).

The impact of increasingly long periods of phytoplankton blooms and the expansion of the area in which they occur has been observed since the early 1990s. This has caused, among other phenomenon, deteriorating light conditions and oxygen supersaturation in the euphotic water layer also in the autumn. Decreasing water transparency limited the euphotic zone to the surface water layer, which worsened conditions for photosynthesis throughout the gulf, even in the shallow regions. This effected the vertical distribution of Angiospermae meadows and thus the habitats of phytophyllic fauna and ichthyofauna. The increase in sedimentation caused the bottom to become muddy, which made the springtime development of macrophytes difficult and overloaded the sediments with large amounts of organic matter and nutrients in the autumn. Reduction conditions appeared in the sediments, and the oxygen balance was disturbed for fairly long periods. The anoxic zone in the Gdańsk Deep reappeared in the 1994–1998 period, and near-bottom hydrogen sulfide occurred (Trzosińska et al. 1999).

There were clear indications in the second half of the 1970s of an increase in the N:P ratio in the upper water layers. Since then, this ratio has increased irregularly although unequivocally. Winter proportions of the magnitude 19.8 indicate phosphorus limitation and are typical of eutrophic environments (Wasmund et al. 2001), while the limiting element might change with the seasons to nitrogen. Between 1984–1989, the average ratio of the annual amplitude of nitrate concentrations to the corresponding amplitude of phosphates in the Gulf oscillated around 16:1 with a standard deviation of about 8. This ratio beneath the halocline is still relatively low. The eutrophication process depends largely on the hydrological conditions of the sea. That is why the eutrophication rate in the 1960s and 1970s differed from that observed in the period 1985–1989, i.e. during an exceptionally long stagnation period even though the inflow nutrients from land-based sources probably did not undergo significant changes. The concentrations that trigger the winter accumulation of phosphates in the surface layers of the Gdańsk Deep have increased since at least the beginning of the 1960s. The rate of this increase was very high until 1976, when it slowed down, and in the 1980s it became negative. Since the beginning of the 1970s, there has been a constant positive trend in nitrate concentrations and a negative one as regards silicate concentrations (Trzosińska and Łysiak-Pastuszak 1996).

In the intermediate layer, which corresponds approximately to the halocline location, nitrates are accumulated in winter at an identical rate to that in the surface waters described above. However, phosphates and silicates exhibited negative trends, which is connected to the sinking of the centre of the halocline (Trzosińska and Łysiak-Pastuszak 1996)

Changes in macrophytes and macrozoobenthos (Ciszewski et al. 1992) has resulted in the disappearance of many commercial fish species, making room for so-called "trash fish" such as *Gasterosteus aculeatus* and *Pungitius pungitius* in the 1980s and 1990s, and *Neogobius melanostomus* since the 1990s. Fishing in the Puck Lagoon has ceased, and in the Puck Bay proper and the Gulf of Gdansk proper, catch efficiency has declined significantly. The level of eutrophication in the Gulf of Gdańsk has been described as high (Łysiak-Pastuszak et al. 2004).

7.5 Conclusions

Due to its hydrologic and biogeochemical system, the Gulf of Gdańsk forms a unique basin in the Baltic Sea, with sources of anthropogenic pollution located nearby in the sedimentation basin of the Gdańsk Deep. Thus, the state of the near-bottom waters of the deep also exemplify conditions in the Gulf of Gdańsk proper. Many alien species occur here, and the decrease in the abundance of commercial fishes is causing a shift in the trophic structure of the gulf.

Although the inflow of pollution from terrestrial sources was significantly curtailed in the 1990s, the status of the gulf has not improved. In comparison with the 1970s, nitrate levels are higher in winter and this means an associated rise in the N:P mole ratio. This, combined with positive trends of phosphate concentrations in the near-bottom layers of the Gdansk Deep that result from anoxia, indicates that the poor state of the gulf is the result of secondary eutrophication.

References

Andrulewicz E, Kruk-Dowgiałło L, Osowiecki A (2004) An expert judgement approach to designating ecosystem typology and assessing the health of Gulf of Gdansk. In: Schernewski G, Löser N (eds) Managing the Baltic Sea. Coastline Reports 2, Rostock, pp 53–61

Andrulewicz E, Witek Z (2002) Anthropogenic pressure and environmental effects on the Gulf of Gdańsk: recent management efforts In: Schernewski G, Schiewer U (eds) Baltic coastal ecosystems, Springer, Berlin, pp 119–139

Bolałek J, Graca B (1996) Ammonia nitrogen at the water-sediment interface in Puck Bay (Baltic Sea). Estuarine Coastal Shelf Sci 43:767–779

Bolałek J, Falkowska L, Korzeniewski K (1993) The hydrochemistry of gulfs (in Polish). In: Korzeniewski K (ed) Puck Bay. IO UG, Gdańsk, pp 222–303

Ciszewski P, Ciszewska I, Kruk-Dowgiałło L, Rybicka D, Wiktor J, Wolska-Pyś M, Żmudziński L, Trokowicz D (1992) Trends of long-term alternations of the Puck Bay ecosystem. Stud i Mat. Oceanol 60, Biologia Morza (8), pp 33–84

Cyberska B, Lauer Z, Trzosińska A (eds) (1996) Environmental conditions in the Polish zone of the southern Baltic Sea during 1995 (in Polish). Institute of Meteorology and Water Management–Maritime Branch in Gdynia, pp 155–163

Cyberski J (1993) The hydrology of lagoons (in Polish). In: Korzeniewski K (ed) Puck Bay. IO UG, Gdańskiego, Gdańsk, pp 40–71

Cyberski J, Szefler K (1993) The climate of gulfs and their lagoons (in Polish). In: Korzeniewski K (ed) Puck Bay. IO UG, Gdańskiego, Gdańsk, pp 14–39

Environment (2002) Environment of the Baltic Sea area 1994–1998. Baltic Sea Environment Proceedings No 82B, pp 129–130

EU WFD (2000) http://www.eucc-d.de/infos/WaterFrameworkDirectice.pdf

Falkowska L (1979) Preliminary studies of the distribution of nutrients in the waters of the Puck Bay Zesz (in Polish). Nauk. Wydz. BiNOZ UG Oceanografia 6:21–35

Graca B (2004) Denitrification in the sediments of the inner Puck Bay–preliminary results. Oceanological Hydrobiol Stud XXXIII:73–91

Grawiński E, Dąbrowski J, Potajałło U (1998) Infection factors isolated from fish caught in the Gulf of Gdańsk and the Vistula Lagoon. Oceanological Stud XXVII:59–77

Jackowski E (2002) Fish of the Puck Bay. Sea Fisheries Inst. Gdynia

Jackowski E (2000) Coastal fish resources on the west coast and conditions for their exploitation (in Polish). Studia i Materiały MIR, ser. B nr 72. Gdynia, pp 5–62

Jackowski E (1998) The state of Puck Bay spawning grounds (in Polish). Studia i Materiały, Seria B nr 71, MIR Gdynia

Jankowska H, Łęczński L (1993) Geology and geomorphology (in Polish). In: Korzeniewski K (ed) Puck Bay. IO UG, Gdańsk, pp 303–338

Korshenko A, Błachowiak-Samołyk K, Opioła R, Shchuka T, Budaeva N, Pöllumäe A (2002) Mesozooplankton. In: Environment of the Baltic Sea area 1994–1998, Baltic Sea, pp 49–50

Kowalczuk M, Żmudziński L (1993) Inventory of the phytobenthos in a region proposed for designation as a protected zone at the foot of redłowski cliff (orłowo cliff) (in Polish). Raporty wewnętrzne CBM PAN, Gdynia (unpublished materials)

Krężel A (1993) The optical properties of water (in Polish). In: Korzeniewski K (ed) Puck Bay. IO UG, Gdańsk, pp 206–222

Kruk-Dowgiałło L (ed) (2004) The impact of chosen pollution sources on the Puck Bay environment (in Polish). Part I, II, III i IV. WW Maritime Institute of Gdańsk Nr 6128

Kruk-Dowgiałło L (ed) (2000a) The natural values of the marine areas protected by HELCOM BSPA of the Pomeranian District, vol 1 (in Polish). Słowinski National Park CRANGON 5, CBM PAN in Gdynia

Kruk-Dowgiałło L (ed) (2000b) The natural values of the marine areas protected by HELCOM BSPA of the Pomeranian District, vol. 3 (in Polish). Nadmorski Landscape Park CRANGON 7, CBM PAN in Gdynia

Kruk-Dowgiałło L (1998) Phytobenthos as indicator of the state of environment of the Gulf of Gdańsk. Oceanological Stud XXVII:105–123

Kruk-Dowgiałło L, Dubrawski R (1998) The state of environment of the Gulf of Gdańsk coastal zone in autumn 1994 and summer 1995. Oceanological Studies XXVI:137–158

Kruk-Dowgiałło L (1996) The role of filamentous brown algae in the degradation of the underwater meadows the Gulf of Gdańsk. Oceanolog. Studies. XXV, pp 125–137

Kruk-Dowgiałło L (1994a) Distribution and biomass of phytobenthos of the inner Puck Bay in the summer 1987 (in Polish). In: Puck Bay – opportunities for reclamation. Wyd IOŚ, Warszawa, pp 109–123

Kruk-Dowgiałło L (1994b) Causes of change in the structural of Puck Bay benthic flora and possibilities of steering its occurrence (in Polish). Materiały Seminaryjne IOŚ Gdańsk, 14 maj 1993. Wyd. IOŚ, Warszawa, pp 47–64

Kruk-Dowgiałło L (1991) Long changes in the structure of under water meadows of the Puck Lagoon. Acta Ichthyol Piscatoria XXI[Suppl]:77–84

Krzymiński W, Kruk-Dowgiałło L, Zawadzka-Kahlau E, Dubrawski R, Kamińska M, Łysiak-Pastuszak E (2004) Typology of Polish marine waters. In: Schernewski G, Wielgat M (eds) Baltic Sea typology. Coastline Reports 4, pp 39–48

Krzymiński W, Łysiak-Pastuszak E, Miętus M (eds) (2001) Environmental conditions in the Polish zone of the southern Baltic Sea during 2000. Institute of Meteorology and Water Management - Maritime Branch in Gdynia

Krzymiński W, Łysiak-Pastuszak E, Miętus M (eds) (2000) Environmental conditions in the Polish zone of the southern Baltic in 1999 (in Polish). Materiały Oddziału Morskiego IMGW, Gdynia

Lakowitz K (1907) Die Algenflora der Danziger Bucht; ein Beitrag zur Kenntnis der Ostseeflora. Engelman, Danzig.

Linkowski T (ed) (2003) Atlas of Baltic catches 2002 (in Polish). MIR, Gdynia

Łysiak-Pastuszak E (ed) (2004) Establishing reference conditions for surface water types, in accordance with Appendix II of the Framework Water Directive 2000/60/WE. (Unpublished materials in Polish)

Łysiak-Pastuszak E, Piątkowska Z (2004) Eutrophication in the Polish coastal zone of the of the Baltic: history, current state, and predicted directions of change (in Polish). In: Środowisko polskiej strefy południowego Bałtyku – stan obecny i przewidywane zmiany w przededniu integracji europejskiej. Gdańskie Towarzystwo Naukowe. Gdańsk, pp 59–71

Łysiak-Pastuszak E, Drgas N, Piątkowska Z (2004) Eutrophication in the Polish coastal zone: the past, present status and future scenarios. Mar Pollut Bull 49:186–195

Majewski A (1994) Natural environmental conditions in the Gulf of Gdańsk and its surroundings (in Polish). In: Błażejowski J, Schuller DUG (eds) Zanieczyszczenie i odnowa Zatoki Gdańskiej – problem o znaczeniu ogólnoeuropejskim. Materiały z Seminarium, Gdynia 1991, pp 22–35

Majewski A (ed) (1990) Gulf of Gdańsk. Wydawnictwa Geologiczne, Warszawa (in Polish)

Matciak M (1998) Transparency of marine coastal waters and environmental conditions – the Puck Bay]. Doctoral thesis. Uniwersytet Gdański, Gdynia (in Polish).

Miętus M, Filipiak J, Owczarek M (2004) The climate of the southern Baltic coast. Current state and perspectives for change (in Polish). In: Cyberski J (ed) Środowisko polskiej strefy południowego Bałtyku – stan obecny i przewidywane zmiany w przededniu integracji europejskiej. GTN, Gdańsk, pp 11–45

Mudryk Z (1999) Microorganisms – pelagic food chains (in Polish). In: Krzymiński W, Kruk-Dowgiałło L (eds) State of the Polish zones of the Baltic Sea (1994-1998). Expert evaluation for the Ministry of the Environment

Niemkiewicz E (2004) Preliminary evaluation of changes in phytoplankton near the Gdańsk-Wschód sewage treatment facility after the undersea collector had been put into operation (in Polish). Inż Morska i Geotechnika 2:76–81

Niemkiewicz E, Wrzołek L (1998) Phytoplankton as eutrophication indicator in the Gulf of Gdańsk water. In: The state of the Gulf of Gdańsk coastal zone (1994, 1995) Oceanological Stud XXVII:77–92

Nowacki J, Jarosz E (1998) The hydrological and hydrochemical division of the surface waters in the Gulf of Gdańsk. Oceanologia 40:261–272

Nowacki J (1993a) Morphology (in Polish). In: Korzeniewski K(ed) Puck Bay. IO UG, Gdańsk, pp 71–79

Nowacki J (1993b) Thermics, salinity and density of water (in Polish). In: Korzeniewski K (ed) Puck Bay. IO UG, Gdańsk, pp 79–112

Opioła R, Błachowiak-Samołyk K (2001) Zooplankton (in Polish). In: Warunki środowiskowe polskiej strefy południowego Bałtyku w 2000r. Materiały Oddz. Morskiego IMGW, Gdynia

Osowiecki A (2000) Directions of changes in the macrophytobenthos of the Puck Bay (in Polish). CRANGON Nr 3, CBM PAN Gdynia

Osowiecki A (1998) Macrozoobenthos distribution in the coastal zone of the Gulf of Gdańsk – autumn 1994 and summer 1995. Oceanological Studies XXVII:123–136

Pempkowiak J (1994) Causes of ecological catastrophes in the vicinity of the Puck Bay (in Polish). In: Błavejowski J, Schuller D (eds) Zanieczyszczenie i odnowa Zatoki. pp 67–73

Pilecki A, Roszman H (2005) Characteristics of terrestrial sources of pollution in the Gulf of Gdańsk (in Polish). Pracownia Ochrony Środowiska, Gdańsk (unpublished materials)

Pliński M (1995) Phytoplankton of the Gulf of Gdańsk in 1992 and 1993 (in Polish). Oceanologia 37:123–135

Pliński M, Florczyk I (1993) Macrophytobenthos (in Polish). In: Korzeniewski K (ed) Puck Bay. IO UG, Gdańsk, pp 416–422

Pliński M (1993) Phytoplankton. In: Korzeniewski K (ed) Puck Bay. IO UG, pp 378–387
Polak-Juszczak L (1996) The correlation between the concentration of heavy metals in the muscle tissue of fish and their habitat. Bull Sea Fisheries Inst 1:35–39
Polak-Juszczak L, Domagała M (1994) Levels of heavy metals in Baltic fish in 1991–1993. Bull Sea Fisheries Inst 3:27–30
Renk H (2000) Primary production in the southern Baltic (in Polish). Studia i Materiały MIR. Ser. A nr 35, Gdynia
Renk H (1993) Primary production in the Puck Bay. In: Korzeniewski K (ed) Puck Bay, Gdańsk, pp 338–366
Rybicka D (2004) Eutrophication in the Vistula Lagoon – actions to improve its state (in Polish). In: Cyberski J (ed) Środowisko polskiej strefy południowego Bałtyku – stan obecny i przewidywane zmiany w przededniu integracji europejskiej. GTN, Gdańsk, pp 85–97
Słomianko P (1974) Physical conditions in the region (in Polish). In: Słomianko P (ed) Puck Bay, Studia i Mat Oceanolog KBM, PAN, 5, pp 7–30
Szostak S, Kuzebski E, Pieńkowska B, Aguirre G (2000) Polish fisheries. Studia i Materiały MIR Ser. E nr 63
Szostak S, Kuzebski E, Pieńkowska B (2001) Polish fisheries. Studia i Materiały MIR Ser. E nr 64
Szostak S, Kuzebski E, Pieńkowska B (2002) Polish fisheries. Studia i Materiały MIR Ser. E nr 66
Szostak S, Kuzebski E, Pieńkowska B (2003) Polish fisheries. Studia i Materiały MIR Ser. E nr 67
Sztobryn M, Kowalska B, Letkiewicz B (2004) Ocena zlodzenia polskiej strefy przybrzeżnej (in Polish). In: Cyberski J (ed) Środowisko polskiej strefy południowego Bałtyku – stan obecny i przewidywane zmiany w przededniu integracji europejskiej. (ed.). GTN, Gdańsk, pp 59–72
Szumilas T, Michalska M, Bartoszewicz M, Nowacki J (2004) Changes in the degree of bacteriological pollution in Gdansk District coastal marine waters in light of local administration actions undertaken to limit pollution loads discharged into these waters (in Polish). In: Cyberski J (ed) Środowisko polskiej strefy południowego Bałtyku – stan obecny i przewidywane zmiany w przededniu integracji europejskiej. Gdańskie Towarzystwo Naukowe, Gdańsk, pp 97–113
Trzosińska A, Kruk-Dowgiałło L, Krzymiński W (eds) (1999) State of the Polish zones of the Baltic Sea (1994–1998) (in Polish). Expert evaluation for the Ministry of the Environment
Trzosińska A, Łysiak-Pastuszak E (1996) Oxygen and nutrients in the southern Baltic Sea. Oceanol Stud 25:41–76
Trzosińska A (1992) Water transparency in the polish zone of the Baltic Sea. Oceanologia 33:203–209
Wandzel T (2000) Round Goby, *Neogobius melanostomus* Pallas, 1811 in the catches of R/V Baltica. Acta Ichthyol Piscatoria 30:81–91
Wasmund N, Andrushaitis A, Łysiak-Pastuszak E, Müller-Karulis B, Nausch G, Neumann T, Ojaveer H, Olenina I, Postel L, Witek Z (2001) Trophic status of the south-eastern Baltic Sea: a comparison of coastal and open areas. Estuarine Coastal Shelf Sci 53:849–864
Wiktor K (1993) Zooplankton. In: Korzeniewski K (ed) Puck Bay. IO UG, Gdańsk, pp 388–395
Witek Z (1995) Biological production and its exploitation in the marine ecosystem of the western part of the Gdansk Basin (in Polish). Sea Fisheries Institute, Gdynia
WIOŚ (2003) Report on the state of the Pomorski Province environment in 2002 (in Polish). WIOŚ, Biblioteka Monitoringu Środowiska, Gdańsk
WIOŚ (1993) Report on the state of the Gdansk Province environment in 1993 (in Polish). WIOŚ, Biblioteka Monitoringu Środowiska, Gdańsk
Wolska-Pyś M (1998) Plankton larvae of invertebrates as an indicator of potential capacity for regeneration of bottom filtrator communities in the Gulf of Gdańsk. Oceanological Stud XXVII:93–105
Wolska-Pyś M (1994) Variation of mezooplankton in the open waters of the southern Baltic as a result of environmental changes (in Polish). In: Kruk-Dowgiałło L (ed) Problemy ochrony Środowiska przyrodniczego morza Bałtyckiego i strefy Materiały z Seminarium Gdańsk 14 maja 1993, IOŚ Warszawa, pp 82–91

Chapter 8
The Vistula Lagoon

B. Chubarenko and P. Margoński

8.1 Environmental Characteristics

8.1.1 Lagoon Morphometry

The Vistula Lagoon is located at the southern coast of the Baltic Sea (Fig. 8.1). It stretches along the shore for ca. 91 km. The width of the lagoon varies between 2 and 11 km, and the water volume and surface area are 2.3 km^3 and 838 km^2, respectively. The average lagoon depth is 2.7 m; the maximum depth excluding the artificially dredged navigable channel is 5.2 m. The state border between the Kaliningrad region (Russia) and Poland divides the lagoon into two parts, which account for 64% and 36% of the water volume and 56.2% and 43.8% of the lagoon area, respectively. The length of the lagoon coastal line is estimated as 270 km (Poland 111 km, Russia 159 km). The single Nasypnoi Island is in the deepest part of the lagoon (Lazarenko and Majewski 1971).

The Vistula Lagoon is separated from the Baltic Sea by a stable sand barrier that is divided into two parts by the lagoon inlet. The southern part (the Vistula Spit) has a length of 55 km and a width of 0.5–2 km. It is covered by forest, and is divided by the state border between Poland and Russia into two segments (55% and 45% of its length). The Baltiysk northern segment of the barrier (11 km) separates the Primorsk Bight from the Baltic Sea.

The width and depth of a single lagoon inlet (Baltiysk Strait, Kaliningrad region) are 400 m and 10–12 m (fairway), respectively; the minimal vertical transect is ca. 4,200 m^2. A navigable channel, the Kaliningrad Marine Canal, connects the Baltiysk Strait and the Pregolia River mouth and passes 43 km along the northern lagoon coast. Its depth varies in the range of 9–12 m. The canal is separated from the lagoon proper by a set of artificial islands. Narrow passes between these islands have a width of 20–50 m and a depth of 1–3 m. The only open segment of the canal has a length of 3.5 km, when it crosses the semi-enclosed Primorsk Bight. The canal was built in 1901 by traders and had a depth of 6 m. It is indeed a hydro-technical construction that requires constant maintenance dredging.

U. Schiewer (ed.), *Ecology of Baltic Coastal Waters. Ecological Studies 197.*

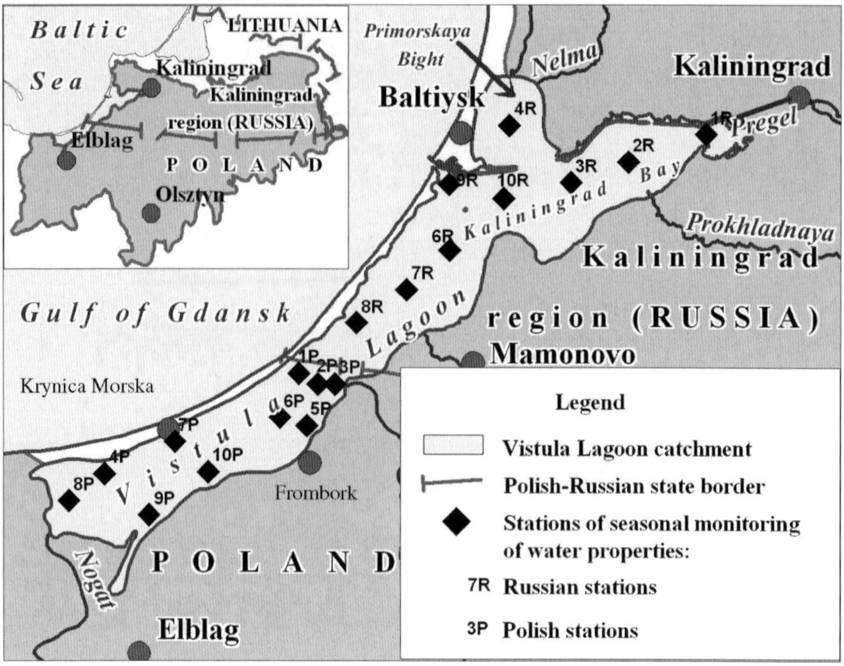

Fig. 8.1 Location of the Vistula Lagoon in the South-Eastern Baltic

8.1.2 Lagoon Hydrography

In former times, the total river runoff towards the lagoon was 11–12 km^3 a^{-1}. Historically, the Vistula Lagoon was formed as part of an estuary of the Vistula River, the total average runoff of which is greater than 30 km^3 a^{-1} (Andrulewicz and Witek 2002). The Vistula River discharged to the lagoon ca 8–9 km^3 a^{-1} water through the Nogat tributary. After regulation in 1916, when the Vistula runoff was directed mostly to the Baltic Sea, the Nogat annual average discharge dropped to the value of 0.7 km^3 a^{-1}. Since that time, the hydrological and sedimentation regimes of the lagoon have changed dramatically: the lagoon has evolved from a freshwater plain estuary toward an estuarine lagoon with significant influence from the Baltic. Lagoon salinity has changed from practically zero to around 3.5 psu on average (see Chaps. 3 and 4 by Schiewer, this volume).

Nowadays, the hydrology of the Vistula Lagoon is controlled by marine water inflow (+17 km^3 a^{-1}) and freshwater gain, which consists of river runoff (+3.68 km^3 a^{-1}), precipitation (+0.5 km^3 a^{-1}), evaporation (–0.65 km^3 a^{-1}) and ground runoff (+0.07 km^3 a^{-1}) (Lazarenko and Majewski 1971). The maximum magnitudes of water gain and loss are observed in winter, with minimums in summer, when the weather is calm and the intensity of water exchange with the Baltic is low.

Marine water inflow is caused by general oscillation of the Baltic water level (±10–20 cm on average) and by local wind surges (±30–50 cm on average). According to Szymkiewicz (1992), the semi-diurnal tidal variations (±1 cm) have only an insignificant influence.

The Pregolia River is nowadays the largest river in the catchment. It opens in the north-eastern corner of the lagoon, just after passing through the biggest city in the watershed, the city of Kaliningrad (formerly Königsberg). The Pregolia River drainage basin is 15,128 km^2, with 50.7% and 48.3% belonging to Poland and Russia, respectively. At ca. 1.53 $km^3\ a^{-1}$ (48 $m^3\ s^{-1}$), its average runoff constitutes ca. 41% of total river runoff to the lagoon. The Nogat River (1,337 km^2, 0.66 $km^3\ a^{-1}$), the Pasleka River (229 km^2, 0.5 $km^3\ a^{-1}$) and the Prokhladnya River (1,170 km^2, 0.3 $km^3\ a^{-1}$) make up another 39% of the lagoon river runoff (Lazarenko and Majewski 1971).

8.1.3 *Transboundary Aspects*

The Vistula Lagoon is a transboundary water pool (Andrulewicz et al. 1994; Andrulewicz and Chubarenko 2004). The Russian and Polish national parts of its surface area are 471 km^2 and 367 km^2, respectively, and free water exchange can occur between them. The single outlet is located in the Russian part of the lagoon. The area of the whole lagoon watershed is 23,871 km^2, 61% belonging to Poland and 39% to Russia. The transboundary lagoon catchment is of parallel-consequence type, i.e. both national lagoon segments have their own catchments (parallel structure), but the catchment entering the lagoon on the Russian side starts in Poland (consequence relation). The drainage area of the exclusively Polish part of the lagoon watershed, which discharges directly to the lagoon, is 6,639 km^2 (27.8%). The remaining lagoon drainage area meets the lagoon in Russian territory. It includes the exclusively Russian part (1,715 km^2, 7.2%), and a transboundary region comprising a Polish upper share (7,922 km^2, 33.2%) and a Russian lower share (7,595 km^2, 31.8%). This transboundary part of the lagoon basin is formed by the catchments of the two big transboundary tributaries of the Pregolia River, namely the Lyna-Lava and Angrapa-Wengorapa rivers, and two small separate rivers, the Prokhladnaya and Momonovka rivers, which flow directly into the lagoon (Chubarenko 2007).

8.1.4 *Lagoon Hydrology*

8.1.4.1 Salinity

The Vistula Lagoon is under both marine and river drain influences, which vary seasonally. An increase in river runoff during March and April leads to minimisation of lagoon salinity in general, and to spatial variations of 0.5–4.5 psu. Then, due to the intensification of periodic marine water inflows, the lagoon is actively saline until

August/September (3.5–6.5 psu). After a smooth desalinisation due to an increase in river discharge in autumn, the ratio between fresh- and saltwater influxes stabilises in late autumn and winter. When ice develops in winter, a significant amount of salt enters the lagoon water from the surface, causing vertical mixing and an increase in lagoon water salinity of 10–25% (I. Chubarenko et al. 2004). The average lagoon salinity of 3.5 psu is only a convenient average value, because salinity varies along the lagoon (Fig. 8.2), decreasing from the lagoon inlet (maximum value) towards the Pregolia River mouth (eastward) and Vistula-Nogat polders (southward). A dramatic salting and transformation of the lagoon into a pure marine enclosed bay is forecast if a second open inlet is constructed (I. Chubarenko and Tchepikova 2001).

Although on average the Vistula Lagoon is usually well mixed vertically due to night convection, Langmuir circulation and storm wind wave mixing, a salinity vertical gradient (of 0.2–0.5 psu m^{-1} on average) is observed 1–3 km from both the inlet and the Pregolia River mouth, especially in spring and late autumn periods. The reason for this is the existence of near-bottom marine water intrusions near the inlet and an upper-layer freshwater flow near the Pregolia River mouth.

8.1.4.2 Water Temperature

The Vistula Lagoon is almost uniform in its thermal structure (Chubarenko et al. 1998a); water temperature differences between the extreme ends of the lagoon are only ca. 0.5–1°C. High gradients, up to 5°C per 100 m (in the horizontal direction) and 5°C per 5 m (in the vertical direction) are observed near the inlet during marine water intrusions. Annual water temperature dynamics (from –0.2 up to 25–26°C) are stipulated by solar heating. The maximum water temperature usually occurs at the end of July or beginning of August, 1 week after the maximum air temperature, and about 2 weeks after the maximum solar irradiation. Spatial and vertical variations of water temperature are insignificant in comparison with temporal distinctions (daily variations during summer are ca. 1–1.5°C on average, with a maximum of 3–4°C). Ice coverage is not stable in the lagoon. In the coldest years permanent ice stays from December until March. In the warmest years this period is very short (1–1.5 months) (Lazarenko and Majewski 1971).

The Vistula Lagoon is characterised by a clockwise-orientated, loop-shaped seasonal temperature–salinity (TS)-diagram (Fig. 8.2). This is typical when the annual dynamics of salinity and water temperature are correlated (B. Chubarenko et al. 2004), and both characterised by one minimum and one maximum shifted in time. For the Vistula lagoon this time-phase shift equals 1–2 months.

8.1.4.3 Retention Time

The main time scale characterising mixing in the reservoir is the retention, or residence or flushing time. In contrast to a water pool fed by a single stream (e.g. a freshwater reservoir), an estuarine lagoon is under the influence of both river drain and marine water inflow. Therefore, one may introduce characteristic flushing times related to these driving streams. Estimated as the ratio of water volume to either river runoff or marine influx, these flushing times equal 198 and 49 days, respectively (B. Chubarenko et al. 2004). An integrated flushing time taking both these driving factors into consideration is ca. 40 days on average for the Vistula Lagoon as a whole. Local flushing time for various lagoon compartments differs from the integrated value for the whole lagoon, varying from half a day for the area just next to the Baltiysk Strait, to 80–150 days for remote corners of the Vistula Lagoon.

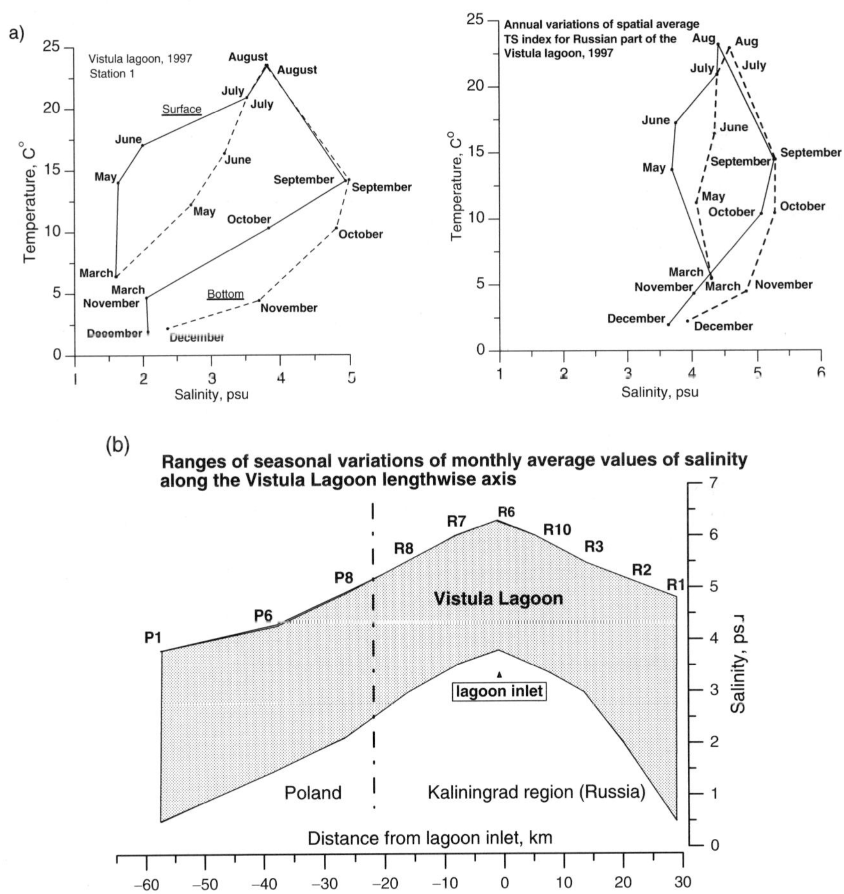

Fig. 8.2a,b Salinity variations in the lagoon. **a** Temperature–salinity (TS) diagrams for the Vistula Lagoon waters. **b** Variations of salinity along the lagoon

8.1.5 Lagoon Water Dynamics

8.1.5.1 Wind Action

The most probable wind directions are in the sector between south and north-west (60%). The strongest winds blow from the south-east and east. The average wind speed for the marine coast (6.1 m s^{-1}) exceeds that observed at the inner lagoon coast: 5.6 m s^{-1} in Tolkmicko and 4.3 m s^{-1} in Momonovo (Bogdanov et al. 2004). The wind wave action is higher on the eastern lagoon coast; the wave energy fluxes calculated at 2 m depth are 10^4–10^5 Ts^{-1} for the onshore component and 10^2–10^3 Ts^{-1} for the alongshore component (1 Ts^{-1} = 10^4 J) (Bogdanov et al. 2004).

8.1.5.2 Currents, Waves, and Vertical Mixing

Waves and currents develop immediately when the wind starts to blow. The average values of advective currents are 0.1–0.2 m s^{-1}, with a maximum of about 0.5 m s^{-1}. The existence of near-bottom compensative currents against the wind direction, and the influence of the Coriolis effect on intermediate and slow currents (≤0.2 m s^{-1}) make it necessary to consider the current pattern in the Vistula Lagoon as fully three-dimensional (I. Chubarenko and B. Chubarenko 2002).

Wind waves are usually limited by depth: a significant wave height for winds of 10 m s^{-1} is 0.8 m, 0.95 m for winds of 15 m s^{-1}, and 1.1 m for winds of 20 ms^{-1}. The waves are rather steep, with a typical length:height ratio of around 7 (Lazarenko and Majewski 1971).

Vertical mixing is ensured by three main mechanisms. Surface cooling leads to night convection, which destroys vertical stratification on a nightly basis. For winds of 3–9 m s^{-1}, Langmuir circulations develop and usually penetrate down to the bottom due to weak stratification. Wind wave mixing becomes significant when winds exceed 6–7 m s^{-1}. Waves cause a resuspension of bottom sediment, which is redistributed over the lagoon by currents. Vertical mixing under ice coverage is caused by salt release upon ice freezing.

8.1.5.3 Water Level

The overall rise in spatial average water level is caused by marine water inflow into the lagoon due to the rise in Baltic water level. Local wind surge causes a level rise at the windward shore and a level drop at the leeward shore. The difference in water levels at the extreme ends of the lagoon in stormy periods is usually around 0.5–0.7 m, but can exceed 1.0–1.7 m (Lazarenko and Majewski 1971). The extreme absolute values of water level rise at the remote corners of the lagoon are ca. 1.3–1.7 m, of which 0.9–1.1 m is contributed by wind-induced local level fluctuation. The induced water level variation at the remote ends of the lagoon starts 3–4 h after an imposed level variation occurs in the inlet, and the lag between occurrence of the maximum (or minimum) levels at the lagoon entrance and its remote ends is ca. 10–12 h.

8.1.5.4 The Role of the Kaliningrad Marine Canal

The existence of a deep navigable canal is of great importance for general lagoon circulation and inner water exchange (B. Chubarenko and I. Chubarenko 2003). The canal contributes in various ways to water movement on different scales.

The canal and the deep lower segment of the Pregolia River (up to the centre of the city of Kaliningrad) form the estuary part of the river. Permanent mixing of marine and river fresh waters occurs here, and the mixing zone seasonally migrates both upstream and downstream over distances of 10–20 km.

These estuary conditions established in the canal play a specific role in the water dynamics of the eastern part of the Vistula Lagoon. The seasonal transport of salty water upstream via the canal toward the Pregolia River mouth supplies an additional flux of salty water directly to the eastern part of the Vistula Lagoon, which intensively increases salinity in this remote corner of the lagoon (up to the end of autumn).

8.1.5.5 Water Exchange Through the Inlet

The average statistics for currents in the Vistula Lagoon inlet are given in Lazarenko and Majewski (1971): 74.9% inflow or outflow uniform current, 11.7% two-layer currents (influx in the bottom layer and outflow at the surface), 13.4% two-stream currents. Two-layer and two-stream regimes are considered intermediate reorganisation processes. Maximal currents of 1.34 m s^{-1} and 1.38 m s^{-1} were observed for inflow and outflow, respectively. The average velocity for unidirectional uniform flow ranges between 0.06 and 0.95 m s^{-1}. The average currents for two-layer or two-stream regimes are usually in the range of 0.1–0.2 m s^{-1}. A historical maximum surface current of 2.5 m s^{-1} was observed in 1894.

8.1.6 Hydrological–Hydrodynamical Typology of the Vistula Lagoon System

The Vistula Lagoon is a complicated hydrodynamic system involving the lagoon itself (shallow spatially extended water pool), the man-made Kaliningrad Marine Canal (semi-isolated deep channel), and an artificially deepened region downstream of the Pregolia River. The lagoon itself could be referred to as a well-mixed estuarine non-tidal lagoon or, in other words, a non-tidal well-mixed bar-built plane estuary parallel to the shore, with significant horizontal variability of characteristics, and predominant lengthwise gradients between the inlet and its remote ends. The canal and deepened region downstream of the Pregolia River could be defined as relatively deep estuary with clearly pronounced along-stream gradients and with seasonal evolution between highly stratified and vertically well mixed conditions.

8.1.7 Sediment Regime

The lagoon bottom deposits consist mainly of three types of sediments: medium and fine grained sands (fractions of 0.1–1.0 mm prevail and cover 30% of the bottom), coarse aleurites (0.05–0.1 mm, 22%) and fine aleurite mud (0.01–0.05 mm, 45%). The remaining 3% is represented by aleuro-pelitic mud (0.005–0.01 mm), shells and pebble-gravel deposits. The muddy sediments cover most of the deep part of the lagoon bottom, with a depth of more than 2–2.5 m, while sandy sediments are mostly found along the dynamically active shallow coastal zone (to a depth of 1.5–2 m). The largest quantity of coarse sand is found in the vicinity (1–1.5 km) of the lagoon inlet, where it actually forms a reversed bar (depth 1.5–2 m) inside the lagoon area (Chechko and Blazchishin 2002).

The concentration of suspended sediments varies in the range 4–230 mg l^{-1}, with an average value of 30 mg l^{-1}, i.e. 10 times more than in the Baltic Sea. On average throughout the year the biotic suspended matter comprises 54% of the total amount, although seasonal variations are evident, e.g. the abiotic component prevails during the windy autumn period, when it comprises more than 60% of the total (Chechko 2002, 2004; Chubarenko et al. 1998b).

Wind wave sediment resuspension (Blazchishin 1998; Chubarenko et al. 2002) is responsible for the stepwise redistribution of sediments throughout the lagoon from its sources – rivers and eroded segments of the coast. Since practically cutting of the flow of the Vistula River in 1916, the overall lagoon sediment budget has changed dramatically. Nowadays, annual loss of suspended matter from the Vistula Lagoon to the Baltic Sea (322,000 t a-1) significantly exceed annual gains from rivers ((88,000 t a-1) and from the Baltic (34,000 t a-1), and the lagoon is a significant source of suspended matter for the Baltic coastal zone. One remarkable feature is that intensive lagoon bottom erosion by resuspension supplies to the Baltic more sediments than the river drain from the lagoon catchment. The Vistula Lagoon is evolving towards a new equilibrium in sediment budget when the lagoon depth increases enough to protect the bottom from active erosion by waves and currents (Chubarenko and Chubarenko 2001).

8.2 Eutrophication

8.2.1 Water Quality

The horizontal distribution of water quality parameters in the Vistula Lagoon is strongly influenced by hydrological and meteorological factors, one of the most important of which is the exchange of water masses between the Gulf of Gdańsk and the lagoon. As a consequence, the area close to the Baltiysk Strait is "washed-out", and the concentrations of nutrients in this area are lower in comparison with those in remote parts of the lagoon (P. Margoński et al., manuscript in preparation). The seasonal dynamics of hydro-chemical parameters in the Vistula Lagoon are presented in Table 8.2.

The high spatial and temporal variation in nutrient concentrations recorded in the Polish part of the lagoon was reported by Renk et al. (2001). During the 1999 vegetative season, and especially during the spring phytoplankton bloom, the concentrations of some nutrients decreased below detectable limits. Phosphate concentrations were relatively high and ranged from 0.25 to 1.96 μmol l^{-1}. The lowest values were observed in April, while winter phosphate concentrations exceeded 4 μmol l^{-1} in February 2000 (M. Zalewski, personal communication). The average concentrations of various forms of inorganic nitrogen oscillated around the following values: nitrites 0.11 μmol l^{-1}; nitrates 0.75 μmol l^{-1}; ammonia nitrogen 0.58 μmol l^{-1} (April–October 1999). However, winter maxima were much higher: nitrites 3.49 μmol l^{-1}; nitrates 130.42 μmol l^{-1}; ammonia nitrogen 16.63 μmol l^{-1} (February 2000, M. Zalewski, personal communication). The ratio of inorganic nitrogen to phosphorus was very low during the vegetative season. It decreased to below 1 from the second half of May to the end of September. The only exception was the average in the second half of July (1.81), which coincided with the blue-green algal bloom. In February 2000, the ratio of inorganic nitrogen to phosphorus even reached a level of 40–50 (M. Zalewski, personal communication). The silicate concentrations were very high; the average value for the Polish part oscillated around 84.5 μmol l^{-1} from April to August, while the average for the September–November period was 107.6 μmol l^{-1}. Winter maxima were higher than 180 μmol l^{-1} (M. Zalewski, personal communication). The average chlorophyll *a* (Chl *a*) concentrations in the 0–1.5 m layer were high at 41.2 mg m^{-3} in 1998 and 43.5 mg m^{-3} in 1999. The highest recorded values exceeded 150 or even 200 mg m^{-3} during blooms in July and August. The light attenuation coefficients varied significantly from 1.5 m^{-1} to 7 m^{-1}.

Monthly water quality averages for the 1981–2001 periods in the Russian part of the lagoon were reported by Senin et al. (2004). Oxygen concentrations were usually very high (9.8–12.7 mg l^{-1}), but biochemical oxygen demand over a 5 day period (BOD_5) ranged between 3.3 and 6.4 mg l^{-1}. Phosphate concentrations varied between 15 and 104 μg l^{-1} (0.48–3.36 μmol l^{-1}) throughout the season, with peaks in spring and summer. The highest nitrate concentrations were observed in early spring, only to drop significantly during the vegetative season (from 865 to 26 μg l^{-1} or from 62 to 1.86 μmol l^{-1}). The Chl *a* concentrations ranged from 36 to 52 mg m^{-3} (41.7 mg m^{-3}, on average).

Water transparency is very low. Although the average Secchi depth is 0.4–0.6 m in the Polish and Russian parts, it very often falls to 0.3–0.4 m during the vegetation period. On cloudless days, the maximum value of photosynthetically active radiation (PAR) penetrating into the water is 2,500–2,600 μmol m^{-2} s^{-1}. Half of this is lost in top 25 cm layer, and about 1% of it reaches to a depth of 1.5 m. The vertical distribution of PAR is well approximated by a simple one-exponential fold, in which a light attenuation coefficient varies seasonally, because of algae vegetation

Table 8.1 Main characteristics of the Vistula Lagoon. *Chl a* Chlorophyll *a*, *dw* dry weight *ww* wet weight, *DIN* dissolved inorganic nitrogen, *DIP* dissolved inorganic phosphorus

	Polish part	Russian part	Lagoon average
a. Climate and hydrology			
Air temperature (°C)	Min: –30.4; max:35.4[a]	Min: –33.3; max: 35.1[a]	Min: –30.4; max: 35.1[a]
Water temperature (°C)	No data	No data	Min: – 0.2; max: 25–26[a]
Ice covering (days a^{-1})	No data	No data	67–75 (max: 140)[a]
Annual global radiation (kW h m^{-2})	No data	No data	898[a]
Average precipitation (mm a^{-1})	No data	No data	597[a]
Mean freshwater inflow ($10^6 m^3 a^{-1}$)	1,662[a]	2,018[a]	3,680[a]
Mean inflow ($10^6 m^3 a^{-1}$) (from the Baltic)	0	17,000[a]	17,000[a]
Mean total outflow ($10^6 m^3 a^{-1}$)	0	20,520[a]	20,520[a]
Residence time (days)	No data	No data	40[b]
b. Morphometry and sediments			
Area (km^2)	367[a]	471[a]	838[a]
Volume ($10^6 m^3$)	828[a]	1,472[a]	2,300[a]
Mean depth (m)	2.3[a]	3.1[a]	2.7[a]
Maximum depth (m)	4.4[a]	5.2[a]	5.2[a]
Catchment area (km^2)	14,561[a]	9,310[a]	23,871[a]
Area/catchment relation	0.0252	0.051	0.0351
c. Pelagic biological components			
Secchi depth (m)	Average 0.4 (0.20–2.00)[c]	Average 0.65 (0.35–2.00)[d]	min: 0.2; max: 2.0 (inflow of marine waters)
Phytoplankton biomass (µg Chl *a* l^{-1})	Average 43, max: 150 (exceptionally 200)[c]	Average 42, max 100[e]	100–200
Dominating phytoplankton species	*Aphanizomenon flos-aquae*, species from genera *Anabaena* and *Merismopedia*[f]	Species from genera *Merismopedia*[f]	
Fish			
Number of species	20		
Dominating species		In terms of biomass: herring; in terms of value: eel, pikeperch, and bream[g]	
Primary production (g C $m^{-2} a^{-1}$)	303.8[c]	180[h]	180–300
Fish catches (t a^{-1})	3,000–15,000[i]		

(continued)

Table 8.1 (continued)

	Polish part	Russian part	Lagoon average
d. Benthic biological components			
Macrophytes			
Biomass (kg m^{-2})	~ 2 (on average)[j]; 1–5 for *Phragmites communis*[j]; 2–3 for *Phragmites communis* (dw)[k]		
Number of species		64[j]	
Dominating species			*Phragmites communis*[j]
Macrozoobenthos			
Biomass (g m^{-2})	80–103 (ww)	20–25 (ww)	20–100 (ww)
Number of species	No data	~180	~180
Dominating species	*Marenzelleria neglecta*[l]	*Marenzelleria neglecta*[m,n]	*Marenzelleria neglecta*
e. Water chemistry, trophic status and pollution			
Salinity (psu)	Average 2.4 (0.9–4.9)[c]	Average 3.8 (monthly averages 0.5–6.9)[d]	0.5–6.9
pH	Average 8.45 (6.8–9.5)[o]	7.9–8.8[e]	6.8–9.5
Oxygen saturation (%)	80–120 (may drop to less than 30% inwinter)[a,o]		
Total nitrogen (μmol l^{-1})	50–400[o]	20–180[p]	20–400
DIN (μmol l^{-1})	0.02–5.30[c] (>140 in winter [q])	No data	0.02–5.30 (>140 in winter)
Total phosphorus (μmol l^{-1})	0.7–16[o]	0.8–20[p]	0.7 20
DIP (μmol l^{-1})	Min: 0.25; max: 1.96[c] (>4 in winter)[q]	0.1–3.1[16]	0.1–3.1 (>4 in winter)
Annual nitrogen input (t a^{-1})	2,700–3,300[f]	12,200[r]	No data
Annual phosphorus input (t a^{-1})	315–330[f]	2,100[r]	No data
Limitation of PP and period	Mainly nitrogen	Phosphorus in early spring only[f]	
Trophic level	Polytrophic/eutrophic status[f]		

[a] Lazarenko and Majewski 1971; [b] Chubarenko et al. 2005; [c]Renk et al. 2001; [d]Chubarenko et al. 1998a and monitoring results of 2000–2005; [e]Senin et al. 2004; [f]P. Margoński et al., manuscript in preparation; [g]Bartel et al. 1996; [h]Aleksandrov 2004; [i]Polish-Russian Commission for the Management of Fish Resources in the Vistula Lagoon; [j]Pliński et al. 1978; [k]Pliński 1995; [l]Żmudziński 2000; [m]Rudinskaya 1999; [n]Ezhova et al. 2005; [o]Inspection Board for Environmental Protection (WIOŚ) 2004 report; [p]Y.M. Senin, personal communication; [q]M. Zalewski, personal communication; [r]Anon 1996

and wind resuspension, between 2.5 (March) and 3.5 (July, November) (Rasmussen 1997; A.F. Kuleshov, personal communication).

Eutrophication processes are influenced not only by riverine loads but also by nutrient exchange between the water and sediments. According to Kwiatkowski

(1996), significant sources of nitrogen and phosphorus are accumulated and released from silty bottom sediments. As a consequence of water mixing, the redistribution of labile inorganic nutrients from the upper layer of bottom sediments to the water column is almost continuous (Ezhova et al. 1999). Therefore, there is a high internal potential for eutrophication. Kwiatkowski (1996) estimated that as much as 138,600 tons nitrogen and 55,800 tons phosphorus have accumulated in the 10 cm sediment layer. Approximately 22% of the nitrogen and 35% of the phosphorus loads are exported to the Gulf of Gdańsk (Kwiatkowski et al. 1996).

8.2.2 Limitation of Primary Production

Light and nutrient availability are among the most important factors controlling primary production. Light limitation is caused mostly by the amount of suspended matter in the water column. Concentrations of suspended matter usually range from 10 to 50 g m^{-3} (or from 5 to 80 g m^{-3} according to Trzosińska 1975). Such large amounts of suspended solids result from frequent resuspension of bottom sediments, which is typical of shallow water bodies exposed to winds, such as the Vistula Lagoon (M. Zalewski et al., manuscript in preparation).

The vast majority of lakes studied within the scope of the OECD project (OECD 1982) fall into the category of phosphorus-limited lakes; however, on the contrary, the Vistula Lagoon rather seems to be a nitrogen-limited water body. Phytoplankton growth limitation estimated with modeling tools confirmed that phosphorus limitation occurs only during early spring. Throughout the rest of the vegetative season, nitrogen is the main limiting factor (Ezhova et al. 1999; Kwiatkowski et al. 1997; M. Zalewski et al., manuscript in preparation).

8.2.3 Trophic State Index

The present trophic status of the Vistula Lagoon has been assessed by P. Margoński et al. (manuscript in preparation). The classic trophic state index (TSI) developed by Carlson (1977) is calculated based on Secchi depth, Chl *a*, and total phosphorus concentrations. Secchi disk transparency is one of the simplest and most frequently collected limnological measurements. Its values are easily understood and appreciated. However, Secchi disk transparency can produce erroneous values in lakes containing high amounts of non-algal particulate matter (Carlson 1977). Under the conditions in the Vistula Lagoon, Secchi depth was influenced heavily by wind-driven resuspension, and phosphorus was not a major limiting factor, so these indices were not useful indicators. Total nitrogen TSI (Kratzer and Brezonik 1981) was designed to be used under nitrogen-limiting conditions. In recent years, the trend line has crossed the threshold between eutrophic and meso-eutrophic types (P. Margoński et al., manuscript in preparation).

The TSI values derived from chlorophyll should be considered as the best indicator, mainly because they reflect the response of the lowest level of biota in the lagoon, which is a good indicator of primary production under given conditions. There were, however, year-to-year oscillations that mostly ranged from 65 to 70. The results indicate that the status is polytrophic/eutrophic (P. Margoński et al., manuscript in preparation).

8.3 Planktonic Communities

8.3.1 *Primary Production*

The average annual production in the Polish part of the lagoon has been estimated to be 303.8 g C m^{-2} a^{-1}, while the average calculated daily production varied from 34 mg C m^{-2} day^{-1} in January to 2,723 mg C m^{-2} day^{-1} in August (Renk et al. 2001). In the mid-1970s, Niedoszytko and Wiktor (1978) obtained much higher values of annual lagoon production of 461.3 g C m^{-2} a^{-1} [see Chaps. 3, 6 (Schiewer) and 7 (Kruk-Dowgiałło and Szaniawska), this volume]. However, they used a different method – the so-called oxygen method. A decrease in the primary production level in the late 1990s might have been caused by lower water transparency, or may have resulted simply from different estimation methods (Renk et al. 2001). The seasonal dynamics of primary production is presented in Table 8.2.

The annual cycle of primary production in the Russian side, which peaks in May and July–August and usually coincides with biomass peaks (Krylova 1985), ranges from 200 (November) to 1,600 (August) mg C m^{-2} day^{-1}. The annual average values calculated from data from 1974–1976 and 2001 varies between 500 and 800 mg C m^{-2} day^{-1} (180 280 g C m^{-2} a^{-1}). The spatial variations are significant: a maximum is usually observed near the Pregolia River mouth while the minimum (for example, 40% less in 2001) is near the Baltiysk Strait (Aleksandrov 2004).

8.3.2 *Bacterioplankton*

Data on bacterioplankton in the lagoon are scarce. Only the *Escherichia coli* titre is regularly monitored in water samples. Since 1997 in the Polish part of the lagoon, the average annual faecal *E. coli* titre has ranged from 1.75 to 1.49 bacteria ml^{-1} (Kopiec 2002).

In the mid-1970s, studies on the quantitative distribution of saprobic and hydrocarbon bacteria in different seasons were conducted in the Polish part of the lagoon

(Maciejowska and Macur 1978). No data are available regarding the Russian part of the Vistula Lagoon.

8.3.3 Phytoplankton

Three phytoplankton groups dominate in terms of abundance in the Polish part of the Vistula Lagoon (Fig. 8.3): Cyanobacteria, green algae, and diatoms. Cyanobacteria comprise over 80% of the total abundance [cf. Chaps. 3, 6 (Schiewer) and 7 (Kruk-Dowgiałło and Szaniawska), this volume]. Species from the genera *Anabaena* and *Merismopedia, Aphanizomenon flos-aquae*, and representatives of the subfamily Gomphosphaerioideae are present among the dominants (P. Margoński et al., manuscript in preparation). The composition of dominating species of Cyanobacteria noted in 1953 was similar (Szarejko-Łukaszewicz 1957). In 1999, the highest biovolume of Cyanobacteria was observed in August, when the mass occurrence of filamentous species from *Anabaena* genus and *Aphanizomenon flos-aquae* was noted. Blooms of these species were observed from June to September in the central and northeastern regions of the Polish part of the Vistula Lagoon (Szarejko-Łukaszewicz 1957; Pliński and Simm, 1978; P. Margoński et al., manuscript in preparation).

Green algae (Fig. 8.3) comprised from 10 to 15% of the total phytoplankton abundance in the 1970s and 1990s (Pliński and Simm 1978; P. Margoński et al.,

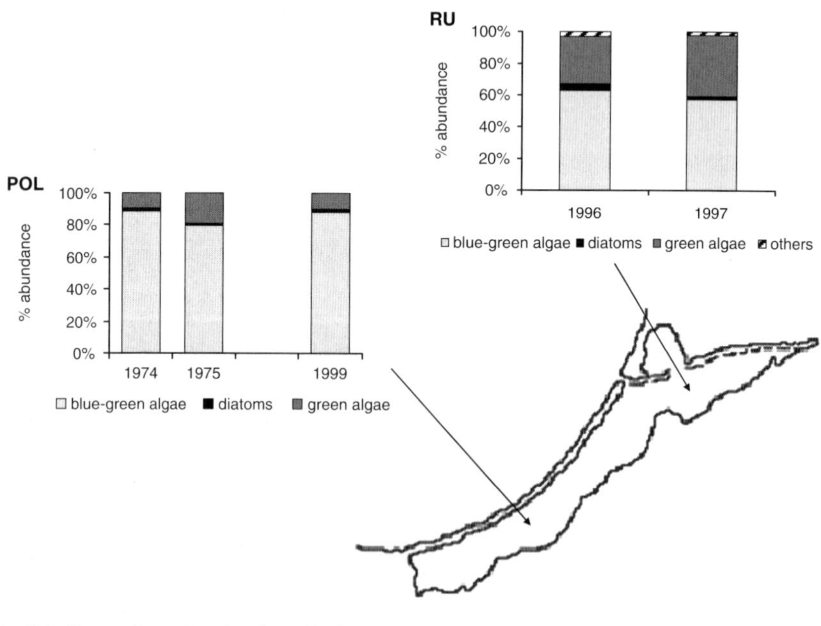

Fig. 8.3 Proportion of main phytoplankton groups in the Vistula Lagoon (P. Margoński et al., manuscript in preparation). *POL* Polish part, *RUS* Russian part

manuscript in preparation). The filamentous green alga *Planktonema lauterborni*, which was not reported in samples taken in 1953, dominated the green algae abundance in spring samples from the 1974–1975 and 1999 periods. On the other hand, representatives of Volvocales – *Volvox aureus*, *Pandorina morum*, and *Eudorina elegans* – were not noted in the 1970s or 1990s (Szarejko-Łukaszewicz 1957; Pliński and Simm 1978; P. Margoński et al., manuscript in preparation).

The share of diatoms in the phytoplankton abundance is lower than 2% (Fig. 8.3). Typically marine species such as *Chaetoceros danicus*, *Ch. holsaticus*, and *Bacillaria paxillifera* were noted among dominants, and the mass occurrence of *Coscinodiscus commutatus* was observed in 1953 only (Szarejko-Łukaszewicz 1957). Although *C. commutatus* and *C. granii* were noted in the 1970s, they did not occur later (Pliński and Simm 1978). On the other hand, *Skeletonema subsalsa* appeared in late 1990s. As this species prefers waters of low salinity, it formed diatom blooms in the central region of the Polish part of the lagoon in 1999 along with *Melosira varians*. The biovolume of diatoms during the mass occurrence of *M. varians* in the western part of lagoon reached values comparable with those of the highest Cyanobacteria blooms close to the Russian–Polish border (P. Margoński et al., manuscript in preparation).

Although cryptophytes were not reported in the phytoplankton taxonomic composition of the Vistula Lagoon in 1953 and 1974–1975, their contribution to phytoplankton abundance was considerable in the 1999 spring season. The lack of cryptophytes might have been due to the applied sampling methodology, i.e. the use of an Apstein net. A high abundance of dinoflagellate species was observed in 1999; among the most abundant species were *Heterocapsa triquetra* and *H. rotundata* (S. Gromisz, personal communication).

As in the Polish part of the lagoon, Cyanobacteria also dominate in the Russian part (Fig. 8.3). The percentage of Cyanobacteria with respect to total phytoplankton abundance is very similar in both parts of the lagoon (P. Margoński et al., manuscript in preparation). Species from the genera *Merismopedia* and *Aphanocapsa*, and representatives of the subfamily Gomphosphaerioideae dominate in terms of abundance. Species from the *Anabaena* genus and *Aphanizomenon flos-aquae* do not occur among the dominants; nevertheless, their share in the total biovolume could be significant. At present, Cyanobacteria blooms are observed annually in the Russian part of the lagoon, while such blooms were noted only sporadically in the 1970s (Krylova and Naumenko 1992).

Significant differences in the phytoplankton composition between the Polish and Russian parts of the lagoon could be observed in the share of dinoflagellates and cryptophytes during the late 1990s. In the Polish part, under conditions of lower salinity, cryptophytes were more abundant, while in the Russian part, with higher salinity, dinoflagellates were more frequent. For example, the percentage of the dinoflagellate *Peridiniella catenata* in total phytoplankton abundance was 15% in May 1996. This species was not found in the western region of the Vistula Lagoon. Likewise, the potentially toxic dinoflagellate *Prorocentrum minimum*, a species alien to the Baltic Sea, was noted only in the Russian part (P. Margoński et al., manuscript in preparation). This species formed blooms in the Gulf of Gdańsk in the late 1990s (Witek and Pliński 2000).

8.3.4 Zooplankton

Significant changes in the zooplankton community occurred after 1915 when the Nogat River was cut off by flood-gates. This caused a three-fold decrease in the freshwater discharge into the lagoon. Until this time, the lagoon had been inhabited mainly by freshwater species (Wiktor and Wiktor 1959). As a consequence of the construction and the subsequent increase in water salinity, the zooplankton became comprised of freshwater euryhaline and brackish water species (Różańska 1967, 1972). The permanent changes in salinity created unfavourable conditions for zooplankton development, thus the majority of species were unable to reach their maximum abundance (Biernacka 1956).

Differences in zooplankton abundance, biomass, and species composition between the Russian and Polish parts of the lagoon are influenced primarily by different salinity regimes (P. Margoński et al., manuscript in preparation). Rotifers play a much more important role in the Polish part than they do in the Russian part, which is dominated by copepods (Fig. 8.4).

The highest species diversity of zooplankton was observed close to the Pregolia River estuary (84 species) and in southwestern area of the Polish part,

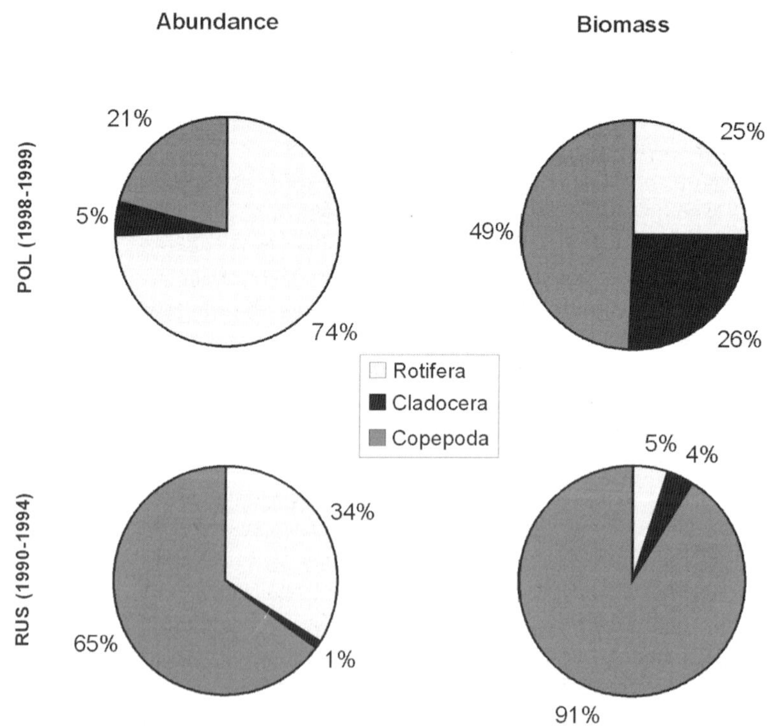

Fig. 8.4 Proportion of main zooplankton groups in the Vistula Lagoon (P. Margoński et al., manuscript in preparation). *POL* Polish part, *RUS* Russian part

while the lowest was close to the Baltiysk Strait (Adamkiewicz-Chojnacka and Majerski 1980; Krylova 1985; Naumenko 1999; Tsybaleva et al. 2000). As regards abundance and species diversity, the dominating group of zooplankton was Rotifera (Adamkiewicz-Chojnacka and Radwan 1989); as they were represented primarily by weakly euryhaline, freshwater species, their abundance and biomass decreased as salinity increased (Różańska 1963; Adamkiewicz-Chojnacka and Leśniak 1985).

The composition of dominating species in the Polish and the Russian parts of the lagoon was similar. The dominant rotifer species in the Russian part in terms of biomass are *Filinia longiseta*, *Brachionus calyciflorus*, *Keratella quadrata*, *Brachionus angularis*, and *Brachionus urceus*. In terms of abundance they are *F. longiseta*, *Keratella cochlearis*, *K. quadrata*, *Brachionus calyciflorus*, and *B. angularis* [see Chaps. 6 (Schiewer), 7 (Kruk-Dowgiałło and Szaniawska), and 9 (Gasiūnaitė et al.), this volume]. Important changes in the dominating cladoceran species have occurred in recent years: *Diaphanosoma brachyurum* has been replaced by *Cercopagis pengoi*. The invader came to dominate with regards to biomass, but in terms of abundance *D. brachyurum* still plays the most important role. Other abundant species include *Bosmina* sp., *Leptodora kindtii*, and *Podon* sp. The dominant copepods are *Eurytemora affinis*, *Acartia tonsa*, *A. bifilosa*, and *Acanthocyclops viridis* (P. Margoński et al., manuscript in preparation).

The dominating rotifer species in the Polish part of the lagoon in terms of biomass are *B. calyciflorus*, *F. longiseta*, *B. angularis*, *Euchlanis dilatata*, and in terms of abundance *F. longiseta*, *B. angularis*, *K. cochlearis*, *B. calyciflorus*, and *K. quadrata*. Among copepods the most important is *Eurytemora affinis*, the much less abundant *A. tonsa* and Cyclopoida. The dominating cladocerans include *Bosmina longirostris*, *D. brachyurum*, and *L. kindtii* (A. Krajewska, personal communication).

The seasonal dynamics of zooplankton abundance and biomass are similar in the two parts of the lagoon (Adamkiewicz-Chojnacka and Różańska 1990), and are characterised by two peaks that are usually coincident in spring (April–May) and summer (August). This is typical of eutrophic waters (Naumenko 2004).

In August 1999, a new predatory Cladocera species, *Cercopagis pengoi* (Ostroumov 1891), appeared in the Vistula Lagoon. It originated from the Ponto-Caspian basin and invaded with ballast waters. The highest abundance, at a density range of 17–533 ind m^{-3}, was observed in the Russian part near the Baltiysk Strait (Naumenko and Polunina 2000a, 2000b; Polunina 2001). In spring 2000, its abundance in the Russian part was ten-fold higher (average value 634 ind m^{-3} and maximum 3,000 ind m^{-3}) (Polunina 2001). In August 1999, it constituted 3.6% of the zooplankton biomass, while by May and June 2000 it had increased its share to 36.8% and 54.4%, respectively (P. Margoński et al., manuscript in preparation). In September 1999, it appeared in the Polish part of the lagoon at an average abundance of 220 ind m^{-3}, which accounted for 0.4% of the zooplankton biomass. In June 2000, with an average abundance of 160 ind m^{-3}, it constituted 1.3% of the total zooplankton biomass (A. Krajewska, personal communication). The zooplankton of the Vistula Lagoon is vulnerable to the salinity gradient. Moreover, as it is subject to the trophic pressure of juvenile

Baltic herring, the naturalisation of the additional predator can negatively affect zooplankton structure as well as initiate significant changes in the trophic chain (Naumenko and Polunina 2000b).

8.4 Benthic Communities

8.4.1 Macrophytes

There are six known belts in the distribution of aquatic flora in the Russian parts of the Curonian and Vistula lagoons (BERNET 2000):

- microphytes belt – includes numerous small algal species and certain tall algae, e.g. *Cladophora*;
- macrophytes belt – pronounced mainly in locations sheltered from wind and dominated by *Chara*, *Tolypellopsis*, *Stelligera*, and *Fontinalis*;
- broad-leaved pondweed belt – comprised of several pondweed species *Potamogeton* sp., water-milfoil *Myriophyllum* sp., and *Batrachium* sp; sometimes stretches a distance of 200–500 m or more inside the lagoon;
- water lily belt – the yellow water-lily *Nuphar lutea*, white water-lily *Nymphaea alba*, fringed water-lily *Nymphoides peltata*, broad-leaved pondweed *Potamogeton natans*; interrelated with reed belt and partially overlapping; particularly widespread in sheltered places at depths of 0.5–1.5 m; width usually ranges from 75 to 200 m;
- rushes belt – dense stands form at depths of 0.5–1.5 m; stretches up to widths of 200 m; at some sites rush stands spread along the shallows far into the lagoon; dominants include common club-rush *Scirpus lacustris*, *Scirpus tabernaemontana*, and in some locations bulrush *Typha latifolia* and lesser bulrush *T. angustifolia*;
- common reed *Phragmites australis* belt – in strips of 20–50 m in length; locally 150–200 m wide; normally proceeds onto the water surface; occasionally a belt of small aquatic vegetation separates it from the coast.

The structure of communities and the spatial distribution of the higher aquatic plants in the Vistula Lagoon are similar to that of the Curonian Lagoon (see Chap. 9 by Gasiunaite et al., this volume). The Vistula Lagoon is distinguishable from the Curonian Lagoon by the additional common reed and common club-rush communities that occupy a major part of the lagoon coast along the Vistula Spit (BERNET 2000).

According to Pliński et al. (1978), the basic phytosociological units in the Polish part of the lagoon included *Scirpo-Phragmitetum* and pure aggregations of *Phragmites communis*, *Myriophyllo-Nupharetum*, *Parvopotamo-Zannichellietum*, and *Potametum lucentis*. The total area covered by plants was 2,197.4 ha, which constitutes 6.86% of the total area of the Polish part of the lagoon. Approximately

40% of the plant cover there were aggregations of *Phragmites communis* and an association of *Scirpo-Phragmitetum* with *Phragmites communis*, which comprised 93% of the overall plant biomass in the lagoon. The same authors compiled a list of recorded species with their abundance and distribution characteristics. Later studies (Pliński 1995) showed that significant changes have occurred, i.e. a decrease in the area covered by vegetation and an especially drastic decrease in the surface covered by narrow-leafed cattail and plants with submerged and floating leaves. Apparently, in some areas, submerged plants and plants growing further away from the coastline were disappearing, probably due to increased turbidity limiting photosynthesis and changes in the structure of bottom sediments. Those changes, however, did not affect homogenous reed aggregations, which remained nearly unchanged.

8.4.2 Macrozoobenthos

The macrozoobenthos of the Vistula Lagoon is dominated by euryhaline organisms of marine and freshwater origin. During the late 1990s, 60 zoobenthos species and groups (Ostracoda and Chironomidae) were identified in the Russian part of the lagoon. They belonged to the following 12 higher taxonomic groups: Hydrozoa, Bryozoa, Nemertini, Turbellaria, Oligocheata, Hirudinea, Polychaeta, Insecta, Malacostraca, Bivalvia, Gastropoda, and Arachnida (Ezhova et al. 2004). The current number of species is significantly lower in comparison with observations from the early twentieth century (Riech 1928) and the 1960s (Aristova 1965, 1973). In addition to the reduction in taxonomic composition over the last 80 years, other changes have also been identified. The majority of species currently noted are of marine origin, and the role of invasive species has increased dramatically (Ezhova et al. 2004) (Fig. 8.5). The reduction in taxonomic composition has been caused by the disappearance of freshwater species and, since 1988, the active colonisation of the lagoon by *Marenzelleria neglecta* (Rudinskaya 2000a, 2000b).

At the end of 1990s the biomass of *M. neglecta* stabilised at a level (4–5 g m^{-1}) lower than that found in the 1989–1996 period (20 g m^{-1}). Polychaeta account for approximately 35% of the total biomass. The highest abundance and biomass of Polychaeta is observed close to the Baltiysk Strait. Freshwater benthic organisms occupy primarily the northeastern part of the lagoon, and, until the 1980s, they comprised 70% of the biomass, whereas that value has now dropped to 46% (Ezhova et al. 2004).

There are three main assemblages in the Russian part of the lagoon: *Macoma*, *Marenzelleria*, and *Chironomidae-Ostracoda* [see Chaps. 7 (Kruk-Dowgiałło and Szaniawska), and 9 (Gasiūnaitė et al.), this volume]. Approximately 20–25 marine and brackish water species constitute the *Macoma* assemblage, which inhabits the area of maximum salinity (5–7 psu) adjacent to the lagoon inlet. This is the richest community in terms of species composition and its average biomass ranges from 32 to 70 g m^{-2} (maximum 376 g m^{-2}). The *Marenzelleria* community is completely

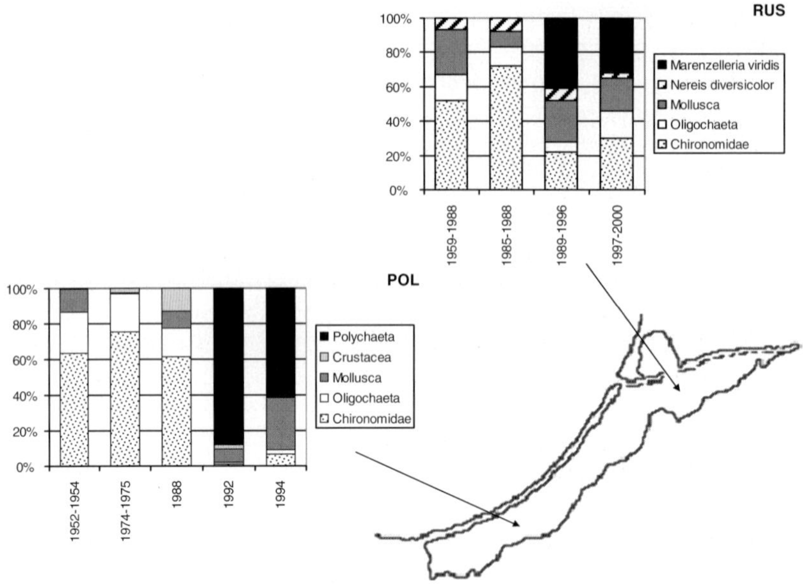

Fig. 8.5 Proportion of main macrozoobenthos groups in the Vistula Lagoon (Brzeska 1995; Cywińska and Różańska 1978; Ezhova et al. 2004; Fall 1993; Marut 1990; Żmudziński 1957). *POL* Polish part, *RUS* Russian part

new to the lagoon. It occupies most of the Russian part and consists of 14–18 groups and species (70% saline and brackish water species; 30% freshwater species). It attains an average biomass range of 8–290 g m^{-2}. Polychaete biomass decreases during the autumn–winter period. The *Chironomidae-Ostracoda* community is typical of more freshwater areas, i.e. close to the Pregolia River mouth and near the Polish–Russian border. Approximately half of the 12–14 species are freshwater species (Ezhova et al. 2004).

During the late 1990s, the zoobenthos biomass in the Russian part of the lagoon ranged from 0.2 to 452 g m^{-2}. The lowest values were calculated for the area close to the Pregolia River mouth and in the eastern part of Primorskaya Bay (0.4–10 g m^{-2}), while the highest levels were noted in the area adjacent to the lagoon inlet (70–452 g m^{-2}). The annual zoobenthos production in the Russian part of the lagoon has decreased in recent years to 23.2 kcal m^{-2} in comparison with results of 35.2 kcal m^{-2} from 1959–1988 and 40.7 kcal m^{-2} from 1989–1996. Although this trend was observed in all groups, it was extremely pronounced with freshwater species (Ezhova et al. 2004).

In the Polish part of the lagoon the vast majority of benthic organisms are freshwater taxa that are characteristic of eutrophic waters. Only polychaetes and most of the crustaceans are euryhaline, marine, or brackish fauna, such as *M. neglecta*, *Neomysis integer*, *Gammarus zaddachi*, *G. salinus*, *Rhithropanopeus tridentatus*, and the rare *Balanus improvisus*. In the open lagoon, the mean macrozoobenthos biomass did not exceed 22 g m^{-2} in 1988, whereas in the early

1990s it increased to 81–103 g m^{-2}. This increase was even more pronounced in the near-shore area where, in the same periods, it rose from 42 g m^{-2} to almost 300 g m^{-2} (Żmudziński 2000).

In the western part of the lagoon, several brackish water species have disappeared since the 1950s: e.g. *Corophium lacustre*, *C. volutator*, and *Potamopyrgus antipodarum*. A drastic decrease of the previous dominants, *Chironomus f. l. semireductus* and oligochaetes, has also been noted (Żmudziński 2000) (Fig. 8.5). Invasive species have played an important role in these changes in taxonomic composition, abundance, and biomass. These include the new gammarid species *Gammarus tigrinus*, *Pontogammarus robustoides*, and *Obesogammarus crassus* (Jażdżewski et al. 2002), and the polychaete *M. neglecta*, which has became a dominant species in nearly the entire area. In some biotopes it has exceeded 90% of the total biomass. This invasion caused a significant increase in the total zoobenthos biomass (Żmudziński 2000).

Żmudziński (1957) described three main assemblages of benthic fauna. The Chironomidae assemblage (*Chironomus f. l. semireductus*, *Procladius* sp., *Microchironomus conjugens* and *Cryptochironomus defectus*), with oligochaetes, and crustaceans (*Corophium volutator* and *Rhitropanopeus harrissi tridentatus*), was characteristic of the muddy bottom. The *Dreissena polymorpha* assemblage, with chironomids, oligochaetes, and crustaceans (*Corophium spp.* and *Rhitropanopeus harrissi tridentatus*), was abundant at the muddy and sandy bottom border. Oligochaetes predominated in the third assemblage, consisting of species associated with the shallow, sandy bottom. Other important taxa were Chironomidae, Nemertini, Bivalvia (*Dreissena polymorpha*), and Crustacea (*Rhitropanopeus harrissi tridentatus*).

8.5 Fish and Fisheries

Due to its productivity, the Vistula Lagoon provides favourable conditions for many fish species. The dominant freshwater species are accompanied by fewer brackish water species. Catch statistics indicate that 30 fish species were caught during the 1889–1920 period (Bartel et al. 1996). These included the following: seven migratory species: Atlantic salmon (*Salmo salar*), sea trout (*Salmo trutta*), sturgeon (*Acipenser sturio*), river lamprey (*Lampetra fluviatilis*), eel (*Anguilla anguilla*), vimba (*Vimba vimba*), and shad (*Alosa fallax*); four marine species: herring (*Clupea harengus*), flounder (*Pleuronectes flesus*), cod (*Gadus morhua*), and eelpout (*Zoarces viviparous*); 19 freshwater species: ruffe (*Gymnocephalus cernuus*), crucian carp (*Carassius carassius*), Prussian carp (*Carassius auratus gibelio*), bream (*Abramis brama*), tench (*Tinca tinca*), perch (*Perca fluviatilis*), roach (*Rutilus rutilus*), pikeperch (*Sander lucioperca*), pike (*Esox lucius*), rudd (*Scardinius erythrophthalmus*), carp (*Cyprinus carpio*), smelt (*Osmerus eperlanus*), burbot (*Lota lota*), sabre carp (*Pelecus cultratus*), bleak (*Alburnus alburnus*), asp (*Aspius aspius*), white bream (*Blicca bjoerkna*), gudgeon (*Gobio gobio*), and wels (*Silurus glanis*); the most important being eel, pikeperch, ruffe, bream, and Prussian carp. Over the last half century, only 20 species were exploited. Those no longer noted in catches are

Table 8.2 Monthly average environmental characteristics for the Vistula Lagoon. *BOD_5* Biochemical oxygen demand over a 5 day period, *PAR* photosynthetically active radiation

Parameter	Period	Russian/ Polish	February	March	April	May	June	July	August	September	October	November	Average	Reference
	1981–2001	Russian		0.7	0.8	0.5	0.6	0.6	0.6	0.6	0.7	0.7	0.6	Senin et al. 2004
Transparency (m)	1994–2003	Russian		0.8	0.7	0.7	0.6	0.6	0.5	0.6	0.6	0.6	0.6	A.F. Kuleshov, personal communication
	1999	Polish			0.75	0.37	0.37	0.49	0.41	0.47	0.33	0.40	0.43	Renk et al. 2001
Temperature (°C)	1981–2001	Russian	0.2	2.6	8.3	14.4	18.2	18.2	20.5	13.7	9.8	4.7	11.1	Senin et al. 2004
Salinty (psu)	1981–2001	Russian		3.4	2.6	3.1	3.6	3.8	4.3	4.2	4.5	4.3	3.8	Senin et al. 2004
Oxygen (mg O_2 l^{-1})	1981–2001	Russian		13.7	14.5	10.8	10.4	9.8	9.8	9.8	10.9	12.1	11.0	Senin et al. 2004
Mineral P (µg l^{-1})	1981–2001	Russian	103.8	24.6	15.0	16.5	45.6	48.2	96.7	39.8	47.9	45.6	48.4	Senin et al. 2004
Nitrates (µg l^{-1})	1981–2001	Russian	152.3	865.3	465.9	92.8	25.8	69.7	25.7	36.8	71.7	207.3	201.3	Senin et al. 2004
BOD_5 (mg O_2 l^{-1})	1981–2001	Russian		5.7	6.4	5.0	4.4	4.2	4.9	3.3	3.6	3.7	4.6	Senin et al. 2004
pH	1981–2001	Russian	7.9	8.4	8.8	8.6	8.7	8.6	8.6	8.4	8.4	8.2	8.5	Senin et al. 2004
Chl *a* (mg m^{-3})	1981–2001	Russian	37.3	36.0	35.8	37.7	48.7	44.8	40.1	43.0	51.8	42.3	41.7	Senin et al. 2004
	1999	Polish			35.9	30.6	26.9	54.3	86.3	36.0	49.0	37.6	43.5	Renk et al. 2001
PO_4-P (µmol l^{-1})	1999–2000	Polish	2.95	0.85	0.21	0.78	0.71	1.32	0.84	0.88	1.16	1.39	0.98	M. Zalewski, personal communication
NO_2-N (µmol l^{-1})	1999–2000	Polish	1.73	0.70	0.20	0.04	0.09	0.06	0.09	0.07	0.23	0.44	0.25	M. Zalewski, personal communication

NO_3-N (µmol l^{-1})	1999–2000	Polish	71.17	84.63	18.96	0.07	0.12	0.13	0.25	0.20	1.73	7.62	9.71	M. Zalewski, personal communication
NH_4-N (µmol l^{-1})	1999–2000	Polish	9.71	2.48	0.17	0.55	0.09	0.39	0.31	0.27	1.32	2.54	1.24	M. Zalewski, personal communication
SiO_2-Si (µmol l^{-1})	1999–2000	Polish	146.91	127.25	92.90	93.71	101.42	88.28	92.28	109.68	108.44	85.48	101.27	M. Zalewski, personal communication
Inorganic N/P ratio	1999–2000	Polish	27.5	103.53	249.27	1.52	0.41	1.29	0.81	0.49	3.20	7.04	34.36	M. Zalewski, personal communication
Light attenuation coefficient (m^{-1})	1994–1995	Russian		2.50	2.60	2.45	3.43	3.67	2.40	3.00	2.90	3.60		Rasmussen 1997; A.F. Kuleshov, personal communication
	1999	Polish			2.32	4.78	3.79	2.43	3.54	3.03	4.67	4.86		Renk et al. 2001
PAR (kJ m^{-2})	1994–1995	Russian			-	-	160 - 520	1,790 - 2,890	540 - 1,840	590 - 2,380	-	-		Rasmussen 1997; A.F. Kuleshov, personal communication
	1999	Polish			8,700	1,438–4,723	5,527–7,494	3,853–9,823	6,458–9,663	4,064–6,633	1,422–2,581	510–1,257		
Primary production (mg C m^{-2} day^{-1})	1999	Polish			645	611–674	636–678	1,473–3,925	1,566–3,870	606–857	303–329	125–146		Renk et al. 2001

Prussian carp, rudd, sturgeon, bleak, gudgeon, eelpout, wels, and shad (Bartel et al. 1996).

The total annual catch in the lagoon fluctuates strongly depending on the abundance of the herring that enters the shallow lagoon waters to spawn. The mass entrance of herring spawners is usually observed in February–March and finishes in May. The period of the beginning of the spawning migration is significantly impacted by hydrological conditions, mainly the disappearance of ice cover. With the exception of the 1957–1964 periods, the share of herring in the catches has always been considerable, exceeding 70% of the total fish biomass in most years. In the last decade (1993–2002) it has oscillated between 68% and 90%. Most herring are caught in the Russian part of the lagoon; this is due primarily to its closer proximity to the Baltiysk Strait. Variation in herring catches (2,000–13,000 tons annually) has had an impact on total catches (3,000–15,000 tons) over the last 30 years. In addition to herring, the most important fish species are eel, pikeperch, and bream. In order to optimise the exploitation of fish productivity of the lagoon, annual catches of pikeperch and bream are regulated by the Polish–Russian Commission for the Management of Fish Resources in the Vistula Lagoon.

8.6 Conclusions and Summary

After the Curonian Lagoon, the Vistula Lagoon is the second largest, semi-enclosed lagoon on the southern coast of the Baltic Sea. It is a transboundary estuarine lagoon under significant marine influence. Hydrological and meteorological factors as well as the existence of the deeper Marine Canal have a strong impact on the horizontal distribution of water quality parameters. The inlet divides this basin into two parts and limits water exchange between them to some extent. The northeastern part has higher river runoff, but more intensive mixing and higher salinity, and the southwestern part is characterised by dynamics that are generally of a low level of intensity. Wind- and thermal-driven water mixing is almost continuous. Although the lagoon is extremely shallow, there is a three-dimensional current pattern and hydrological structure, which is especially visible in areas near the inlet and river mouths. Ongoing maintenance – dredging and the deepening of the lagoon inlet – increases the marine water influx, which, although it is one of the most important positive factors affecting water quality (nutrient concentrations are much lower near the inlet), increases salinity. The high internal potential for eutrophication is caused by significant sources of nitrogen and phosphorus, which have accumulated in the sediments and are released from them. The lagoon is a remarkable source of sediment to the Baltic Sea, and currently it discharges more than it collects from the catchment area. By releasing stored sediments, the tendency in the lagoon is towards a new sediment equilibrium.

The present trophic state has been evaluated as polytrophic/eutrophic. Light and nutrient availability are the most important parameters controlling primary production. The average annual production in the Polish and the Russian parts of the lagoon was estimated at 303.8 and 180 g C m^{-2} $year^{-1}$, respectively. Phytoplankton growth is limited mainly by nitrogen. Phosphorus limitation is observed only during early spring.

Three phytoplankton groups dominate in terms of abundance: Cyanobacteria, green algae, and diatoms. Blooms of the *Anabaena* genus and *Aphanizomenon flos-aquae* have been observed from June to September in the central part of the lagoon. Cyanobacteria blooms have also been observed regularly in the Russian part, while during the 1970s they were noted only sporadically. Diatom blooms have been recorded in the western part of lagoon.

Differences in zooplankton abundance, biomass, and species composition between the Russian and the Polish parts are influenced mostly by the different salinity regimes. Rotifers play a much more important role in the Polish part as compared to the Russian part, where copepods dominate.

There are pronounced differences in macrozoobenthos communities in the two parts of the lagoon. In the Russian part, the majority of species observed are of marine origin, whereas the western part is dominated by freshwater taxa. However, the reduction of taxonomic composition as well as the increasing role of invasive species is characteristic of both parts. The Polychaeta *Marenzelleria neglecta* has become a dominant species in almost the entire area, in some places exceeding 90% of the total macrozoobenthos biomass.

Due to its productivity, the Vistula Lagoon provides favourable conditions for many fish species. The dominant freshwater species are accompanied by fewer brackish water species. Variation in herring catches (2,000–13,000 tons annually) has had an impact on total catches (3,000–15,000 tons) over the last 30 years. In addition to herring, the most important fish species are eel, pikeperch, and bream.

Acknowledgements The authors would like to thank all of the researchers cited for the advances they have made in the investigations of the Vistula Lagoon ecosystem and for the results reviewed in this chapter. International projects like the Vistula Lagoon Project (Denmark-Poland-Russia) 1994–1997 (Rassmussen 1997) and MANTRA-East (EU 5FP) 2002–2004 (P. Margoński et al., manuscript in preparation) deserve recognition for their significant contributions to recent investigations of the Vistula Lagoon.

References

Adamkiewicz-Chojnacka B, Leśniak D (1985) Evaluation of biomass of the rotifers (*Rotatoria*) in salty water of the Vistula Bay in 1975, 1977 and 1978 (in Polish). Studia i Materiały Oceanologiczne PAN, Biologia Morza 46:273–308

Adamkiewicz-Chojnacka B, Majerski J (1980) Zooplankton of the Vistula Bay during summer season (in Polish). Zesz Nauk ART Olsztynie 10:85–94

Adamkiewicz-Chojnacka B, Radwan S (1989) Vertical distribution of *Rotatoria* biomass in brackish waters of the Vistula Lagoon (in Polish). Pol Arch Hydrobiol 36:105–111

Adamkiewicz-Chojnacka B, Różańska Z (1990) Seasonal variations of the biomass of *Copepoda* and *Rotatoria*: permanent components of the zooplankton in the Vistula Lagoon. Limnologica (Berlin) 20:71–73

Aleksandrov SV (2004) Primary production in the Vistula Lagoon. In: Alimov AF, Ivanova MB (eds) Regularities of hydrobiological regime in water bodies of different types (in Russian). Moscow: "Scientific World" pp 139–141

Andrulewicz E, Chubarenko B (2004) Case study: Vistula Lagoon (Poland/Russia) – transboundary management problems and an example of modelling for decision making. In: Gonenc IE, Wolflin J (eds) Coastal lagoons: ecosystem processes and modelling for sustainable use and developments. CRC, Boca Raton, pp 423–439

Andrulewicz E, Witek Z (2002) Anthropogenic pressure and environmental effects on the Gulf of Gdansk: recent management efforts. In: Schernewski G, Schiewer U (eds) Baltic coastal ecosystems. Springer, Berlin, pp 119–139

Andrulewicz E, Chubarenko B, Żmudziński L (1994) Vistula Lagoon – a troubled region with a great potential. WWF Baltic Bull 1:16–21

Anon (1996) Prioritizing hot spot remediations in the Vistula Lagoon catchment. Environmental Assessment and Planning for the Polish and Kaliningrad Parts of the Lagoon. Ministry of Environment, Denmark

Aristova GI (1965) Benthos of Curonian and Vistula lagoons (in Russian). Atlantic Institute for Fishery and Oceanography, Kaliningrad, pp 40–49

Aristova GI (1973) Benthos of Curonian and Vistula lagoons of the Baltic Sea and its significance in fish food (in Russian). PhD Thesis. Kaliningrad Technical Institute for Fishery and Fish Industry

Bartel R, Wilkońska H, Borowski W (1996) Changes of the ichthyofauna in the Vistula Lagoon. In: Volskis R (ed) Proceedings of the International Meeting on Baltic Network of Biodiversity and Productivity of Selected Species in Coastal Ecosystems, Nida, (Lithuania), 4–8 October 1995, pp 12–24

BERNET (2000) Wetland management. Bernet Theme Report, Attachment. Fyn Country, Odense, Denmark

Biernacka I (1956) Przyczynek do znajomości pierwotniaków Zalewu Wiślanego (in Polish). Pol Arch Hydrobiol 3:43–68

Blazchishin AI (1998) Zur Geoökologie des Kurischen und des Frischen Haffs. Schriftenr Geowiss 3:39–57

Bogdanov NA, Vorontsov AA, Morozova LN (2004) Tendencies of chemical pollution and dynamics of the Kaliningrad bay (in Russian). Russ Water Resour 31:576–590

Brzeska A (1995) Zoobentos of the Vistula Lagoon (in Polish). MSc Thesis. Wyższa Szkoła Pedagogiczna w Słupsku. Zakład Ekologii i Ochrony Środowiska. Słupsk

Carlson RE (1977) A trophic state index for lakes. Limnol Oceanogr 22:361–369

Chechko VA (2002) Analysis of spatial–temporal variability of suspended matter of the Kaliningrad Bight of the Baltic Sea (in Russian). Russ Water Resour 4:425–432

Chechko VA (2004) Characteristic features of sedimentation in estuaries of lagoon type (by the example of Vistula and Curonian lagoons of the Baltic Sea). In: Alimov AF, Ivanova MB (eds) Regularities of hydro biological regime in water bodies of different types (in Russian). Moscow: "Scientific World", pp 172–175

Chechko VA, Blazchishin AI (2002) Bottom deposits of the Vistula Lagoon of the Baltic Sea. The Baltica: an International Yearbook on Geology, Geomorphology and Palaeogeography of the Baltic Sea 15:13–22

Chubarenko B (2007) Shared watersheds in the South-Eastern Baltic. In: Chubarenko B (ed) Transboundary waters and basins in the south-eastern Baltic. Terra-Baltica, Kaliningrad, pp 5–12

Chubarenko BV, Chubarenko IP (2001) New way of natural geomorphological evolution of the Vistula Lagoon due to crucial artificial influence. In: Emeliayanov EM (ed) Geology of the Gdansk Basin, Baltic Sea. Yantarny Skaz, Kaliningrad, pp 372–375

Chubarenko BV, Chubarenko IP (2003) The transport of Baltic water along the deep channel in the Gulf of Kaliningrad and its influence on fields of salinity and suspended solids. In: Dahlin H, Dybern B, Petersson S (eds) ICES Cooperative research report, N 257

Chubarenko BV, Chubarenko IP, Kuleshov AF (1998a) The results of 1994–1997 investigations of hydrological structure and transparency of water in the Russian part of the Vistula lagoon. In: Proceedings of the Symposium on Freshwater fish and the herring population in the coastal lagoons: environment and fisheries. 6–7 May 1998, Gdynia, Poland. Sea Fishery Institute, Gdynia, pp 45–59

Chubarenko BV, Kuleshov AF, ChechkoVA (1998b) Field study of spatial–temporal variability of suspended substances and water transparency in Russian part of Vistula lagoon. In: Monographs in system ecology, vol 2. Proceedings of the second international workshop on interdisciplinary approaches in ecology. System research in Ecology: linking watershed, riverine and marine processes. Klaipeda, Lithuania, 25–27 August 1997, pp 12–17

Chubarenko B, Lund-Hansen LCh, Beloshitskii A (2002) Comparative analysis of potential wind-wave impact on bottom sediments in the Vistula and Curonian lagoons. The Baltica: an International Yearbook on Geology, Geomorphology and Palaeogeography of the Baltic Sea 15:30–39

Chubarenko BV, Koutitonski VG, Neves R, Umgiesser G (2004) Modelling concept. In: Gonenc IE, Wolflin J (eds) Coastal lagoons: ecosystem processes and modelling for sustainable use and developments. CRC, Boca Raton, pp 231–306

Chubarenko IP, Chubarenko BV (2002) General water dynamics of the Vistula Lagoon. Environ Chem Phys 24:213–217

Chubarenko IP, Tchepikova IS (2001) Numerical modelling analysis of artificial contribution to salinity increase into the Vistula Lagoon (Baltic Sea). Int J Ecol Modelling Syst Ecol 138:87–100

Chubarenko IP, Esiukova EE, Hutter K (2004) Impact of the ice formation on seasonal salinity increasing the Vistula Lagoon. In: Proceedings of the VI scientific workshop on high resolution operational model of the Baltic Sea. 8–10 September 2003, Saint Petersburg, Russia. MORZASCHITA and Russian State Hydrometeorological University, St. Petersburg, pp 58–68

Cywińska A, Różańska Z (1978) Zoobenthos of the Vistula Lagoon (in Polish). Studia i Materiały Oceanologiczne, Biologia Morza 21:145–160

Ezhova EE, Chubarenko BV, Blazhchishin AI, Chubarenko IP (1999) Sediment–water interactions in the Vistula Lagoon – some preliminary results. In: Heiskanen AS, Lundsgaard C, Registadt M, Olli K, Floderus S (eds) Sedimentation and recycling in aquatic ecosystems. Finn Environ 263:78–83

Ezhova EE, Rudinskaya LV, Pavlenko-Lyatun MV (2004) Macrozoobenthos of the Vistula Lagoon (in Russian). In: Alimov AF, Ivanova MB (eds) Regularities of hydrobiological regime in water bodies of different types. Moscow: "Scientific World", pp 146–164

Ezhova EE, Żmudziński L, Maciejewska K (2005) Long-term trends in the macrozoobenthos of the Vistula Lagoon, southern Baltic Sea. Species composition and biomass distribution. Bull Sea Fisheries Inst 1:55–73

Fall A (1993) Macrozoobenthos of the Vistula Lagoon (in Polish). M.Sc. Thesis. Uniwersytet Gdański, Zakład Genetyki, Gdańsk

Jażdżewski K, Konopacka A, Grabowski M (2002) Four Ponto-Caspian and one American gammarid species (*Crustacea, Amphipoda*) invading Polish waters. Contrib Zool 71:115–122

Kopiec J (2002) The Vistula Lagoon (in Polish). Report of the surface waters research carried out in the Polish part of the Vistula Lagoon in 2001. Inspection for Environmental Protection, Elblg

Kratzer CR, Brezonik PL (1981) A Carlson-type trophic state index for nitrogen in Florida lakes. Water Res Bull 17:713–715

Krylova OI (1985) Functioning of plankton and benthos of the Curonian and the Vistula Lagoons of the Baltic Sea in relation to their ecological differences (in Russian). Atlant NIRO, Kaliningrad, Dep. in CNIREIRH, N 714-RX-D-85, 21.10.1985

Krylova OI, Naumenko EN (1992) Phytoplankton and primary production of the Vistula Bay (in Russian). Ecological fisheries research in the Vistula Bay of the Baltic Sea. Trudy AtlantNIRO, pp 14–33

Kwiatkowski J (1996) Evaluation of possibilities of ecological quality improvement in the Vistula Lagoon (in Polish). GEOMOR, Sopot

Kwiatkowski J, Chubarenko B, Rasmussen EK (1996) The Vistula Lagoon as a trap of nutrient pollution of the Baltic Sea. In: Proceedings of the Conference: Functioning of Coastal Ecosystem in Various Geographical Regions, 5–7 September 1996, Gdańsk

Kwiatkowski J, Rasmussen EK, Ezhova EE, Chubarenko BV (1997) The eutrophication model of the Vistula Lagoon. Oceanological Stud 26:5–33

Lazarenko NN, Majewski A (eds) (1971) Hydrometeorological regime of the Vistula Lagoon (in Russian). Hydrometeoizdat, Leningrad

Maciejowska M, Macur SW (1978) Bacteriological analysis of the Vistula Lagoon waters from the point of view of hydrocarbon pollution (in Polish). Stud Mater Oceanol KBM PAN 21:37–52

Margoński P, Bielecka M, Horbowa K, Gromisz S, Witek Z, Zalewski M, Wielgat M, Naumenko EN, Senin YM, Semenova SN, Warzocha J (in prep.) Nutrient loads and their effect: The problems from a natural science perspective. In: Bielecka M, Lewandowski A, Margoński P (eds) Functioning of the Vistula Lagoon. Sea Fisheries Institute, Gdynia

Marut S (1990) Macrozoobenthos of the Vistula Lagoon (in Polish). MScThesis. Wyższa Szkoła Pedagogiczna w Słupsku, Instytut Biologii, Słupsk

Naumenko EN (1999) Species composition, seasonal and long-term dynamics of zooplankton abundance and biomass in the Vistula Lagoon of the Baltic Sea. In: Proceedings of the Symposium on freshwater fish and the herring populations in the coastal lagoons – environment and fisheries. 6–7 May 1998, Gdynia, Poland, pp 161–173

Naumenko EN (2004) Zooplankton of the Vistula Lagoon (in Russian). In: Alimov AF, Ivanova MB (eds) Regularities of hydrobiological regime in water bodies of different types. Moscow "Scientific World", pp 141–146

Naumenko EN, Polunina YuYu (2000a) *Cercopagis pengoi* (Ostroumov, 1891) (*Cladocera, Podonidae*) a new species introduced to the Vistula Lagoon (Baltic Sea) (in Russian). In: Introduced species in the European Seas of Russia. Kolskij Nauchnyj Tsentr RAN, Apatity, pp 121–129

Naumenko EN, Polunina YuYu (2000b) New *Cladocera* species *Cercopagis pengoi* (Ostroumov, 1891) (*Crustacea*) in the Vistula Lagoon on the Baltic sea. ICES CM 2000/U 14:8

Niedoszytko K, Wiktor K (1978) Primary production in the Vistula Lagoon (in Polish). Stud Mater Oceanol KBM PAN, 21:95–122

OECD (1982) Eutrophication of waters. Monitoring, assessment and control. OECD, Paris

Pliński M (1995) Vascular plants of the northern part of the Vistula Lagoon. Bull Mar Inst 22:81–87

Pliński M, Simm A (1978) Seasonal fluctuations in the composition, distribution and quantity of phytoplankton in the Vistula Lagoon in 1974 and 1975 (in Polish). Studia i Materiały Oceanologiczne, Biologia Morza 21:53–80

Pliński M, Kreńska B, Wnorowski T (1978) Floristic relationships and biomass of vascular plants of the Vistula Lagoon (in Polish). Studia i Materiały Oceanologiczne, Biologia Morza 21:161–196

Polunina YuYu (2001) Population biology of the Ponto-Caspian invader *Cercopagis pengoi* (*Cladocera*) in the Vistula Lagoon of the Baltic Sea. Baltic Sea Science Congress 2001. Past, present and future – a joint venture. Stockholm University, Sweden, 25–29 November, Abstract, Poster No. 125:276

Rasmussen EK (1997) Summary report on International project "Prioritising hot spot remediation in the Vistula Lagoon catchment: environmental assessment and planning for the Polish and Kaliningrad parts of the Lagoon", Water Quality Institute (DK); Danish Hydraulic Institute (DK); GEOMOR (Poland); P.P. Shirshov Institute of Oceanology, Atlantic Branch (Russia), 2nd edn

Renk H, Ochocki S, Zalewski M, Chmielowski H (2001) Environmental factors controlling primary production in the Polish part of the Vistula Lagoon. Bull Sea Fisheries Inst 1:77–95

Riech F (1928) Beiträge zur Kenntnis der litoralen Lebensgemeinschaften in der poly- und mesohalinen Region des Frischen Haffes. Schr Phys Ökonom Ges 65:32–47

Różańska Z (1963) The Vistula Firth zooplankton (in Polish). Zesz Nauk WSR w Olsztynie 16:41–57

Różańska Z (1967) Zooplankton i krótka charakterystyka hydrologiczna Zalewu Wiślanego i Zalewu Szczecińskiego w świetle badań XX- lecia (in Polish). Stud Mater Mor Inst Ryb Gdynia Series A 4:27–37

Różańska Z (1972) Zooplankton zalewów przybałtyckich (in Polish). Studia i Materiały Oceanologiczne 3:141–148

Rudinskaya LV (1999) Water salinity impact upon bottom invertebrates structure in the Vistula Lagoon. In: Proceedings of symposium on freshwater fish and the herring populations in the coastal lagoons – environment and fisheries. 6–7 May 1998, Gdynia, Poland, pp 202–219

Rudinskaya LV (2000a) The influence of water salinity on the structure of benthic invertebrates in the Vistula Lagoon of the Baltic Sea (in Russian). Hydrobiological Studies in the basin of the Atlantic Ocean. V.1.: Freshwater hydrobiology. Atlantic Institute for Fishery and Oceanography, Kaliningrad, pp 50–58

Rudinskaya LV (2000b) Dynamics of biomass and abundance of the *Marenzelleria viridis* and its influence upon the structure of benthic community in the Vistula Lagoon. In: Invading species in the European Seas of Russia. Apatity: KNC RAN, pp 193–202

Senin YuM, Smyslov VA, Khlopnikov MM (2004) General characteristics of the Vistula Bay. In: Alimov AF, Ivanova MB (eds) Regularities of hydrobiological regime in water bodies of different types. Moscow: "Scientific World": pp 17–18

Szarejko-Łukaszewicz D (1957) Qualitative investigations of phytoplankton of Firth of Vistula in 1953 (in Polish). Prace MIR, Gdynia 9:439–451

Szymkiewicz R (1992) A mathematical model of storm surge in the Vistula Lagoon, Poland. Coastal Eng 16:181–203

Trzosińska A (1975) Suspended particulate matter (in Polish). In: Lazarenko NN, Majewski A (eds) The hydrometeorological system of the Vistula Lagoon. IMGW, Wydawnictwo Komunikacji i Łączności, Warsaw

Tsybaleva GA, Andronov VN, Polunina YuYu (2000) Structural characteristics of zooplankton from the Pregola River–Kaliningrad Sea Channel system. Proceedings of the International Conference KGTU. Kaliningrad, Part 1, pp 94–96

Wiktor J, Wiktor K (1959) Study on biology of the Baltic Sea coastal lagoons in Poland (in Polish). Przyroda Polski Zachodniej 3:7–28

Witek B, Pliński M (2000) The first recorded bloom of *Prorocentrum minimum* (Pavillard) in the coastal zone of the Gulf of Gdańsk. Oceanologia 42:29–36

Zalewski M, Witek Z, Wielgat M, Margoński P, Senin YM (in prep.) Ecological modeling of the Vistula Lagoon. In: Bielecka M, Lewandowski A, Margoński P (eds) Functioning of the Vistula Lagoon. Sea Fisheries Institute, Gdynia

Żmudziński L (1957) The Firth of Vistula zoobenthos. Reports of the Sea Fisheries Institute in Gdynia 9:453–500

Żmudziński L (2000) Long-term changes in macrozoobenthos of the Vistula Lagoon. Sea and Environment (Jura ir Aplinka), Klaipeda University, 1:46–50

Chapter 9
The Curonian Lagoon

Z.R. Gasiūnaitė, D. Daunys, S. Olenin, and A. Razinkovas

9.1 Environmental Characteristics

Curonian lagoon (Fig. 9.1) is a large (1,584 km^2) (Žaromskis 1996) coastal water body connected to the south-eastern Baltic Sea by a narrow (0.4–1.1 km) strait (Klaipėda port area). The mean depth of the lagoon is approximately 3.8 m (Žaromskis 1996). The strait is ca. 11 km long, with artificially deepened waterways down to 14 m depth. The eastern side of the lagoon (mainland shore) represents a shallow plain, gently sloping westward down to a depth of 1–2 m, whereas the western side (the Curonian Spit shore) reaches up to 4 m depth (Table 9.1).

The lagoon water circulation is generally determined by the wind and by the Nemunas river discharge [see Chaps. 3, 4 (Schiewer), and 8 (Chubarenko and Margoński), this volume]. From the hydrological point of view, the Curonian lagoon can be divided into two parts: a northern part influenced by the Nemunas river, and a southern part where the wind is the main driving factor (Razinkovas et al. 2005). Approximately 23 km^3 fresh water gained in the form of river runoff pass the lagoon annually. More than 40% of this amount is discharged into the sea during the spring months, whereas 5 km^3 incoming seawater are mixed in the lagoon mostly in the autumn months (Pustelnikovas 1998). Episodic inflows of seawater cause irregular rapid (hours–days) salinity fluctuations in the range of 0–7 psu in the northern part of the lagoon (Daunys 2001). Seawater inflows of 1–6 days are most common (Gasiūnaitė 2000); the seawater intrusions are usually restricted to the northern part of the lagoon, rarely propagating more than ca. 40 km.

In the southern part of the lagoon, the current regime is driven mainly by the wind which, depending on its direction and speed creates different circulation sub-systems. In most cases the system evolves a dominant gyre, with anti-clockwise (wind from west) or clockwise (wind from south-east) direction, and some smaller gyres. In a south-western wind, the circulation pattern is characterised by a two-gyre system (Razinkovas et al. 2005).

Due to Nemunas water input and water level differences between the sea and the lagoon, a south–north current moves from the Nemunas delta to the north towards the Klaipėda strait (Žaromskis 1996). This current is usually more evident during the spring flood period, when the Klaipėda strait is characterised by a unidirectional

U. Schiewer (ed.), *Ecology of Baltic Coastal Waters. Ecological Studies 197.*

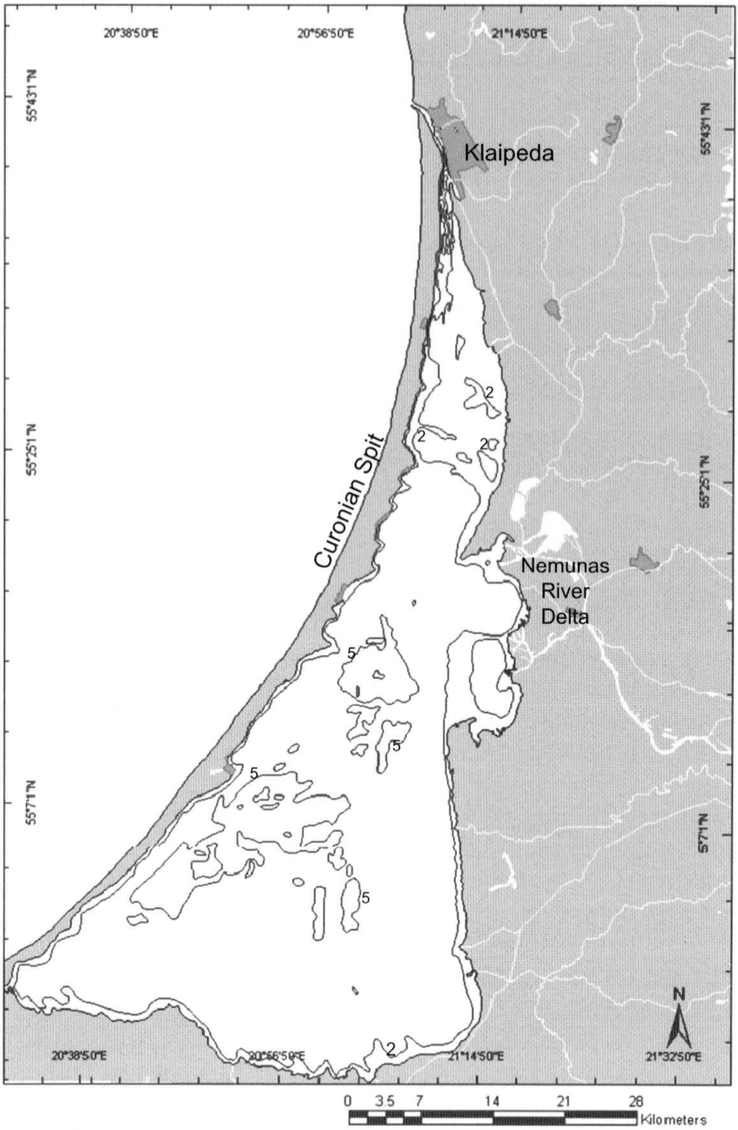

Fig. 9.1 Map of the Curonian lagoon

flow of fresh water from the lagoon to the Baltic Sea with a current speed of 1.5–2 m s^{-1} (Pustelnikovas 1998).

The seasonal water temperature dynamics are typical of shallow temperate water bodies, with an annual amplitude of up to 25–29°C (Žaromskis 1996). Temperature stratification of the water column is weak and unstable. The temperature of the

Table 9.1 Main characteristics of the Curonian Lagoon. *dw* Dry weight, *ww* wet weight, *Chl a* chlorophyll *a*, *POC* particulate organic carbon, *DOC* dissolved organic carbon, *DIN* dissolved inorganic nitrogen, *DIP* dissolved inorganic phosphorus

a. Climate and hydrology	
Air temperature (°C)	Annual average: 6.5; monthly averages: –2.8–16.8[a]
Water temperature (°C)	Monthly averages: 0.1–19.3; up to 24–25 in summer[b]
Ice covering (days a^{-1})	110 (min: 12; max: 169)[b]
Annual global radiation (kW h m^{-2})	0.07–0.53[c]
Average precipitation (mm a^{-1})	740[a]
Mean freshwater inflow ($10^6 m^3 a^{-1}$)	23,100[d]
Mean inflow ($10^6 m^3 a^{-1}$) (from Baltic)	5,100[d]
Mean total outflow ($10^6 m^3 a^{-1}$)	26,500[d]
Residence time (days)	81[c]
b. Morphometry and sediments	
Area (km^2)	1,584[e]
Volume ($10^6 m^3$)	6,000[e]
Mean depth (m)	3.8[b]
Maximum depth (m)	5.8 in the southern part (up to 14 in artificially deepened Klaipeda port area)[e]
Catchment area (km^2)	100,458[e]
Area/catchment relation	0.02[e]
Organic matter content (% dw)	< 1% in fine sand, > 5–15% in accumulation areas[e]
Sediment fraction < 63 µm (% dw)	Up to 4% in fine sand; 40–95% in accumulation areas[q]
Sediment distribution	Fine sand predominate, muddy areas in sheltered or deeper parts[d]
c. Pelagic biological components	
Secchi depth (m)	0.3–2.2[h]
Bacteria ($10^6 ml^{-1}$)	0.8–39.3[d]
Phytoplankton biomass (µg Chl *a* l^{-1})	45[9]; spring max: 302, autumn max: 458[g]
Dominating phytoplankton species	Brackish water: *Heterocapsa rotundata, Achnanthes taeniata, Thalassiosira levanderi, Skeletonema costatum, Heterocapsa triquerta, Nodularia spumingena, Coscinodiscus granii, Skeletonema costatum* and *Teleaulax* spp.[j] Fresh water: *Stephanodiscus hantzschii, Aphanizomenon flos-aquae* and *Microcystis aeruginosa*[j]
Mesozooplankton	
Biomass (mg l^{-1})	< 8 mg l^{-1} (ww)[e]
Dominating species	Brackish water: *Acartia bifilosa, Temora longicornis, Eurytemora hirundoides, Podon polyphemoides and Evadne nordmanni*[e,k] Fresh water: *Bosmina* spp., *Chydorus sphaericus, Cyclops strenuus, Acanthocyclops* sp., *Daphnia* spp., *Diaphanosoma brachyurum, Leptodora kindti, Eudiaptomus graciloides and Mesocyclops leuckarti*[e,k]

(continued)

Table 9.1 (continued)

c. **Pelagic biological components**	
Fish	
Number of species	57
Dominating species	Roach (*Rutilus rutilus*), perch (*Perca fluviatilis*), redeye (*Scardinius erythrophalmus*), white bream (*Blicca bjoerkna*) and common bream (*Abramis brama*)[a]
POC (mg C l^{-1})	0.4–2.4[c]
DOC (mg l^{-1})	2.2–4.5[c]
Fish catches (t a^{-1})	700 (only Lithuanian part)[m]
d. **Benthic biological components**	
Macrophytes	
Biomass (g m^{-2})	No data
Number of species	18 submerged macrophyte species, no data on total species number[n]
Dominating species	*Phragmites australis*, *Schoenoplectus lacustris*, *Potamogeton perfoliatus*, *Potamogeton pectinatus*[o]
Macrozoobenthos	
Biomass (g m^{-2})	5.2 (Oligochaet–Chironomid-dominated community); 1,140 (*Dreissena*-dominated community)
Number of species	85 species and higher taxa
Dominating species	*Marenzelleria neglecta*, *Dreissena polymorpha*, *Valvata*, *Bithynia* spp., *Radix* spp., *Potamopyrgus antipodarum*, *Theodoxus fluviatilis*, *Chaetogammarus*, *Pontogammarus*[p]
e. **Water chemistry, trophic status and pollution**	
Salinity (psu)	0–8[e]
pH	8.1–9.2[f]
Oxygen saturation (%)	20–115[b]
Total nitrogen (µmol l^{-1})	119.7 (min: 15; max 453.2); winter max: 210.6[g]
DIN (µmol l^{-1})	23.5 (min: 0.8; max: 127) [g]
Total phosphorus (µmol l^{-1})	3.5 (min: 0; max: 21.6); winter max: 7.7[g]
DIP (µmol l^{-1})	1.0 (min: 0; max: 6.5); winter max: 4.1[g]
Annual nitrogen input (t a^{-1})	33,000–64,000[r]
Annual phosphorus input (t a^{-1})	1,200–4,000[b]
Limitation of PP and period	P-limitation in spring; N-limitation in summer[i]
Main pollutants	N, P[b]
Trophic level	Eutrophic[j]

[a]Bukantis (1994); [b]Žaromskis (1996); [c]A. Razinkovas (unpublished); [d]Pustelnikovas (1998); [e]Gasiūnaitė (2000); [f]Pilkaitytė (2003); [g]CHARM database; [h]Razinkovas and Pilkaitytė (2002); [i]Pilkaitytė and Razinkovas (2006); [j]Olenina (1997); [k]Gasiūnaitė and Razinkovas (2004); [l]Repečka et al. (1996); [m]Virbickas and Repečka (1996); [n]Z. Sinkevičienė (unpublished); [o]Plokštienė (2002); [p]Daunys (2001); [q]Gulbinskas (1995); [r]Četkauskaitė et al. (2001)

near-bottom water layer is lower than that in the upper layer by 1–2°C (Pustelnikovas 1998). The Klaipėda strait is always ice free, while in the rest of the lagoon ice cover is present for 110 days on average (Žaromskis 1996).

The oxygen concentration fluctuates spatially and temporally (both diurnally and seasonally) (Jurevičius 1959). Low concentrations down to 1.8 ml l^{-1} were found during the ice cover period; local anoxia may take place in summer at night.

Nutrient concentration dynamics are typical of temperate and boreal transitional waters with strong riverine inputs. The highest concentrations of nutrients are observed in winter and early spring. The concentration of phosphate decreases rapidly in April and starts to increase in early summer due to decomposition of organic material. The nitrogen concentration can decrease to analytical zero in May; nitrate concentration tends to increase from midsummer, whereas ammonium concentrations have no pronounced seasonal pattern. The silica concentration is lowest during spring after the diatom bloom. It remains low throughout the summer and starts to increase again in early autumn (Razinkovas and Pilkaitytė 2002).

The main bottom sediments in the lagoon are sand and silt. The northern part of the lagoon is transitory with regards to sediment transport, while the central part is most heterogeneous in respect to bottom geomorphology and sediment type. Here, the prevailing type is fine sand, mixed with gravel and pebbles, peat and moraine. Muddy bottoms occur in local depressions in the deeper western part of the Lagoon along the Curonian Spit (Pustelnikovas 1998).

9.2 Historical Overview

Investigations into the Curonian lagoon plankton communities started nearly 80 years ago with species inventories (Szidat 1926; Willer 1935), followed by analysis of selected species distribution (Schmidt-Ries 1940), dynamics of abundance and biomass as well as production (Kiselytė 1959; Ūselytė 1959). In the 1980s, studies also focused on the impact of anthropogenic pollution on the phytoplankton, plankton crustaceans, rotifers, protozoans and mollusc larvae (e.g. Mažeikaitė 1978a, 1978b; Jankevičius and Baranauskienė 1978; Jankavičiūtė; and Jankevičius 1978; Jankevič ius et al. 1979; Klimašauskienė 1979).

Recent studies have been directed towards both the descriptive and functional characteristics of the lagoon plankton communities (e.g. Olenina 1997; Gasiūnaitė and Olenina 1998; Olenina 1998; Gasiūnaitė; and Didžiulis 2000; Gasiūnaitė 2000; Gasiūnaitė and Razinkovas 2000, 2002, 2004; Naumenko 1996; Razinkovas and Pilkaitytė 2002; Žiliukienė 2003; Pilkaitytė 2004; Pilkaitytė and Razinkovas 2006). Bacterioplankton and protozooplankton in the Curonian lagoon are still poorly studied.

Continuous ichthyological investigations in the lagoon were started in 1949. The first detailed list of lamprey and fish species (Maniukas 1959) was further supplemented by additional information on species ecology, migration and production

(Bružinskienė and Virbickas 1988; Repečka and Milerienė 1991; Gaigalas et al. 1992; Repečka 1995, 1999; Gaigalas 2001; Repečka 2003a). The monitoring of ichthyoplankton began only recently (Repečka 1996; Repečka et al. 1996).

Studies on bottom macrofauna diversity and biology in the Curonian lagoon started in the early 1920s (Szidat 1926; Willer 1931; Lundbeck 1935). The most comprehensive inventory of the Curonian Lagoon bottom macrofauna was carried out in the 1950s (Gasiūnas 1959). Several later studies focused on the ecology of introduced species (e.g. Razinkov 1990; Daunys et al. 2000), evaluation of food sources for commercial fishes (Bubinas 1983; Lazauskienė et al. 1996), and the accumulation of heavy metals and cytogenetic damage in benthic animals (e.g. Jagminienė 1995; Baršienė 2002). The main structural characteristics of benthic communities and trophic groups were also investigated (Aristova 1965, 1971; Bubinas 1983; Olenin 1987, 1997; Rudinskaya 1994). Salinity was analysed as an important factor for distribution of some benthic species (Gasiūnas 1959; Bubinas 1983; Olenin 1987; Daunys 2001) and their reproductive success (Gasiūnas1959; Daunys et al. 2000). The role of sediment organic carbon was also shown to be important for the macrofauna in the Lagoon (Daunys 2001). The ecological effects of invasive alien species were described by Olenin and Leppäkoski (1999), and a quantitative evaluation at the ecosystem scale was carried out (Daunys et al. 2006). Recently, the diversity of selected bottom macrofauna groups was revised (Zettler et al. 2005) and long-term changes evaluated (M.L. Zettler and D. Daunys, manuscript submitted).

9.3 Phytoplankton

In total, 438 phytoplankton species [see Chaps. 3, 4 (Schiewer) and 5 (Radziejewska and Schernewski), this volume] are found in the lagoon (Olenina 1997; Gasiūnaitė et al. 2005). According to phytoplankton community characteristics, three areas could be distinguished: (1) salinity gradient in the northern part of the lagoon, (2) the freshwater open area in central and southern parts, and (3) the river–lagoon transition.

The overall phytoplankton abundance in the Curonian lagoon decreases markedly with increasing salinity (Fig. 9.2). A brackish-water phytoplankton community (see Chap. 4 by Schiewer, this volume) enters the lagoon during inflows of Baltic Sea water and is dominated by *Heterocapsa rotundata*, *Achnanthes taeniata* and *Thalassiosira levanderi* in spring, *Skeletonema costatum*, *Heterocapsa triquerta* and *Nodularia spumingena* in summer, and *Coscinodiscus granii*, *Skeletonema costatum* and *Teleaulax* spp. in autumn (Olenina 1997).

The freshwater phytoplankton assemblage structure and dynamics are typical of eutrophic waters (see Chap. 5 by Radziejewska and Schernewski, this volume). Phytoplankton succession starts with a prevalence of diatoms from January until June, and continues with dominance of Cyanobacteria or co-dominance of Cyanobacteria and diatoms, until the biomass peak is reached in

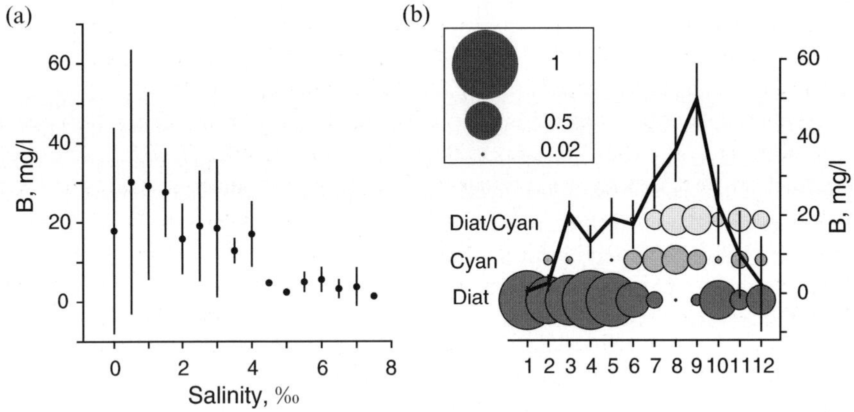

Fig. 9.2 **a** Phytoplankton biomass (*B*; mean ±1 SD) along the salinity gradient (data from CHARM database). **b** The sequence of freshwater phytoplankton seasonal succession (modified from Gasiūnaitė et al. 2005). The size of the *spheres* represents the monthly ratio of number of samples, representing particular seasonal assemblage and total number of samples (*inset*). *Solid line* Monthly means (± 1 SD) of the total phytoplankton biomass [mg wet weight (ww) l^{-1}; right-hand *y*-axis]

August–September (Fig. 9.2). Diatoms are again relatively abundant in late autumn. *Stephanodiscus hantzschii* is the dominant species both in spring and late autumn–winter; the potentially toxic species *Aphanizomenon flos-aquae* and *Microcystis aeruginosa* are responsible for the biomass peak in summer (Olenina 1998; Gasiūnaitė, et al. 2005, Pilkaitytė and Razinkovas 2006). Nitrogen-fixing Cyanobacteria are present in the phytoplankton community throughout the year, but the maximum concentrations are observed when favourable weather conditions are coupled with low inorganic N/P ratios (Pilkaitytė and Razinkovas 2006).

Bacillariophyceae (mostly *Centrales* and *Penales*) dominate in the Nemunas river low reach phytoplankton community from April to July, Chlorophyceae are most abundant during May–September; dominance of Cyanophyceae is characteristic only of August. The variation of the Cyanophyceae/Bacillariophyceae ratio throughout the seasons is markedly lower in the river (0.1–0.8) than in the lagoon (0.01–>40). No important changes in phytoplankton abundance were found in the medium-scale (ca. 7 km) river lagoon gradient; however, a quantitative gradient could be observed on a larger spatial scale, especially during Cyanobacteria blooms. Structural changes along the gradient are well expressed during the Cyanobacteria-dominated period in the lagoon (June–October): a high Cyanophyceae/Bacillariophyceae ratio was typical for the central part of the lagoon, whereas a low ratio is characteristic of Nemunas river distributaries and lagoon stations situated closer to the river mouths (Z.R. Gasiūnaitė et al., manuscript in prep.).

9.4 Zooplankton

As with phytoplankton, three areas important for zooplankton community characteristics could be distinguished according to the level of physical forcing: (1) a spatially and temporally unstable salinity gradient in the northern part of the lagoon; (2) the least hydrodynamically active limnetic part of the lagoon; (3) the Nemunas river-lagoon transition with abrupt changes in current velocity. The large vegetated littoral also represents a rather heterogeneous environment for plankton assemblages, structured by complex predator–prey interactions.

A brackish-water zooplankton assemblage enters the lagoon only during sea water intrusions and is dominated by *Acartia bifilosa*, *Temora longicornis*, *Eurytemora hirundoides*, *Podon polyphemoides* and *Evadne nordmanni* (Gasiūnaitė 2000). The Ponto-Caspian invader *Cercopagis pengoi* was first recorded in 1999 in northern part of the lagoon during seawater intrusions (Gasiūnaitė and Didžiulis 2000).

The spatially and temporally unstable salinity gradient is unfavourable for both freshwater and brackish-water zooplankton species (see Chap. 8 by Chubarenko and Margoński, this volume). The total abundance of zooplankton varies as a function of salinity, and highest values were always found at salinity corresponding to the distribution of limnetic assemblage (Fig. 9.3). The salinity gradient in the Curonian lagoon includes the critical salinity range between 5 and 8 psu, commonly characterised by minimum species diversity; however, no significant salinity-associated

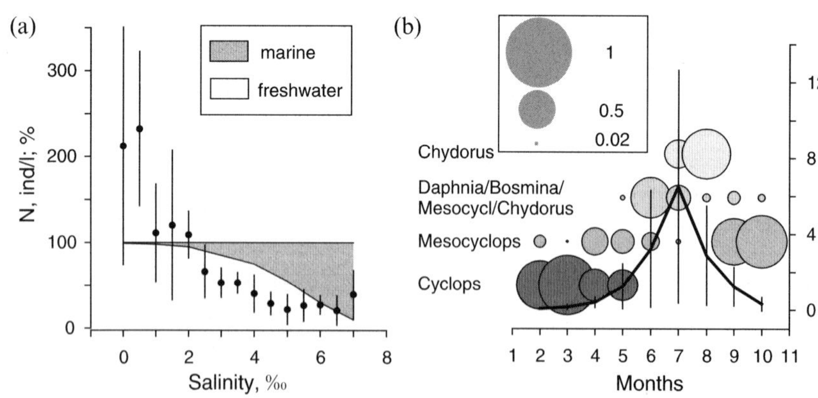

Fig. 9.3 **a** Plankton crustacean abundance (*N*; mean ±1 SD) along the salinity gradient and relative abundance (%) of freshwater and marine species (modified from Gasiūnaitė 2000). **b** The sequence of freshwater plankton crustacean seasonal succession (modified from Gasiūnaitė and Razinkovas 2004). The size of the *spheres* represents the monthly ratio of number of samples, representing particular seasonal assemblage and total number of samples (*inset*). *Solid line* Monthly means (± 1 SD) of the total zooplankton biomass (mg ww l^{-1}; right-hand *y*-axis)

biodiversity decrease was found. Changes in species diversity could be explained rather by stochastic than by deterministic factors: certain species composition and quantitative patterns at certain salinity values seems to depend mostly on those of the "source" assemblages (Gasiūnaitė 2000).

Freshwater assemblage is permanently present in the lagoon and is dominated by *Bosmina* spp., *Chydorus sphaericus*, *Cyclops strenuus*, *Daphnia* spp., *Diaphanosoma brachyurum*, *Leptodora kindti*, *Eudiaptomus graciloides* and *Mesocyclops leuckarti* (Gasiūnaitė; and Razinkovas 2004). Rotifers are dominated mainly by *Keratella* spp. and *Brachionus* spp.

Four successional assemblages of limnetic crustacean zooplankton were derived (Gasiūnaitė and Razinkovas 2004). Cyclopoids dominate in the zooplankton community until May, while large *Daphnia* specimens appear only at the beginning of summer. The shift to small-bodied *Chydorus* is observed later at midsummer and coincides with the dominance of Cyanobacteria in phytoplankton. Cyclopoida usually dominate in September–October (Fig. 9.3). These changes are consistent throughout the pelagic part of the lagoon and generally follow the pattern predicted for eutrophic freshwaters except for the pronounced domination of cyclopoids instead of daphniids in spring and autumn. This could be explained by comparatively short water residence times at these periods, which are unfavourable for development of large cladocerans (Gasiūnaitė; and Razinkovas 2004).

The Nemunas river zooplankton in the low reach is dominated by copepods (60–100% of total density) over the year, cladocerans generally dominate in the lagoon from June to September (up to 90%) (Gasiūnaitė and Razinkovas 2004; Z.R. Gasiūnaitė et al., manuscript in preparation). The mean Copepoda/Cladocera ratio in the lagoon varies from 20 to 50 in early spring, decreases to 2–11 in May, does not exceed 0.5 in summer, and increases again to 1–3 in autumn. This ratio varies from 1.5 to 7 in the Nemunas river over the year, except for some occasions in spring when cladocerans are completely absent. Compared to the lagoon, the overall zooplankton abundance in the river is roughly ten times lower throughout the year. The total zooplankton abundance increases abruptly towards the lagoon. Simultaneous structural changes along the gradient are clearly expressed during both cladoceran- and copepod-dominated periods. A high Copepoda/Cladocera ratio is typical of river channels, with a low ratio for the lagoon. In the case of a uniform Copepoda/Cladocera ratio along the gradient, structural changes are expressed by the domination of adult cyclopoids in the lagoon and copepodit stages in the river (Z.R. Gasiūnaitė et al., manuscript in preparation).

The large littoral zone with macrophyte beds, dominated by *Potamogeton perfoliatus* and *P. pectinatus* (Plokštienė 2002), could have an essential structuring effect on zooplankton structure and distribution, providing spatial refuges for zooplankton and shifting food-web interactions.

The littoral zooplankton assemblage in the lagoon is dominated by *Mesocyclops leuckarti*, *Eudiaptomus graciloides*, *Daphnia* spp., *Chydorus sphaericus*, *Bosmina coregoni*, *Thermocyclops* sp., and *Acanthocyclops* sp. (Z.R. Gasiūnaitė and E. Grinienė, manuscript in preparation). In general, both depth and macrophyte distribution influence the changes in crustacean community structure along the littoral–pelagic gradient. During the day, larger plankton crustaceans (*Daphnia*, *Eudiaptomus* and *Mesocyclops*) tend to avoid vegetated littoral, whereas small-bodied crustaceans (e.g. *Chydorus*) are associated with shallow vegetated areas. During the night, crustacean abundance increases up to 2–8 times in the shallow littoral area (both vegetated and open) and decreases approximately 2.5 times in the upper and middle layers of the pelagic zone. In small-scale shallow patchy littoral habitats, several patterns of zooplankton distribution can be observed. Quantitative and structural zooplankton characteristics can be temporally and spatially uniform through open and vegetated littoral patches, or the total zooplankton density and density of some dominant species (*Daphnia*, *Mesocyclops*) can be significantly higher in open areas during both night and day, with *Acanthocyclops* and *Chydorus* specimens usually being associated with macrophyte beds.

9.5 Fish

Among the 57 fish species recorded in the Curonian lagoon, 11 are of marine origin. The most common are roach (*Rutilus rutilus*), perch (*Perca fluviatilis*), redeye (*Scardinius erythrophalmus*), white bream (*Blicca bjoerkna*) and common bream (*Abramis brama*) (Repečka et al. 1996). The main commercial fishes are roach, perch and bream; their commercial catches are between 700 and 1,680 t a^{-1} (Virbickas et al. 1996).

The migratory fishes in the Curonian lagoon are represented by Atlantic salmon (*Salmo salar*), sea trout (*Salmo trutta trutta*), smelt (*Osmerus eperlanus*), twaite shad (*Alosa fallax*), whitefish (*Coregonus lavaretus*), vimba (*Vimba vimba*), eel (*Anguilla anguilla*), and river and marine lampreys (*Petromyzon marinus* and *Lampetra fluviatilis*) (Repečka 2003b). The migration of salmon starts in mid-August and reaches a peak at the end of September–October. During the last 10 years the yearly abundance of migrating salmon ranged from 3,400 to 7,800 individuals (4,818 ind. on average) (Kesminas et al. 2003). The most intense sea trout migration is observed in September–October, when the abundance of migrating sea trout reaches 5,500–11,500 individuals. Smelt is the most abundant migratory fish in the coastal zone of the Baltic sea. Smelt enter the lagoon at the end of October; the most intensive migration is observed in December–February. Commercial smelt catches in the lagoon vary from ca. 75 to 210 t a^{-1}. Vimba enter the lagoon in September–November and March–May. Stocks of vimba increased recently, reaching up to 50 t a^{-1} in recent years (Repečka 2003a).

The juvenile fish assemblage is dominated by smelt and pikeperch (*Sander lucioperca*) in the lagoon pelagic habitats, while roach, perch, three-spined stickleback (*Gasterosteus aculeatus*) and gudgeon (*Gobio gobio*) prevail in the littoral (Žiliukienė 1998; Žiliukas 2003). Fish fry dynamics have a pronounced seasonal pattern: three-spined stickleback dominate in the littoral in spring, roach and gudgeon in summer and perch in autumn. Overall fish fry abundance varies from 66 to 514 ind 100 m^{-3} (Žiliukas 2003).

9.6 Benthic Communities

Salinity is the main factor determining benthic species distribution in the lagoon (Fig. 9.4) (Daunys 2001). Communities dominated by freshwater species are characterised by higher species diversity and relatively large biomasses, while the contribution of widespread oligochaet and chironomid species is usually constant (Fig. 9.5). The zonation of benthic macrofauna in the lagoon is based on the peculiarities of the salinity regime (Olenin and Daunys 2004): (1) Klaipeda strait, affected by seawater inflows and highest anthropogenic pressure; (2) the northern part of the lagoon influenced by episodic inflows of seawater; (3) the freshwater area in front of the Nemunas River delta in the central part of the lagoon, strongly influenced by the river outflow [cf. Chap. 3 (Schiewer), this volume; Sect. 3.4 Schlei].

The western side of the Klaipeda Strait is characterised by the variety of its bottom substrates: fine and coarse sand, gravel and pebble bottoms, moraine–clay and stones, and patches of mud as well as artificial substrates such as concrete embankments, submerged wood, etc. The area is not exposed to oxygen deficiency and is inhabited by a rather diverse benthic fauna that is able to withstand rapid environmental fluctuations and essential anthropogenic pressure. The number of species, abundance and biomass vary within a wide range and are subject to rapid changes. Communities are dominated by *Nereis diversicolor*, *Marenzelleria neglecta*, oligochaets and chironomids, *Balanus improvisus*, *Cordylophora caspia*, *Mya arenaria*, *Macoma balthica* and *Mytilus edulis*.

Muddy bottoms comprise inlets on the eastern side of the Klaipeda strait belonging to the port area. The main bottom sediments are black mud on sites with a mixture of sand and gravel, containing also human litter. The sediments are polluted with organic material, heavy metals and oil products. Only the most tolerant species can survive in this heavily disturbed biotope, oligochaetes and chironomids are the main forms [see Chaps.

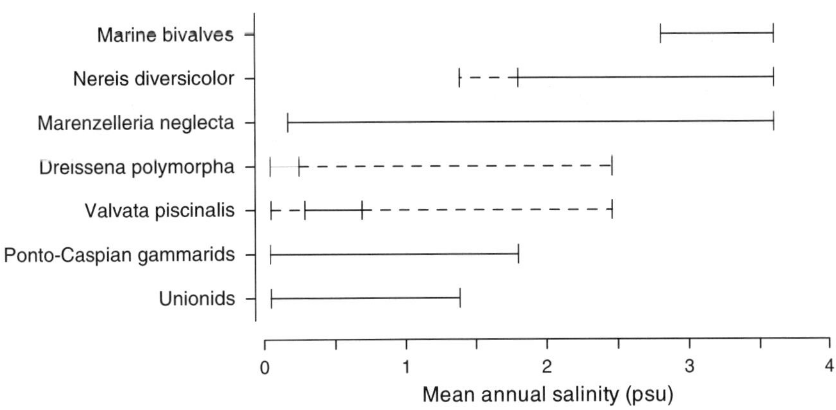

Fig. 9.4 Occupied salinity range (*solid lines*) and tolerance limits (*dashed lines*) of dominant bottom macrofauna species with respect to mean annual salinity

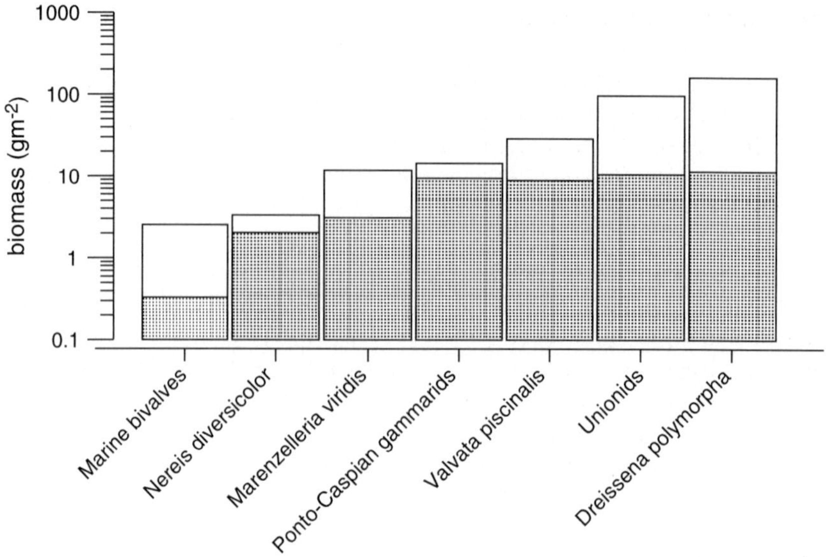

Fig. 9.5 Biomass distribution among main bottom macrofauna communities showing the contribution of oligochaets and chironomids (*shaded*)

7 (Kruk-Dowgiałło and Szaniawska) and 8 (Chubarenko and Margoński), this volume], while *Nereis diversicolor* and *Marenzelleria neglecta* may be found in comparatively less polluted locations. Benthic macrofauna is absent in the most polluted sites.

In the northern part of the Curonian lagoon, two main zones can be distinguished with respect to bottom macrofauna communities: a large eastern shallow (depth <1.5 m) area with fine sand as the prevailing bottom substrate, and the deeper (depth between 1.5 and 4 m) western area along the Curonian Spit covered mainly with muddy sediment.

The sandy bottoms on the eastern side of the lagoon may be sub-divided into a variety of biotopes: fine sand with macrophytes (e.g. Potamogeton species); fine sand with large native unionids (*Unio tumidus*); fine sand and silt with oligochaets and chironomids as well as biotopes dominated by the alien invasive species *Dreissena polymorpha*, *Marenzelleria neglecta* and amphipods of Ponto-Caspian of North American origin (mainly *Obesogammarus crassus*, but also *Pontogammarus robustoides* and *Gammarus tigrinus*). The biotope occupied by large numbers of mobile amphipods is present the a very narrow (<20 m) uppermost part of the underwater slope (depth <0.5 m) and may be distinguished only during the warm period of the year when dense communities of gammarids develop (Daunys and Olenin 1999). All other biotopes alternate with each other on a scale of 10–100 m. Invasive benthic macrofauna constitutes an important part of the biotope-forming species, contributing up to 95% of total community biomass. Even in locations where unionids are the predominant species, approximately 65% of them are fouled by zebra mussels (Olenin and Leppäkoski 1999).

The main community in the muddy bottom biotope is dominated by oligochaets and chironomids. A comparatively large part of the muddy bottoms is covered by shell deposits formed mainly by *Valvata* species with an admixture of *Bithynia* spp., *Radix* spp., *D. polymorpha* and *Potamopyrgus antipodarum*. In areas less exposed to saline water inflows, clumps of living zebra mussels may also be found.

In the central part of the lagoon, muds and sands form two principally different biotopes, which alternate with each other on a scale of 100 m. The muddy bottom is dominated by the zebra mussel *Dreissena polymorpha*. In contrast to the northern part of the lagoon, the sediment is formed mainly by zebra mussel shell deposits. Due to the habitat engineering activity of *D. polymorpha*, the community of co-occurring species is rich in species number (up to 29 per sample, and about 50 in total). The total biomass (up to 11 kg m^{-2}) and abundance (up to 100,000 ind. m^{-2}) are the highest of the entire Curonian lagoon.

Sandy bottoms (mainly fine sand and aleurite) in the central part of the lagoon are dominated by a variety of oligochaetes and chironomids (Olenin 1987, 1988; Daunys 2001). Approximately half of the lagoon bottom macrofauna species are found in these bottoms; however, none of them is constant (occurrence >40%). The species number varies from 2 to 16 per sample, and the total biomass from 10 to 40 g m^{-2}. Fine sand is mixed with mud on sites situated close to local organic pollution sources (Nida, Juodkrantė, etc.). In such places, only oligochaets and chironomids are found in benthic samples.

9.7 Biological Processes and Interactions

A food web structure based on already known interactions and trophic compartments in the pelagic part of the lagoon was compiled by Razinkovas and Zemlys (2000).

The littoral part of the lagoon, covering approximately 10–12% of the total area is, however, not yet investigated sufficiently to quantify food web interactions. In the littoral part of the Curonian lagoon both invertebrate and vertebrate predators could have an effect on the distribution of zooplankton. The main littoral planktophagous invertebrates are the Ponto-Caspian mysids *Paramysis lacustris* and *Limnomysis benedeni*, associated with the open sandy/aleurytic bottom at a depth of 1–3 m and submerged vegetation, respectively (Razinkovas 1997). The planktophagous fish in the littoral are represented by juveniles of roach (*Rutilus rutilus*) and perch (*Perca fluviatilis*) (Žiliukienė 1998; Žiliukas 2003). Macrophyte beds in the littoral zone of the lagoon are dominated mainly by *Potamogeton* (*P. pectinatus* and *P. perfoliatus*) and *Cladophora* species (Plokštienė 2002). Among the main consumers of macrophytes in the littoral zone are the introduced Ponto-Caspian amphipod species.

The Curonian lagoon food web comprises 15 compartments (Fig. 9.6). A general feature of the web is the very high phytoplankton biomass (see Chaps. 3 and 4 by Schiewer, this volume) matched only by the benthic compartments. Consequently, most of the primary production in the system is transferred to the detritus food chain or even exported to the Baltic Sea rather than grazed by zooplankton. The low abundance of pelagic fish also implies that zooplankton is recycled within its two compartments rather than transferred to higher trophic levels. The surprisingly high biomass of benthic suspension feeders is due mainly to *D. polymorpha*-dominated biotopes in the central part of the lagoon (approximately 20% of the total area). However, due to low water residence time (Daunys et al. 2006) and the comparatively low bivalve metabolism rates, the role of zebra mussels in organic matter transformation is less pronounced in comparison to that of deposit feeders including small oligochaets and chironomids. The top predators (birds) have a considerable impact on fish populations. This impact is comparable to that of fishing efforts and may increase due to the growth of cormorant colonies.

9.8 Protection

Some parts of the Curonian lagoon are protected as they are included in both a regional and a national park. Curonian Spit National Park, covering the Lithuanian part of the Curonian spit, also includes the narrow westernmost portion of the Curonian lagoon (up to the navigation channel). However, a similar national park in

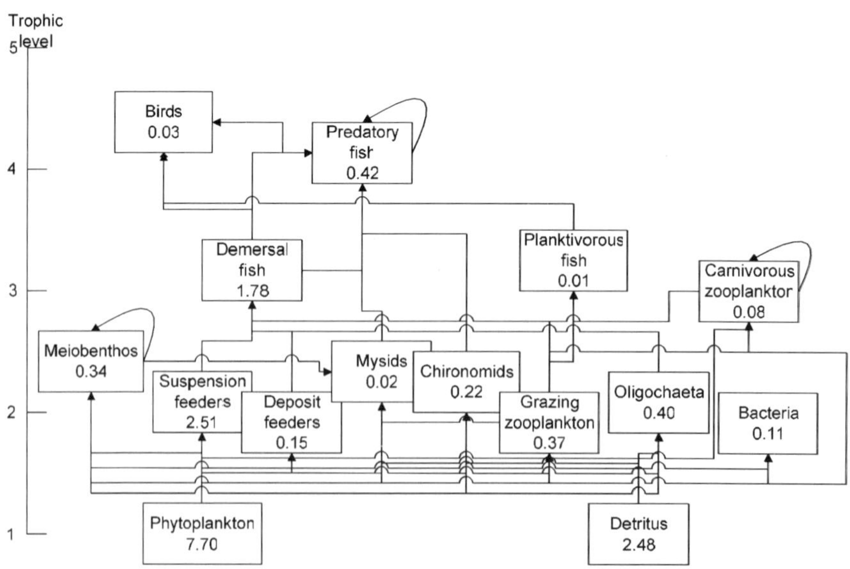

Fig. 9.6 Curonian lagoon food web. The biomass (g C m^{-2}) of each compartment is indicated

the Russian part of the Curonian spit does not cover lagoon aquatory. The Nemunas Delta Regional Park covers the Nemunas Delta as well as several bays and aquatory around the Ventes Ragas peninsula. Additionally, seasonal restrictions on fishing activity protect the spawning and migration periods of important fish species.

9.9 Summary

The Curonian lagoon is a large, highly eutrophied, mainly freshwater coastal water body, connected to the south-eastern Baltic Sea. The lagoon water circulation is determined by the wind and river discharge; seawater intrusions (up to nearly 7 psu) are frequent and irregular. The main bottom sediments in the lagoon are sand and silt. The northern part of the lagoon acts as a transitory area of sediment transportation, while the central part is quite heterogeneous with respect to bottom geomorphology and sediment type. Muddy bottoms occur in local depressions in the deeper western part of the lagoon.

Both phytoplankton and zooplankton are dominated by freshwater species; marine species enter the lagoon only during seawater intrusions. Seasonal succession of plankton communities generally follows a pattern typical of the eutrophic ecosystems: phytoplankton blooms are dominated by *Stephanodiscus hantzschii* in spring, and *Aphanizomenon flos-aquae* and *Microcystis aeruginosa* in summer.

The most common fish species are roach, perch, redeye, white bream and common bream. Migratory fishes include Atlantic salmon, sea trout, smelt, twaite shad, whitefish, vimba, eel and lampreys. The main commercial fishes are roach, perch and bream.

Salinity is the main factor determining benthic species distribution in the lagoon, whereas sediment type and amount of organic material are responsible for relatively local variations rather than large scale shifts. The communities are dominated mainly either by euryhaline species (e.g. *Nereis diversicolor*, *Marenzelleria neglecta*) in the zone influenced by seawater inflows or by tolerant (oligochaete and chironomid species), usually alien, species (e.g. *Dreissena polymorpha*, Ponto-Caspian amphipods) in the freshwater part of the lagoon.

A general feature of the Curonian lagoon food web is the very high phytoplankton biomass. Most of the primary production in the system is transferred to the detritus food chain or even exported to the Baltic Sea rather than being grazed by zooplankton. The top predators (mainly cormorants) have a considerable impact on fish populations, comparable to that of fishing efforts.

References

Aristova GI (1965) Benthos of the Kurskij Gulf (in Russian). Proc. AtlantNIIRO, Kaliningrad, 14:19–39

Aristova GI (1971) Biocenosis of *Dreissena polymorpha* in the Curonian Lagoon (in Russian). Proc AtlantNIIRO, Kaliningrad, 25:9–16

Baršienė J (2002) Genotoxic impacts in Klaipėda Marine Port and Būtingė Oil terminal areas. Mar Environ Res 54:475–479

Bružinskienė A, Virbickas J (1988) Commercial and sport fishery (in Lithuanian). Mokslas, Vilnius

Bubinas A (1983) Zoobenthos. In: Biogeochemistry of the Curonian Lagoon (in Russian). Mokslas, Vilnius, pp 44–60

Bukantis A (1994) The Lithuanian climate (in Lithuanian). Vilnius University

Četkauskaitė; A, Zarkov D, Stoškus L (2001) Water quality control, monitoring and wastewater treatment in Lithuania 1950–1999. Ambio 30:297–305

Daunys D (2001) Patterns of the bottom macrofauna variability and its role in the shallow coastal Lagoon. Summary of PhD dissertation, Klaipėda University

Daunys D, Olenin S (1999) Littoral bottom communities of the northern part of the Curonian Lagoon (in Lithuanian, with English summary). Ekologija 2:19–27

Daunys D, Schiedek D, Olenin S (2000) Species strategy near its boundary: the *Marenzelleria* cf. *viridis* (Polychaeta, Spionidae) case in the south-eastern Baltic Sea. Int Rev Hydrobiol 85:639–651

Daunys D, Zemlys P, Olenin S, Zaiko A, Ferrarin C (2006) Impact of the zebra mussel *Dreissena polymorpha* invasion on the budget of suspended material in a shallow lagoon ecosystem. Helgoland Mar Res 60:113–120

Gaigalas K (2001) Fish in the Curonian Lagoon basin, their resources and fishery (in Lithuanian). Eglė, Klaipėda

Gaigalas K, Gerulaitis A, Kesminas V, Milerienė E, Repečka R, Valušienė V, Žiliukas V, Žiliukienė V (1992) Rare fishes of Lithuania (in Lithuanian). Academia, Vilnius

Gasiūnaitė ZR (2000) Coupling of the limnetic and brackishwater plankton crustaceans in the Curonian lagoon (Baltic Sea). Int Rev Hydrobiol 85:653–662

Gasiūnaitė ZR, Didžiulis V (2000) Ponto-Caspian invader *Cercopagis pengoi* (Ostroumov, 1891) in Lithuanian coastal waters. Sea Environ 2:97–101

Gasiūnaitė ZR, Olenina I (1998) Zooplankton–phytoplankton interactions: a possible explanation of the seasonal succession in the Kursiu Marios lagoon. Hydrobiologia 363:333–339

Gasiūnaitė ZR, Razinkovas A (2000) The role of plankton and nektobenthic crustaceans in the pathways of the primary production transformation in the Curonian lagoon (in Lithuanian, with English summary). Ekologija 3:22–27

Gasiūnaitė ZR, Razinkovas A (2002) The salinity tolerance of two cladoceran species from the Curonian lagoon: an experimental study. Sea Environ 2:28–32

Gasiūnaitė ZR, Razinkovas A (2004) Temporal and spatial patterns of the crustacean zooplankton dynamics in transitional lagoon ecosystem. Hydrobiologia 514:139–149

Gasiūnaitė ZR, Cardoso AC, Heiskanen A-S, Henriksen P, Kauppila P, Olenina I, Pilkaitytė R, Purina I, Razinkovas A, Sagert S, Schubert H, Wasmund N (2005) Seasonality of coastal phytoplankton in the Baltic Sea: influence of salinity and eutrophication. Estuarine Coastal Shelf Sci 65:239–252

Gasiūnas I (1959) Forage macrozoobenthos of the Kuršių; marios (in Russian with German summary). In: Jankevičius K, Gasiūnas I, Gediminas A, Gudelis V, Kublickas A, Maniukas I (eds) Kuršių Marios. Pergalė, Vilnius, pp 191–292

Gulbinskas S (1995) Distribution of recent bottom sediments in the Curonian Lagoon–Baltic Sea sedimentation area (in Lithuanian). Geogr Yearbook 28:296–314

Jagminienė I (1995) The dynamics of the structure and productivity of benthocenosis in the northern part of the Kuršių Marios Lagoon (in Lithuanian with English summary). Fishery and Aquaculture in Lithuania 1:97–102

Jankavičiūtė GJ, Jankevičius KK (1978) Quantitative and qualitative changes of phytoplankton in the northern part of the Curonian lagoon under anthropogenic impact (in Russian). In: Baranauskienė A, Budrienė S, Dušauskienė-Duž R, Malamienė B, Petrauskas B, Raziulytė R, Jankevičius K, Jankavičiūtė G (eds) Physiologo–biochemical conditions of the development of plankton organisms in the northern part of the Curonian lagoon. Vilnius, pp 33–61

Jankevičius K, Baranauskienė AJ (1978) Hydrobiological research in the Curonian lagoon (in Russian). In: Baranauskienė A, Budrienė S, Dušauskienė-Duž R, Malamienė B, Petrauskas B, Raziulytė R, Jankevičius K, Jankavičiūtė G (eds) Physiologo–biochemical conditions of the development of plankton organisms in the northern part of the Curonian lagoon. Vilnius, pp 3–7

Jankevičius KK, Antanynienė AS, Baranauskienė JJ, Budrienė SF, Mažeikaitė SI, MalomienėVA, Šlapkauskaitė GV, Šalienė RI, Jankavičiūtė GJ (1979) Anthropogenic eutrophication of the northern part of the Curonian lagoon in 1974 and 1975 (in Russian). Works of the Academy of Science of Lithuanian SSR 87:49–61

Jurevičius R (1959) Hydrochemical characteristic of the Kuršių Marios Lagoon (in Russian with German summary). In: Jankevičius K, Gasiūnas I, Gediminas A, Gudelis V, Kublickas A, Maniukas I (eds) Kuršių Marios. Pergalė, Vilnius, pp 69–108

Kesminas V, Virbickas T, Repečka R (2003) The present state of salmon (*Salmo salar* L.) in Lithuania. Acta Zool Lithuanica 13:176–187

Kiselytė T (1959) Zooplankton of the Kuršių Marios lagoon (in Russian with German summary). In: Jankevičius K, Gasiūnas I, Gediminas A, Gudelis V, Kublickas A, Maniukas I (eds) Kuršių Marios. Pergalė, Vilnius, pp 169–189

Klimašauskienė VP (1979) The impact of the sewage on the zooplankton in Nemunas delta, central and northern parts of the Kuršių marios lagoon. Works of the Academy of Science of Lithuanian SSR 85:1–55

Lazauskienė L, Bubinas A, Klimašauskienė V, Jagminienė I, Vaitonis G (1996) Assessment of fish forage resources in Kuršių Marios Lagoon, Nemunas Delta and Baltic Sea nearshore areas. Fishery and Aquaculture in Lithuania 2:46–60

Lundbeck J (1935) Über die Bodenbevölkerung besonders der Chironomidenlarven des Frischen und Kurischen Haffes. Int Rev Ges Hydrobiol Hydrograph 32:265–284

Maniukas J (1959) The ichthyofauna, the state of resources and commercial fishing in the Curonian Lagoon (in Russian with German summary). In: Jankevičius K, Gasiūnas I, Gediminas A, Gudelis V, Kublickas A, Maniukas I (eds) Kuršių Marios. Pergalė, Vilnius, pp 293–401

Mažeikaitė SI (1978a) Zooplankton species composition, abundance and biomass in the northern part of the Curonian lagoon under anthropogenic pollution (in Russian). In: Baranauskienė A, Budrienė S, Dušauskienė-Duž R, Malamienė B, Petrauskas B, Raziulytė R, Jankevičius K, Jankavičiūtė G (eds) Physiologo–biochemical conditions of the development of plankton organisms in the northern part of the Curonian lagoon. Vilnius, pp 87–111

Mažeikaitė SI (1978b) Zooplankton of the northern part of the Curonian lagoon in 1974 and 1975 (in Russian). Works of the Academy of Science of Lithuanian SSR 64:55–56

Naumenko E (1996) Seasonal and long-term dynamics of zooplankton community composition, abundance and biomass in the Curonian lagoon of the Baltic Sea. Monogr Syst Ecol 1:23–35

Olenin S (1987) Zoobenthos of the Curonian Lagoon: results of biological monitoring, 1980–1984 (in Russian). In: Chemistry and biology of seas. Proceedings of the State Oceanographic Institute, Hydrometeoizdat, Moscow, pp 175–191

Olenin S (1988) Some aspects of biocenotic and trophological structure of the bottom fauna in the Curonian Lagoon (in Russian with English summary). Reg Hydrometeorol 12:46–55

Olenin S (1997) Comparative study of the south-eastern Baltic coastal zone and the Curonian Lagoon bottom communities. In: Andrushaitis A (ed) Proceedings of the 13th Baltic Marine Biologists Symposium, Institute of Aquatic Ecology, University of Latvia, Riga, pp 151–159

Olenin S, Daunys D (2004) Coastal typology based on benthic biotope and community data: the Lithuanian case study. In: Schernewski G, Wielgat M (eds) Baltic Sea typology. Coastline Rep 4:65–83

Olenin S, Leppäkoski E (1999) Non-native animals in the Baltic Sea: alteration of benthic habitats in coastal inlets and Lagoons. Hydrobiologia 393:233–243

Olenina I (1997) Phytoplankton development in the Curonian lagoon and south-eastern Baltic Sea coastal area. PhD Thesis, Institute of Botany, Vilnius

Olenina I (1998) Long-term changes in the Kuršių Marios lagoon: eutrophication and phytoplankton response. Ekologija 1:56–65

Pilkaitytė R (2003) Phytoplankton seasonal succession and abundance in the eutrophic estuarine lagoons. Summary of Doctoral Dissertation, Klaipeda

Pilkaitytė R (2004) Phytoplankton community response to the nutrient enrichment: an example from N-limited ecosystems. Jūra ir aplinka 1:76–82

Pilkaitytė R, Razinkovas R (2006) Factors controlling phytoplankton blooms in a temperate estuary: nutrient limitation and physical forcing. Hydrobiologia 555:41–48

Plokštienė D (2002) Development and changes of the vegetation of the Kuršių marios lagoon (in Lithuanian). Sea Environ 2:33–41

Pustelnikovas O (1998) Geochemistry of Sediments of the Curonian Lagoon (Baltic Sea). Institute of Geography, Vilnius

Razinkov A (1990) Production processes in the opossum shrimps populations of water basins of the river Nemunas basin (in Russian). PhD Thesis. Institute of Zoology, Minsk

Razinkovas A (1997) Spatial distribution and migration patterns of the mysids in the Curonian Lagoon. In: Andrushaitis A (ed) Proceedings of the 13th Baltic Biologists Symposium. Riga, pp 113–116

Razinkovas A, Pilkaitytė R (2002) Factors limiting phytoplankton development in the Curonian lagoon (in Lithuanian). Sea Environ 1:39–46

Razinkovas A, Zemlys P (2000) Balance of the organic matter in the ecosystem of the Curonian lagoon. Sea Environ 1:32–37

Razinkovas A, Bliudziute L, Erturk A, Ferrarin C, Lindim C, Umgiesser G, Zemlys P (2005) Curonian lagoon: a modelling study – Lithuania. In: Russo RC (ed) Modeling nutrient loads and response in river and estuary systems. Report No. 271., Committee on the Challenges of Modern Society, North Atlantic Treaty Organization, Brussels, pp 194–222

Repečka R (1995) Investigations of the fish fauna in the Lithuanian part of the Curonian lagoon. Coastal Conservation and the Management in the Baltic Region. In: Proceedings of the EUCC-WWF Conference. Riga-Klaipeda-Kaliningrad, pp 117–126

Repečka R (1996) Assessment and monitoring of coastal fish resources, Lithuania. In: Assessment and monitoring of coastal fish resources in Estonia, Latvia and Lithuania. Sweedmark Consultancy Report, pp 1–23

Repečka R (1999) Biology and resources of the main commercial fish species in the Lithuanian part of the Curonian lagoon. In: Freshwater fish and the herring population in the coastal lagoons. Environment and Fisheries, Symposium Proceedings, Sea Fisheries Institute, pp 185–195

Repečka R (2003a) The species composition of the ichthyofauna in the Lithuanian economic zone of the Baltic Sea and the Curonian lagoon and its changes in recent years Acta Zool Lithuanica 13:149–157

Repečka R (2003b) Changes in biological indices and abundance of salmon, sea trout, smelt, vimba and twaite shad in the coastal zone of the Baltic sea and the Curonian lagoon at the beginning of spawning migration. Acta Zool Lithuanica 13:195–216

Repečka R, Milerienė E (1991) Species composition and abundance in shore-zone fish communities of the Curonian lagoon during 1985–1989 (in Russian). Ekologija 4:68–80

Repečka R, Žiliukas V, Stankus S (1996) The abundance of the commercial fish larvae and fry in the coastal zone of the Baltic Sea and the Curonian lagoon (in Lithuanian). Fishery and Aquaculture in Lithuania 2:95–108

Rudinskaya LV (1994) Long term benthos dynamics in the Curonian Lagoon, Baltic Sea (in Russian). In: Hydrobiological Studies in the Atlantic Ocean and Baltic Sea Basin. AtlantNIIRO, Kaliningrad, pp 41–50

Schmidt-Ries H (1940) Untersuchungen zur Kenntnis des Pelagials eines Strandgewässers. Z Fischerei 36:213–291

Szidat L (1926) Beiträge zur Faunistik und Biologie des Kurischen Haffs. Schriften der Physical-ökonom Gesellschaft Königsberg in Pr 65:5–31

Ūselytė S (1959) Phytoplankton of the Curonian lagoon and its seasonal dynamics (in Russian with German summary). In: Jankevičius K, Gasiūnas I, Gediminas A, Gudelis V, Kublickas A, Maniukas I (eds) Kuršių Marios. Pergale, Vilnius, pp 139–167

Virbickas J, Repečka R (1996) Fishes and fishery in the Lithuanian coastal zone of the Baltic Sea and Curonian lagoon. In: Proceedings of the 1st Conference of Lithuanian Oceanologists, Klaipeda University, Klaipeda, 235–250

Virbickas J, Domarkas A, Pašukonis J (1996) On exploitation of fish stocks in the Baltic Sea (Lithuanian economic zone) and Lithuanian aquatory of the Kurshiu Lagoon. Fishery and Aquaculture in Lithuania 2:61–68

Willer A (1931) Vergleichende Untersuchungen an Strandgewässern. Verh Int Ver Theor Angew Limnol 5:197–231

Willer A (1935) Das Kurische Haff als Grenzgewässer. Schriften der Physical-ökonom Gesellschaft Königsberg in Pr 38:5–31

Zettler ML, Zettler A, Daunys D (2005) Bemerkenswerte Susswassermollusken aus Litauen. Aufsamlungen vom September 2004. Malakol Abh 23:27–40

Žaromskis R (1996) Oceans, Seas, Estuaries (in Lithuanian). Debesija, Vilnius

Žiliukas V (2003) Assessment of the dynamics of main ecological parameters of the fish fry community in the coastal zone of the Curonian Lagoon near Vente Cape. Acta Zool Lithuanica 13:167–175

Žiliukienė V (1998) Ichthyoplankton of the Curonian lagoon (in Lithuanian). Fishery and Aquaculture in Lithuania 3:215–225

Žiliukienė V (2003) Quantitative structure, abundance and biomass of zooplankton in the Lithuanian part of the Curonian lagoon in 1996–2002. Acta Zool Lithuanica 13:97–105

Chapter 10
Gulf of Riga and Pärnu Bay

J. Kotta, V. Lauringson, G. Martin, M. Simm, I. Kotta, K. Herkül, and H. Ojaveer

10.1 Environmental Characteristics

The Gulf of Riga (Fig. 10.1) is a relatively shallow and isolated water-body with a surface area of 16,330 km^2. On its eastern and southern sides it is flanked by the Estonian and Latvian mainlands, and on the northern side by the islands of Saaremaa and Muhu. The Gulf of Riga is connected to the Baltic Proper via the Irbe Strait, and to the Väinameri Archipelago Sea by the Suur Strait. Annual river inflow (Table 10.1) ranges between 18 and 56 km^3 (on average 32 km^3), while the volume of the gulf is 424 km^3. The residence time of water masses is 2–4 years (HELCOM 1996). The Gulf of Riga receives fresh water from a huge drainage area (134,000 km^2), mostly entering the southern part of the basin (Andrushaitis et al. 1995). In general, the bottom relief of the area is quite flat, with gentle slopes towards deeps. The northern part of the Gulf is characterised by a wide coastal zone with diverse bottom topography and extensive reaches of boulders. The southern part of the Gulf of Riga is more exposed; steep and soft substrate prevails. In the deeper parts of the Gulf, silty sediments prevail.

The average salinity of the Gulf of Riga varies from 0.5–2.0 psu in surface layers in its southern and northeastern areas to 7 psu at the Irbe Strait. In most parts of the Gulf the salinity is 5.0–6.5 psu. During the ice-free season the salinity is higher in the bottom layers and lower in the surface layer. However, due to its shallowness, the Gulf of Riga lacks a permanent halocline (Berzinsh 1995; Raudsepp 2001).

Because the Gulf of Riga is a shallow water basin, changes in air temperature have a direct influence on the dynamics of both surface and deep water. In a "typical" year the water is cold and no clear thermocline occurs until May. Later, the surface water temperature rises to about 17–20°C and a thermocline builds up. The water temperature below 30 m remains relatively stable at 3°C. The thermocline reaches a depth of 25 m in August and disintegrates in September–October due to intensive wind mixing. In the course of autumn storms the surface water cools down and the deep water temperature rises to 5–10°C (Raudsepp 2001).

The duration of the ice season has a large interannual variability. The range of variation in the number of ice days is several months. The average number of ice days varies spatially from 80 days in the Irbe Strait and open Gulf of Riga to 150

U. Schiewer (ed.), *Ecology of Baltic Coastal Waters. Ecological Studies 197.*

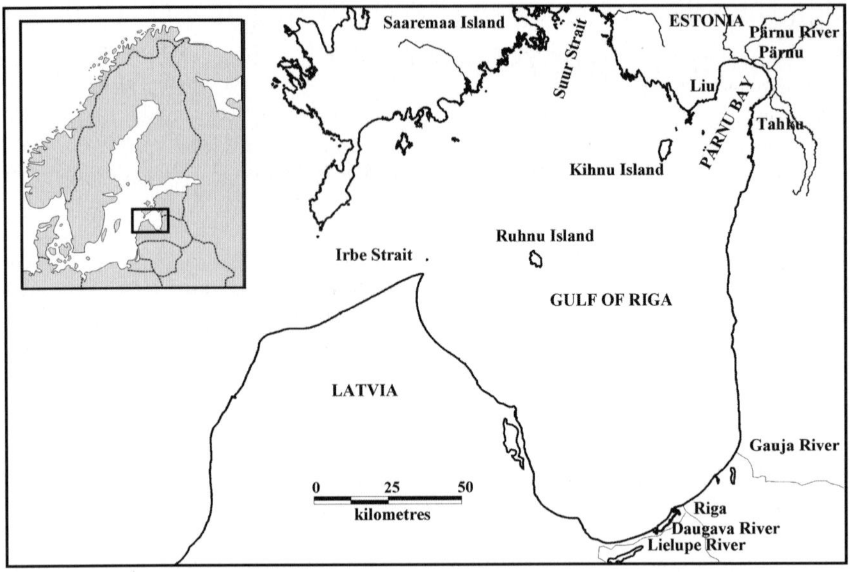

Fig. 10.1 Map of the Gulf of Riga and Pärnu Bay

days in Pärnu Bay. The number of ice days has decreased by 5–7 days in the last century (Jevrejeva 2000).

The sea level of the Gulf of Riga was modelled based on realistic meteorological forcing and historical data. Extremely low levels (–1.25 m below mean sea level) generally do not occur locally, whereas high levels (up to 2.75 m above the mean as measured in Pärnu Bay in 2005) are short term and localised. The shallow and narrow bays exposed to the direction of the strongest possible storm winds (SW and W) are prerequisites for these high values (Suursaar et al. 2003a, 2003b). The most effective forcing function for the system is the wind stress above the straits, which has a considerable role in sea level motions with a time scale of between 1 year and 1–2 decades. In winter, water flows and exchange depend strongly on the existence of ice-cover in the Gulf of Riga. The seasonality of the water exchange process is governed by seasonal changes in large-scale atmospheric circulation above the North Atlantic (Otsmann et al. 1997, 2001). The Gulf of Riga is connected with the Baltic Proper mainly through Irbe Sound, which covers about 80–85% of the total exchange. The average measured velocity is 12 cm s^{-1}, with maximum values of up to 72 cm s^{-1}. The water flow in the Suur Strait is uniform but the temporal variability of the currents can be quite complex, with the average measured velocity being 19 cm s^{-1} and maximum values up to 1 m s^{-1}. The total volume of the water running through the strait in both directions is about 100 km^3 s^{-1} (Suursaar et al. 1995). The average flow velocities in Pärnu Bay are estimated at 4–11 cm s^{-1}, with a maximum value of 90 cm s^{-1} (Suursaar et al. 2002).

The oxygen regime of the Gulf of Riga is relatively good due to its shallowness and strong vertical mixing. In most areas oxygen concentrations are higher than

Table 10.1 Main characteristics of the Gulf of Riga. *ww* Wet weight, *dw* dry weight, *Chl a* chlorophyll *a*, *DIN* dissolved inorganic nitrogen, *DIP* dissolved inorganic phosphorus

a. Climate and hydrology	
Air temperature (°C)	min: –32; max: +36
Water temperature (°C)	min: 0; max: 20 (surface)
	min: 0; max: 6.5 (bottom)
Ice covering (days a^{-1})	Irbe Strait: 80
	Pärnu Bay: 150
Annual global radiation (kW h m^{-2})	1,116
Average precipitation (mm a^{-1})	590
Mean freshwater inflow ($10^6 m^3 a^{-1}$)	3,000 (min: 18,000; max: 56,000)
Mean inflow ($10^6 m^3 a^{-1}$) (from Baltic)	113,000
Mean total outflow ($10^6 m^3 a^{-1}$)	147,000
Residence time (days)	min: 730; max: 1,460
b. Morphometry and sediments	
Area (km^2)	16,330
Volume (km^3)	424
Mean depth (m)	27
Maximum depth (m)	60
Catchment area (km^2)	134,000
Area/catchment relation	0.122
Organic matter content (% dw)	4.65
C/N-ratio (0–10 cm depth)	8.53
Phosphate release (t a^{-1})	9,400
Ammonia release (t a^{-1})	0
Sediment distribution	Sand and mixed on coastal areas, clay–silt at depth
c. Pelagic biological components	
Secchi depth (m)	Min: < 3; max: >5
Bacteria ($10^6 ml^{-1}$)	8.1 (coastal area)
	0.57 (central gulf)
Phytoplankton biomass (µg Chl *a* l^{-1})	Spring max: 77.17
	Autumn max: 13
Phytoplankton biomass ($mm^3 l^{-1}$)	Spring max: 14
	Autumn max: 2.28
Dominating phytoplankton species	*Achnanthes taeniata, Thalassiosira baltica, Chaetoceros wighamii, Peridiniella catenata, Ch. holsaticus, Melosira nummuloides*
Flagellates	
Biomass (mg C l^{-1})	Max: 0.5 (ww)
Dominating species	*Dinophysis acuminate, D. baltica, Ebria tripartite, Peridiniella catenata, Plagioselmis prolonga, Teleaulax* spp, *Chrysochromulina* spp
Ciliates	
Biomass (mg C l^{-1})	1.39 (ww)

(continued)

Table 10.1 (continued)

Dominating species	*Strombidium* spp, *Tintinnopsis* spp, *Lohmanniella* spp
Mesozooplankton	
Biomass (mg C l^{-1})	0.1–1.3 (ww)
Dominating species	*Eurytemora affinis*, *Acartia bifilosa*, *Bosmina longispina*
Fish	
Number of species	>50
Dominating species	*Clupea harengus membras*, *Zoarces viviparus*, *Platichthys flesus*, *Gasterosteus aculeatus*, *Perca fluviatilis*, *Sander lucioperca*, *Blicca bjoerkna*, *Rutilus rutilus*, *Abramis brama*
Primary production (g C m^{-2} a^{-1})	290
Bacterial production (µg day^{-1})	10–330 (open gulf in summer)
Mesozooplankton production (g C m^{-2} a^{-1})	27
Fish catches (t a^{-1})	48,000
d. Benthic biological components	
Macrophytes	
Biomass (g dw^{-1} m^{-2})	Average: 216; max: 3,700
Number of species	69
Dominating species	*Fucus vesiculosus*, *Pilayella littoralis*, *Ectocarpus siliculosus*, *Furcellaria lumbricalis*, *Cladophora glomerata*, *Ulva intestinalis*
Macrozoobenthos	
Biomass (g m^{-2})	35 (dw)
Number of species	145
Dominating species	*Macoma balthica*, *Mya arenaria*, *Cerastoderma glaucum*, Oligochaeta, *Monoporeia affinis*
e. Water chemistry, trophic status and pollution	
Salinity (psu)	5.5 (min: 0.5; max: 7)
pH	8
Oxygen saturation (%)	94–130 (surface) 52–94 (bottom)
Total nitrogen (µmol l^{-1})	28–54 Winter max: 54
DIN (µmol l^{-1})	1–13.5
Total phosphorus (µmol l^{-1})	0.78–1.5 Winter max: 1.5
DIP (µmol l^{-1})	0.4–0.8
Annual nitrogen input (t a^{-1})	141,741
Annual phosphorus input (t a^{-1})	3,084
Limitation of PP and period	N, P or Si depending on season and area
Main pollutants	N, P (the main pollutant is the Daugava River)
Trophic level	Eutrophic

5 ml l^{-1}. From April to the middle of October seasonal stratification may restrict vertical water exchange, thus promoting oxygen depletion and storage of nutrients in the bottom water until the water column is remixed in autumn. Concentrations below 2 ml l^{-1} have occasionally been found in the deepest part of the gulf (>45 m). Since the mid-1960s, a statistically significant decreasing trend in the concentration of oxygen has been observed in the study area (Berzinsh 1995; Yurkovskis 2004).

The gulf is on average twice as eutrophicated as the Baltic Proper, and the outflow of nutrients through the straits is higher than the inflow (Mägi and Lips 1998). The total inputs of nitrogen (N) and phosphorus (P) are 141,741 t a^{-1} and 3,084 t a^{-1}, respectively. The seasonal and vertical regime of nutrients in the Gulf differs somewhat from that in the open sea. The nutrient concentrations are set by the occasional inflows of saline and nutrient-rich deep water from the Gotland Basin via the Irbe Strait, and year-to-year variations in river inflows. The rivers play a crucial role in the total input of nutrients and exceed the combined contribution from atmospheric deposition, point emission from cities and industries along the coast, as well as nitrogen fixation by marine organisms. Higher concentrations of nutrients are found in the southern and northeastern parts of the Gulf, i.e. adjacent to the mouths of the Daugava, Lielupe, Gauja and Pärnu rivers. Because of its shallowness, the Gulf of Riga has no clear chemocline. Although higher concentrations of nutrients are often observed in deeper water in May–October, this pool is accessible to the surface layers through occasional mixing events. Particularly strong vertical mixing processes in autumn and winter result in a high nutrient content in the upper layer in January and February. Both dissolved inorganic nitrogen and phosphate pools of the upper mixed layer are exhausted by mid-May, except at river mouths, where nutrient concentrations decline only in July. In summer and early autumn the concentrations remain low. After November a gradual increase takes place due to the higher intensity of vertical mixing. Nitrate and phosphate contents both increased during the period 1974–1988 and decreased in recent years. However, the decreasing trend for total P was not so clear. The obvious depletion of the silicate-Si pool in 1985–1991 reversed after 1995 (Suursaar 1995; Astok et al. 1999; Stålnacke et al. 1999; Põder at al. 2003; Yurkovskis 2004). The sedimentation rate is roughly 2 mm $year^{-1}$ and accumulation of carbon is 5 g C m^{-2} (Danielsson et al. 1998).

Pärnu Bay (Fig. 10.1) is a shallow semi-enclosed water basin in the northeastern part of the Gulf of Riga. The surface area of the bay is about 700 km^2 and its volume is 2 km^3. The depth increases gradually from 7.5 m in its inner part (NE of the Liu–Tahku line) to 23 m in the southwestern part. The hydrological conditions of the bay are formed under the complex influence of meteorological processes, river discharge (Pärnu River, freshwater inflow 2 km^3 annually), and water exchange with the open part of the Gulf of Riga. The currents are generally weak in this area and are mainly wind induced.

Pärnu Bay suffers from extensive anthropogenic eutrophication. The town of Pärnu, with its 70,000 inhabitants, and the Pärnu River are the major sources of pollution in the bay. The contents of total N, total P, and silicate in the seawater have increased two-fold on average, and the primary production of phytoplankton increased substantially during the 1970s and 1980s (Ojaveer 1995; Tenson 1995). Since 1990, wastewater from the town of Pärnu has been mechanically and biologically treated. However, the Pärnu River, which is responsible for about 10% of the total riverine inflow to the Gulf of Riga, bringing about 40–50 t total P and >4,000 t total N into Pärnu Bay annually, is still a significant source of nutrients (Suursaar 1995).

10.2 Plankton Communities

The maximum water transparency (Secchi depth >5 m) occurs in the open Gulf of Riga and in the Irbe Strait. Transparency is lowest (<3 m) in the shallower parts of Pärnu Bay and the southern Gulf. The transparency has declined significantly only in summers due to an increase in biological production (Berzinsh 1995). Since 1991 the pollution load in the Gulf has decreased steadily and the water transparency has increased (Tenson 1995). The maximum chlorophyll *a* (Chl *a*) content reached 77.2 μg l^{-1} (Table 10.1c).

In the Gulf of Riga the vernal succession of phytoplankton starts at the end of March and ends in the first half of June. The arctic and arctic boreal diatoms (mainly *Achnanthes taeniata*, *Thalassiosira baltica*, *Chaetoceros wighamii*, *Ch. holsaticus*, *Melosira nummuloides*, *Navicula vanhoeffenii*) and dinoflagellates (*Peridiniella catenata*) are the most common groups during this period [see Chaps. 4 (Schiewer) and 7 (Kruk-Dowgiałło and Szaniawska), this volume]. During the growth phase the diatoms *A. taeniata* and *T. baltica* are dominant by biomass; during the peak phase, the same species plus *Ch. wighamii* are dominant, while *T. baltica* and *P. catenata* dominate the phase of decline. Summer succession starts in the middle of June and ends in the middle of September. In June eurythermal species still play an important role. In July and August polythermal Cyanobacteria (*Snowella lacustris*, *Aphanizomenon flos-aquae*, *Nodularia spumigena*), dinoflagellates (*Dinophysis acuminata*, *D. baltica*, *Ebria tripartita*), diatoms (*Actinocyclus octonarius*) and green algae (*Oocystis* spp.) dominate. In this period the summer bloom of Cyanobacteria is observed. In the second half of September most polythermal species disappear and are again replaced by eurythermal species. The species structure of phytoplankton varies between years. In the Augusts of 1979 to 1994, *A. octonarius*, *Dinophysis* spp., *Snowella* spp., *A. flos-aquae* and *Oocystis* spp. comprised most of the biomass, whereas since 1990 heterotrophic species have increased in importance. Interannual spring phytoplankton dynamics are determined mainly by climatic conditions. Depending on the harshness of the winter, either *A. taeniata*, *T. baltica* or *P. catenata* dominate the phytoplankton communities. In addition, the type of nutrient limitation affects the structure of phytoplankton communities. The long-term changes in phytoplankton in summer are likely determined by nutrient availability (Yurkovskis et al. 1999). Besides nutrient availability, salinity and the depth of the mixed layer explain the variability in phytoplankton biomasses (Seppälä and Balode 1999).

The phytoplankton of Pärnu Bay has been investigated since the early 1960s. As compared to adjacent areas [Gulf of Riga 222 taxa (Kalveka 1983); Haapsalu Bay 311 taxa (Piirsoo 1984)], Pärnu Bay is rich in phytoplankton species (442 taxa) and biomasses are higher [see Chaps. 3 (Schiewer) and 8 (Chubarenko and Margoński), this volume]. In very early spring, mostly at the end of March, the diatom *A. taeniata* blooms just below the ice cover at biomasses of up to 500 g m^{-3}. Following the breakup of ice, light conditions improve and a more diverse association of phytoplankton builds up. The prevailing species are *A. taeniata*, *Melosira arctica*, *Nitzschia frigida*, *Dactylococcopsis rhaphidioides*, *Gonyaulax catenata* and *Thalassiosira* sp. In early spring, *Chlamydomonas* sp. may be dominant over a period of several hours up to 1 day. In summer, the development of the cyanobacterium *Anabaena flos-aquae* is typical for Pärnu Bay. The frequency and duration of cyanobacterial blooms increased from 1985 to 1993. In the past decade the situation has partially reversed. Additionally, recent changes in phytoplankton involve the abundant occurrence of a new species, *Prorocentrum scutellum*, and the mass development of *Chlamydomonas* sp. in Pärnu Bay, where the production rate has increased from 1968 to 1987. Since then, primary productivity has again decreased (Tenson 1995). Phytoplankton production was estimated at 130–140 g C a^{-1} in the 1970s and 1990s, and at 190–200 g C a^{-1} in the 1980s.

In the Gulf of Riga zooplankton is represented by brackish water, eurythermal, oligothermal and polythermal species. The copepods *Eurytemora affinis* and *Acartia bifilosa* and the cladoceran *Bosmina coregoni maritima* dominate in summer samples [see Chaps. 6 (Schiewer), 8 (Chubarenko and Margoński), and 12 (Telesh et al.), this volume]. Interannual changes in mesozooplankton are related to phytoplankton biomass. Together with increasing Chl *a* values, the density of mesozooplankton increased between 1970 and 1990. Basic changes were due to an increasing share of *E. affinis* (until 1985) and *B. coregoni maritima* (after 1988). From 1991, following a decline in phytoplankton biomass, a decrease in zooplankton biomass was observed (Yurkovskis et al. 1999). The mean biomass value is between 0.1 and 1.3 mg wet weight (ww) l^{-1}. Since the 1990s, two non-indigenous species – the cladoceran *Cercopagis pengoi* and the polychaete *Marenzelleria neglecta* – have invaded the Gulf of Riga and their numbers have increased exponentially. In recent years, however, their densities have notably decreased (Ojaveer et al. 2004). This is in agreement with earlier observations that, in many cases, invading species achieve a peak population density and then decline due to interactions between the invader and its resources or enemies (Carlton 1996; Williamson and Fitter 1996).

Rotifers, cladocerans, copepods and meroplankton form mesozooplankton in Pärnu Bay. Although net samples may not give fully reliable estimates of rotifer density, this group has traditionally been determined in the mesozooplankton of the Baltic Sea (Simm 1995). Rotifers and copepods dominate the zooplankton community in the bay. A total of 17 species of rotifers has been observed. *Synchaeta baltica* and *Keratella quadrata* are the dominating species. The share of other genera is 0.22% of the total abundance. Cladocerans are represented by 11 species with *Bosmina coregoni maritima* as the dominant species of this group. Copepods are represented by 8 species. The most important copepods are *Acartia bifilosa*, *A. tonsa* and *Eurytemora affinis*. Other copepod species contribute only 0.33% of the total abundance. The North-American cirriped *Balanus improvisus* is the prevailing species in meroplankton. Mesozooplankton communities are not spatially different in Pärnu Bay. As expected, freshwater species, e.g. *Leptodora kindti* and *Diaphanosoma brachyurum*, are more frequent on coasts or in the estuary of the Pärnu River, whereas the brackish water species that comprise the majority of the mesozooplankton have similar abundances and biomasses in the inner and outer parts of the bay.

In general, the density of rotifers and copepods declined from the 1970s to the mid-1980s and then increased until 2004. For biomasses, the same trend holds for copepods only, as the ratio of smaller to larger rotifers (i.e. *Keratella* vs *Synchaeta*) has substantially increased in recent decades. The abundance and biomass of cladocerans have significantly declined since the early 1990s. Although low densities of cladocerans were recorded temporarily in 1974–1977 and 1986–1987, such low numbers were not previously measured. Prior to 1993, the density of cladocerans was on average 11,000 individuals (ind) m^{-3} and since then only 3,000 ind m^{-3}. The decline in cladoceran abundance coincides with, and can be partly explained by, the invasion of the predatory cladoceran *C. pengoi* (Kotta et al. 2004; Ojaveer et al. 2004). Meroplankton abundances are low but highly variable. Since the 1990s, the contribution of meroplankton among other zooplankton taxa has increased significantly, due mainly to the addition of the larvae of the North-American polychaete *M. neglecta* (Ojaveer and Lumberg 1995; Kotta and Kotta 1998).

As a result of these long-term trends, mesozooplankton seasonal dynamics in the 1970s/1980s differed substantially from those in the 1990s/2000s. Rotifers had highest abundances in April–June, cladocerans in August and copepods in September–November during the 1970s/1980s. In recent decades, rotifers dominated in May–August, copepods in April and September, and meroplankton in November. Based on biomasses, rotifers and copepods had an equal share in April–June, cladocerans dominated in July–August and copepods dominated again in September–November during the 1970s/1980s. In recent decades, copepods dominated practically the whole seasonal cycle, except for a meroplankton peak in November. Changes in the seasonal dynamics of mesozooplankton were due to an increase in the densities of *Eurytemora affinis*, *Acartia bifilosa*, *A. tonsa* and *Bosmina improvisus*, and the decreasing densities of *B. coregoni maritima* and *Pleopsis polyphemoides* in plankton assemblages.

The relationships between environmental variables and mesozooplankton are weak in the study area. Among abiotic variables, water temperature and solar radiation best explain the variability in mesozooplankton content. Trophic interactions strongly modify the effect of these abiotic variables (Kotta et al. 2004). For example, the dynamics of *E. affinis* and *Acartia* spp. is controlled by abiotic environment, site-specific competition for food, and predation by other invertebrates and fish. Both species are eurythermic and euryhaline but *Acartia* spp. prefer somewhat higher salinities than *E. affinis*. In spring the values of temperature and salinity in Pärnu Bay are low. Thus, *E. affinis* has a competitive advantage over *Acartia* spp.

and the peak of *E. affinis* is observed in June. In July, with higher temperature and salinity, *Acartia* spp. outcompete *E. affinis*. By this time fish larvae have reached the length that allows them to prey on adult copepods. In addition, the densities of other predators (e.g. mysids, *Cercopagis pengoi*, *Leptodora kindti*, Baltic herring *Clupea harengus membras* and gobies Gobiidae) increase notably. The females of *E. affinis*, which carry the egg sac, are a relatively easy prey for the latter predators. Consequently, the density of *E. affinis* decreases abruptly and that of *Acartia* spp. increases. The peak of *Acartia* spp. is observed in September, when temperature and salinity conditions are favourable for the species and predation by fish has decreased. Thus, we can conclude that the development of mesozooplankton depends on the interactions of many abiotic and biotic variables, and that the probability of finding strong correlations with a single environmental variable is rather low.

Over 50 fish species and two cyclostomes – the anadromous river lamprey (*Lampetra fluviatilis*) and the sea lamprey (*Petromyzon marinus*) – live in the basin. The river lamprey is common, whereas the sea lamprey is a rare species. The ichthyofauna of the Gulf of Riga consists mainly of marine boreal species. Here is found the most abundant and important commercial species, the Baltic herring (*Clupea harengus membras*), but also eelpout (*Zoarces viviparus*), flounder (*Platichthys flesus*), three-spined stickleback (*Gasterosteus aculeatus*), sprat (*Sprattus sprattus balticus*) and cod (*Gadus morhua callarias*). Migratory fish are of lesser importance. Among anadromous and catadromous fish, the most important are smelt (*Osmerus eperlanus*), whitefish (*Coregonus lavaretus*), vimba bream (*Vimba vimba*), salmon (*Salmo salar*) and historically also sturgeon (*Acipenser sturio*). The number of freshwater fish species is relatively high and they are important in coastal fisheries over the entire basin. The most common species are perch (*Perca fluviatilis*), pikeperch (*Sander lucioperca*) and several cyprinid species such as whitebream (*Blicca bjoerkna*), roach (*Rutilus rutilus*), and bream (*Abramis brama*). The Gulf of Riga is also known as an important refuge for glacial relict species. In addition to historical introductions [e.g. rainbow trout (*Oncorhynchys mykiss*), gibel carp (*Carassius gibelio*), common carp (*Cyprinus carpio*) Siberian salmon (*Oncorhynchys keta*), humpback salmon (*O. gorbusha*), sterlet (*Acipenser ruthenus*), Siberian sturgeon (*A. baeri*), and Russian sturgeon (*A. gueldenstaedti*)], two new alien species have been recorded recently: the round goby (*Neogobius melanostomus*) and the bighead (*Aristichthys nobilis*). Among aliens, however, only gibel carp is currently forming a self-reproducing population and increasing its abundance as well as colonising new areas (M. Vetemaa, personal communication).

At about 90% of total values, herring strongly dominates in commercial catches. Herring is exploited by trawls and trapnets. Autumn spawners are also caught by nets. Sprat and smelt may also occur in herring trawl catches. Amongst non-commercial fish, sticklebacks (three-spined stickleback and the nine-spined stickleback *Pungitius pungitius*) are highest in abundance. Currently, the abundance of piscivorous fish is low in open and deeper areas but relatively higher in coastal areas. Values were substantially higher in previous decades; for instance, abundant stocks of cod and four-horned sculpin (*Triglopsis quadricornis*) were recorded.

The Gulf of Riga herring has slow growth and one of the smallest age-specific length and weight values in the Baltic Sea. The stock is mostly stationary. Only a minority of the older herring leaves the Gulf after the spawning season in the summer–autumn period and afterwards returns to the Gulf (ICES 2004). There are two herring populations in the basin: spring- and autumn-spawning herring. Their growth dynamics are roughly similar and depend mainly on food availability, temperature and salinity. The two populations differ from each other in morphology and behaviour. Spring herring spawns in shallow coastal areas over almost the entire coast except for some areas in the south and south-west. Autumn herring spawns in relatively deeper areas, mainly in the northern and eastern part of the basin, including sea areas adjacent to Ruhnu Island. The Gulf of Riga is an important spawning area for open sea herring (Ojaveer and Gaumiga 1995, and references therein). In general, herring catches in the Gulf of Riga were relatively low and stable during the late 1970s and the 1980s (around 15,000 t annually), but increased since the early 1990s to their historically highest value in 2003 (ca. 41,000 t). The spawning stock biomass was relatively stable during the 1970s and the 1980s (ca 40,000–60,000 t), but has increased since late 1980s, reaching its highest recorded level in 2003 (ca. 124,000 t). The dynamics of recruitment (age 1) is rather similar to that of the spawning stock biomass (ICES 2004). The autumn spawning herring decreased from 45% of the total herring abundance in the mid-1970s down to 1–3% in the late 1980s and the 1990s (Ojaveer and Gaumiga 1995).

The abundance of sprat is usually relatively low. Sprat do not form a separate unit in the basin and migrate into the gulf during periods of high stock values in the open Baltic. The same is valid for cod, as the stock size in the Gulf of Riga is dependent on the dynamics of the external populations, which in turn depend on e.g. hydrological conditions in the main spawning grounds in the open Baltic Sea.

In Pärnu Bay, pikeperch is the second most important commercial fish species. As a result of over-fishing, the pikeperch catch in Pärnu Bay decreased by more than 10-fold during the second half of the 1990s and a total fishing ban was implemented during the early 2000s. Prior to 2000, a total of 194 t pikeperch was caught annually in the whole Gulf of Riga, but since then the catch is only 50 t a^{-1}. In the period 1960–1992, the fishing mortality rate was close to optimal but since then growth-overfishing of the stock has led to an annual loss of at least 38% of the potential long-term catch (Eero 2004). Year-class strength of the pikeperch is determined largely by the water temperature in May–June and by the abundance of sand goby (*Pomatoschistus minutus*) larvae (Erm 1976). The fish are characterised by size-dependent winter mortality at age 0, although longer durations of ice cover result in decreased size-dependent winter mortality (Lappalainen et al. 2000).

Three forms of whitefish can be distinguished in the basin: (1) sparsely-rakered sea-spawning whitefish (spawns near Ruhnu Island), (2) sparsely rakered anadromous whitefish (spawns in the Pärnu River), and (3) higher raker count whitefish of the eastern Baltic Sea (probably spawns in the southern part of the basin). Catches of whitefish were relatively high in the Gulf of Riga in the mid-twentieth century, but are currently low (14 t a^{-1} prior to 1992; 5 t a^{-1} since 1992). The stocking of fingerlings for first and second forms is not sufficiently effective to notably increase commercial catches (Sõrmus and Turovski 2003).

The somatic growth rate, the absolute fecundity, and the number of vertebrae of eelpout differ significantly between the eurytherm (the shallow Pärnu Bay) and the cold stenotherm (Ruhnu Deep) biotopes, suggesting the existence of different eelpout populations in the basin. Adaptation of the originally cold stenotherm eelpout enabled the fish to occupy more productive eurytherm biotopes and thus facilitate the increase in species abundance. As a result of such adaptations, the phenotype of the species has changed compared to the original (stenotherm) stock. Adaptation of the fish to new conditions has not caused changes in intraovarian growth (Ojaveer et al. 2004).

The permanent fronts at the Pärnu River estuary and in the Irbe Sound area, as well as at the seasonal thermocline, continuously attract large fish aggregations. From spring to autumn, pelagic fish (mainly herring but also sticklebacks, sprat and smelt) closely follow the location of the thermocline as it descends from near-coastal areas towards the open part of the basin. Higher indices for pelagic fish stomach fullness and a lower percentage of empty stomachs were recorded in and near the thermocline (Ojaveer 1997).

Clupeids feed mainly on zooplankton (but also on nektobenthic invertebrates in autumn), whereas the diet of smelt and sticklebacks is more diverse. The diet of pelagic fish has high spatio-temporal variability depending on prey availability and fish size. The highest dietary overlap was observed for herring and sprat. Sticklebacks are important food competitors for clupeids. The diet of smelt was in general different from that of other pelagic fish (excluding the diet of young smelt in Pärnu Bay; Ojaveer 1997).

Demersal and bentho-pelagic fish [like eelpout, fourhorned sculpin, flounder, eel (*Anguilla anguilla*), vimba bream and sea-snail (*Liparis liparis*)] exhibit clear selective feeding behaviour that is governed by the morphological features and physiological state of the predator. These fish prefer mysids, *Saduria entomon* and amphipods as prey. Feeding electivity varies by sub-region, season and fish size, and is determined mainly by prey availability, which is in turn dependent on the density and size of a given prey and interrelationships between various prey groups (Kostrichkina 1970). The diet of several commercially less-important fish like white bream, ruffe (*Gymnocephalus cernuus*) and eelpout is very similar to that of commercially more valuable species (e.g. eel, large perch, flounder; Schukina 1970).

The great cormorant (*Phalacrocorax carbo sinensis*) is the most important fish-consuming bird in the Gulf of Riga, inhabiting small islets with little human disturbance. The first nest was recorded in 1989 (Lilleleht 1995) and since then the abundance of cormorants has increased continuously, with a total stock of about 4,500 birds counted in 1999. A total of 920 successfully breeding couples and 420 destroyed nests (fishermen commonly destroy the nests of cormorants) were reported in 1999 (Veber 2001). In subsequent years the number of breeding couples increased gradually to 1,258 pairs in 2002 (T. Veber personal communication). Traditionally, cormorants are presumed to exhaust the resources of fisheries. However, in the Gulf of Riga in 1999, only 195 t of fish were consumed by cormorants compared to commercial catches of about 8,333 t. The diet of cormorants in the Gulf of Riga consisted mainly of commercially less important species such as the eelpout (66% of yearly consumption by weight) and also the numerically greater sandeel (*Hyperoplus lanceolatus*) and gobies, except in spring, when vimba bream was the most consumed species (11.4% of yearly consumption by weight). Perch, pikeperch, herring and roach were of less importance in the diet. The share of herring, as well as that of sticklebacks, was probably underestimated due to methodological difficulties. Feeding cormorants avoid fish with body width over 8–10 cm. Flounder was absent in the food samples while ruffe, perch, herring and pikeperch were consumed less than would be expected based on their relative abundances in test catches (Veber 2001).

10.3 Benthic Communities

The Gulf of Riga has a low phytobenthos diversity compared to the Swedish Baltic coast [see Chaps. 14 (Hill and Wallström) and 15 (Kautsky), this volume]. Among 69 listed phytobenthos species, 12 belong to Bangiophyceae, 16 to Fucophyceae, 21 to Chlorophyceae, 8 to Charophyceae and 12 to phanerogams. As is typical of most brackish water systems, the number of marine species decreases with decreasing salinity. In general, the number of species belonging to Bangiophyceae and Fucophyceae declines and that of Chlorophyceae increases along the falling salinity gradient (Martin 2000).

The overall quantitative characteristics of phytobenthic communities in the Gulf of Riga are comparable to those in other parts of the Baltic Sea with a similar salinity regime (Kautsky 1995; Kautsky et al. 1999). In general, the width of the *Fucus* belt is much smaller in the Gulf of Riga than in the Baltic Proper. At the same time, the maximum biomass is extremely high at 3.7 kg dry weight (dw) m^{-2} as compared to the biomass in the Baltic proper (Kautsky et al. 1999). Similar high values have been found only near the Island of Gotland, where the average phytobenthic biomass reached 1.4 kg dw m^{-2} within the *Fucus* belt (Kautsky 1989) or 460 g dw m^{-2} within the *Cladophora* belt in eutrophied areas adjacent to fish-farms in the Finnish archipelago (Mäkinen and Autio 1986).

It is estimated that almost half of the total phytobenthic biomass in the Gulf of Riga is composed of Fucophyceae (Martin 2000). About one-quarter of the total biomass is formed by *Fucus vesiculosus* and another quarter by the filamentous brown algae *Pilayella littoralis* and *Ectocarpus siliculosus*. The filamentous algal species *Ceramium tenuicorne*, *Polysiphonia fucoides*, *E. siliculosus* and *Sphacelaria arctica* have the highest constancy (Martin 1999, 2000).

The formation of phytobenthic communities in the Gulf of Riga is controlled by physical disturbances (ice, water level fluctuations, wave exposure) in the shallowest part of the phytobenthic zone, and by the availability of light and suitable substrate near the lowest depth limit of distribution of phytobenthos (Martin 2000; Kautsky et al. 1999).

Both historical and recent data demonstrate similar patterns in the depth distribution of the phytobenthic biomass in the Gulf of Riga (Martin 1999; Kautsky et al. 1999). In general, the maximum biomass is observed at the depth of 1–3 m. The species forming this biomass change between locations, in most cases being *Fucus vesiculosus* and its epiphytes. The rapid decrease of phytobenthic biomass below the *Fucus* belt is typical for the northern Baltic Proper, which lacks other algae with large thalli. On soft substrates, and also in shallow areas due to higher turbidity, phanerogams and Charophytes are usually light-limited. This situation can be observed in Pärnu Bay (Kautsky et al. 1999; Kukk and Martin 1992).

The net primary production of dominating phytobenthic species in the northern part of the Gulf of Riga has been estimated (Paalme and Kukk 2003; Paalme 2005). The ephemeral green alga *Enteromorpha intestinalis* had a markedly higher net primary production rate (26 mg O_2 g dw^{-1} h^{-1}) as compared to other opportunistic algae. The ephemeral brown alga *P. littoralis* and the ephemeral green alga *Cladophora glomerata* were characterised by relatively high net primary production rates (up to 13 mg O_2 g dw^{-1} h^{-1}). The perennial algal species *F. vesiculosus* and *Furcellaria lumbricalis* had net primary production rates about 2–5 times lower.

Macrofauna in the Gulf of Riga is characterised by a small number of benthic invertebrate species [see Chaps. 14 (Hill and Wallström) and 15 (Kautsky), this volume] due to low salinity and uniformity of habitat, especially in the southern part of the Gulf. Järvekülg (1979) found 139 zoobenthic species in the Gulf of Riga (including oligochaetes, harpacticoid copepods, ostracods and insect larvae). Other authors (Kotta et al. 1998a and references therein; this chapter) added six other species, including two alien species. Freshwater species are prevalent in the near-shore areas and euryhaline marine species in the open part of the Gulf. More than 50% of the Gulf is occupied by the *Macoma balthica* association or that of *Cerastoderma glaucum*, *Mya arenaria* and *M. balthica*. Besides the latter bivalves, Oligochaeta and *Monoporeia affinis* are frequent benthic invertebrates.

The density and biomass of macrozoobenthos is lower in the southern part of the Gulf of Riga than in other regions. Within the coastal range, and between the surface and 15 m depth, the macrobenthic biomass is currently estimated at 16 g dw m^{-2} in the southern gulf and 37 g dw m^{-2} in the northern gulf. This is due to several reasons. First, unstable sediments coupled with intensive currents and wave action make the colonisation of *M. arenaria* and *C. glaucum* difficult; secondly, lack of hard substrate resulting in unfavourable conditions for *Mytilus trossulus*, *Dreissena polymorpha* and *Balanus improvisus*; and finally, poor bottom vegetation resulting in low densities of phytophilous species.

Less eutrophicated regions of the Gulf of Riga have relatively diverse macrozoobenthic assemblages. The main determinants of species composition, abundance and biomass of macrozoobenthos in the area are type of substrate and vegetation. Crustaceans, oligochaetes and polychaetes are most abundant, and burrowing bivalves have the highest biomass, in unvegetated areas on soft substrate. The highest densities are observed in less eutrophicated areas and the lowest densities in more eutrophicated areas. Vascular plant communities are numerically dominated by insect larvae and gastropods of freshwater origin while the biomass is dominated by burrowing clams. Mainly phytophilous crustaceans and suspension-feeding bivalves are found in algal communities on hard substrates. Along with the type of substrate and the vegetation, a third important variable in explaining the structure of benthic invertebrate assemblages in the Gulf of Riga is salinity. Higher proportions of brackish and fresh water species were found in the north-eastern and southern parts of the Gulf, in areas subjected to larger freshwater inflow (Kotta 2000; Kotta et al. 2000a).

In the relatively clean coastal areas, the total abundance and biomass of macrozoobenthos have not changed significantly in recent decades. However, there is an indication that, due to the mass occurrence of filamentous algae, the abundance and the biomass of epibenthic macrofauna have increased in macrovegetation (Kotta et al. 2000b). On the other hand, species diversity has declined markedly in unvegetated areas (Kotta et al. 2000a). Several species, e.g. *Idotea chelipes*, *Asellus aquaticus*, *Jaera albifrons* and *Lymnaea peregra*, which were common in the 1970s, have become rare in extensive areas. All these species are considered to be phytophilous. Hence, it is likely that the decline in species diversity reflects the overall impoverishment of the benthic vegetation in the area.

More eutrophicated regions, such as Pärnu Bay and the sea areas adjacent to the Daugava River, have significantly lower benthic diversity than the areas described above. The abundance and biomass structure of macrozoobenthos are determined mainly by nutrient load, type of substrate and salinity. Suspension feeding bivalves dominate on the hard bottoms, *D. polymorpha* at salinities less than 5 psu, and *M. trossulus* in more saline environments. Occasionally, these two species are found together in the same samples. Soft bottom assemblages are characterised by the deposit-feeding bivalve *M. balthica* and partly by the deposit-feeding amphipod *Corophium volutator* (Kotta and Kotta 1995; Kotta 2000).

In eutrophicated areas, the biomass of macrozoobenthos has increased notably during the last few decades. Similarly to the cleaner coastal sea, many phytophilous species have reduced their distribution area considerably. The most uniform assemblages have developed in localities where the highest load of nutrients was recorded (Kotta and Kotta 1995, 1997).

To conclude, the phytobenthos may be considered a factor of prime importance for the development of benthic invertebrate assemblages in the cleaner coastal sea whereas the sediment type and nutrient load affect macrozoobenthos mostly in more eutrophicated and deeper regions. The species diversity is significantly lower in more eutrophicated regions (e.g. Pärnu Bay and the estuary of the Daugava River). Notable changes in the structure of macrozoobenthic assemblages have taken place both in less and more eutrophicated regions during recent decades. This is expressed by a reduction in species diversity and an increase in the dominance of suspension-feeding bivalves.

The Ponto-Caspian bivalve *Dreissena polymorpha*, or zebra mussel, was first recorded in the Gulf of Riga (Pärnu Bay) in the middle of the nineteenth century. Today, permanent populations occur in the northeastern, eastern and southern parts of the Gulf of Riga (Kotta et al. 1998b). In the northeastern part of the Gulf of Riga, the species inhabits depths between 1.5 and 5 m. In the southern and eastern parts of the gulf the zebra mussel occurs between 4 and 10 m. In these areas, the lack of hard substrate and the high turbidity prevent the mussels from inhabiting the shallow waters. Generally, the abundances and biomasses of this species are low (around 50 ind m^{-2} and 5 g m^{-2}), with the exception of the southernmost part of the Gulf of Riga, where abundances and biomasses reach as high as 8,400 ind m^{-2} and 1,463 g m^{-2}, respectively. This could be explained by the inflow of fresh water from the Daugava River, resulting in lower salinity and high nutrient concentrations, i.e. high phytoplankton biomass. During the spring and summer of 1996, salinity and the concentrations of Chl *a* in the southern Gulf of Riga varied between 3.6–4.5 ppm and 8.6–30.1 μg l^{-1}, respectively, whereas these values were 5.4–5.6 ppm and 8.3–22.0 μg l^{-1} in the northern part of the Gulf. The current dispersion area of *D. polymorpha* coincides with the area previously documented in the literature (see references in Kotta et al. 1998a) with the exception that Shurin (1961) also found a population of zebra mussels living on a flowering plant in the southern coastal sea off Saaremaa Island. *Dreissena* has an upper tolerance of salinity at ca. 4 psu and is therefore excluded from vast areas of the Gulf.

The detritus-feeding polychaete *Marenzelleria neglecta* was introduced into the Gulf of Riga in 1988 (Lagzdins and Pallo 1994). The polychaete has spread quickly to most parts of the Gulf. In shallower areas, the species prefers to live in sand or gravel substrates. Its abundance is higher in more densely vegetated areas. Deeper down, *M. neglecta* is confined to silty clay bottoms. Only a small proportion of the total polychaete population consists of *M neglecta*, except for the deepest parts of the Gulf of Riga, where it may dominate the total abundance. Concurrent with the invasion of *M. neglecta*, the density of the native polychaete *Hediste diversicolor* has dropped significantly (Kotta and Kotta 1998). After an increase during the first decade of invasion, the density of *M. neglecta* has since declined. Nevertheless, the larvae of this polychaete are today the most prevalent taxa in meroplankton and have, in general, doubled the total meroplankton abundance.

Nektobenthic mysid shrimps are an important component in the ecosystem of the Gulf of Riga. They link benthic and pelagic environments as well as primary and secondary production to higher trophic levels. The distribution of mysid species in the whole Gulf of Riga has been studied periodically since the 1950s (see references in Kotta et al. 1998a; Kotta 2003). However, most of these studies each represent only a short period of the productive season (e.g. June–July) and are limited in their spatial scale. The quantitative estimates of mysid distribution in the central and southern parts of the Gulf were based on traditional zooplankton and zoobenthos sampling methods. Hence, the presented values are likely to be biased and do not reflect the actual conditions in the sea. The only comprehensive large-scale study on the distribution of mysids dates back to the years 1974–1975 (Kotta and Kotta 1999) and allows us to make some generalisations about the past distribution of mysids at a basin scale and to demonstrate the relationship between mysid distribution and ambient environment. Long-term data on the abundances and biomasses of mysids are available only from the northeastern part of the Gulf of Riga and Pärnu Bay.

A total of five mysid species have been found in the Gulf of Riga – *Neomysis integer*, *Mysis mixta*, *M. relicta*, *Praunus flexuosus* and *P. inermis*. The mysid *N. integer* is the most prevalent species in the Gulf of Riga and practically the only mysid species found in Pärnu Bay. This species inhabits the whole Gulf (see Kotta 2003 and references therein) and is occasionally found in rivers at 1 km distance upstream from river estuaries (Sanina 1961). On average, it dominates at depths above 10 m but is also abundant at deeper study sites. *M. mixta* and *M. relicta* inhabit depths below 5 m. *M. mixta* prefers intermediate depths (10–40 m), whereas *M. relicta* is confined mostly to the deepest part of the Gulf (20–55 m). Both species are particularly abundant on steeper slopes where strong temperature gradients were observed. This can be explained by the higher productivity of frontal areas supporting high densities of pelagic consumers, among them the mysids. The abundances of *M. relicta* are relatively low in the deepest parts of the Gulf of Riga, which is most likely related to the occasionally unfavourable oxygen conditions measured in the area. *P. flexuosus* and *P. inermis* inhabit only coastal areas at depths of down to about 20 m. *P. inermis* has a wider distribution area, and higher abundances and biomasses than *P. flexuosus*. Both species are phytophilous and are found in areas with abundant benthic vegetation. These species spend their entire lifecycle within the phytobenthic zone and, unlike *M. relicta*, do not perform extensive seasonal migrations to deeper areas in the Gulf of Riga.

Most of the variability in the distribution of *M. mixta*, *M. relicta* and *N. integer* can be explained by temperature. *M. relicta* is confined to the deeper parts of the gulf where the temperature is constantly low. *N. integer* migrates to the coastal areas in late spring. Its production increases with water temperature in summer. In the course of the summer, as the depth of the thermocline increases, *Neomysis* migrates down to deeper areas but stays above the thermocline. An opposite migration takes place on the other side of the thermocline, where *M. mixta* avoids the expanding warm water. The distribution of the different mysid populations becomes more homogeneous after storms in autumn (Kotta and Kotta 1999).

On the bank slopes, the distribution of mysids is aggregated and related to the temperature conditions. Higher densities of mysids coincide with areas where the thermocline boundaries touch the sea floor. In biotopes with strong temperature gradients, mysids are often observed outside their natural temperature preferences. *M. relicta* was recorded above the thermocline at 10°C and *N. integer* below the thermocline at 2°C (Kotta and Kotta 2001a).

All three mysid species perform diel vertical migrations. These migrations are more intense in late summer and autumn than in spring and early summer. The vertical migration of mysids is related to light levels. The proportion of juveniles (<11 mm) in the vertical migrations is higher than that of adults, and the migration of adults is delayed about 3–4 h. Usually, *N. integer* appears in the water column first in the evening, followed later by *Mysis* spp (Kotta and Kotta 2001b). During the low-light-induced upward migrations, mysids prey on phytoplankton and zooplankton while in daytime they feed mainly on detritus (Chekhova 1961; Sanina 1961). Vertical migrations of mysids in stratified waters allow them to save their energy for reproduction (food is less sparse in warmer waters, the consumption of energy is lower at lower temperatures) and better escape from predators.

In winter and spring, the density of overwintering generations of *N. integer* is low. The increase in the biomass of the cohorts is due to the growth in length.

N. integer breeds continuously through the summer and has two recruitment peaks. The mortality of the species increases abruptly from September onwards. *M. mixta* releases its young in April. The abundance of juveniles peaks in mid-June. In the course of the summer and autumn, the abundance of *M. mixta* decreases steadily while the average length of individuals increases. The highest mortality is observed through August until October. This decline is likely a result of herring and smelt predation on the mysid population.

In the northeastern Gulf of Riga, the highest abundance and biomass values of *M. mixta* and *M. relicta* are recorded after severe winters, whereas a high abundance of *N. integer* is often expected during warm summers. The latter can probably be explained by a positive correlation between temperature and the density of zooplankton (Kotta 2003). The significant correlations between the density of copepods, rotifers and mysids suggest that zooplankton is the main diet of the mysids in the study area (Kotta et al. 2004).

10.4 Benthic–Pelagic Coupling and Eutrophication

In order to describe the processes in different subsystems of the Gulf of Riga, the popularity of experimental studies as a tool to resolve these problems has increased notably in recent decades. Such studies involve, for example, the development of methods to detect human-induced changes relative to natural variability, or the interactions and flows of matter and energy between different components of the system by quantifying habitat and feeding choice of various invertebrate species (e.g. Kotta 2000; Orav-Kotta 2004).

It was found that in hydrologically active areas or upwelling areas, and in sites adjacent to municipal discharges and river estuaries, macrofaunal biomasses and abundances were consistently higher than in the adjacent sea despite similar bottom topography and sediment characteristics (Kotta et al. 2003). Two different types of hydrologically active areas may be distinguished in the Gulf of Riga: upwelling areas and the fronts between different sub-basins. The former type is represented by, for example, Kihnu Shallow. This area is characterised by a steep bank slope, which forms a barrier to the prevalent near-bottom currents and, hence, results in daily upwelling. The latter type is found at the Irbe Strait and the Suur Strait. As a result of the mixing of different water masses, deep-water nutrients are transported to the surface layer supporting higher pelagic (Ojaveer and Kalejs 1974) and, consequently, benthic production (Järvekülg 1979). Both types of hydrologically active areas are ranked as important fishery grounds, reflecting their high productivity (Ojaveer and Kalejs 1974). In general, macrozoobenthos does not respond to changing nutrient concentrations in the upwelling areas. In the Irbe Strait, however, the biomass of macrozoobenthos was positively correlated with total N in the near-bottom water. Neither the concentration of nutrients nor the biomass correlated with the species number of macrozoobenthos. The deposit feeder *Macoma balthica* comprised the bulk of the biomass. The abundance of nektobenthic mysids was very high in the studied strait. A similarly high benthic biomass was observed only in the heavily polluted Pärnu Bay and in the sea adjacent to the Daugava River.

Due to its shallowness and isolation, the Gulf of Riga is strongly influenced by riverine input. Increasing riverine input, intense agriculture, and the rapid development of industry and urbanization has resulted in high loads of nutrients in the Gulf of Riga since the 1960s. In the 1970s and 1980s the nutrient pool in the Gulf has increased significantly. In the 1990s, however, nutrient loadings decreased as a consequence of the decline in freshwater inflow, progress in sewage treatment and economic recession in the countries around the Gulf. The load of nitrogen and phosphorus was estimated at 100,000 and 2,000 t a^{-1} in the 1970s, 125,000 and 2,400 t a^{-1} in the 1980s and 100,000 and 2,100 t a^{-1} in the 1990s (HELCOM 1996; Yurkovskis 2004). Following the changes in nutrient loads, abrupt changes in phytoplankton and zooplankton communities were observed. In response to high direct nutrient loading, and the corresponding changes in biogeochemical processes, the ecosystem of the Gulf of Riga moved from N limitation (1974–1978) to P limitation (1979–1990). The shift in zooplankton community structure towards the dominance of herbivores in 1970–1998 indicates an increase in eutrophication in the Gulf of Riga. During the 1990s, the concentration of total P increased and that of total N decreased. Consequently, total P lost its role as the limiting nutrient. At the same time, Si deficiency started to occur during the spring bloom. As a result, the role of dinoflagellates, as well as that of green, blue-green and heterotrophic algae increased and more intense competition between herbivorous species of zooplankton developed. The nutrient limitation may also have significant spatial variability. In spring 1999 the limiting nutrient appeared to be total P at the entrance of the Daugava River and Si in the central Gulf. There was no correlation between limiting nutrient and spring phytoplankton community structure. In summer there was a good correlation between phytoplankton biomass and dissolved organic P. Both land-based loading and internal processes have contributed to the long-term change in the nutrient pool. The dynamics of total N was determined by riverine loading (N leaching from watersheds). The variation in total P was due to internal processes, i.e. P outflux from sediments. The eutrophication effects of river discharge were reduced by the high pelagic buffering capacity of the Gulf, which was related to the variability in pelagic–benthic coupling and efficiency of pelagic nutrient cycling (Floderus et al. 1999; Reigstad et al. 1999; Tamminen and Seppälä 1999; Wassmann and Tamminen 1999; Yurkovskis et al. 1999; Põder et al. 2003; Yurkovskis 2004).

Phytobenthos has higher biomass in areas away from local eutrophication sources and larger fresh water inflows, and these communities are dominated by *Fucus vesiculosus*. The overall high nutrient concentrations in the Gulf of Riga favour the development of high phytobenthos biomasses in remote areas of the Gulf. In more eutrophicated areas (e.g. Pärnu Bay) the biomass of phytobenthos is lower and the communities are dominated by Chlorophyceae. The biomass of these communities is confined to the shallowest part of the areas and decreases rapidly with depth. Low biomasses are due mainly to substrate limitation and high water turbidity (Kautsky et al. 1999; Martin 2000).

Information on the structure of phytobenthic communities in the Gulf of Riga has been available since the late nineteenth century and quantitative data since the 1960s (see references in Martin 2000; Martin et al. 2004). These studies show that many species that inhabited the Gulf of Riga at the beginning of the twentieth century, no longer occur there. *Cladophora fracta*, *Anfeltia pilicata*, *Aglaothamnion roseum*, *Dictyosiphon chordaria*, *Chorda filum* and *Elachista fucicola* have disappeared in the southern part of the Gulf. Several species of Charophyceae, Fucophyceae and Bangiophyceae have disappeared in a given area or in the whole Gulf, whereas the occurrence of Chlorophyceae species has been constantly rising. In Pärnu Bay, the filamentous green alga *Cladophora glomerata* has expanded its distribution and biomass compared to the beginning of the 1960s. The extensive decline of *F. vesiculosus* and *Furcellaria lumbricalis* can be related partly to increased industrial and municipal pollution. However, this decline was delayed about 10–15 years in the Gulf of Riga as compared to other parts of the Baltic Sea.

The dynamics of algal communities has high spatial variability. On the southern coast of Saaremaa Island the phytobenthic biomass has decreased significantly at the depth interval of 2–3 m since 1987. No significant changes have been observed at other depths. The changes have occurred on both mixed sediment bottoms and sand-dominated substrates. More recently, a continuous decline process was described in Kõiguste Bay during the years 1995–1998 (Kotta et al. 2000b; Martin et al. 2003). During this time period, *F. vesiculosus* was replaced by the filamentous brown alga *Pilayella littoralis*. However, in recent years the *Fucus* belt has partly recovered, as also observed in many parts of the Baltic Sea.

In the open Gulf of Riga the biomass of soft bottom macrofauna increased substantially in the 1970–1980s and decreased down to its initial levels in the 1990s. The main changes were due to the bivalve *M. balthica*. The abundance of the prevailing amphipods *Monoporeia affinis* and *Pontoporeia femorata* increased in the shallower parts and decreased in the deeper parts of the Gulf. To date, there has been no recovery of this species in these areas. The annual production of macrozoobenthos was estimated at 5.2 g C m^{-2} during the 1980s and 2.6 g C m^{-2} during the 1990s (Cederwall et al. 1999).

In the late 1950s the effect of Pärnu River on macrozoobenthos was unclear. Highest biomasses were observed in the outer part of Pärnu Bay (Järvekülg 1960). Since 1960 the macrozoobenthos biomass in the bay has gradually increased. Kotta and Kotta (1995) reported a 1.5- to 8-fold increase in the total biomass of macrozoobenthos, as well as the disappearance of several oligosaprobic and some mesosaprobic species. The densest communities have shifted from open areas towards the mouth of Pärnu River where the most intense influx of pollutants occurs. In 1991, maximum densities were estimated at 20,500 ind m^{-2} and biomass at 200 g dw m^{-2}. The amphipod *Corophium volutator* and the bivalve *Macoma balthica* accounted for more than 90% of macrozoobenthos density. *M. balthica* dominated in biomass. Since the late 1980s, the density of the nectobenthic mysid *Neomysis integer* in Pärnu Bay has increased many times over (Kotta et al. 2004). These changes were likely triggered by increased contamination of the bay with nutrients originating from municipal discharges and agricultural run-off. As a result, the content of total

N, total P and silicate in the seawater increased two-fold on average, and the primary production of phytoplankton increased substantially in the 1970s and 1980s (Ojaveer 1995; Tenson 1995). Following the economic recession of Estonia in the 1990s, macrobenthic biomass declined significantly. During the recession the intensity of agriculture was substantially reduced, resulting in a decline in the nutrient content of the seawater (Suursaar 1995). However, since 2000 nutrient concentrations and macrozoobenthos biomass have been gradually increasing. The nutrient dependence of macrozoobenthos biomass is indicated by the significant positive correlation between benthic species biomass and total N and total P values in the water column. With the increase in nutrient concentrations, the densities of *C. volutator*, *M. balthica* and Oligochaeta increased while those of *Prostoma obscurum*, *Hydrobia ulvae* and *H. ventrosa* decreased. On soft substrate, the diffuse leakage and the point source nutrient supply had a similar effect on macrozoobenthic species composition and dominance structure. However, functional diversity was higher in areas subjected to diffuse nutrient input in comparison to areas receiving a point source nutrient discharge (Kotta et al. 2000a). Dredging and dumping activities resulted in an increased concentration of organic matter in both the water column and the upper layer of sediment. Thus, as with eutrophication, large-scale dredging increases the abundance and biomass values of opportunistic invertebrates, especially in areas adjacent to dredging activities. In Pärnu Bay these species are *Macoma balthica*, *Corophium volutator* and *Hediste diversicolor*. In most cases, benthic communities reach equilibrium within 2–3 years. Recolonisation and stabilisation may take longer period at dredging and dumping sites.

Shurin (1953) demonstrated the impact of the Daugava River on the species composition of macrozoobenthos in the southern part of the Gulf of Riga. Relatively diverse zoobenthic communities with a high proportion of freshwater species were found down to a depth of 3 m. Freshwater species disappeared gradually between a depth of 3 and 8 m. Deeper down (8–20 m) only three benthic species were recorded: *Macoma balthica, Saduria entomon* and *Monoporeia affinis*. Järvekülg (1979) stated that the density of zoobenthos was lower (242 ind m^{-2}) at the mouth of the Daugava River as compared to the surrounding areas. Similarly, Lagzdins (1975) found a 4- to 7-fold difference in the number of species between the mouth of the Daugava River and the open sea. Strikingly, the biomass and abundance values of zoobenthos in the mouth of the Pärnu River were considerably higher than in the adjacent areas (Järvekülg 1960). Gaumiga and Lagzdins (1995) reported that biomass and abundance of zoobenthos in the Gulf of Riga have increased considerably over the course of the last 30 years.

On hard substrate, the effect of eutrophication is expressed as excessive growth of filamentous algae. As a consequence of the higher coverage of filamentous algae, the stock of the most prevalent algal species *Fucus vesiculosus* diminished notably in wide areas in the mid-1990s. The most prevalent benthic herbivore, the isopod *Idotea baltica*, switched to *Furcellaria lumbricalis* as a habitat and *Pilayella littoralis* as a food. The highest densities of *I. baltica* were observed in shallower areas where the proportion of the filamentous epiphyte *P. littoralis* on *F. lumbricalis* was highest. Removing fast-growing epiphytes, isopods protected the slow-growing *F. lumbricalis* against the nuisance alga *P. littoralis* (Kotta 2000; Kotta et al. 2000b). At very high eutrophication levels drift algal mats replaced attached algal communities. Benthic mesograzers significantly consume drift algae. Grazing by the dominant mesograzers *I. baltica* and *Gammarus oceanicus* is primarily a function of the photosynthetic activity of the algae. When the algae are photosynthetically active, their consumption is almost negligible. Grazing increases exponentially prior to their decomposition. In the presence of herbivores, the complete breakdown of drift algae was reached already within 30–40 days. Without the mesograzers, algal decomposition took place within 70 days (Paalme et al. 2002). Benthic suspension feeders increase algal decomposition indirectly. When algae were photosynthetically active, *Mytilus trossulus*, through the fertilising effect of biodeposition, promoted algal growth and, hence, retarded algal decomposition. Since the early 2000s, *F. vesiculosus* has partly recovered and again provides both food and shelter to *I. baltica* (Orav-Kotta 2004).

The bivalves *Dreissena polymorpha* and *Mytilus trossulus* are the most important suspension feeders on the hard bottoms in the coastal area of the Gulf of Riga, and contribute significantly to benthic–pelagic coupling. As a result of increasing eutrophication, the development of dense populations of suspension feeding mussels has been observed. The biodeposition and clearance of the bivalves increase with ambient temperature. In more eutrophicated regions biodeposition and clearance capacity rates increase curvilinearly with the ambient concentrations of Chl *a*, and level off at high food concentrations. In less eutrophicated conditions a linear model gives the best fit, suggesting that a saturation level is not obtained. Additional variation in biodeposition and clearance is explained by the interaction of water temperature, current velocity and Chl *a*. Salinity has a significant effect on the *D. polymorpha* biodeposition and clearance rate. On average, the population of suspension feeders clears from 3 to 2,426% of overlaying water in the littoral area daily, constituting an important sink for primary production. The impact of benthic suspension feeders on pelagic communities is smaller in eutrophicated systems relative to cleaner environments (Kotta et al. 2005).

The introduction of a successful predator may destabilise the whole ecosystem. Prior to 1990, bottom-up effects dominated in the pelagic food web in Pärnu Bay, whereas top-down effects were more common in subsequent years. The population dynamics of mysids showed low and relatively stable abundances from the early 1970s to the late 1980s, with high and fluctuating abundances thereafter. Zooplankton abundance and biomass, and the severity of winter, were the best predictors of mysid abundance and biomass. Since 1991, the maximum density of mysids has declined, with concurrent invasion of the predatory cladoceran *Cercopagis pengoi*. Strong predation pressure of *C. pengoi* on other zooplankton taxa has partly released phytoplankton from zooplankton grazing and resulted in the food limitation of mysids (Kotta et al. 2004).

Dreissena polymorpha and *Marenzelleria neglecta* are ranked among the most influential benthic exotics in the Gulf of Riga (Kotta 2000). The high filtering capacity gives *D. polymorpha* the potential to influence phytoplankton communities and, via phytoplankton, control the stock of zooplankton and planktivorous fish communities. Unconsumed proportions (pseudofaeces) and faeces are deposited on the benthos, supporting also abundant communities of deposit feeders. Field experiments have indicated that *M. neglecta* enhances the content of sediment Chl *a* and reduces the growth and survival of the polychaete *Hediste diversicolor* and the growth of the amphipod *Monoporeia affinis*. Adult specimens of the bivalve *Macoma balthica* have a negative effect on the *Marenzelleria* population. Competitive interactions between *M. neglecta* and native fauna may explain why polychaete densities are low when densities of *M. balthica* are high, and why *M. affinis* has not recovered in areas with high density of *M. neglecta*. According to these experiments, *M. neglecta* should prevail only in such biotopes where it can avoid competitive interactions with *M. balthica*. Such areas appear to be adjacent to river mouths, where the food for deposit feeders is in excess and bivalves are stressed by low salinity, but likely also in deeps where the amphipods *M. affinis* and *Pontoporeia femorata* prey on the bivalve spat. Because *M. balthica* is one of the most common species in the soft bottom sediments in the Gulf of Riga, competitive interactions between *M. neglecta* and *M. balthica* appear to be a key factor in keeping the densities of *M. neglecta* low in the study area. The most recent newcomers in the Gulf of Riga are the gammarid amphipod *Gammarus tigrinus* and the round goby *Neogobius melanostomus*. In 2003, the amphipod invaded a mesocosm experiment in which we studied the susceptibility of different communities to the addition of species from neighbouring areas. The species was associated mainly with artificial communities that contained the cockle *Cerastoderma glaucum*. Densities ranged between 100 and 800 ind m^{-2}. In 2004 *Gammarus tigrinus* was already a dominating crustacean species in more sheltered parts of the bay (Herkül et al. 2006). The first report of round goby was in Pärnu Bay in 2002 (Ojaveer and Špilev 2003). Since then no other findings of the species have been confirmed.

Smelt catches decreased drastically during the 1980s and the first half of the 1990s. The most probable reasons for this were the prevalence of unfavourable meteorological conditions and the continuous pollution of river spawning grounds. This resulted in a drastic drop in the abundance of smelt larvae on the spawning grounds in Pärnu Bay. In recent years, the larval abundance of smelt has again increased and certain signs of recovery of landings are evident (Špilev and Turovski 2003).

As a result of industrial and residential developments, conditions in the Gulf of Riga had deteriorated considerably by the end of the 1950s. In order to fight against pollution, effective measures had to be launched, but it was not until the 1970s and 1980s that central water treatment plants (mechanical-chemical) were put into operation. Since the middle of the 1980s, the public has taken an active approach to the protection of the natural environment, putting pressure on administrative authorities and improving water protection related activities. In the 12 years since independence, Estonia and Latvia have achieved satisfactory results in implementing water use and protection measures, and in drafting and implementing water-related legal acts (Velner 2005).

10.5 Conclusions and Summary

The Gulf of Riga with its wide and shallow basin represents an unique system in the Baltic Sea with a well-aerated and biologically productive benthal. Consequently, many benthos-related invertebrate and fish species have formed strong populations in the Gulf of Riga. Despite its wide sublittoral areas, the gulf is characterised by a low phytobenthic diversity compared to the Swedish Baltic coast (cf. Chap. 15 by Kautsky, this volume). Although high nutrition rates enable remarkably high phytobenthic biomasses locally, the whole system undoubtedly relies on planktic primary production. Naturally poor conditions for phytobenthos involve low salinity in the whole Gulf, sandy littoral with very low stability in the southern part of the Gulf and turbulence-induced poor water transparency in the open coastal areas. Eutrophication-related processes in recent decades have further impoverished phytobenthic communities and depleted the related phytophilous macrofauna.

Pärnu Bay is a shallow and partly exposed area with high fresh water inflow and pollution load. The low biomass of macrophyte communities in its shallowest regions is in marked contrast to the high phytoplankton diversity and biomass in the bay. As a consequence of increasing eutrophication, the share of marine macrofauna has increased. This is in contrast with other areas of similar salinity in the Northern Baltic Sea where human-induced pollution has resulted in a collapse of marine species and a full dominance of small opportunistic freshwater forms.

Atmospheric processes in the northern Atlantic, together with human activities, explain the interannual variations in nutrient pools and primary productivity. Coincident trends in climate and human activities prevent their clear quantitative separation. Trophic interactions control the development of pelagic and benthic communities at higher trophic levels in the Gulf of Riga. The introduction of an effective intermediate predator may partly explain the major shift in the functioning of the pelagic food web in Pärnu Bay. Declines in anthropogenic inputs and nutrient pools point to an improvement in the water quality of the Gulf of Riga in the 1990s and 2000s.

References

Andrushaitis A, Seisuma Z, Legzdina M, Lenshs E (1995) River load of eutrophying substances and heavy metals into the Gulf of Riga. In: Ojaveer E (ed) Ecosystem of the Gulf of Riga between 1920 and 1990. Estonian Academy, Tallinn, pp 32–40

Astok V, Otsmann M, Suursaar Ü (1999) Water exchange as the main physical process in semi-enclosed marine systems: the Gulf of Riga case. Hydrobiologia 393:11–18

Berzinsh V (1995) Dynamics of hydrological parameters of the Gulf of Riga. In: Ojaveer E (ed) Ecosystem of the Gulf of Riga between 1920 and 1990. Estonian Academy, Tallinn, pp 8–31

Carlton JT (1996) Pattern, process, and prediction in marine invasion ecology. Biol Conserv 78:97–106

Cederwall H, Jemakovs V, Lagzdins G (1999) Long-term changes in the soft-bottom macrofauna of the Gulf of Riga. ICES J Mar Res 56 [Suppl]:41–48

Chekhova VA (1961) Vertical migration of *Mysis mixta* and *M. oculata f. relicta* in the Gulf of Riga (in Russian). Tr Nauchno-Isseld Inst Rybn Khoz SNKH Latv SSR 3:307–327
Danielsson A, Carman R, Rahm L, Aigars J (1998) Spatial estimation of nutrient distributions in the Gulf of Riga sediments using cokriging. Estuarine Coastal Shelf Sci 46:713–722
Eero M (2004) Consequences of management of pikeperch (*Stizostedion lucioperca* L.) stock in Pärnu Bay (Baltic Sea) under two different economic regimes, 1960–1999. Fish Res 68:1–7
Erm V (1976) On the factors determining pike-perch year-class strength in Pärnu Bay of Pärnu (in Russian). Rybokhoz Issled Basseine Balt Morya 12:51–58
Floderus S, Jähmlich S, Ekebom J, Saarso M (1999) Particle flux and properties affecting the fate of bacterial productivity in the benthic boundary layer at a mud-bottom site in South-Central Gulf of Riga. J Mar Syst 23:233–250
Gaumiga R, Lagzdins G (1995) Macrozoobenthos. In: Ojaveer E (ed) Ecosystem of the Gulf of Riga between 1920 and 1990. Estonian Academy, Tallinn, pp 198–211
HELCOM (1996) Third periodic assessment of the state of the marine environment of the Baltic Sea. 1989–1993: Background document. Baltic Sea Environ Proc 64B
Herkül K, Kotta J, Kotta I, Orav-Kotta H (2006) Effects of physical disturbance, isolation and key macrozoobenthic species on community development, recolonisation and sedimentation processes. Oceanologia 48S:267–282
ICES (2004) Report of the Baltic Fisheries assessment working group. ICES CM 2004/ACFM:22
Järvekülg A (1960) Materjale Pärnu lahe põhjaloomastiku kohta. Proc Estonian Acad Sci Biol 9:199–214
Järvekülg A (1979) Bottom fauna in the eastern part of the Baltic Sea (in Russian). Valgus, Tallinn
Jevrejeva S (2000) Long-term variability of sea ice and air temperature conditions along the Estonian coast. Geophysica 36:17–30
Kalveka BJ (1983) Special development of phytoplankton in the Gulf of Riga during 1978–1980 (in Russian). Rybokhoz Issled Basseine Balt Morya 18:3–9
Kautsky H (1989) Quantitative distribution of plant and animal communities of the phytobenthic zone in the Baltic Sea. Contributions from the Askö laboratory, University of Stockholm, 53:1–80
Kautsky H (1995) Quantitative distribution of sublittoral plant and animal communities in the Baltic Sea gradient. In: Elefteriou A, Ansell A, Smith C Jr (eds) Biology and ecology of shallow coastal waters. Olsen and Olsen, Fredensborg, pp 23–31
Kautsky H, Martin G, Mäkinen A, Borgiel M, Vahteri P, Rissanen J (1999) Structure of phytobenthic and associated animal communities in the Gulf of Riga. Hydrobiologia 393:191–200
Kostrichkina EM (1970) Feeding electivity of fishes of the Gulf of Riga (in Russian). Rybokhoz Issled Basseine Balt Morya 5:97–109
Kotta I (2003) Population ecology of mysids (Crustacea, Mysidacea) in the Gulf of Riga, NE Baltic Sea. Dissertations on Natural Sciences, Tallinn Pedagogical University, TPU Press 7:1–147
Kotta I, Kotta J (1997) Changes in zoobenthic communities in Estonian waters between the 1970s and 1990s. An example from the southern coast of Saaremaa and Muuga Bay. In: Ojaveer E (ed) Proceedings of the 14th Baltic Marine Biologists Symposium. Estonian Academy Publishers, Pärnu, Estonia, pp 70–79
Kotta I, Kotta J (1999) Distribution and migration of mysids in the Gulf of Riga (Northern Baltic). Proc Estonian Acad Sci Biol Ecol 48:284–295
Kotta I, Kotta J (2001a) Distribution of mysids on the bank slopes in the Gulf of Riga. Proc Estonian Acad Sci Biol Ecol 50:14–21
Kotta I, Kotta J (2001b) Vertical migrations of mysids in the Gulf of Riga. Proc Estonian Acad Sci Biol Ecol 50:248–255
Kotta I, Orav-Kotta H, Kotta J (2003) Macrozoobenthos assemblages in highly productive areas of the Estonian coastal sea. Proc Estonian Acad Sci Biol Ecol 52:149–165
Kotta J (2000) Impact of eutrophication and biological invasions on the structure and functions of benthic macrofauna. Dissertationes Biologicae Universitatis Tartuensis 63:1–160

Kotta J, Kotta I (1995) The state of macrozoobenthos of Pärnu Bay in 1991 as compared to 1959–1960. Proc Estonian Acad Sci Ecol 5:26–37

Kotta J, Kotta I (1998) Distribution and invasion ecology of *Marenzelleria viridis* (Verrill) in the Estonian coastal waters. Proc Estonian Acad Sci Biol Ecol 47:210–217

Kotta J, Kotta I, Martin G, Kukk H (1998a) A survey of data published on the littoral zoobenthos of the Gulf of Riga. Proc Estonian Acad Sci Biol Ecol 47:83–97

Kotta J, Orav H, Kotta I (1998b) Distribution and filtration activity of Zebra mussel, *Dreissena polymorpha* (Pallas) in the Gulf of Riga and the Gulf of Finland. Proc Estonian Acad Sci Biol Ecol 47:32–41

Kotta J, Kotta I, Viitasalo I (2000a) Effect of diffuse and point source nutrient supply on the low diverse macrozoobenthic communities of the northern Baltic Sea. Boreal Environ Res 5:235–242

Kotta J, Paalme T, Martin G, Mäkinen A (2000b) Major changes in macroalgae community composition affect the food and habitat preference of *Idotea baltica*. Int Rev Hydrobiol 85:693–701

Kotta J, Simm M, Kotta I, Kanošina I, Kallaste K, Raid T (2004) Factors controlling long-term changes of the eutrophicated ecosystem of Pärnu Bay, Gulf of Riga. Hydrobiologia 514:259–268

Kotta J, Orav-Kotta H, Vuorinen I (2005) Field measurements on the variability in biodeposition and grazing pressure of suspension feeding bivalves in the northern Baltic Sea. In: Dame R, Olenin S (eds) The comparative roles of suspension feeders in ecosystems. Springer, Dordrecht, pp 11–29

Kukk H, Martin G (1992) Long-term dynamics of the phytobenthos in Pärnu Bay, the Baltic Sea. Proc Estonian Acad Sci Ecol 2:110–118

Lagzdins G, Pallo P (1994) *Marenzelleria viridis* (Verrill) (Polychaeta, Spionidae) – a new species for the Gulf of Riga. Proc Estonian Acad Sci Biol 43:184–188

Lagzdins GS (1975) Benthic fauna in the Gulf of Riga as the indicator of pollution. In: Virbitskas Yu, Krotas R, Manyukas I (eds) Fundamentals of bioproductivity of the inner water bodies of the Baltic region (in Russian). In: Proceedings of XVIII scientific conference of the research of the inner waterbodies of the Baltic States. Institute of Zoology and Parasitology, Academy of Sciences of Lithuanian SSR, Vilnius, pp 429–430

Lappalainen J, Erm V, Kjellman J, Lehtonen H (2000) Size-dependent winter mortality of age-0 pikeperch (*Stizostedion lucioperca*) in Pärnu Bay, the Baltic Sea. Can J Fish Aquat Sci 57:451–458

Lilleleht V (1995) Veel kormoranidest (in Estonian). Eesti Loodus 2:44–46

Martin G (1999) Distribution of phytobenthos biomass in the Gulf of Riga (1984–1991). Hydrobiologia 393:181–190

Martin G (2000) Phytobenthic communities of the Gulf of Riga and the inner sea of the West-Estonian Archipelago. Dissertationes Biologicae Universitatis Tartuensis 64:1–140

Martin G, Torn K, Kotta J, Orav-Kotta H (2003) Estonian marine phytobenthos monitoring programme: preliminary results and future perspectives. Proc Estonian Acad Sci Biol Ecol 52:112–124

Martin G, Kukk E, Kukk H, Kotta J (2004) Historical review of the literature on phytobenthic investigations in the Gulf of Riga. Proc Estonian Acad Sci Biol Ecol 53:236–250

Mägi L, Lips U (1998) Budget components of water, salt, and nutrients in the Gulf of Riga in 1993–95. Proc Estonian Acad Sci Biol Ecol 47:14–31

Mäkinen A, Autio K (1986) *Cladophora glomerata* (Chlorophyta) as an indicator of coastal eutrophication. In: Proceedings of the third Finnish-Swedish seminar of the Gulf of Botnia. Vesihallituksen Julkaisuja – Water Res Inst Finland 68:160–163

Ojaveer E (1995) Large-scale processes in the ecosystem of the Gulf of Riga. In: Ojaveer E (ed) Ecosystem of the Gulf of Riga between 1920 and 1990. Estonian Academy, Tallinn, pp 268–277

Ojaveer E, Gaumiga R (1995) Cyclostomes, fishes and fisheries. In: Ojaveer E (ed) Ecosystem of the Gulf of Riga between 1920 and 1990. Estonian Academy, Tallinn, pp 212–267

Ojaveer E, Kalejs M (1974) On some oceanographic factors determining the abundance and distribution of pelagic fish in the Baltic Sea (in Russian). Okeanologiya 14:544–554

Ojaveer E, Špilev H (2003) Round goby, *Neogobius melanostomus* (Pallas). In: Ojaveer E, Pihu E, Saat T (eds) Fishes of Estonia. Estonian Academy, Tallinn, pp 336–337

Ojaveer H (1997) Composition and dynamics of fish stocks in the Gulf of Riga ecosystem. Dissertationes Biologicae Universitatis Tartuensis 31:1–138

Ojaveer H, Lumberg A (1995) On the role of *Cercopagis (Cercopagis) pengoi* (Ostroumov) in Pärnu Bay and the NE part of the Gulf of Riga ecosystem. Proc Estonian Acad Sci Ecol 5:20–25

Ojaveer H, Simm M, Lankov A (2004) Population dynamics and ecological impact of the non-indigenous *Cercopagis pengoi* in the Gulf of Riga (Baltic Sea). Hydrobiologia 522:261–269

Orav-Kotta H (2004) Habitat choice and feeding activity of benthic suspension feeders and mesograzers in the northern Baltic Sea. Dissertationes Biologicae Universitatis Tartuensis 89:1–117

Otsmann M, Astok V, Suursaar Ü (1997) A model for water exchange between the Baltic Sea and the Gulf of Riga. Nord Hydrol 28:351–364

Otsmann M, Suursaar Ü, Kullas T (2001) The oscillatory nature of the flows in the system of straits and small semi-enclosed basins of the Baltic Sea. Cont Shelf Res 21:1577–1603

Paalme T (2005) Nuisance brown macroalga *Pilayella littoralis*: primary production, decomposition and formation of drifting algal mats. Dissertations on Natural Sciences, Tallinn Pedagogical University, TPU Press, 10:1–104

Paalme T, Kukk H (2003) Comparison of net primary production rates of *Pilayella littoralis* (L.) Kjellm. and other dominating macroalgal species in Kõiguste Bay, northeastern Baltic Sea. Proc Estonian Acad Sci Biol Ecol 52:125–133

Paalme T, Kukk H, Kotta J, Orav H (2002) "In vitro" and "in situ" decomposition of nuisance macroalgae *Cladophora glomerata* and *Pilayella littoralis*. Hydrobiologia 475/476:469–476

Piirsoo K (1984) Phytoplankton of the Haapsalu Bay and its seasonal dynamic (in Russian). Gidrobiol Reshim Balt Morya, Tallinn, pp 24-33

Põder T, Maestrini SY, Balode M, Lips U, Béchemin C, Andrushaitis A, Purina I (2003) The role of inorganic and organic nutrients on the development of phytoplankton along a transect from the Daugava River mouth to the open Baltic, in spring and summer 1999. ICES J Mar Sci 60:827–835

Raudsepp U (2001) Interannual and seasonal temperature and salinity variations in the Gulf of Riga and corresponding saline water inflow from the Baltic Proper. Nord Hydrol 32:135–160

Reigstad M, Heiskanen AS, Wassmann P (1999) Seasonal and spatial variation of suspended and sedimented nutrients (C, N, P) in the pelagic system of the Gulf of Riga. J Mar Syst 23:211–232

Sanina LV (1961) About the biology of *Neomysis vulgaris* Thompson in the Gulf of Riga (in Russian). Tr Nauchno-Isseld Inst Rybn Khoz SNKH Latv SSR 3:293–306

Schukina IN (1970) Feeding of some demersal fish in Pärnu Bay (in Russian). Rybokhoz Issled Basseine Balt Morya 5:110–121

Seppälä J, Balode M (1999) Spatial distribution of phytoplankton in the Gulf of Riga during spring and summer stages. J Mar Syst 23:51–67

Shurin AT (1953) Benthic fauna in the Gulf of Riga (in Russian). Tr Latvijskovo Odteleniya VNIRO I:77–113

Shurin AT (1961) The grouping of benthic fauna in the Gulf of Riga (in Russian). Tr NIIRH SNH Latv SSR III: 343–368

Simm M (1995) Zooplankton in the northeastern part of the Gulf of Riga. In: Ojaveer E (ed) Ecosystem of the Gulf of Riga between 1920 and 1990. Estonian Academy, Tallinn, pp 169–175

Sõrmus I, Turovski A (2003) European whitefish, *Coregonus lavaretus* (L.) s.l., Baltic Sea forms. In: Ojaveer E, Pihu E, Saat T (eds) Fishes of Estonia. Estonian Academy, Tallinn, pp 121–130

Špilev H, Turovski A (2003) Smelt, *Osmerus eperlanus* (L.). In: Ojaveer E, Pihu E, Saat T (eds) Fishes of Estonia. Estonian Academy, Tallinn, pp 143–147

Stålnacke P, Vagstad N, Tamminen T, Wassmann P, Jansons V, Loigu E (1999) Nutrient runoff and transfer from land and rivers to the Gulf of Riga. Hydrobiologia 410:103–110
Suursaar Ü (1995) Nutrients in the Gulf of Riga. In: Ojaveer E (ed) Ecosystem of the Gulf of Riga between 1920 and 1990. Estonian Academy, Tallinn, pp 41–50
Suursaar Ü, Astok V, Kullas T, Nõmm A, Otsmann M (1995) Currents in the Suur Strait and their role in the nutrient exchange between the Gulf of Riga and the Baltic Proper. Proc Estonian Acad Sci Ecol 5:103–123
Suursaar Ü, Kullas T, Otsmann M (2002) Flow modelling in the Pärnu Bay and the Kihnu Strait. Proc Estonian Acad Sci Eng 8:189–203
Suursaar Ü, Kullas T, Otsmann M (2003a) A model study of the sea level variations in the Gulf of Riga and the Väinameri Sea. Cont Shelf Res 22:2001–2019
Suursaar Ü, Kullas T, Otsmann M, Kõuts T (2003b) Extreme sea level events in the coastal waters of western Estonia. J Sea Res 49:295–303
Tamminen T, Seppälä J (1999) Nutrient pools, transformations, ratios, and limitation in the Gulf of Riga, the Baltic Sea, during four successional stages. J Mar Syst 23:83–106
Tenson J (1995) Phytoplankton of the Pärnu Bay. In: Ojaveer E (ed) Ecosystem of the Gulf of Riga between 1920 and 1990. Estonian Academy, Tallinn, pp 104–126
Veber T (2001) Interactions between cormorants and fishery in coastal waters of Estonia (in Estonian). MSc Thesis, University of Tartu, Tartu, 1–66
Velner H-A (2005) Water protection of the Gulf of Finland and Estonian waterbodies. Tallinn University of Technology, Tallinn
Wassmann P, Tamminen T (1999) Pelagic eutrophication and sedimentation in the Gulf of Riga: a synthesis. J Mar Syst 23:269–283
Williamson MH, Fitter A (1996) The characters of successful invaders. Biol Conserv 78:163–170
Yurkovskis A (2004) Long-term land-based and internal forcing of the nutrient state of the Gulf of Riga (Baltic Sea). J Mar Syst 50:181–197
Yurkovskis A, Kostrichkina E, Ikauniece A (1999) Seasonal succession and growth in the plankton communities of the Gulf of Riga in relation to long-term nutrient dynamics. Hydrobiologia 393:83–94

Chapter 11
Haapsalu and Matsalu Bays

J. Kotta, A. Jaanus, and I. Kotta

11.1 Environmental Characteristics

Haapsalu and Matsalu Bays (Fig. 11.1) are situated in the Väinameri area (the "Sea of Straits") – a low-water region between continental Estonia and its western islands. The surface areas of Haapsalu and Matsalu Bays are small: 50 and 67 km^2, respectively. These values are gradually decreasing owing to landlift and riverine sediment discharge. The base rock is formed by the limestone, marl and dolomites of the Ordovician and Silurian deposits. The bed sediments are sands, gravels and clays of various structures. The bays are relatively shallow, with a maximum depth of less than 5 m and an average depth of 1.5–2 m (Lutt and Kask 1980).

Haapsalu Bay has a moderate freshwater inflow. The Taebla River, which has a catchment area of only 107 km^2, is the most important source of fresh water to the bay (Table 11.1). Thus, the hydrological conditions of Haapsalu Bay are influenced mainly by the exchange of water with the Väinameri area. The eastern and central parts of the bay are separated by peninsulas and water exchange is limited. Salinity values are relatively stable at 6–7 psu in the open western part of the bay and 2.5–4 psu in the central and eastern parts of the bay. On the other hand, the Kasari River with its catchment area of 3,214 km^2 contributes significantly to the hydrological conditions of Matsalu Bay. The average amount of annually inflowing fresh water is seven times greater than the volume of the bay. The shallow eastern part of the bay is characterised by fresh water with salinity below 0.5 psu, whereas in the western part of the bay the salinity may exceed 6 psu. Inflowing fresh water may quickly dilute the whole bay. Similarly, sea water may reach the mouth of the Kasari River as a result of strong westerly winds (Mardiste and Kaasik 1985; Porgasaar and Simm 1985).

The mean annual air temperature is 5.3–5.6°C. The hydrochemical regime of the bays varies within different seasons and years. Vertical gradients in water temperature are lacking during the ice-free season due to the shallowness of the bay. The water temperature of the region is determined by its geographical position, water level and water exchange with the sea. Rapidly increasing water temperatures during the spring period, high temperatures in summer, and a long period of subzero temperatures in winter are typical of Haapsalu and Matsalu Bays. The mean annual temperature in the surface layer is about 7°C, the absolute

U. Schiewer (ed.), *Ecology of Baltic Coastal Waters. Ecological Studies 197.*

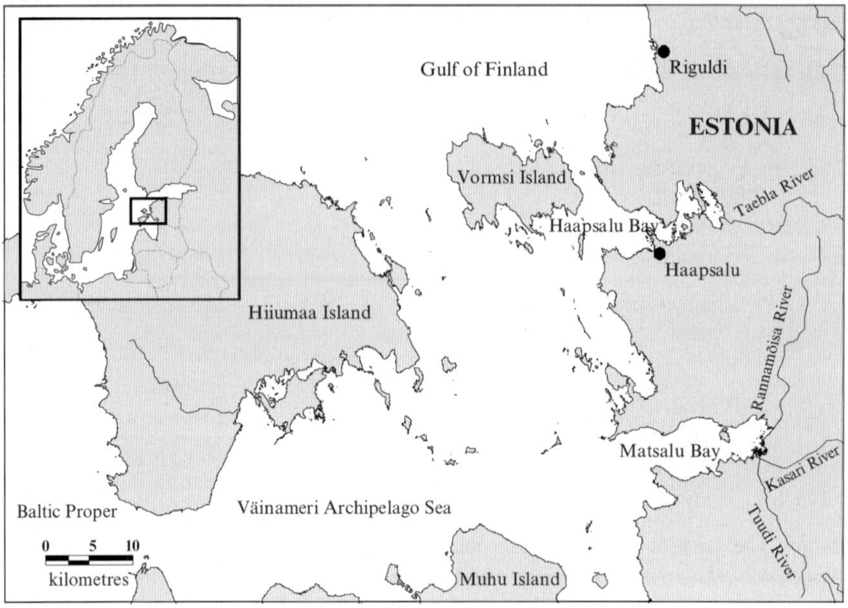

Fig. 11.1 Map of Haapsalu and Matsalu Bays

Table 11.1 Main characteristics of Haapsalu and Matsalu Bays. *ww* Wet weight, *dw* dry weight, *Chl a* chlorophyll *a*, *DIN* dissolved inorganic nitrogen, *DIP* dissolved inorganic phosphorus

	Haapsalu Bay	Matsalu Bay
a. Climate and hydrology		
Air temperature (°C)	Min: –33; max: +34	Min: –33; max: +34
Water temperature (°C)	Min: –0.4; max: +28	Min: –0.4; max: +28
Ice covering (days a^{-1})	140–145	140–145
Annual global radiation (kW h m^{-2})	Max: 1,116	Max: 1,116
Average precipitation (mm a^{-1})	<550	<550
Mean freshwater inflow ($10^6 m^3 a^{-1}$)	No data	950
Mean inflow ($10^6 m^3 a^{-1}$)	No data	No data
Mean total outflow ($10^6 m^3 a^{-1}$)	No data	No data
Residence time (days)	No data	<52
b. Morphometry and sediments		
Area (km^2)	50	67
Volume ($10^6 m^3$)	75	135
Mean depth (m)	2.0	1.5
Maximum depth (m)	5.0	4

(continued)

Table 11.1 (continued)

	Haapsalu Bay	Matsalu Bay
Catchment area (Km²)	>107	3,400
Area/catchment relation	<0.47	0.02
Sediment distribution	Silty sand, clay–silt	Silty sand
c. Pelagic biological components		
Secchi depth (m)	0.8–4.2	0.2–3.5
Bacteria ($10^6 ml^{-1}$)	No data	1.0–7.3
	0.7–18.6	–16.6
Phytoplankton biomass (μg Chl a l^{-1})	Spring max: 7.4	Spring max: 16.6
	Autumn max: 17.4	Autumn max: 6.0
	Spring max: 3.2	Spring max: 3.2
Phytoplankton biovolume ($mm^3 l^{-1}$)		
	Autumn max: 9.9	Autumn max: 1.2
Dominating phytoplankton species	*Chaetoceros wighamii*, *Achnanthes taeniata*, *Diatoma tenuis*, *Fragilaria* spp., *Cylindrotheca closterium,* and *Aphanocapsa* spp. *Merismopedia* spp. *Microcystis aeruginosa*	*Diatoma elongatum*, *Chaetoceros holsaticus*, *Chaetoceros wighamii*, *Pseudanabaena limnetica*, *Merismopedia punctata*, *Aphanizomenon flos-aquae, Nodularia spumigena*
Flagellates		
Biomass (mg l^{-1})	0.03–0.80	
Dominating species	*Plagioselmis prolonga*, *Teleaulax* spp., *Heterocapsa rotundata*, *Peridiniella catenata*, *Pyramimonas* spp. Heterocapsa triquetra	*Peridiniella catenata*, *Cryptomonadales*, *Pyramimonas* spp., *Eutreptiella* sp
Ciliates		
Biomass (mg C l^{-1}) (ww)	0.0–0.2	0–0.002
Dominating species	*Mesodinium rubrum*	*Mesodinium rubrum*
Mesozooplankton		
Biomass (mg l^{-1})	0–0.8 (ww)	0–0.4 (ww)
Dominating species	*Keratella quadrata*, *Eurytemora affinis*	*Keratella quadrata*, *Eurytemora affinis*
Fish		
Number of species	>17	45

(continued)

Table 11.1 (continued)

	Haapsalu Bay	Matsalu Bay
Fish catches (t a^{-1})	90–410	110–776
d. Benthic biological components		
Macrophytes		
Biomass (g m^{-2})	0–250 (dw)	0–750 (dw)
Number of species	31	75
Dominating species	*Potamogeton pectinatus*, loose *Fucus vesiculosus*, *Zostera marina*, *Zannichellia palustris*	*Cladophora glomerata*, *Chara aspera*
Macrozoobenthos		
Biomass (g m^{-2})	0.4–50 (dw; dominating species)	3.5–30 (dw)
Number of species	106 (including chironomids and ostracods)	175 (including chironomids and ostracods)
Dominating species	Chironomidae, *Mya arenaria*, *Macoma balthica*	*Macoma balthica*, *Asellus aquaticus*, Chironomidae
e. Water chemistry, trophic status and pollution		
Salinity (psu)	1.28–6.38	0.5–7.5
pH	7.2–9.0	7.0–9.2
Oxygen saturation (%)	9–159	No data
Total nitrogen (μmol l^{-1})	9–157; winter max: 1400	8–37 winter max: 200
DIN (μmol l^{-1})	0–65	No data
Total phosphorus (μmol l^{-1})	0.2–2.8; winter max: 27.5	0.3–2.8; winter max: 10.9
DIP (μmol l^{-1})	0.0–9.6	0.0–2.0
Annual nitrogen input (t a^{-1})	192	1500–3000
Annual phosphorus input (t a^{-1})	3.6	25–82
Limitation of PP and period	N-limited (May–July)	No data
Main pollutants	Haapsalu town, Taebla River	Kasari River
Trophic level	Moderately eutrophicated	Moderately eutrophicated

minimum is –0.4°C and the absolute maximum is 28°C. The coldest month is February and the warmest July. The first ice usually comes during the second half of November and the region becomes free of ice in the second half of April. The mean number of days of ice-cover in winter varies between 140 and 145.

Annual precipitation is less than 550 mm. Fluctuations in water level and water currents depend on the direction and velocity of the wind. The mean current speed in exposed parts of the bays is 15–25 cm s^{-1}, but maximum values over 60 cm s^{-1} have been observed. The waves are low due to the limited extension and depth of the bays. Waves of up to 0.3 m can bring muddy bottom sediments into suspension but flushing of the bay occurs only during storm surges. As the bays are shallow, wind-induced surges cause floods, which have been observed at 270 cm above average water level. Such events occur infrequently and mainly in autumn. However, the bays' currents can neither effectively remove muddy bottom deposits nor import sand from nearby regions (Eipre and Pärn 1982; Suursaar et al. 2003).

Oxygen conditions in Haapsalu and Matsalu Bays are good or satisfactory, although short periods of hypoxia may occur, especially under ice-cover in the shallower parts of the bay. The water quality in the central and western parts of Haapsalu Bay depends greatly on the sewage inflow from the town. Although the seasonal cycles of total phosphorus and nitrogen are rather irregular, concentrations of total nitrogen in Haapsalu Bay are higher during the summer and autumn period. We have no information on the winter concentrations of inorganic nitrogen. During the vegetation period in 2000, nitrates and nitrites stayed mostly below the detection limit. The nutrient load in Matsalu Bay is very variable, and depends on the seasonality in discharge and nutrient content of the river water. Early spring accounts for about 30–40% of the annual total phosphorus and nitrogen loads. The summer period (June-August) provides only 4% of the annual nutrient load. The concentrations of total phosphorus are higher in winter (maximum 11 µmol P l^{-1} in the eastern bay, and 1.6–2 µmol P l^{-1} in the central and western parts of the bay). The concentrations of phosphates and the percentages of inorganic compounds of phosphorus are also higher in the winter period. Total nitrogen declines steeply in midsummer; however, in the western part the concentrations show no clear trend throughout the vegetation period. Winter concentrations of total nitrogen as high as 200 µmol N l^{-1} have been measured in the eastern bay, and levels reach 91–114 µmol N l^{-1} in the central and western parts (Eipre and Pärn 1982; Porgasaar 1993).

11.2 Plankton Communities

Water transparency – measured as the disappearance depth of Secchi disk – is relatively good in the western parts of Haapsalu and Matsalu Bays during the summer and autumn periods. In most cases the water is transparent to the bottom. The main cause of turbidity is wind-induced mixing. In the eastern and central parts of the bays, resuspension of mud and organic waste from the sea bottom, and the exceptionally high density of phytoplankton communities, are the main factors aggravating water transparency during late summer (Jaanus 2003).

The ecosystem of Haapsalu Bay is characterised by very high biological variability. The concentration of chlorophyll *a* (Chl *a*) and total phytoplankton biomass may differ by 15- to 20-fold between the western open and eastern shallow parts of the bay. Typically for eutrophicated sea areas, the most intensive development of phytoplankton is shifted to the late summer period in the central and eastern parts of the bay. Nowadays, the maximum Chl *a* concentrations are estimated at 10–15 μg l^{-1}, indicating strong eutrophication of the site. However, in recent decades phytoplankton biomass has declined significantly, as concentrations measured in August 1989 reached as high as 66 μg l^{-1}. The succession of phytoplankton in the western part is similar to that in non-eutrophicated coastal waters. Chl *a* concentrations remain below 5 μg l^{-1} and do not exceed 1–2 μg l^{-1} during the summer and autumn periods (Jaanus 2003). Species composition and phytoplankton biomass differ notably between the open western and the semi-enclosed central-eastern parts of Haapsalu Bay. The maximum biomass values of the inner parts are approximately 20 times as high as those in the exposed parts of the bay. In the open parts, the dominating species during the spring maximum period is the arctic diatom *Achnanthes taeniata*, and the total phytoplankton biomass does not exceed 0.8 mg wet weight (ww) l^{-1}. The summer phytoplankton biomass remains relatively low at 0.1–0.3 mg ww l^{-1}, except for short and irregular periods in July–August. There is no clear domination of any algal group during the summer period, probably due to the dominance of macrophytes in such a shallow area. In the semi-enclosed parts of the bay, *Diatoma tenuis* and other pennate diatoms are responsible for the highest biomass values in April–May. The colonial Cyanobacteria *Merismopedia* spp., *Microcystis* spp and cf. *Aphanocapsa planctonica* prevail from July to October. Maximum biomasses during summer–autumn blooms are measured at 18 mg ww l^{-1}. Although a high percentage of Cyanobacteria was observed already in the 1980s, the ranges of total phytoplankton biomass exhibit an increasing trend. The biomass in the eastern part of Haapsalu Bay in August 1985, 1989 and 2000 varied within the ranges 0.8–8.2, 1.8–12.9 and 5.2–13.3 mg ww l^{-1}, respectively (Jaanus 2003).

The seasonal fluctuation in salinity and nutrient load strongly affects the development of phytoplankton in Matsalu Bay [see Chaps. 4 (Schiewer) and 9 (Gasiūnaitė et al.), this volume]. The fresh water inflow and nutrient load are high during spring floods. The most diverse phytoplankton groups in Matsalu Bay were diatoms, Cyanobacteria and green algae. The total number of nanoplanktonic flagellates decreased towards the open sea and the dominant species also changed. The abundance of Cyanobacteria increased towards the mouth area of the bay. During the 1980s, the maximum spring biomasses observed at salinity 2 psu consisted mainly of the diatom *Diatoma tenuis* (up to 95% of total biomass). In the western bay, centric diatoms from the genus *Chaetoceros* and the dinoflagellate *Peridiniella catenata* co-dominated. The latter species are typical for the brackish Baltic Sea waters. Kasari River inflow was small during the summer months and nutrients were consumed mainly by the green alga *Cladophora glomerata* in the central part, and by reed beds in the inner part of the bay. In the summer period, a horizontal patchiness was clearly observed, where different microalgal species prevailed. In June, the phytoplankton was formed by cryptomonads, euglenophytes, prasinophytes and other nanoplankton. In August, the abundance of Cyanobacteria increased. The average summer phytoplankton biomass did not exceed 1 mg ww l^{-1}. The total phytoplankton biomasses sporadically reached 2–3 mg ww l^{-1}, but in some summer periods they remained markedly lower (0.3–0.5 mg ww l^{-1}) (Piirsoo 1979, 1986, 1996; Piirsoo and Porgasaar 1985).

Mesozooplankton has high seasonality in terms of abundance and biomass in Haapsalu and Matsalu Bays. Zooplankton densities are very low at 150 ind m^{-3} during winter due to the unfavourable oxygen regime under ice cover. Species diversity and densities, especially of rotifers, increase rapidly during spring. In June, the density of zooplankton may already exceed 750,000 ind m^{-3}. In August, a second biomass maximum occurs due to the development of copepods and cladocerans. Later, densities gradually decrease until the next spring bloom. Mesozooplankton biomasses are relatively low compared to those of other areas of the Baltic Sea. Typical open Baltic communities are found in the mouth area of the bays, whereas freshwater species prevail in more diluted waters. In general, both abundances and biomasses increase with decreasing salinity. The communities are very unstable, probably due to strong water movement. Among mesozooplankton, *Eurytemora affinis*, *Acartia bifilosa* and *Mesocyclops* spp. dominate in Haapsalu Bay. Their maximum densities reach 12,000, 16,000 and 19,000 ind m^{-3}, respectively. *E. affinis* is frequently found throughout the bay, *A. bifilosa* in the western and more saline parts and *Mesocyclops* spp. in the eastern more diluted parts of the bay. The freshwater cladoceran *Chydorus sphaericus* is more abundant in the easternmost Haapsalu Bay. Among rotifers *Keratella quadrata* is found throughout the entire bay, *Synchaeta baltica* in the western parts and *Notholca caudata* and *Brachionus* spp. in the eastern parts of the bay. In autumn, the abundance of polychaetes is very high in western parts of the bay. Mesozooplankton species diversity is very high in Matsalu Bay due to the occurrence of many semiplanktic, phytophilous and freshwater species. Rotifers are the most common and abundant zooplankton taxa, especially in the central and eastern parts of the bay. Their maximum densities have been estimated at 270,000 ind m^{-3}. The density of cladocerans exceeds 1,000 ind m^{-3} in reedbeds but likely also in the central and western parts of the bay. Copepods are found mainly in exposed parts of the bay with densities over 35,000 ind m^{-3} (Remm 1986).

Haapsalu and Matsalu Bays are important reproduction areas for many fishes of the Estonian western archipelago sea. A total of 43 fish and 2 lamprey species (sea and river lamprey) were recorded in this area in the 1980s. Among fish species, 15 were marine, 14 freshwater and 16 migratory or semi-migratory species (Erm and Kangur 1987; Erm et al. 2002). Bottom-feeding fishes prevail in the area. In the more saline western part of Matsalu Bay, the most important and common food of fishes are different bivalve, gastropod and amphipod species. In the eastern parts insect larvae prevailed as fish food. The growth rate of fishes is higher in the bay than in freshwater biotopes (Erm et al. 2002; Saat and Eschbaum 2002).

The total catch has decreased from 1,200 t a^{-1} in the 1980s to 200 t a^{-1} in the early 2000s. Baltic herring (*Clupea harengus membras*) formed 70% of the total catch in the 1980s. The abundance of predatory fish [pike (*Esox lucius*), pikeperch (*Sander lucioperca*)] was comparatively high as they constituted more than 10% of the total catch. Other important commercial fishes were ruffe (*Gymnocephalus cernuus*), smelt (*Osmerus eperlanus*) and roach (*Rutilus rutilus*). The structure of spawning stocks of commercial species was normal. The main fishing gears used were fyke nets, gill nets and herring pond nets. Fishery was concentrated in the reproduction period of spring spawning fishes (April–June) (Erm et al. 2002). Due to increased fishing intensity, catch of most species increased in the late 1980s and early 1990s. Too intensive a catch in the 1990s has led to overfishing and the catch has declined significantly in recent years. The stock of perch (*Perca fluviatilis*) collapsed and recent catch has been less than 1% of the long-term average. The spawning stock consisted mostly of young fish. Concurrent with the overexploitation of more valuable species, roach became commercially important. In addition to fishery, pike stock suffered from low-water springs in the 1980 and early 1990s (Saat and Eschbaum 2002).

11.3 Benthic Communities

Isolation, low exposure, shallowness, substrate availability and moderate-to-strong inflow of nutrients explain the development of rich phytobenthic communities in Haapsalu and Matsalu Bays. In Haapsalu Bay the most important species is the higher plant *Potamogeton pectinatus*, which dominates the entire bay regardless of eutrophication level. The cyanobacterium *Lyngbya aestuarii* occurs frequently at the inner parts of the bay. Loose-lying dwarf forms of the brown alga *Fucus vesiculosus* are found in wide areas. Its maximum biomasses may exceed 1 kg dw m^{-2}. Concurrent with this species, the cyanobacterium *Gloeotrichia pisum* and the higher plant *Myriophyllum spicatum* are found in high numbers. In more exposed areas, the higher plants *Zostera marina* and *Zannichellia palustris* prevail. However, their biomasses are low as compared to total biomass values in other parts of the bay (Trei 1982). Salinity and nutrient levels are the main determinants of the spatial distribution of macrophytes within Matsalu Bay. The reed *Phragmites australis* is the most common species at the estuary of the Kasari River, the charophyte *Chara aspera* prevails in the northeastern part of the bay and the red algae *Furcellaria lumbricalis*, *Polysiphonia nigrescens* and *Ceramium tenuicorne* dominate in the western parts of the bay (Trei 1985; Torn and Martin 2003). Benthic vegetation is an important consumer of nutrients in Matsalu and Haapsalu bays as more than 150-fold difference in nitrates is found between riverine water and bay water in summer.

The macrozoobenthos of Haapsalu Bay is mainly of freshwater origin. The most numerous groups are chironomid larvae, ostracods and gastropods. In the easternmost parts of the bay, where ostracods and chironomid larvae predominate, abundance and biomass have strong seasonal fluctuations (total abundance and biomass values: 6,000–40,000 ind m^{-2}, 0.4–7.0 g dw m^{-2}). In the middle parts of the bay, *Mya arenaria* dominates in biomass and oligochaetes and ostracods prevail numerically (total abundance and biomass values: 9,000–64,000 ind m^{-2}, 23–305 g dw m^{-2}). In the more exposed parts of the bay, the communities are characterised by high biomasses of *Macoma balthica*, *Cerastoderma glaucum* and *Hydrobia ulvae* and high abundances of ostracods and *H. ulvae* (total abundance and biomass values: 17,500–41,500 ind m^{-2}, 20–50 g dw m^{-2}). Seasonal fluctuations are not very marked in that part of the bay. Macrozoobenthos of Matsalu Bay is characterised by high species diversity. Freshwater species prevail numerically (13,000 ind m^{-2}) but their biomasses are low (8 g dw m^{-2}). A great number of freshwater invertebrates (Oligochaeta, leeches, ostracods, water mites, chironomid larvae and snails) live in Matsalu Bay that are not found in the other regions of Estonian coastal waters. Due to the low and largely varying salinity, high temperatures in summer and subzero water temperature in winter, many arctic-boreal, boreal and all relicts of the arctic fauna except *Saduria entomon* are absent from Matsalu Bay. As the salinity is highest in the western part of Matsalu Bay, the bottom fauna of that region is of marine appearance, with moderate biomasses (30 g dw m^{-2}) and high densities (12,000 ind m^{-2}). The marine bivalve *Macoma balthica* dominates in biomass and the marine amphipod *Corophium volutator* in abundance. The species diversity of zoobenthos is highest in the central part of the bay where marine fauna is replaced by freshwater fauna. Due to fluctuating ecological conditions, macrozoobenthic communities in that part of the bay have very low biomasses (< 2 g dw m^{-2}), very high abundances (> 17,000 ind m^{-2}) and varying species composition. The eastern part of the bay is inhabited mainly by freshwater fauna. The bottom fauna outwith reed-beds is characterised by a small number of species, very low biomass (2 g dw m^{-2}) and high density (8,000 ind m^{-2}). The snail *Bithynia tentaculata* and the isopod *Asellus aquaticus* dominate in biomass, and Chironomidae larvae and *A. aquaticus* in abundance. The bottom fauna in reed-beds is species-rich, and has low biomass (<10 g dw m^{-2}), with *A. aquaticus* and the bivalve *Unio pictorum* dominating in the biomass and at a very high density (19,000 ind m^{-2}).

In the Estonian coastal sea, most of the evidence about various benthic processes (e.g. production, matter and energy flows) is circumstantial. This is because researchers concentrated primarily on the issues of spatial distribution or temporal trends of the biota whereas experimental studies were almost non-existent. It was not until the mid-1990s that process studies gained more attention; however, so far no such studies have been performed in Haapsalu and Matsalu Bays.

11.4 Benthic–Pelagic Coupling and Eutrophication

Municipal sewage is considered the main source of eutrophication in Haapsalu Bay. The central part of Haapsalu Bay receives municipal and industrial sewage water from the town of Haapsalu (13,000 inhabitants). A mechanical wastewater treatment plant was reconstructed in 1981. The next reconstruction of the existing plant began in 1995 and was completed in 1997. The construction works for nitrogen removal and sand separation were finalised by 2001. As a result, the pollution load in Haapsalu Bay has decreased markedly in the past decade. In the late 1970s the maximum winter concentrations of total phosphorus were 93 µmol P l^{-1} in the western and central part of the bay. Currently, the maximum values of total nitrogen and phosphorus are estimated at 136 µmol N l^{-1} and 3.8 µmol P l^{-1}. However, the concentrations of total nitrogen were not measured during the 1970s and 1980s and we also lack information on the winter contents of total phosphorus in recent decades. By the time field work began in 2000, phytoplankton had reached a biomass level corresponding to the spring maximum, and the concentrations of phytoplankton in the central and eastern parts of Haapsalu Bay were relatively high also during later measurements in October. Therefore, the concentrations of dissolved inorganic nutrients in the water column remained low as well. However, the exceptionally high summer concentrations of total phosphorus recorded in 1979 and 1989 (2.9 and 3.8 µmol P l^{-1}, respectively) were not observed in 2000. The concentration of dissolved inorganic phosphorus also decreased during the last few decades (maximum 2.2–2.3 and 0.9 µmol P l^{-1} in central Haapsalu Bay in 1979 and 2000, respectively; HELCOM 2001; Jaanus 2003).

The discharge of dissolved nutrients and toxic substances by the Kasari River affects the biota of Matsalu Bay. The central and western parts of Matsalu Bay are weakly eutrophicated and the eastern part of the bay is highly eutrophicated. At the estuary of the Kasari River the winter values of total nitrogen and phosphorus exceeded 95 µmol N l^{-1} and 3 µmol P l^{-1} in the 1970s and 1980s. Current values of the winter nutrients in the area are lacking (Porgasaar 1982, 1993; Porgasaar and Simm 1985; Jaanus 2003). In autumn, the total phosphorus concentration increases in the western parts of Haapsalu and Matsalu Bays and decreases in the eastern parts of the Bays. This pattern may be associated with the influence of surface waves and turbidity during stormy weather, which can cause an increase in phosphorus concentrations of about 2- to 3-fold (Suursaar et al. 2001). Analyses of the spatial variability of total phosphorus and nitrogen show a decreasing gradient from river water to estuary. The bulk of the nutrients are probably accumulated by lush benthic vegetation and phytoplankton. As the phosphorus and nitrogen load in the vegetation period makes up about 25–30% of their annual input, and the assimilation of nutrients by vegetation is active, the outflow of inorganic compounds from Matsalu Bay is small during the vegetative period. Although relatively small, the amount of nutrients accumulated has a decisive impact on the state of Matsalu Bay (Porgasaar 1993).

Oxygen concentrations are generally high during ice-free seasons owing to shallowness of the area and exposure to winds. Spatial and temporal differences are due to macrophytes and their decomposition. The high biomass of macrophytes in eastern parts of the bays results in oversaturation of oxygen in summer (>125%, pH 9) and its depletion during autumn months (<80%, pH 7.6). In summer months, however, the oxygen concentration in the near-bottom layer may occasionally be depleted in the whole bay due to the intensive decomposition of organic matter. Oxygen saturation is stable at 90–96% in the western and more exposed parts of the bays. Anoxia may temporally occur under ice-cover, especially at the innermost edges of the bays (Porgasaar and Simm 1985).

Despite the relatively small area of Haapsalu Bay, the differences in nutrient concentrations and phytoplankton characteristics between its open western and other parts are enormous. The main reason is the limited water exchange between these areas, which leads to advanced eutrophication of the eastern shallow bays. Although the pollution load was reduced after reconstruction of the municipal wastewater treatment plant, a large amount of phosphorus has accumulated in the sediments, functioning as an additional nutrient supply for phytoplankton under favourable environmental conditions – vertical mixing of water masses and oxygen deficiency. Ice cover – an annual phenomenon in Haapsalu Bay – is also a very important regulating factor. As water exchange under ice cover is extremely low, anoxic conditions are common during late winter, and the release of phosphorus from bottom sediments is likely. Consequently, concentrations of total phosphorus are at their highest under ice cover. Anoxic conditions with possible nutrient release from sediments may develop infrequently during the intensive decomposition of organic matter in summer. Thus, the concentration of total phosphorus plays a decisive role in the functioning of the Haapsalu Bay ecosystem. Moderate or strong positive correlation was found between this parameter and phytoplankton biomass development during different observation periods. The effect of nitrogen load on phytoplankton productivity is probably less important (Jaanus 2003). Comparisons of the dominant species of phytoplankton and biomass calculations are problematic in most cases. Methods of sampling and analysis are continually developing and numerous taxonomic changes have also occurred during the last decades. Nevertheless, we can say that the general succession of phytoplankton communities in Haapsalu and Matsalu Bays has not altered and the biomass values are similar to those found a few decades ago.

Following the increasing load of nutrients, the density of mesozooplankton rose by up to 20 times in Haapsalu Bay between the 1960s and 1970s. The differences were due mainly to the massive development of the rotifers *Keratella quadrata*, *K. cochlearis*, *Brachionus quadrata* and *Filina longiseta*, and the cladoceran *Chydorus sphaericus* (Järvekülg et al. 1981). As data on mesozooplankton is spatially and temporally very scattered and different methods have been used, we cannot make any sound comparisons about long-term changes. There are some indications, however, that mesozooplankton communities have become more stable and that the abundance of rotifers has declined in recent decades.

Benthic systems have responded to increased anthropogenic eutrophication in Haapsalu and Matsalu Bays since the 1960s. Sandy sediments have been replaced by silty–sand or silty sediments over extended areas (Järvekülg 1982). Phytobenthic communities have changed in terms of species composition, distribution area and biomass. Macrophytobenthic communities adjacent to municipal sewage outlets disappeared in the 1970s. The epiphyte *Ceramium tenuicorne*, which prevailed in the study area in the 1960s, has significantly reduced its distribution area and biomasses. In the transition area between reeds and open water a new community

Erm V, Kangur M (1987) Matsalu märgala põhjatoiduliste kalade toitumisest. In: Mägi E (ed) Loodusvaatlusi 1985, vol 1. Valgus, Tallinn, pp 104–118

Erm V, Kangur M, Saat T (2002) Matsalu märgala kaladest ja kalapüügist 1980 aastatel. In: Saat T (ed) Väinamere kalastik ja kalandus. Tartu Ülikooli Kirjastus, Tartu, pp 122–158

HELCOM (2001) Thematic reports on HELCOM PITF regional workshops held in the Baltic republics. Baltic Sea Environ Proc 83:87–91

Jaanus A (2003) Water environment of Haapsalu Bay in retrospect (1975–2000). Proc Estonian Acad Sci Biol Ecol 52:91–111

Järvekülg A (1982) Läänemere antropogeenne eutrofeerumine. In: Järvekülg A (ed) Eesti NSV rannikumere kaitse. Valgus, Tallinn, pp 15–27

Järvekülg A, Piirsoo K, Porgasaar V, Remm K, Seire A, Vilbaste S (1981) Inimtegevuse mõju Haapsalu lahe vee omadustele, elustikule ja ökoloogilisele seisundile. In: Inimtegevus ja keskkonnakaitse. Eesti NSV Teaduste Akadeemia, Tallinn, pp 73–76

Lutt J, Kask J (1980) Matsalu lahe põhjasetted. In: Mägi E (ed) Loodusvaatlusi 1978, vol 1. Valgus, Tallinn, pp 166–178

Mardiste H, Kaasik T (1985) Matsalu lahe ja Kasari jõe hüdroloogiline režiim. In: Kumari E (ed) Matsalu – rahvusvahelise tähtsusega märgala. Valgus, Tallinn, pp 15–25

Piirsoo K (1979) On the summer phytoplankton of the West-Estonian coastal waters (in Russian). Proc Estonian Acad Sci Biol 28:56–67

Piirsoo K (1986) Species composition and seasonal dynamics of the phytoplankton in Matsalu Bay (in Russian). Proc Estonian Acad Sci Biol 35:61–71

Piirsoo K (1996) Environmental factors regulating the phytoplankton biomass in Estonian coastal waters. In: Andrushaitis A (ed) Proceedings of the 13th Symposium of the Baltic Marine Biologists, Riga, Zinatne, pp 73–77

Piirsoo K, Porgasaar V (1985) Fütoplankton ja klorofüllisisaldus Matsalu lahes. In: Kumari E (ed) Matsalu – rahvusvahelise tähtsusega märgala. Valgus, Tallinn, pp 36–43

Porgasaar V (1982) Hapniku-, fosfori- ja klorofüllisisaldus Eesti NSV rannikumere eutrofeerunud piirkondades. In: Järvekülg A (ed) Eesti NSV rannikumere kaitse. Valgus, Tallinn, pp 40–48

Porgasaar V (1993) Content and distribution of phosphorus and nitrogen in the coastal waters of West Estonia. Proc Estonian Acad Sci Ecol 3:166–180

Porgasaar V, Simm H (1985) Matsalu lahe hüdrokeemiline režiim. In: Kumari E (ed) Matsalu – rahvusvahelise tähtsusega märgala. Valgus, Tallinn, pp 26–35

Remm K (1986) Matsalu lahe zooplanktoni koosseis ja dünaamika. Valgus, Tallinn

Saat T, Eschbaum R (2000) Väinamere kalastiku muutused viimastel aastakümnetel. In: Frey T (ed) Kaasaegse ökoloogia probleemid, vol 8. Eesti Ökoloogiakogu, Tartu, pp 265–269

Saat T, Eschbaum R (2002) Väinamere kalastik ja selle muutumine viimastel aastakümnetel. In: Saat T (ed) Väinamere kalastik ja kalandus. Tartu Ülikooli Kirjastus, Tartu, pp 9–45

Suursaar Ü, Kullas T, Otsmann M (2001) The influence of currents and waves on ecological conditions of the Väinameri. Proc Estonian Acad Sci Biol Ecol 50:231–247

Suursaar Ü, Kullas T, Otsmann M (2003) Modelling of flows, sea level variations and bottom stresses in the coastal zone of West Estonia. In: Brebbia CA, Almorza D, López-Aguayo F (eds) Coastal engineering VI. Computer modelling and experimental measurements of seas and coastal regions. WIT, Southampton, pp 43–52

Torn K, Martin G (2003) Changes in the distribution of charophyte species in enclosed sea bays of western Estonia. Proc Estonian Acad Sci Biol Ecol 52:134–140

Torn K, Martin G, Kukk H, Trei T (2004) Distribution of charophyte species in Estonian coastal water (NE Baltic Sea). Sci Mar 68[Suppl 1]:129–136

Trei T (1982) Inimtegevuse mõju põhjataimestiku levikule Kuressaare ja Haapsalu lahes. In: Järvekülg A (ed) Eesti NSV rannikumere kaitse. Valgus, Tallinn, pp 60–67

Trei T (1985) Long-term changes in the bottom macroflora of the coastal waters of Estonia. In: Trei T (ed) Problems concerning bioindication of the ecological condition of the Gulf of Finland. Valgus, Tallinn, pp 117–122

Chapter 12
The Neva Estuary Ecosystem

I.V. Telesh, S.M. Golubkov, and A.F. Alimov

12.1 Environmental Characteristics

Neva Estuary is located in the eastern part of the Gulf of Finland. The western boundary of this estuary is not clearly defined geographically, due to the lack of a comprehensive definition of an estuary in general (Elliott and McLusky 2002). In the past, several attempts at subdividing Neva Estuary have been undertaken by researchers. One of the most recent classifications (Golubkov et al. 2003a) distinguishes certain parts of Neva Estuary that are well separated by natural and/or artificial boundaries. Some publications, however, present schemes of this estuarine water system in which Neva Bay is falsely excluded from the estuary subdivision, whereas in fact it should be considered as the head of Neva Estuary.

Based on both hydrobiological data and critical analyses of current definitions and classifications of estuaries (see reviews: Elliott and McLusky 2002; Telesh 2004), the present chapter aims to generally distinguish the inner and outer parts of Neva Estuary, the former being sub-divided into two sections: upper and lower (Fig. 12.1). The upper part of the inner estuary (i.e. the upper inner estuary) is the shallow freshwater semi-enclosed Neva Bay, which can be also called the head of Neva Estuary. The lower part of the inner estuary (i.e. lower inner estuary) is (according to most Russian publications) the slightly brackish-water eastern Gulf of Finland located between the island of Kotlin and longitude ca. 29°E. Intrusions of brackish water from the lower inner estuary to its upper part, Neva Bay, also happen often. The outer estuary is the region between ca. 29°E and the longitude of the island of Moshniy. Recent hydrobiological investigations have demonstrated that these areas differ in abiotic parameters as well as in structural and functional ecosystem characteristics.

In this chapter we consider the characteristics of the upper and lower parts of the inner Neva Estuary, i.e. Neva Bay and the eastern Gulf of Finland, respectively.

Neva Estuary is generally characterised by a number of features common to other major Baltic estuaries. As for most Baltic estuaries, Neva Estuary is (1) brackish-water, non-tidal, and shallow, (2) strongly affected by wind-mixing, (3) affected by stochastic water exchange with the Baltic Proper, (4) subject to

U. Schiewer (ed.), *Ecology of Baltic Coastal Waters. Ecological Studies 197.*

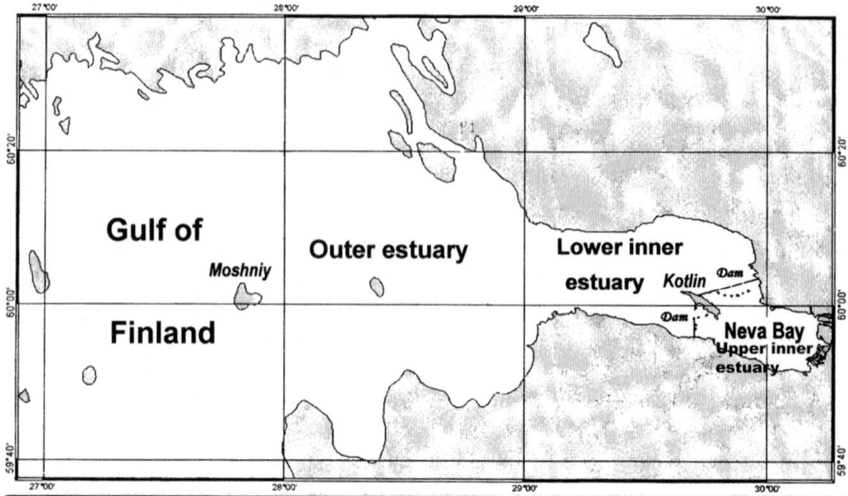

Fig. 12.1 Map of Neva Estuary and its sub-divisions

fluctuations of ecosystem parameters caused mainly by physical factors, (5) characterised by horizontal gradients of salinity, nutrients and plankton abundance, with (6) intensive benthic-pelagic coupling, (7) sensitive to nutrient loads (eutrophication), (8) characterised by intensive accumulation of humic substances, (9) dominated by eurytopic species, and (10) known for high biological diversity and productivity (Telesh 2004).

The general environmental characteristics (morphology, hydrology, and water chemistry) of Neva Estuary are presented in Table 12.1 (a,b,e). The Neva River (74 km long) flows out of Lake Ladoga, the largest lake in Europe, and discharges 76 km^3 water annually into the head of Neva Estuary, with an average concentration of total particulate organic matter (POM) in water ranging between 4.6 and 12.0 mg l^{-1} (Orlova et al. 2004). The head of the estuary (Neva Bay) is a very shallow water body (mean depth 4 m, maximum depth 12 m, volume of water 1.2×10^6 m^3) with an area/catchment ratio of 1.17 and a water residence time of 5.5 days. The share of allochthonous POM in Neva Bay is high. According to data from 2002, the mid-summer total phosphorous load from the Neva River to Neva Estuary is ca. 2,090 t a^{-1}, which supports gross primary production in the open waters of the estuary ranging from 1.43 to 2.32 g C m^{-2} day^{-1} (Orlova et al. 2004). Yearly global radiation in the area is 3,049 MJ m^{-2}; average precipitation is 600 mm a^{-1} (Smirnova 1997). The organic matter content of the sediments is 3–5 % dry weight (dw) (Emelyanov 1997); nitrogen input to the estuary is 140,000 t a^{-1}, phosphorus input 7,400 t a^{-1} (Leppänen et al. 1997).

At present, the three major ecological problems in Neva Estuary are generally classed as eutrophication, pollution and invasion by non-indigenous species (Telesh 2004). These three processes act within ecosystems, thus impacting them by causing changes in natural biological diversity, in the number and structure of

Table 12.1 Main characteristics of the Neva Estuary (Neva Bay and eastern Gulf of Finland). *dw* Dry weight, *ww* wet weight, *Chl a* chlorophyll *a*, *POC* particulate organic carbon, *DIN* dissolved inorganic nitrogen, *DIP* dissolved inorganic phosphorus

	Neva Bay (upper inner estuary)	Eastern Gulf of Finland (lower inner estuary)
a. Climate and hydrology		
Air temperature (°C)	Min: –10; max: –1 (December–March) Min: + 2; max: + 10 (April–May) Min: + 12; max: + 18 (June–Sept) Min: + 1; max: + 8 (October–November)[a]	
Water temperature (°C)	ca. 0 (December–March) mean: + 3 to + 10 (April–May) ca. + 18 (July–August) ca. + 5 (September–November)[b]	
Ice covering (days a^{-1})	120–130[b]	
Annual global radiation (MJ m^{-2})	3,049[a]	
Average precipitation (mm a^{-1})	600[a]	
Mean freshwater inflow ($10^6 m^3$ a^{-1})	78,600[a]	81,600[c]
Mean inflow ($10^6 m^3$ a^{-1}) (from Baltic)	0[c]	52,300[c]
Mean total outflow ($10^6 m^3$ a^{-1})	77,740[d]	134,100[c]
Residence time (days)	5.5	No data
b. Morphometry and sediments		
Location	Between Neva River mouth and island Kotlin	Between island Kotlin and ca. 29° E (cape Stirsudden)
Area (km^2)	329[e]	1,146[c]
Volume ($10^6 m^3$)	1,316	22,920
Mean depth (m)	4	20
Maximum depth (m)	12[c]	45[c]
Catchment area (km^2)	281,000[a]	350,400[a]
Area/catchment relation	0.0012	0.0033
Organic matter content (% dw)	4–6[f]	3–5[7]
Sediment fraction < 10 µm (% dw)	15–25[f]	94–96[f]
Sediment oxygen demand (mmol O_2 m^{-2} day^{-1})	12.44[h]	
Phosphate release (mg P m^{-2} day^{-1})	0.11[aa]	5.40[aa]
Ammonia release (mg N m^{-2} day^{-1})	0.61[aa]	1.65[aa]
Denitrification rate (mg N m^{-2} day^{-1})	0.1–0.2[h]	0.6–15.7[h]
Sediment composition	Sand, silt, clay[f]	Silt, clay, sand, boulders, gravel, pebbles[f]

(continued)

Table 12.1 (continued)

	Neva Bay (upper inner estuary)	Eastern Gulf of Finland (lower inner estuary)
c. **Pelagic biological components**		
Secchi depth (m)	1.28[i]	1.84[i]
Bacteria (10^6 ml^{-1})	2.5–5.0[j]	1.49–6.29[j]
	3.86[k]	7.37[i]
Phytoplankton biomass (μg Chl *a* l^{-1})	Spring max: 19.7	Spring max: 14.7
	Autumn max: 16.0[l,m]	Autumn max: 15.3[l,m]
Biovolume (spring/autumn max) (mm^3 l^{-1})	16.5/8.1 (average 3.0)[k]	14.0/11.1 (average 4.5)[k]
Dominating phytoplankton species	*Aulacosira islandica, Diatoma elongatum, Gonyaulux catenata, Achnanthes taeniata*[k]	
Ciliates		
Biomass (mg C l^{-1})	0.19–0.60 (max: 3.7) (ww)[o]	
Dominating species	*Tintinnidium fluviatile, Tintinnopsis cratera, Strombidium mirabile, S. viride, Coleps hirtus, Stokesia vernalis*	
Mesozooplankton		
Biomass (mg ww l^{-1})	0.02–1.65[l]	0.29–1.63[l]
Dominating species	Rotifera: *Keratella cochlearis, Synchaeta stylata, S. pectinata, S. grandis, Polyarthra vulgaris, P. dolichoptera, Asplanchna priodonta*; Cladocera: *Bosmina longirostris, B. longispina, B. coregoni, Chydorus sphaericus, Leptodora kindti;* Copepoda: *Mesocyclops leuckarti, Thermocyclops oithonoides, Eurytemora lacustris, E. affinis, Eudiaptomus gracilis*	
Fish		
Number of species	6[o]	60[o]
Dominating species	*Abramis brama, Sander lucioperca, Rutilus rutilus, Perca fluviatilis, Gymnocephalus cernuus, Gasterosteus aculeatus, Esox luceus, Coregonus albula, C. lavaretus*	
Seston (mg l^{-1})	7.25 (dry weight)[i]	3.79 (dry weight)[i]
POC (mg C l^{-1})	6.2–35.5[p]	4.1–24.9[p]
Primary production (g C m^{-2} a^{-1})	348[i]	
Bacterial production (μg C l^{-1} h^{-1})	32–270[j]	
Mesozooplankton production (g C m^{-2} a^{-1})	90–255[q]	30–100[r]

(continued)

Table 12.1 (continued)

	Neva Bay (upper inner estuary)	Eastern Gulf of Finland (lower inner estuary)
Fish catches (t a^{-1})	1,892[s]	14,969–31,536[o]
d. Benthic biological components		
Macrophytes		
Biomass (g dw m^{-2})	130–2,110[t]	50–300[u]
Number of species	68[v]	
Dominating species	*Phragmites australis, Scirpus lacustris, Nuphar lutea, Potamogeton perfoliatus, Cladophora glomerata*[v]	
Macrozoobenthos		
Biomass (g ww m^{-2})	10–17 (max: 200)	
Number of species	210	
Dominating species	*Potamothrix hammoniensis, Limnodrillus hoffmeisteri, Chironomus plumosus* (larvae), *Dreissena polymorpha*	
e. Water chemistry, trophic status and pollution		
Salinity (psu)	0.06–0.11[w]	1.0–5.0; (bottom: > 7.0)[b]
pH	7.29–7.99[p]	
Oxygen saturation (%)	84–100[p]	80–105 (surface); 29–63 (bottom)[p]
Total nitrogen (µmol l^{-1})	24.9–30.4[p]	23.5–31.1[p]
DIN (µmol l^{-1})	1.5–12.9[h]	No data
Total phosphorus (µmol l^{-1})	0.7–2.2[p]	0.73–3.0[p]; winter max: 1.3[x]
DIP (µmol l^{-1})	No data	1.3–6.0[x]
Annual nitrogen input (t a^{-1})	45,338[d]	140,000[y]
Annual phosphorus input (t a^{-1})	2,089[d]	7,400[aa]
Limitation of PP and period	Light[h]	Phosphorus[h]
Main pollutants	Oil products, phenols, heavy metals (Hg, Cd, Pb etc.), PCB, chlororganic substances, N, P[z]	
Trophic level	Mesotrophic to eutrophic	

[a]Smirnova (1997); [b]Mikhailov (1997); [c]Ostov (1971); [d]Kondratyev et al. (1997); [e]Nezhikhovsky (1981); [f]Spiridonov et al. (1999); [g]Emelyanov (1997); [h]Alimov et al. (2001); [i]Golubkov et al. (2004); [j]Drabkova et al. (1999); [k]V.N. Nikulina (personal communication); [l]Telesh et al. (1999); [m]Nikulina (2003); [n]Khlebovich (1987); [o]Kudersky (1999); [p]Frumin and Kryuchkov (1999); [q]Telesh (1987); [r]Shishkin et al. (1989); [s]Ruzhin (1987); [t]Belavskaya (1987); [u]Bäck et al. (2005); [v]Orlova et al (2005); [w]Alimov (1997); [x]Pitkänen and Välipakka (1997); [y]Leppänen et al. (1997); [z]Frumin & Susareva (1997); [aa]Ignatieva (1997)

trophic webs, and in the overall productivity and functioning of coastal ecosystems. The consequences of these alterations can be demonstrated by the negative effect on water quality and the decrease in production at the higher trophic level (fish and mammals).

12.2 Planktonic Communities

Pelagic communities of Neva Estuary have been studied intensively since the early 1980s. These investigations permit evaluation of changes that have occurred in plankton during two recent stages of the "technogenic period" in the estuary (Alimov et al. 1996). In the 1980s, the first stage was characterised by phasing-in two wastewater treatment plants in Neva Bay, and the construction of a major part of the St. Petersburg flood protection barrier – a hydraulic system and a number of dams at the border between Neva Bay and the eastern Gulf of Finland. The latter caused serious alteration of the hydrological regime in the estuary, which impacted significantly on the structure and functions of pelagic communities during the second stage of the "technogenic period" in the 1990s.

12.2.1 Phytoplankton

In the 1980s, several areas with different characteristic pelagic communities could be distinguished in Neva Bay: central, northern and southern zones, and near-shore areas within macrophyte associations (Telesh 1987). In the central and northern parts of Neva Bay, species composition of plankton communities was determined by the pelagic flora and fauna of the southern Lake Ladoga and the Neva River. The phytoplankton species list for Neva Bay obtained during a 3-year study (1982–1984) comprises 323 species of algae [see Chaps 3, 4 (Schiewer), 5 (Radziejewska and Schernewski), 7 (Kruk-Dowgiałło and Szaniawska), 8 (Chubarenko and Margoński), 9 (Gasiūnaitė et al.), and 10 (Kotta et al.), this volume], among them 214 planktonic and 171 species in periphyton (Nikulina 1987). The maximum algal biomass was observed in spring due to diatoms (Nikulina 1987; Nikulina and Anokhina 1987). The mean phytoplankton biomass in Neva Bay in the vegetation seasons of 1982–1984 was 1.03±0.84 mg wet weight (ww) l^{-1} (Nikulina 1987). During summer and autumn, Cyanobacteria, green algae and diatom species were dominant. In the southern zone of Neva Bay, however, the phytoplankton communities were typical of eutrophic conditions already in the 1980s. The average concentrations of chlorophyll-*a* (Chl-*a*) for the vegetation periods of 1982–1984 were 9.41 ± 3.81 µg l^{-1} in the southern part of the bay, and 6.34 ± 1.89 µg l^{-1} in the northern Neva Bay (Umnova 1987). Primary production in the central part of Neva Bay in the 1980s was limited by high turbidity and short water residence times. The decomposition of organic matter in Neva Bay during that period averaged ca. 0.50 g C m^{-2} day^{-1} while mean primary production was low (0.26 g C m^{-2} day^{-1}) because of the extremely high concentration of inorganic particulate matter due to dredging activities in the bay and the high load of allochthonous organic matter (Pavelieva 1987; Gutelmakher et al. 1987; Nikulina 1996).

In the 1990s, the summer phytoplankton community was dominated by Cryptophytae (34% of total biomass), and diatoms (31%); green algae contributed on average 15.6% to the community biomass, Cyanobacteria only 12.3%; the total phytoplankton biomass in Neva Bay varied within the range 0.2–1.94 mg ww l^{-1}; the Chl-*a* concentration was 2.07–19.71 µg l^{-1}, averaging 9.74 ± 5.18 µg l^{-1}; the primary production of plankton was 0.72 ± 0.22 g C m^{-2} day^{-1} (Telesh et al. 1999).

In the eastern Gulf of Finland, the average seasonal phytoplankton biomass in the 1980s was 1.5–2.5 mg ww l^{-1} (Alimov et al. 1994). Among the Cyanobacteria, the genera *Aphanothece*, *Microcystis* and *Planktothrix* accounted for 80–90% of the biomass, suggesting accumulation of organic substances and nutrients in the water. In the shallowest waters of the eastern Gulf of Finland, the diatom *Skeletonema subsalsum* (A.Cl.) Bethge appeared during the 1980s, accounting for 50% of the total phytoplankton biomass (Alimov et al. 1994). This alga, especially when highly abundant, can serve as an indicator of changing conditions due to pollution. Seasonal dynamics of phytoplankton density and biomass in the eastern Gulf of Finland during that period was represented by a bimodal curve, with maxima in May and August, and with average mid-summer–autumn biomass reaching 2 mg ww l^{-1} (Nikulina 2003). In late August through September, a Cyanobacteria bloom caused by *Aphanizomenon flos-aquae* (L.) Ralfs, *Microcyctis aeruginosa* Kutzing and *Anabena flos-aquae* (Lyngb.) Breb. was typical of this area and characteristic of mesotrophic waters (see Chap. 6 by Schiewer, this volume); the algae *Limnothrix planktonica* (Wolosz.) Meff. and *Planktothrix agardhii* (Gom.) of the *Oscillatoria*-complex were permanently present but did not form predominant assemblages (Nikulina 2003). Annual primary production in the eastern Gulf of Finland during the 1980s was typical of mesotrophic waters, averaging ca. 100 g C m^{-2} a^{-1} (Shishkin et al. 1989).

In the 1990s, the *Oscillatoria*-complex became predominant in the eastern Gulf of Finland during summer and autumn; the period of high biomass values (average 2–7, maximum 8–11 mg ww l^{-1}) typical of eutrophic waters became more extensive, the proportion of Cyanobacteria and cryptophytes in the total phytoplankton biomass increased while the ratio of bacillariophytes decreased (Nikulina 2003). Average phytoplankton biomass increased in the deeper areas of the inner Neva Estuary while, on the contrary, concentrations of Chl-*a* decreased with depth (Fig. 12.2). Cyanobacteria blooms in the eastern Gulf of Finland are now defined mainly by nitrogen-fixing species (*Anabaena* spp., *Aphanizomenon* spp., *Nodularia spumigena* Mertens); as a rule, such blooms occur when the ratio of dissolved mineral N to P is low, i.e. < 20 (Nikulina 2003). These data are in agreement with the fact that, during the past decade, the nutrient loads to the eastern Gulf of Finland from agriculture and point sources have been steadily declining. For example, in 1998 the discharges (in relation to maximum values) were 30% for total phosphorus, 62% for total nitrogen, 53% for ammonium nitrogen, 76% for nitrites, and 89% for nitrates (Kondratyev et al. 1997). Presently, the Cyanobacteria *P. agardhii*, *A. flos-aquae*, *L. planktonica*, *Anabaena* spp., *Aphanizomenon* spp., and *N. spumigena* dominate the phytoplankton community.

Thus, changes in the phytoplankton community of the eastern Gulf of Finland reflect modifications of the nutrient regime in the ecosystem; however, the altered hydrological situation in the basin has had little impact on the community structure and abundance of phytoplankton. Although the average mid-summer and autumn

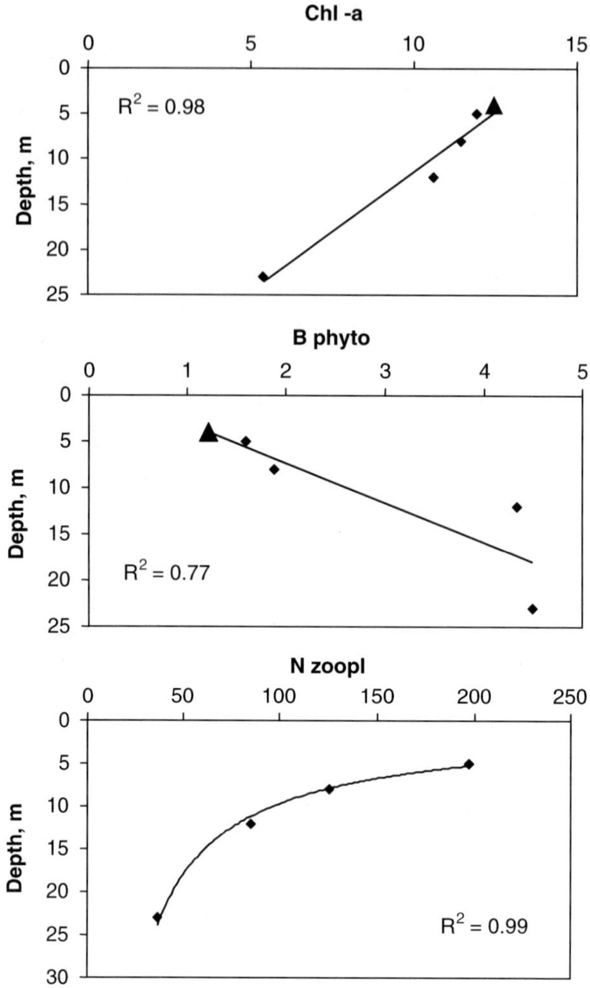

Fig. 12.2 Concentration of chlorophyll *a* (*Chl-a*; μg l^{-1}), phytoplankton biomass (*B phyto*; mg l^{-1}), and zooplankton density (*N zoopl*; ind l^{-1}) at locations with different depths in the inner Neva Estuary (data for the 1990s, from Telesh et al. 1999). *Triangles* Data for Neva Bay, *squares* data for the eastern Gulf of Finland; means of 3–6 data sets

biomass of planktonic algae had nearly doubled compared to the 1980s, the composition of phytoplankton changed only insignificantly during the last decade of the twentieth century, and a tendency to return to the community structure known from earlier decades has been recorded in recent years (2001 and 2002; Nikulina 2003). Moreover, variation in the ratio of maximum to minimum phytoplankton biomass during one vegetation season has decreased 9- to 12-fold from 1997 to 2000, compared to 9- to 22-fold in the 1980s, which witnesses a reduction in pollution and anthropogenic eutrophication (Alimov 2003).

12.2.2 Bacterioplankton

Neva Estuary is a heavily nutrient-loaded water system where the external load exceeds significantly the internal production of organic matter. In the 1980s, gross primary production in Neva Bay was ca. 30 times lower than the amount of allochthonous organic substances entering the estuary (Winberg and Gutelmakher 1987), while in the eastern Gulf of Finland this difference was only 6.2 tines (Shishkin et al. 1989).

A high external load of organic matter favours populations of heterotrophic microorganisms–decomposers; their density in Neva Bay averaged 68,000 cells ml^{-1} (Pavelieva et al. 1987). The total average bacterioplankton density in Neva Bay in the 1980s varied within the range 2.5–5.0 × 10^6 cells ml^{-1} in the central area of the bay to 7–10 × 10^6 cells ml^{-1} in the littoral zone (Table 12.2). Average bacterial biomass in the pelagic zone of the bay in those years was 0.74–1.53 g C m^{-2} (see Chap. 3 by Schiewer, this volume), which was comparable to the biomass of mesozooplankton and three times higher than the protozoan biomass, constituting 18–26% of the POM concentration in the water column (Pavelieva et al. 1987).

Microbial production in Neva Bay reaches a maximum in late summer and usually decreases westwards from ca. 240 mg C m^{-2} day^{-1} in the Neva River mouth to 130 mg C m^{-2} day^{-1} near the island of Kotlin (Pavelieva et al. 1987). On average, gross bacterial production for the vegetation season (May–September) in Neva Bay was 45.3 t C day^{-1}, which was 55% of primary production of phytoplankton (Pavelieva et al. 1987). Decomposition of organic matter by bacteria in the summer seasons of 1982, 1983 and 1984 was 150, 190 and 200 t C day^{-1}, respectively, for the whole Neva Bay, and the ratio of microbial decomposition to primary production of phytoplankton for the same periods was 1.6, 2.1 and 3.2, respectively (Pavelieva et al. 1987).

Data on bacterioplankton in Neva Estuary in the 1990s is available only for the autumn months: September and October. The microbial densities in the autumns of 1994–1996 varied within the range 1.49–6.29 × 10^6 cells ml^{-1} and averaged 3.5–4.3 × 10^6 cells ml^{-1} (Drabkova et al. 1999). These values are similar to data for the 1980s (Table 12.2) and correspond to meso- and slightly eutrophic conditions in the water body.

Table 12.2 Bacterioplankton density (N × 10^6 cells ml^{-1}) in Neva Bay

Year (months)	Mean	Range	Reference
1982 (May–September)	2.5	1.3–3.2	Pavelieva et al. 1987
1983 (May–September)	3.2	2.4–4.4	
1984 (May–September)	5.0	3.1–10.0	
1994 (September)	3.7	2.6–4.8	Drabkova et al. 1999
1995 (September)	4.3	1.5–6.3	
1996 (September)	3.5	1.9–5.9	

In the eastern Gulf of Finland, the parameters of pelagic microbial communities are comparable to those in Neva Bay; although bacterial production in the lower inner Neva Estuary demonstrates a lesser degree of inter-annual variation than in the upper inner estuary, indicating a higher level of ecosystem stability in the former (Drabkova et al. 1999).

12.2.3 Zooplankton

The history of zooplankton research in Neva Estuary was reviewed recently by Telesh (2001). Since the earliest studies at the beginning of the twentieth century (Skorikov 1910), the zooplankton composition in Neva Estuary has been shown to depend on the planktonic fauna of southern Lake Ladoga inflowing into the estuary with Neva River waters (Telesh 1986; 1987). In total, ca. 400 species, including Protozoa, Rotifera, Cladocera, and Copepoda, have been registered in the estuary since the early twentieth century. Thus, the present-day zooplankton communities in Neva Estuary are characterised by very high species diversity due to heterogeneity of biotopes and enrichment of the estuarine fauna by species from numerous small freshwater bodies (lakes, ponds, rivers and creeks) located close to the shores of the estuary (Telesh 2004). More than 50% of the ca. 280 zooplankton species registered nowadays in Neva Estuary can also be found in the communities of other shallow coastal waters of the southern and eastern Baltic Sea (Telesh and Heerkloss 2002, 2004).

The zooplankton community structure in Neva Bay and the eastern Gulf of Finland is generally different; however, in each of these sub-basins it has been relatively stable in terms of species composition, numbers and biomass.

A characteristic feature of zooplankton in Neva Estuary, and particularly in its upper inner part (Neva Bay), is the substantial contribution of microzooplankton: protozoans and rotifers. Ciliates form the major component of the protozooplankton community in this estuary (see Chap. 3 by Schiewer, this volume). Khlebovich (1987) identified 55 species of planktonic ciliates in Neva Bay, among which 12 taxa define the community structure while 4 species form a group of dominants that are permanent throughout the vegetation season: *Tintinnidium fluviatile* Stein, *Tintinnopsis cratera* Hada, *Strombidium mirabile* Penard and *S. viride* Stein. In spring, ciliates can contribute up to 80% of total zooplankton biomass. The average density of ciliates in Neva Bay is 3,000 ind. l^{-1}, biomass 0.36 mg l^{-1}; however, these parameters can reach values twice as high in the shallow near-shore zones of the estuary (Khlebovich 1987). The average production of ciliate populations in Neva Bay is 50 mg C m^{-2} day^{-1}, which is 19% of the primary production of plankton and ca. 30% of bacterial production; the latter speaks for an important role of ciliates in the estuarine trophic web as consumers of microorganisms responsible for the considerable self-purification in the ecosystem of Neva Estuary (Khlebovich 1987).

The net zooplankton in Neva Bay is dominated by rotifers [see Chaps. 3 (Schiewer), 8 (Chubarenko and Margoński) and 9 (Gasiūnaitė et al.), this volume], which contribute on average 86% to the total density and 84% to the total zooplankton biomass, and a comparatively high abundance of juvenile copepods, mainly from the genera *Mesocyclops*, *Thermocyclops*, and *Eurytemora* is typical of this area (Telesh 1987; Telesh et al. 1999). Among rotifers, species from the genera *Keratella* (*K. cochlearis* Gosse, *K. quadrata* O.F.M.), *Synchaeta* (*S. stylata* Wierz., *S. pectinata* Ehrbg, *S. grandis* Zach., *S. lakowitziana* Lucks), and *Polyarthra* (*P. vulgaris* Carlin, *P. dolichoptera* Idelson, *P. remata* Skorikov) usually dominate in Neva Bay; high abundance of a facultative predatory species *Asplanchna priodonta* Gosse may be registered at certain times. Brackish water intrusions to the upper inner Neva Estuary along the shipping channel in the southern Neva Bay, as proved by permanent populations of some marine Harpacticoida species, were rather common events until the late 1980s (Telesh 1987). Since then, following the construction of a storm-surge barrier in Neva Estuary, no marine zooplankton species have been found in the upper estuary.

Although demonstrating high spatial heterogeneity within the estuary, mean May–September values of zooplankton biomass in the open waters of Neva Bay in the 1980s varied slightly around 1.0 mg ww l^{-1}, while in macrophyte associations, zooplankton biomass was 10- to 30-fold higher (Telesh 1987). Productivity of zooplankton was highest in near-shore macrophyte stands, reaching 0.52–1.2 g C m^{-2} in the vegetation season. Decomposition of organic matter by zooplankton was highest in reed beds: 5.3–36.3 mg C m^{-2} day^{-1}, which accounted for 0.3–2.2% of the total decomposition of organic substances by invertebrate communities among coastal vegetation (Telesh 1987).

In the 1990s, zooplankton biomass in the bay was lower, averaging 0.42 mg ww l^{-1} and ranging between 0.02 and 1.65 mg ww l^{-1} (Telesh et al. 1999). In Neva Bay, zooplankton density in the 1980s and 1990s generally showed a 20% increase in the western direction (Telesh 2004).

In the eastern Gulf of Finland, zooplankton density and biomass decrease in deeper areas (Fig. 12.2). Currently, the zooplankton biomass in the eastern Gulf of Finland varies within the range 0.29–1.63 mg ww l^{-1} (mean 0.71) while in the 1980s it was on average around 0.44 mg ww l^{-1} (Shishkin et al. 1989). The community is dominated by copepods (70.3% of total zooplankton biomass); cladocerans contribute 14.4%, rotifers 15.5% (Telesh et al. 1999). Among copepods, the most abundant species are usually *Mesocyclops leuckarti* Claus, *Eurytemora lacustris* Poppe, and *Eurytemora affinis* (Poppe). Among cladoceran species, predatory forms from the genera *Leptodora*, *Bythotrephes*, and the most recent Ponto-Caspian invaders *Evadne anonyx* Sars and *Cercopagis pengoi* (Ostroumov), are relatively abundant. The impact of invasive predators on the native zooplankton community (Fig. 12.3) has already initiated a tendency to population decrease in aboriginal cladoceran filtrators, mainly daphniids (Telesh et al. 2001). Our results and calculations show that the permanent population of *C. pengoi* (in case of high densities) can facilitate alteration of pelagic trophic webs, matter cycles, and energy flows through the ecosystem that might be followed by changes in the trophic status and water quality in the ecosystem (Telesh et al. 2001; Telesh and Ojaveer 2002; Litvinchuk and Telesh 2006).

Phytoplankton and zooplankton are pelagic components that are strongly functionally inter-related in all water bodies, and in this respect Neva Estuary is no exception. Knowledge of pelagic community structure can provide explanations for such phenomena as, for example, the lack of correlation between zooplankton and phytoplankton biomasses in the eastern Gulf of Finland compared to Neva Bay, where this correlation is very strong (Fig. 12.4). Thus, the dominance of Cyanobacteria in the phytoplankton assemblages of the eastern

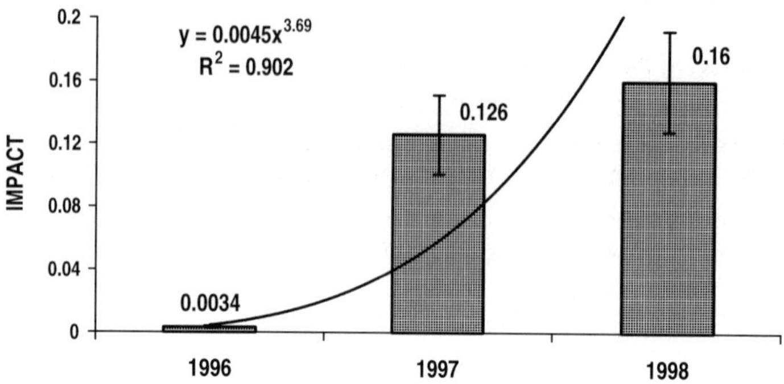

Fig. 12.3 Impact of the invasive predatory cladoceran *Cercopagis pengoi* on the zooplankton community in the eastern Gulf of Finland (1996–1998) calculated according to Telesh et al. 2001

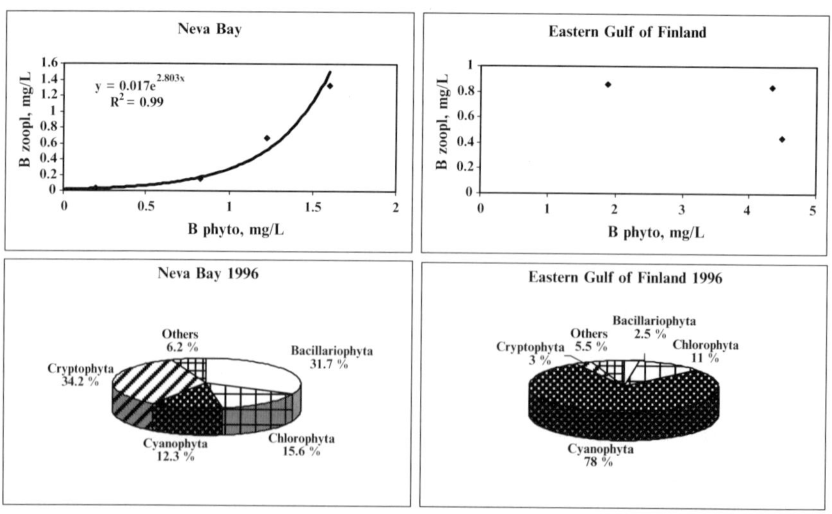

Fig. 12.4 Relationship between zooplankton biomass (*B zoopl*) and phytoplankton (*B phyto*) in Neva Bay and the eastern Gulf of Finland. *Data points* are averages for four zones in Neva Bay and three zones in the Gulf of Finland; *n*=20 for total data set; data from Telesh et al. (1999)

Gulf of Finland favour the existing zooplankton community structure, with a prevalence of obligatory and facultative predators and a minor role for phytophagous filtrators.

12.3 Fish Communities

Of the 60 species of fishes and cyclostomes recorded in Neva Estuary, 54 are known in both the outer estuary and lower inner estuary (the eastern Gulf of Finland), and 6 inhabit only the upper inner estuary, Neva Bay (Kudersky 1997; 1999).

Less than half of these 60 species (namely, 25–28) can be considered as commercial fishes, of which only 15 species form the basis of fisheries in the area. According to the statistical data, fish catches in Neva Estuary in 1946 through 1995 fluctuated from 4,035 to 42,597 t a^{-1}, or from 3.4–8.0 to 20–35 kg ha^{-1} (Kudersky 1999).

The fish population in the estuary is heterogeneous and can be divided into three groups: (1) marine (20 species), (2) anadromous and semi-anadromous (11 species), and (3) freshwater (29 species). The characteristics of fish groups and species given below follow the publications of Kudersky (1997, 1999, and references therein).

The marine group includes only five commercial species (Baltic herring *Clupea harengus membras* L., sprat *Sprattus sprattus* (Schneider), cod *Gadus morhua callaris* L., eelpout *Zoarces viviparus* (L.), and flounder *Pleuronectes flesus trachurus* Dunker), the rest being scarce and non-commercial. The Baltic herring, eelpout and flounder are the permanent inhabitants of Neva Estuary, while sprat and cod enter the estuary periodically from the western areas of the Baltic Sea.

The group of anadromous and semi-anadromous fishes and cyclostomes includes eight commercial species, of which river lamprey *Lampetra fluviatilis* (L.), salmon *Salmo salar* L., sea trout *Salmo trutta* L., vendace *Coregonus albula* L., whitefish *Coregonus lavaratus* (L.), *Vimba vimba* (L.) are anadromous, eel *Anguilla anguilla* (L.) is catadromous, and the semi-anadromous species is smelt *Osmerus eperlanus* (L.).

The freshwater group is most numerous in species number but most of these fishes are scanty or even scarce. Within this group, commercial statistical data is available for pike *Esox lucius* L., bream *Abramis brama* (L.), pikeperch *Sander lucioperca* (L.), burbot *Lota lota* (L.), three-spine stickleback *Gasterosteus aculeatus* L., and nine-spine stickleback *Pungitius pungitius* (L.). Besides the species listed, there are also *Leuciscus idus* (L.), white bream *Blicca bjorkna* (L.), and some others. Small-sized freshwater fishes with low commercial value include roach *Rutilus rutilus* (L.), bleak *Alburnis alburnis* (L.), perch *Perca fluviatilis* L., and ruffe *Gymnocephalus cernua* (L.).

The total number of 60 species mentioned above, however, does not include some fishes (ca. 10 species) that appear in the estuarine waters occasionally due to acclimatisation and fishery activities, e.g. *Acipenser ruthenus* L., *Acipenser baerii* Brandt, *Salmo gairdneri* Rich., *Coregonus autumnalis migratorius* Georgi, *C. nasus*

Pallas, *C. muksun* Pallas, *C. peled* Gmelin, *Catostomus rostratus* Tilesius, *Cyprinus carpio* L., *Perccottus glenii* Dyb. (Kudersky 1999). However, the latter species, the amur sleeper *Perccottus glenii*, was recently shown to significantly affect the native invertebrate communities, as well as the other fish populations (e.g. roach), in Neva Bay through predation – either directly or indirectly through competition for food (Orlova et al. 2006).

Fish populations and, consequently, fish catches in the estuary are presently impacted by both natural and man-induced factors. The latter is seriously affecting quality of water and sediments (e.g. chemical and organic pollution, heavy metals, etc.) and thus fish spawning and nursery grounds are deteriorating, especially in the vast coastal areas with macrophyte stands in Neva Bay. The exceptionally broad coastal zone of the shallow upper inner estuary is favourable for macrophytes [dominants: *Phragmites australis* (Cav.) Trin. ex Stend., *Scirpus lacustris* L., *Nuphar lutea* (L.) Smith]. In the 1980s, macrophytes covered an area of ca. 7.0 km^2 at the head of Neva Estuary, which accounted for 2% of the total area of the bay (Belavskaya 1987). Recent results have demonstrated the high diversity of aquatic plant associations (68 species in total) and the tendency towards expansion of the macrophyte-covered zone in Neva Estuary (Orlova et al. 2005).

12.4 Benthic Communities

Historical data show that, at the beginning of twentieth century when the state of ecosystem was determined primarily only by natural processes, the glacial relict crustaceans *Pallasea quadrispinosa* Sars and *Mysis relicta* Loven dominated strongly in the benthic communities of the freshwater Neva Bay, whereas in the brackish inner and outer parts of the estuary, dominants were *Monoporeia affinis* Lindstrom and *Saduria entomon* (Linne) (Skorikov 1910). The composition of zoobenthic communities of Neva Bay changed greatly in the 1920s. Since that time, worms (Oligochaeta, mainly Tubificiidae) and small filtering molluscs (Pisidiidae, Bivalvia) became the dominant groups. Only one glacial relict crustacean species, *Pallasea quadrispinosa*, was present in small numbers in the north-western part of the bay, which is characterised by a relatively high water flow rate. In the mid-1980s, the biomass of Pisidiidae molluscs in the eastern part of Neva Bay reached very high values: 0.5–1.0 kg ww m^{-2}. These molluscs were filtering out ca. 60% of suspended organic matters brought by Neva River waters and were responsible for the extremely high rate of decomposition of organic matter by macrozoobenthos in this area (Alimov and Golubkov 1987).

In the 1990s, the composition of benthic communities in Neva Bay changed significantly due to an increase of dominance of Oligochaeta, an increase in importance of chironomid larvae, and a decrease in the share of bivalve molluscs (Fig. 12.5). Especially dramatic changes have happened in the eastern part of Neva Bay, where the biomass of zoobenthos declined from 200–1,500 mg ww m^{-2} in the 1980s to 1–17 mg ww m^{-2} in the early 2000s. As a result, the average biomass of

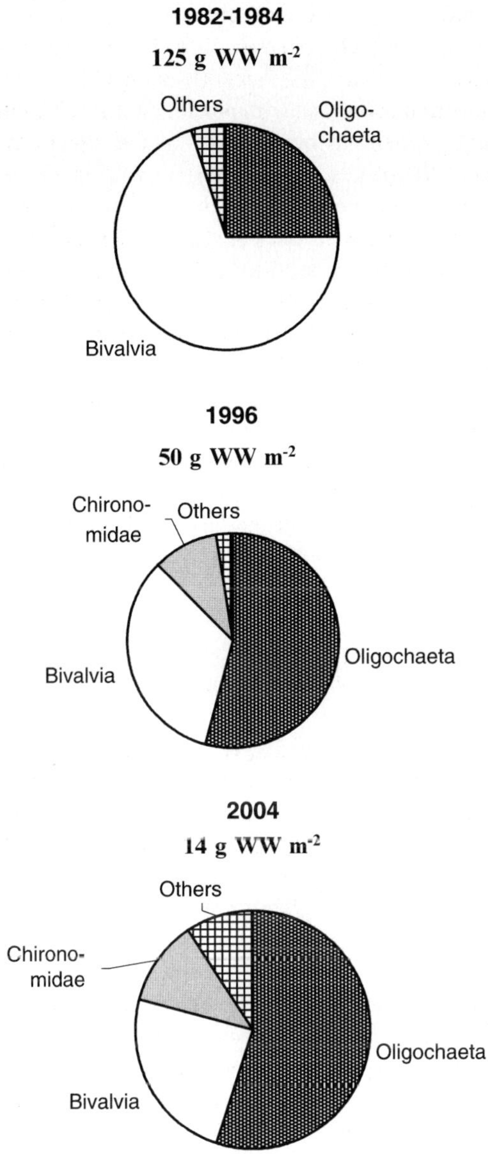

Fig. 12.5 Community structure (% of biomass) and average biomass of zoobenthos in Neva Bay in 1982–1984, and in mid-summer of 1996 and 2004

benthos in Neva Bay has decreased from 100–150 g ww m^{-2} in the 1980s to ca. 50 g ww m^{-2} in the 1990s, and to 10–14 g ww m^{-2} in the early 2000s (Fig. 12.5). The species composition of bivalve molluscs has also changed. Large numbers of the small Pisidiidae that formed high biomasses in the eastern part of Neva Bay have disappeared; instead, rather dense beds of large-sized molluscs, Unionidae, have appeared in southern and north-western areas of the bay. The biomass of these mussels can reach 200 g ww m^{-2} in some local beds in the north-western part of Neva Bay.

These changes in composition and abundance of zoobenthic communities in the north-western part of the bay may be explained by the construction of a storm-surge barrier (dam), which diminishes water exchange with the lower part of the inner estuary. The great decline of zoobenthos in the eastern part of Neva Bay during the last decade is probably related to the construction of the Northern Waste Water Treatment Plant, which has decreased inflow of dissolved and suspended organic matters to the easternmost part of Neva Estuary.

The biodiversity of macrozoobenthos in the lower part of the inner estuary (the eastern Gulf of Finland) is rather low and includes only four groups of bottom animals: Oligochaeta (18 species), Polychaeta (1 species), Crustacea (2 species), and Insecta–Chironomidae (3 species). Two species of Oligochaeta, *Potamothrix hammoniensis* (Mich.) and *Limnodrillus hoffmeisteri* Clap. and larvae of *Chironomus plumosus* L. (Insecta) dominated in the bottom animal communities in the eastern part of the lower inner estuary. The mean biomass of zoobenthos in this part of the estuary was about 17 g ww m^{-2} in the early 2000s (Golubkov et al. 2003a).

Comparing present data to the results of earlier studies (Shishkin et el. 1989) revealed slight changes in zoobenthos in the open water areas during the last two decades. However, the composition of macrofauna was quite different at the beginning of the twentieth century, when the relict crustaceans *Monoporeia affinis*, *Saduria entomon*, and *Mysis relicta* largely dominated the zoobenthos (Skorikov 1910). The first two species are still dominants in the outer Neva Estuary (Golubkov et al. 2003a; Maximov 2003). The most important change in zoobenthos of the open waters in the lower inner estuary is the disappearance of marine bivalve molluscs *Macoma balthica* (L.) from benthic communities in the 1990s. Before that time, in the mid-1980s, rather dense populations of this species with biomass up to 50 g ww m^{-2} were observed at depths of around 20 m. In the late 1980s, this population gradually declined, and in the early 1990s these molluscs were almost entirely eliminated from the zoobenthic community (Maximov 2003).

An important factor affecting the distribution and dynamics of the bottom fauna in the deep western part of Neva Estuary is the periodical near-bottom intrusions of saline oxygen-poor waters from the deeper western part of the Gulf of Finland. These intrusions lead to an impoverishment of macrobenthos in considerable areas of the western part of the estuary (Maximov 2003). Especially drastic changes were observed in 1996, when, after the intrusion of near-bottom saline waters from the western Gulf of Finland, macrobenthos entirely disappeared from the deep western part of the outer estuary. In the years following, a gradual recovery of bottom communities was observed, but even in 1999–2001 macrofauna in previously azoic areas in the eastern Gulf of Finland was still poor, and its biomass was less than 1 g ww m^{-2} (Maximov 2003). Thus, at present, a declining gradient of benthic biomass exists from the eastern (Neva Bay) to the western (outer estuary) part of Neva Estuary (Fig. 12.6).

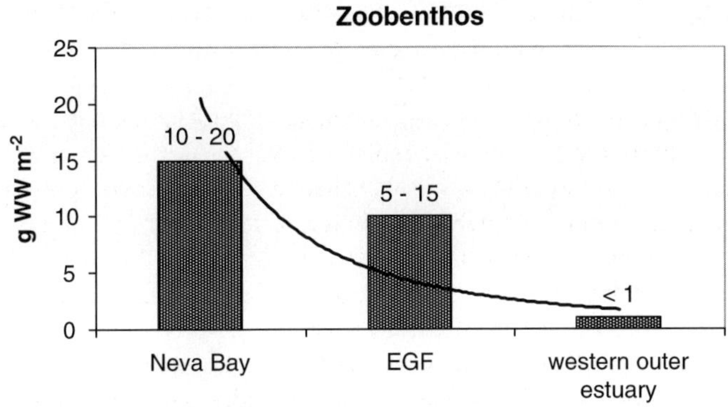

Fig. 12.6 Present-day spatial dynamics of average values (*columns*) and range (*numbers above columns*) of zoobenthos biomass (g ww m^{-2}) in Neva Estuary. *EGF* Eastern Gulf of Finland

Benthic animal communities accumulate information about their habitat conditions, i.e. chemical characteristics of bottom sediments and water masses above them. Moreover, benthic invertebrates respond to the changing quality of their environment by alterations in community structure. These modifications in species composition of benthic communities reflect ecosystem transformations that have occurred over a relatively long time-span. For this reason, studies of the benthic communities in Neva Estuary in recent decades have contributed to elaboration of the Integrated Index (IP index) for water quality assessment (Balushkina 1997). The IP index of Balushkina (1997) allows water quality evaluation based on different indices, and often providing conflicting results, to be combined in an integrated parameter giving a more complete characterisation of the animal community as a whole and reflecting specific features of human impact on the entire estuarine ecosystem.

Along with the influences of pollution and eutrophication on the ecosystem, significant changes in zoobenthic communities have occurred in Neva Estuary following the invasions of alien species: *Dreissena polymorpha* Pallas (Bivalvia), *Marenzelleria neglecta* (formerly *Marenzelleria viridis* Verrill) (Polychaeta), and amphipods *Gmelinoides fasciatus* Stebbing and *Pontogammarus robustoides* (Sars) (Orlova et al. 2005, 2006). *Gmelinoides fasciatus* has practically replaced the aboriginal amphipod species *Gammarus lacustris* Sars in the littoral communities of Neva Estuary (Berezina 2004; Berezina and Panov 2004).

By the early 2000s, *D. polymorpha* had established dense populations in all nearshore localities in the lower inner and outer Neva Estuary, where hard substrates are available and the salinity varies from fresh water up to 5.5 psu; these molluscs foul hydro-technical constructions and contribute greatly to nutrient cycling and the production of organic matter in the coastal waters of Neva Estuary (Golubkov et al. 2003b; Orlova et al. 2004).

Another important event was the establishment and expansion of the worm *M. neglecta* in the lower inner and in the outer estuary in the late 1990s. The biomass of this species has increased progressively in recent years, and in 2004 it reached 8 g ww m^{-2} in the southern part of the inner estuary (Maximov 2003). At high densities, this species affects ecosystem functioning considerably and can stimulate eutrophication processes due to perturbation of bottom sediments (Golubkov 2004).

12.5 Benthic–Pelagic Coupling, "Biological Pollution" and Eutrophication

Neva Estuary is one of the most eutrophied areas in the eastern Baltic Sea, and is constantly under the impact of intensive anthropogenic stress of different types, including pollution by waste discharges from St. Petersburg and its surroundings. "Biological pollution" i.e. invasive species introduced with ballast waters of trans-oceanic ships is another serious problem in this area (Panov et al. 1999; Orlova et. al 1999, 2006). Formation and successful development of self-sustainable populations of alien species in the coastal waters of Neva Estuary can be considered as a manifestation and/or result of an anthropogenic stress followed by a decrease in biological diversity, modification of trophic webs and, consequently, alteration of the functional characteristics of an estuarine ecosystem (Telesh and Orlova 2004).

Biological pollution of Neva Estuary by invasive species can have a negative impact on the structure and functions of the estuarine zooplankton community, which is presently responsible for ca. 40% of total decomposition of organic matter in this water system (Golubkov et al. 2003a). For example, the recent (1995) invasion of the predatory planktonic Ponto-Caspian cladoceran *Cercopagis pengoi* to Neva Estuary has already initiated a tendency towards a population decrease in aboriginal cladoceran filtrators as a result of predation by *Cercopagis* during the peak of its development (Telesh et al. 2001; Litvinchuk and Telesh 2006).

The zebra mussel – the bio-fouling bivalve mollusc *Dreissena polymorpha* – is another Ponto-Caspian invader to the eastern Gulf of Finland that can cause considerable ecosystem effects in these coastal waters. It is well known that, due to the planktonic larval stage in its life cycle, this species can successfully colonise vast areas of recipient water bodies (Ovchinnikov 1954; Mills et al. 1993; Starobogatov 1994). Being an edificatory species, and one of the major freshwater filtrators, this invader can influence not only the structure of benthos but also impact plankton communities (Starobogatov 1994; MacIsaac 1996; Karatayev et al. 1997). Zebra mussel beds are known to facilitate benthic–pelagic coupling by excretion of bio-available nutrients and by clarifying the water of particulate matter, thereby improving light conditions in the deeper water layers (Orlova et al. 2004).

Plankton interactions in Neva Estuary, including relations between invasive and aboriginal species, have been intensively studied in recent years (see the reviews: Telesh 2004, 2006; Krylov et al. 2004, and references therein). However, until recently the relationships between native planktonic organisms and benthic invasive species with pelagic larvae in their life cycle were not a specific focus of research in this area.

There is evidence from the literature for two types of benthic–pelagic relationship between *Dreissena* and microzooplankton in Neva Estuary: (1) competition for food, and (2) direct consumption of microzooplankton (rotifers) by the molluscs, i.e. predation (Telesh and Orlova 2004). A recent study aimed to verify the proposed hypotheses by comparative analysis of the spatial distribution of planktonic rotifers, *Dreissena* larvae, and adult molluscs of this species in the coastal waters of the eastern Gulf of Finland (Telesh and Orlova 2004).

The high diversity of rotifers registered during this study allowed conclusions to be drawn about favourable environmental conditions for microzooplankton in the coastal waters of the eastern Gulf of Finland. The similarity of species composition of rotifer assemblages in the open and coastal waters of Neva Estuary is explained by its shallowness and the significant water exchange between the littoral and central areas of this water body (Telesh 1987). The large variability of numerical data on rotifers at different locations in this estuary is not surprising due to the highly dynamic hydrological regime. Previous investigations have demonstrated that irregularity in plankton distribution in the littoral zone of Neva Estuary is due predominantly to the variable water regime, which is in turn defined largely by wind-induced water movements (Praeobrazhenskij et al. 1987).

Nevertheless, analysis of the vertical distribution of microzooplankton in the water column of the littoral zone down to a depth of 5 m revealed certain correlations between estuarine community components. In particular, a tendency to share niches between rotifers inhabiting presumably the upper water layers, and larvae of *D. polymorpha*, which were concentrated mostly near bottom, was revealed (Fig. 12.7). Similarity in body size (200–250 μm), lifestyle (planktonic), and comparable size of food particles available for consumption (1–12 μm) allow speculation about possible competitive interactions for food between microzoo-

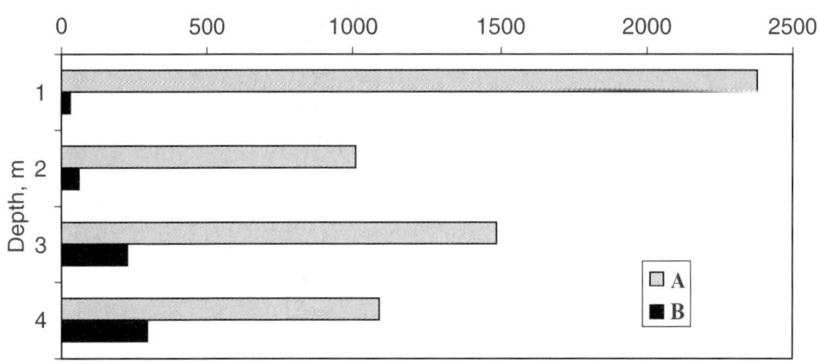

Fig. 12.7 Density of rotifers (*A*, ind l^{-1}) and meroplanktonic larvae of *Dreissena polymorpha* (*B*, × 10 ind l^{-1}) at stations at different depths in the littoral zone of the eastern Gulf of Finland (modified from Telesh and Orlova 2004)

plankton (rotifers) and the bivalve mollusc *D. polymorpha* at the larval stages of their life cycle (Telesh and Orlova 2004). The reliable negative correlation ($R^2 = 0.67$, $n = 15$) between rotifer biomass in the near-bottom water layer and biomass of adult *D. polymorpha* can be explained either by out-competing of rotifers by molluscs due to exploitation of common food resources, or by the direct elimination of rotifers by *Dreissena* during filtration (Telesh and Orlova 2004).

The results of this study prove that there is no reason to neglect the possibility of intensive benthic–pelagic interactions in Neva Estuary, especially in late summer and during the autumn months (August and September, when peak reproduction and release of larvae into plankton is expected) in warm years favourable for *D. polymorpha* recruitment (Telesh and Orlova 2004). In general, the high filtration and sedimentation rates of *D. polymorpha* and of microzooplankton confirm the intensive process of water self-purification in the littoral zone of the lower inner Neva Estuary.

Besides, recent results have demonstrated that the contribution of mussel beds to benthic–pelagic coupling in this estuary is sufficient to consider them as one of the major sources of dissolved inorganic phosphorus (DIP) in near-shore waters. Calculations carried out by Orlova et al. (2004) showed that, in the 17-km-long coastal zone (down to 5 m depth) located in the resort district along the northern shore of the eastern Gulf of Finland, the zebra mussel population can release ca. 514 kg DIP daily, whereas the inflow from contributing brooks to the same area is only 7 kg DIP daily, and the load from the Neva River 25 kg day^{-1}. Thus, it appears reasonable to assume the development of *D. polymorpha* beds in the lower inner Neva Estuary as a factor supporting persistent vegetation of nuisance benthic algae [e.g. the green filamentous alga *Cladophora glomerata* (L.) that can detach and form drifting algal mats], and favouring eutrophication of the estuary through the permanent release of high amounts of bio-available nutrients (Orlova et al. 2004).

12.6 Functional Response of Aquatic Communities to Anthropogenic Stress

Historical data show a gradual deterioration of the environment in Neva Estuary during the last few decades, which has made this estuary one of the Baltic Sea "hot spots" (Telesh et al. 1999). However, until recently the ecosystem of Neva Estuary had demonstrated a high degree of sustainability and was considered a marginal filter zone that detains and decomposes a significant fraction of the incoming pollutants due to the very high intensity of self-purification processes (Golubkov et al. 2001). Nevertheless, recent ecosystem alterations have stimulated an increase in primary production in the estuary, caused a reduction in the organic matter decomposition activity of pelagic communities, and given rise to filamentous algal blooms and intensive secondary pollution in the coastal zone of Neva Estuary (Golubkov et al. 2003a, 2003b).

Publications from the early twentieth century lack information about primary production (PP) and decomposition (D) of organic matter in Neva Estuary 100 years ago when the waters of this area were classified as oligotrophic. It is known, however, that PP does not usually exceed 30 g C m^{-2} a^{-1} in oligotrophic waters (Håkanson and Boulion 2002), which yields on average 0.2 g C m^{-2} day^{-1} for a 150-day vegetation season. At present, the PP rate in the inner Neva Estuary is 2.32 ± 0.45 g C m^{-2} day^{-1} (Golubkov et al. 2003a), which is ca. 10 times higher than the threshold value. PP rates in the Gulf of Finland in general and in the Baltic Sea as a whole average 148 and 139 g C m^{-2} a^{-1}, respectively (Elmgren 1984), equivalent to 1.2 and 1.0 g C m^{-2} day^{-1}. Thus, present-day primary productivity in Neva Estuary (especially in its inner part) is considerably higher than in most of the open waters of the Baltic Sea (Fig. 12.8a). Moreover, the PP/D ratio in Neva Estuary has increased from 0.16 in 1984–1988 (Shishkin et al. 1989) to 1.15 in 2001 (Golubkov et al. 2003a) (Fig. 12.8b). Values of PP/D > 1 indicate the prevalence of PP over

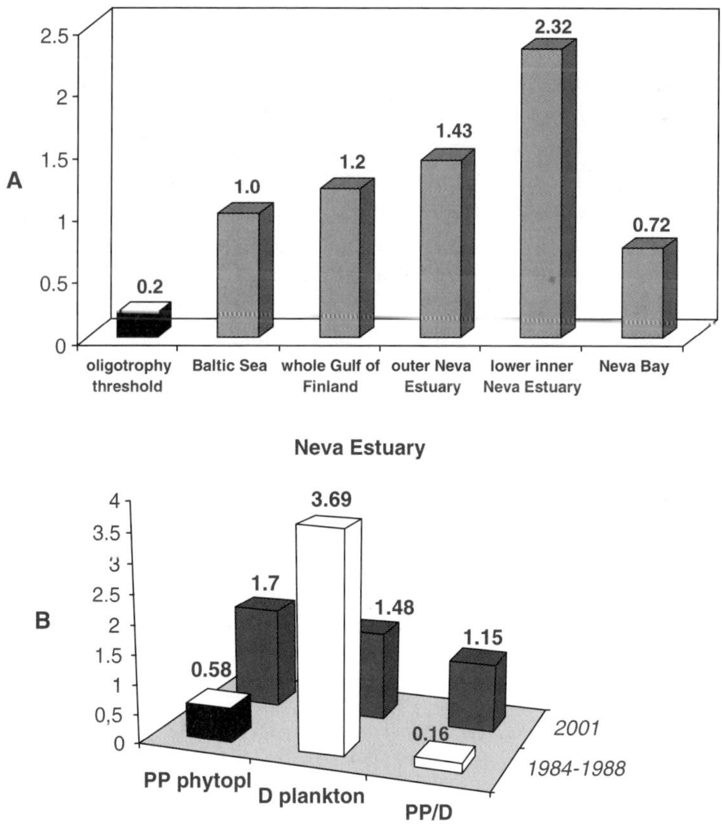

Fig. 12.8 Average primary production (*PP*) of phytoplankton (g C m^{-2} day^{-1}) in different areas of the Baltic Sea (**A**, see text), and dynamics of relationship between PP and decomposition of organic matter by plankton (D) in Neva Estuary (**B**) (data from Golubkov et al. 2003a)

D processes, and indicate continuous on-going eutrophication. This functional response of aquatic communities to eutrophication has been exposed, in particular, in the large-scale growth of the attached green filamentous alga *Cladophora glomerata* along a major part of the coastline in the inner Neva Estuary (Table 12.3), which has a detrimental effect on the nearshore environment and spoiling the sandy beaches in the resort area due to storm casts.

12.7 Conclusions

Pelagic and benthic communities in Neva Estuary form important components that participate in producing and structuring the organic matter cycles and energy fluxes within this ecosystem. Long-term data on biodiversity and abundances of phytoplankton, bacterioplankton and zooplankton in Neva Estuary have demonstrated a considerable level of ecosystem stability. The pronounced alterations in the community structure and biomass of macrozoobenthos in the estuary have been caused mainly by strong exploitative competition between native and invasive benthic species (in the littoral zone) and the negative impact of periodical near-bottom intrusions of saline oxygen-depleted waters (in the deep western areas).

However, the 3-fold increase (since the 1980s) in gross PP, and the 7-fold increase in the PP/D ratio, which has exceeded 1.15 since 2001, demonstrates the increasing level of eutrophication in Neva Estuary. Presently, this is illustrated by prolonged intensive Cyanobacteria blooms, expansion of macrophytes stands, and the large-scale growth of the filamentous green alga *Cladophora glomerata*, which can detach from the underlying substrate, form floating mats and thus have a detrimental effect on the beaches of the resort area. Interactions between benthic and pelagic components provide strong benthic–pelagic coupling in Neva Estuary, which is enhanced by the significant release of nutrients by the dense *Dreissena* beds, as well as by the shallowness of the water body and intensive water mixing.

Thus, comparative analyses and quantitative estimates of long-term changes in biodiversity, productivity, and other characteristics of aquatic communities contribute to the assessment of the functional response of an estuarine ecosystem to anthropogenic stress, and provide biological indicative parameters (such as PP/D ratio) for evaluation of this process.

Table 12.3 Average projective coverage (%), mid-summer biomass (*B*) and gross primary production (*PP*) of the filamentous alga *Cladophora glomerata* at stations at different depths along the northern coast of the inner Neva Estuary (data from Golubkov et al. 2003a)

Depth	0.5 m	1.5 m	3.0 m
Projective coverage (%) of hard substrates by *C. glomerata*	90	60	40
B (g dw m^{-2})	103 ± 0.29	53 ± 0.18	16 ± 0.09
PP (g C m^{-2} day^{-1})	5.33	2.10	0.12

Knowledge of ecosystem structure and understanding of an ecosystem's functional mode are the basic prerequisites for any kind of management and environmental policy-making in the area under consideration. The intensive long-term hydrobiological research that has been conducted in Neva Estuary since the early 1980s provides a good opportunity to evaluate the ecosystem's response to the large-scale management of an estuary under intensive anthropogenic pressure. Recent results have demonstrated that different levels of eutrophication and pollution in certain areas of Neva Estuary influence ecosystem sustainability and define variations in the functional mode of this important marginal filter between the eastern-most part of the Baltic coastal zone and the rest of the Baltic Sea.

Acknowledgements The authors acknowledge the financial support of the Russian Foundation for Basic Research (grants 02-04-48646, 04-04-49207, 05-04-90588, 08-04-00486), grant for the Leading Scientific School on Production Hydrobiology from the Russian Ministry of Education and Science (H-5577.2006.4), and Russian Academic Programs "Biological Resources" and "Biodiversity".

References

Alimov AF (2003) Towards a theory of the functioning of aquatic ecosystems. Backhuys, Leiden

Alimov AF, Golubkov SM (1987) Functional role of zoobenthos in the ecosystem of Neva Bay (in Russian). In: Winberg GG, Gutelmakher BL (eds) Neva Bay: hydrobiological investigations. Nauka, Leningrad, pp 170–174

Alimov AF, Nikulina VN, Panov VE, Telesh IV, Finogenova NP (1994) Hydrobiological characteristics of the Neva Bay of the Gulf of Finland. Hydrobiol J 30:14–26

Alimov AF, Golubkov SM, Panov VE (1996) Regularities of functioning and strategy for management of the ecosystems in the estuary of the Neva River (in Russian). In: Alimov AF, Frolov AK (eds) Ecological state of water bodies and water flows of the Neva River drainage basin. SPb Scientific Centre RAS, pp 187–203

Balushkina EV (1997) New integrated index for water quality evaluation based on structural characteristics of zoobenthos. In: Proceedings of the Final Seminar of the Gulf of Finland Year 1996, 17–18 March 1997, Helsinki, pp 177–201

Belavskaya AP (1987) Characteristics of near-shore plant associations (in Russian). In: Winberg GG, Gutelmakher BL (eds) Neva Bay: hydrobiological investigations, Nauka, Leningrad, pp 66–69

Berezina NA (2004) Causes, characteristics and consequences of non-indigenous amphipod species dispersal in aquatic ecosystems in Europe (in Russian). In: Alimov AF, Bogutskaya NG (eds) Biological invasions in aquatic and terrestrial ecosystems. KMK, Moscow, pp 254–268

Berezina NA, Panov VE (2004) Distribution, population structure and salinity resistance of the invasive amphipod *Gmelinoides fasciatus* (Stebbing) in the Neva Estuary (Gulf of Finland, Baltic Sea). Hydrobiologia 514:199–206

Drabkova VG, Kapustina LL, Kaurova ZG, Tamatorina NL (1999) Bacterioplankton and its heterotrophic activity (in Russian). In: Rumyantsev VA, Drabkova VG (eds) Gulf of Finland in the conditions of anthropogenic impact. St. Petersburg, pp 138–161

Elliott M, McLusky DS (2002) The need for definitions in understanding estuaries. Estuarine Coastal Shelf Sci 55:815–827

Elmgren R (1984) Trophic dynamics in the enclosed, brackish Baltic Sea. Rapp P-V Réun Cons Int Explor Mer 183:152–169

Emelyanov EM (1997) Distribution of chemical elements in the bottom sediments (in Russian). In: Davidan IN, Savchuk OP (eds) Ecosystem models. Assessment of the present state of the Gulf of Finland. Part II. Gidrometeoizdat, St. Petersburg, pp 166–174

Frumin GT, Kryuchkov AM (1999) Hydrochemical characteristics (in Russian). In: Rumyantsev VA, Drabkova VG (eds) Gulf of Finland in the conditions of anthropogenic impact, St. Petersburg, pp 48–66

Golubkov SM (2004) Impact of invasive species on the functioning of aquatic ecosystems (in Russian). In: Alimov AF, Bogutskaya NG (eds) Biological invasions in aquatic and terrestrial ecosystems. KMK, Moscow, pp 243–253

Golubkov SM, Balushkina EV, Anokhina LE, Nikulina VN, Orlova MI, Panov VE, Umnova LP (2001) The role of biological active zones in the organic pollution and purification of the Neva Estuary. Proc Zool Inst Russ Acad Sci 289:95–100

Golubkov SM, Alimov AF, Telesh IV, Anokhina LE, Maximov AA, Nikulina VN, Pavel'eva EB, Panov VE (2003a) Functional response of midsummer planktonic and benthic communities in the Neva Estuary (eastern Gulf of Finland) to anthropogenic stress. Oceanologia 45:53–66

Golubkov SM, Bäck S, Nikulina VN, Orlova MI, Anokhina LE, Umnova LP (2003b) Effect of eutrophication and invasion of *Dreissena polymorpha* in the coastal zone of the eastern Gulf of Finland. Proc Estonian Acad Sci Biol Ecol 52:218–223

Gutelmakher BL, Pavelieva EB, Nesterenko BV (1987) Primary production of plankton (in Russian). In: Winberg GG, Gutelmakher BL (eds) Neva Bay: hydrobiological investigations. Nauka, Leningrad, pp 29–39

Håkanson L, Boulion VV (2002) The lake foodweb – modeling predation and abiotic/biotic interactions. Backhuys, Leiden

Karatayev AY, Burlakova LE, Padilla DR (1997) The effect of *Dreissena polymorpha* invasion on aquatic communities in Eastern Europe. J Shellfish Res 16:187–203

Khlebovich TV (1987) Planktonic Infusoria (in Russian). In: Winberg GG, Gutelmakher BL (eds) Neva Bay: hydrobiological investigations. Nauka, Leningrad, pp 77–82

Kondratyev S, Yefremova L, Sorokin I, Alyabina G, Chernykh O (1997) External load on Russian part of the Gulf of Finland: Seasonal dynamics of riverine inputs and atmospheric deposition in 1995–1996. In: Proceedings of the Final Seminar of the Gulf of Finland Year 1996, 17–18 March 1997, Helsinki, pp 19–29

Krylov PI, Bolshagin PV, Panov VE, Bychenkov DE, Naumenko EN, Polunina YuYu (2004) Invasions of predatory Cladocera and possible reasons for their success (in Russian). In: Alimov AF, Bogutskaya NG (eds) Biological invasions in aquatic and terrestrial ecosystems. KMK, Moscow, pp 268–275

Kudersky LA (1997) Changes of the fish crop in the eastern part of the Gulf of Finland as a result of natural and man-induced factors' impact. In: Proceedings of the Final Seminar of the Gulf of Finland Year 1996, 17–18 March 1997, Helsinki, pp 261–277

Kudersky LA (1999) Composition and long-term changes in fish population of the Neva Bay and the eastern Gulf of Finland (in Russian). In: Rumyantsev VA, Drabkova VG (eds) Gulf of Finland in the conditions of anthropogenic impact. St. Petersburg, pp 257–303

Leppänen J-M, Pitkänen H, Savchuk OP et al (1997) Eutrophication and its effects in the Gulf of Finland. In: Proceedings of the Final Seminar of the Gulf of Finland Year 1996, 17–18 March 1997, Helsinki, pp 31–49

Litvinchuk LF, Telesh IV (2006) Distribution, population structure, and ecosystem effects of the invader *Cercopagis pengoi* (Polyphemoidea, Cladocera) in the Gulf of Finland and the open Baltic Sea. Oceanologia 48[Suppl]:243–257

L'vova AA, Makarova GE, Karatayev AYu, Kirpichenko MYa (1994) Planktonic larvae (in Russian). In: Starobogatov YaI (ed) Dreissena, *Dreissena polymorpha* (Pall.) (Bivalvia, Dreissenidae). Systematics, ecology, and practical importance. Nauka, Moscow, pp 149–156

MacIsaac HJ (1996) Potential abiotic and biotic impacts of zebra mussel on the inland waters of North America. Am Zool 36:287–299

Maximov AA (2003) Changes of bottom macrofauna in the eastern Gulf of Finland in 1985–2002. Proc Estonian Acad Sci Biol Ecol 52:378–393

McLusky DS, Elliott M (2004) The estuarine ecosystem: ecology, threats, and management. Oxford University Press, Oxford, UK

Mills EL, Leach JH, Carlton JT, Secor CL (1993) Exotic species in the Great Lakes: a history of biotic crises and anthropogenic introductions. J Great Lakes Res 19:1–54

Nikulina VN (1987) The dynamics of numbers and biomass of phytoplankton (in Russian). In: Winberg GG, Gutelmakher BL (eds) Neva Bay: hydrobiological investigations. Nauka, Leningrad, pp 20–29

Nikulina VN (1996) Planktonic algae as indicators of the ecological state of water bodies (in Russian). In: Alimov AF, Frolov AK (eds) Ecological state of water bodies and water flows of the Neva River drainage basin. Scientific Centre RAS, St. Petersburg, pp 13–35

Nikulina VN (2003) Seasonal dynamics of phytoplankton in the inner Neva Estuary in the 1980s and 1990s. Oceanologia 45:25–39

Nikulina VN, Anokhina LE (1987) Species composition of phytoplankton and periphyton (in Russian). In: Winberg GG, Gutelmakher BL (eds) Neva Bay: hydrobiological investigations, Nauka, Leningrad, pp 14–20

Orlova M, Golubkov S, Kalinina L, Ignatieva N (2004) *Dreissena polymorpha* (Bivalvia: Dreissenidae) in the Neva Estuary (eastern Gulf of Finland, Baltic Sea): is it biofilter or source for pollution? Mar Pollut Bull 49:196–205

Orlova MI, Panov VE, Krylov PI, Telesh IV, Khlebovich VV (1999) Changes in planktonic and benthic communities of the eastern part of the Gulf of Finland concerned with bioinvasions (in Russian). Proc Zool Inst RAS, St. Petersburg 279:305–325

Orlova MI, Antsulevich AE, Telesh IV, Berezina NA, Maximov AA, Zhakova LV, Litvinchuk LF, Kovalchuk NA, Kostygov AYu (2005) Methodical recommendations for the collection and treatment of data for conducting the monitoring of biological pollution in the Gulf of Finland (in Russian). St. Petersburg

Orlova MI, Telesh IV, Berezina NA, Antsulevich AE, Maximov AA, Litvinchuk LF (2006) Effects of nonindigenous species on diversity and community functioning in the eastern Gulf of Finland (Baltic Sea). Helgoland Mar Res 60:98–105

Ovchinnikov IF (1954) Dreissena in Rybinsk Reservoir. Abstracts of the 3rd Ecological conference. Kiev, pp 107–109

Panov VE, Krylov PI, Telesh IV (1999) The St. Petersburg harbour profile. In: Gollasch S, Leppäkoski E (eds), Initial risk assessment of alien species in nordic coastal waters. Nord: 8, Nordic Council of Ministers, Copenhagen, pp 225–244

Pavelieva EB (1987) Decomposition rate of organic matter in water and functional role of bacterioplankton (in Russian). In: Winberg GG, Gutelmakher BL (eds) Neva Bay: hydrobiological investigations, Nauka, Leningrad, pp 45–52

Pavelieva EB, Chernova GB, Belova MA (1987) Numbers and biomass of bacterioplankton (in Russian). In: Winberg GG, Gutelmakher BL (eds) Neva Bay: hydrobiological investigations, Leningrad, Nauka, pp 40–45

Praeobrazhenskij LYu, Telesh IV, Khlebovich TV (1987) Influence of hydrological factors on the formation of littoral zooplankton communities (in Russian). In: Winberg GG, Gutelmakher BL (eds) Neva Bay: hydrobiological investigations, Nauka, Leningrad, pp 105–111

Shishkin BA, Nikulina VN, Maximov AA, Silina NI (1989) Main characteristics of the biota of the Gulf of Finland and its role information of water quality (in Russian). Gidrometeoizdat, Leningrad

Skorikov AS (1910) On the fauna of Neva Bay and waters adjacent to island Kotlin (in Russian). Zool Museum Acad Sci Annu Rep 15:474–489

Smirnova AI (1997) Climate characteristics. River inflow (in Russian). In: Davidan IN, Savchuk OP (eds) Ecosystem models. Assessment of the present state of the Gulf of Finland. Part II. Gidrometeoizdat, St. Petersburg, pp 175–193

Starobogatov YaI (ed) (1994) Dreissena, *Dreissena polymorpha* (Pall.) (Bivalvia, Dreissenidae). Systematics, ecology, and practical importance (in Russian). Nauka, Moscow

Telesh IV (1986) Transformation of the lake zooplankton in rivers. Dok Akad Nauk SSSR 291:495–498

Telesh IV (1987) Planktonic rotifers and crustaceans (in Russian). In: Winberg GG, Gutelmakher BL (eds) Neva Bay: hydrobiological investigations, Nauka, Leningrad, pp 81–103

Telesh IV (2001) Zooplankton studies in the Neva Estuary (Baltic Sea): a brief excursion into history. Proc Estonian Acad Sci Biol Ecol 50:200–210

Telesh IV (2004) Plankton of the Baltic estuarine ecosystems with emphasis on Neva Estuary: a review of present knowledge and research perspectives. Mar Pollut Bull 49:206–219

Telesh IV (2006) Impact of biological invasions on the diversity and functioning of zooplankton communities in the Baltic estuarine ecosystems (a review) (in Russian). Izvestiya Samarsk. Scientific Center RAS 8:220–232

Telesh IV, Heerkloss R (2002) Atlas of estuarine zooplankton of the southern and eastern Baltic Sea. Part I: Rotifera. Naturwissenschaftliche Forschungsergebnisse, Bd. 68, Kovač, Hamburg, 90 pp (with CD)

Telesh IV, Heerkloss R (2004) Atlas of estuarine zooplankton of the southern and eastern Baltic Sea. Part II: Crustacea. Naturwissenschaftliche Forschungsergebnisse, Bd. 72. Kovač, Hamburg, 118 pp (with CD)

Telesh IV, Ojaveer H (2002) The predatory water flea *Cercopagis pengoi* in the Baltic Sea: invasion history, distribution and implications to ecosystem dynamics. In: Leppäkoski E, Gollasch S, Olenin S (eds) Invasive aquatic species of Europe. Distribution, impacts and management, Kluwer, Dordrecht, pp 62–65

Telesh IV, Orlova MI (2004) Relations between alien species *Dreissena polymorpha* (Pallas) and microzooplankton in the coastal waters of Neva Estuary (Gulf of Finland, Baltic Sea). In: Alimov AF, Bogutskaya NG (eds) Biological invasions in aquatic and terrestrial ecosystems (in Russian). Nauka, St. Petersburg, pp 268–275

Telesh IV, Alimov AF, Golubkov SM, Nikulina VN, Panov VE (1999) Response of aquatic communities to anthropogenic stress: a comparative study of Neva Bay and the eastern Gulf of Finland. Hydrobiologia 393:95–105

Telesh IV, Bolshagin PV, Panov VE (2001) Quantitative estimation of the impact of the alien species *Cercopagis pengoi* (Crustacea: Onychopoda) on the structure and functioning of plankton community in the Gulf of Finland, Baltic Sea. Dok Biol Sci 377:157–159

Umnova LP (1987) Particulate organic and inorganic matter and chlorophyll (in Russian). In: Winberg GG, Gutelmakher BL (eds) Neva Bay: hydrobiological investigations, Nauka, Leningrad, pp 7–13

Winberg GG, Gutelmakher BL (1987) Present state of the Neva Bay ecosystem (in Russian). In: Winberg GG, Gutelmakher BL (eds) Neva Bay: hydrobiological investigations, Nauka, Leningrad, pp 198–205

Chapter 13
The Gulf of Finland

H. Pitkänen, J. Lehtoranta, and H. Peltonen

13.1 Introduction

In this chapter, we describe the unique physical and biogeochemical conditions of the Gulf of Finland. Although morphologically a direct continuation of the Baltic Proper, several features distinguish the Gulf markedly from other parts of the Baltic Sea. Relative to its surface area, the Gulf of Finland has the largest catchment area and highest freshwater inflow within the Baltic Sea. The external loading of nitrogen (N) and phosphorus (P) relative to the surface area of the Gulf are about two and three times, respectively, those of the Baltic Sea. The Gulf of Finland has a distinct but unstable salinity stratification, which makes the sea area very sensitive to external perturbations. At times, this may lead to e.g. deep water anoxia/hypoxia, elevated benthic release of P and extensive cyanobacterial blooms. The poor oxygen conditions in the Gulf are caused primarily by the voluminous spring bloom and preceding sedimentation of detrital plankton material, which in turn is caused by high wintertime nutrient concentrations. Poor oxygen conditions and intensified sediment release of P have led to strongly increasing deep-water and winter concentrations of inorganic P over the last 10 years. In contrast, N concentrations have decreased since the late 1980s, suggesting the effect of decreased external N loading at the same time. This latter phenomenon probably contributed to the decreased magnitude of the phytoplankton spring blooms in the 1990s. The prevailing poor oxygen conditions have weakened the populations of benthic invertebrates and further decreased the nutrient retention and removal capacity of the sediments. The most important controlling factor in the recent adverse changes in fish stocks has been hydrographical changes. Especially in coastal waters, eutrophication and its associated consequences has shifted the biomass dominance from percids towards cyprinids. Summertime eutrophication has been going on in the Gulf of Finland since at least the 1970s. Since the mid-1990s the main reason for this development, both in the open Gulf and in the coastal waters, seems to have been the general development of eutrophication caused by internal loading, rather than any changes in land-based nutrient loads. The further reduction of the external nutrient load is, however, the only permanent way to counteract both sediment accumulation of labile organic matter and benthic release of nutrients, as well as summertime eutrophication of the Gulf of Finland.

U. Schiewer (ed.), *Ecology of Baltic Coastal Waters. Ecological Studies 197.*

13.2 An Infinite Estuary Sensitive to Physical Forcing

Two principal factors control the hydrodynamics and the density stratification in the Gulf of Finland. In the east, the largest river of the Baltic Sea catchment, the River Neva, enters the Gulf. In the west, the Gulf is connected with the Baltic Proper without any clear geographical borderline or underwater ridge. The Gulf of Finland constitutes a continuum between a freshwater and a marine system, and can, in a broad sense, be considered as an infinite estuary of the River Neva (see also Chap. 12 by Telesh et al., this volume). This is shown by the fact that the effects of the River Neva can be seen in both the physical, chemical and biological nature of the whole Gulf (e.g. Elken et al. 2002). About 75% of the freshwater inflow into the Gulf comes from Neva, and a clear horizontal salinity gradient is observable throughout the length of the Gulf. However, the strongest horizontal salinity gradient is naturally found in the easternmost part of the Gulf, and usually only the water area between the Island of Seskar and the City of St. Petersburg is regarded as the estuary of the River Neva (Fig. 13.1; Pitkänen et al. 1993).

These two factors influence the physical and biogeochemical conditions in the westernmost and easternmost ends of the Gulf of Finland, respectively, and make it difficult to exactly separate the natural eutrophication of the Gulf from anthropogenically induced effects. The both spatially and temporally variable salinity stratification of this shallow sea area also makes the benthic ecosystem of the Gulf very sensitive to external perturbations via changes in oxygen conditions.

Before the 1990s only sparse scientific information covering the whole Gulf of Finland was available. In particular, the biogeochemical behaviour of the Neva Estuary, which receives most of the external nutrient loading of the whole Gulf, was relatively poorly known – at least outside the Soviet Union. Anthropogenic

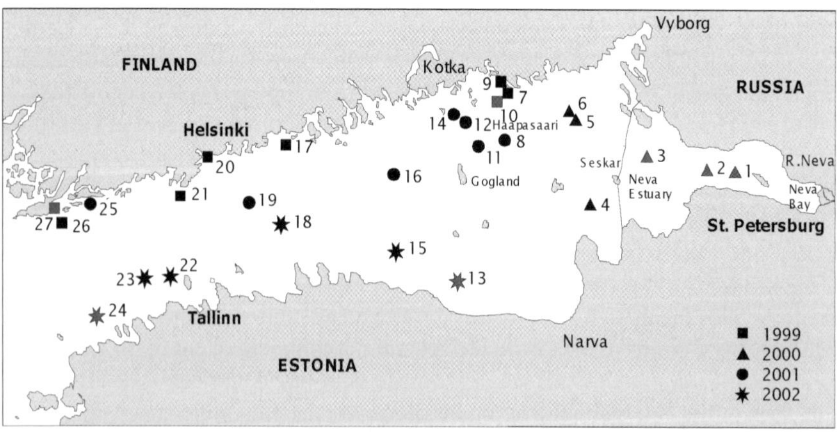

Fig. 13.1 The study area: the Gulf of Finland and the sampling stations for the sediment flux studies in Fig. 13.6

loading from the Soviet part of the catchment, covering about 75% of the whole catchment, was reported in, e.g., HELCOM documents, but the reliability of this information was questionable (Velner 2005).

The situation changed drastically at the beginning of the 1990s. Since that time, Estonian–Finnish–Russian trilateral cooperation has produced a variety of joint studies on, e.g., nutrient loading and eutrophication, the causes of algal blooms, benthic communities and alien species, the sources and occurrence of harmful substances in sediments and fish, as well as on physical and ecosystem modelling (Sarkkula 1997; Elken et al. 2003).

The most severe and complex environmental problem of the Gulf is obviously eutrophication and its consequences (see also Chap. 12 by Telesh et al., this volume). Although the external load of both N and P into the Gulf decreased by about one-third between the late 1980s and the early 2000s (Kiirikki et al. 2003), the summer eutrophication has continued (Kahru et al. 2000; Kauppila and Lepistö 2001). At the same time, benthic release of nutrients – internal loading – has intensified and partly counteracted the positive effects of nutrient load reductions (Pitkänen et al. 2001a; Raateoja et al. 2005).

The aim of this chapter is to describe the nutrient dynamics and trophic status of the Gulf of Finland, as well as their interrelationships and dependencies on external factors. Special emphasis is placed on the role of sediment–water interactions – a feature that strongly influences the state of the Gulf. Additionally, some biological characteristics describing recent changes in the state of Gulf are discussed.

13.3 Geomorphology and Hydrodynamics

The area of the Gulf of Finland totals 30,000 km^2. The Gulf receives river waters from an area of 421,000 km^2 (Fig. 13.1, Table 13.1), and is geomorphologically a direct continuation of the Baltic Proper (no sills at the boundary). Hydrographically, the Gulf can be considered as a partially mixed estuary. This is caused by its topography and by the high freshwater inflow of about 2,500 m^3 s^{-1} (mean) from the River Neva, both of which cause a clear but unstable salinity stratification (Pitkänen et al. 1993; Alenius et al. 1998). The open waters of the Gulf can be divided into deeper western (west of Gogland) and shallower eastern (east of Gogland) parts. The estuarial nature is very prominent, especially east of the Island of Seskar (Fig. 13.1), where most of the land-based nutrients also enter via the River Neva and from the city of St. Petersburg (Kiirikki et al. 2003).

The average residence time of water in the Gulf is 2–3 years (Alenius et al. 1998; Table 13.1). A considerable salinity gradient can be found in the surface layer of the Gulf from the Neva Bay in the east (0 psu) to the westernmost Gulf (7 psu) due to the inflow of the River Neva at the eastern end of the Gulf. Near the bottom, the salinity increases from 5 psu in the east to about 10 psu in the west. The strength of the halocline varies both temporally and spatially, being strongest and most stable in the deep western Gulf. During the late 1990s and early 2000s there has been a tendency towards stronger haloclines both in the Baltic Proper and in the Gulf of Finland (Elken et al. 2002; Lips et al. 2002). This has changed the physical conditions, especially in the eastern Gulf, where the halocline was absent in the early 1990s (Pitkänen and Tamminen 1995), but developed again in the mid 1990s (Pitkänen et. al 2003).

Due to the relatively stagnant deep water layer, the oxygen conditions are usually poorest in the middle and western Gulf (Andersin and Sandler 1991; Perttilä et al. 1995; Pitkänen et. al. 2003). In these areas, well-built and linear correlations can be found between near bottom salinity, oxygen and P (Perttilä et al. 1995). In the shallower eastern Gulf, as well as in the coastal waters, substantial deviations from this relationship can be observed. This indicates that, in these areas, the role of local conditions is more important than the hydrographical conditions generated by the immediate connection with the Baltic Proper (Pitkänen et al. 2001a, 2003; Lehtoranta 2003).

Table 13.1 Main characteristics of the Gulf of Finland. *dw* Dry weight, *DIN* dissolved inorganic nitrogen, *DIP* dissolved inorganic phosphorus, *DOC* dissolved organic carbon, *POC* particulate organic carbon, *POM* particulate organic matter, *GoF* Gulf of Finland

a. Climate and hydrology	
Air temperature (°C)	+3.7[a]
Water temperature (°C)	0 to +15[b]
Ice covering (days a^{-1})	40–130[c]
Average precipitation (mm a^{-1})	643[a]
Mean freshwater inflow ($10^6 m^3 a^{-1}$)	114,000[d]
Mean inflow ($10^6 m^3 a^{-1}$) (from Baltic)	480,000[d]
Mean total outflow ($10^6 m^3 a^{-1}$)	600,000[d]
Residence time (days)	700–1,000[d]
b. Morphometry and sediments	
Area (km^2)	29,600[e]
Volume (km^3)	1,100[e]
Mean depth (m)	38[e]
Maximum depth (m)	123[e]
Catchment area (km^2)	412,900[f]
Area/catchment relation	0.072[f]
Organic matter content (% dw)	17.6 (sediment surface)[g]
Sediment fraction < 63 μm (% dw)	75–90[h]
C/N-ratio (0–10 cm depth)	10.5 (sediment surface)[g]
Phosphate release (mg P m^{-2} day^{-1})	5.5[g]
Ammonia release (mg N m^{-2} day^{-1})	23 (DIN, sediment surface)[g]
Denitrification rate (mg N m^{-2} day^{-1})	6.3[h]
Sediment composition	Organic rich silt and clay
c. Pelagic biological components	
Secchi depth (m)	4.0 (summer)[i]
Bacteria ($10^6 ml^{-1}$)	0.4–6.6[j]
Phytoplankton biomass (μg Chl *a* l^{-1})	Summer: 3–10[k]
Phytoplankton biomass (μg Chl *a* l^{-1})	Spring max: 20–80

(continued)

Table 13.1 (continued)

	Autumn max: 10–20[k]
Dominating phytoplankton species	Spring: *Achnanthes taeniata, Scrippsiella hangoei, Skeletonema costatum, Peridinella catenata Chaetoceras* spp.; Autumn/late summer: *Nodularis* sp., *Aphaniozomenon* sp., *Planktothrix agardhii, Dinophysis acuminata, Heterocapsa triquetra*[k]
Flagellates biomass (mg C l^{-1})	0.005–0.04 (ww)[l]
Ciliates biomass (mg C l^{-1})	0.0–0.3 (ww)[l]
Mesozooplankton biomass (mg C l^{-1})	0.03–0.3 (ww)[l]
Fish	
Number of species	52[m]
Dominating species	*Clupea harengus, Sprattus sprattus*[m]
POC (mg C l^{-1})	0.1–0.8[n]
DOC (mg C l^{-1})	3.6–8.3[n]
C/N in POM	4.2–6.9 (w/w)[o]
Primary production (g C m^{-2} a^{-1})	74–111[p]
Bacterial production (μg C l^{-1} h^{-1})	0.1–15.3 (variation of seasonal averages)[q]
Fish catches (t a^{-1})	50,000 (mean 1980–2003)[r]
d. Benthic biological components	
Macrophytes	
Species number	97[s]
Dominating species	*Fucus vesiculosus, Cladophora glomerata, Pilayella littoralis*
e. Water chemistry, trophic status and pollution	
Salinity (psu)	0–7 (surface)[a]
	5–10 (deep water)[a]
pH	8.0–8.4[t]
Oxygen saturation (%)	0–80 (deep water)[u]
Total nitrogen (μmol l^{-1})	20–40[v] (winter, surface)[v]
DIN (μmol l^{-1})	7–20 (winter, surface)[v]
Total phosphorus (μmol l^{-1})	0.8–1.6[v] (winter, surface)[v]
DIP (μmol l^{-1})	0.7–1.3[v] (winter, surface)[v]
Annual nitrogen input (t a^{-1})	120,000–130,000[w,x]
Annual phosphorus input (t a^{-1})	6,000–7,600[w,x]
Limitation of P and period	N or N/P, P in estuaries[l,y]
Main pollutants	Nutrients, heavy metals, organochlorines
Trophic level	Mesotrophic (Western GoF) to eutrophic (Eastern GoF)

[a]HELCOM (2002); [b]Haapala and Alenius (1994); [c]Seinä and Peltola (1991); [d]Alenius et al. (1998); [e]HELCOM (1996); [f]HELCOM (1998); [g]Lehtoranta (2003); [h]Vallius (1999); [i]Laamanen et al. (2004); [j]Heinänen (1992); [k]Kauppila and Lepistö (2001); [l]Kivi et al. (1993); [m]H. Peltonen (personal communication); [n]Leppänen and Tamelander (1981); [o]Laakkonen et al. (1981); [p]Lignell (1993); [q]Heinänen (1992); [r]Haahti and Kangas (2004); [s]Nielsen et al. (1995); [t]Perttilä et al. (1980); [u]Lips et al. (2002); [v]Pitkänen et al. (2001b); [w]Pitkänen and Räike (2004); [x]HELCOM (2004); [y]Pitkänen and Tamminen (1995)

13.4 External Nutrient Loading

The annual land-based inputs into the Gulf of Finland have been estimated at around 6,000–7,600 t total P and 110,000–120,000 t total N (HELCOM 2004; Pitkänen and Räike 2004; Table 13.1). The annual atmospheric precipitation increases the load by 16,000–17,000 t annum $(a)^{-1}$ N (Bartnicki et al. 2002; Pitkänen and Räike 2004). The inputs of P and N are about 15% and 12%, respectively, of the combined land-based and atmospheric total inputs into the Baltic Sea (HELCOM 2004). Relative to its surface area, the external N and P loading of the Gulf of Finland is about 2- to 3-fold, respectively, compared with the average for the whole Baltic Sea (Fig. 13.2).

Roughly 70% of total P and 50% of total N enter the Gulf in its easternmost part, and it has been estimated that about 40% of the P input (2,800 t a^{-1}) and 60%

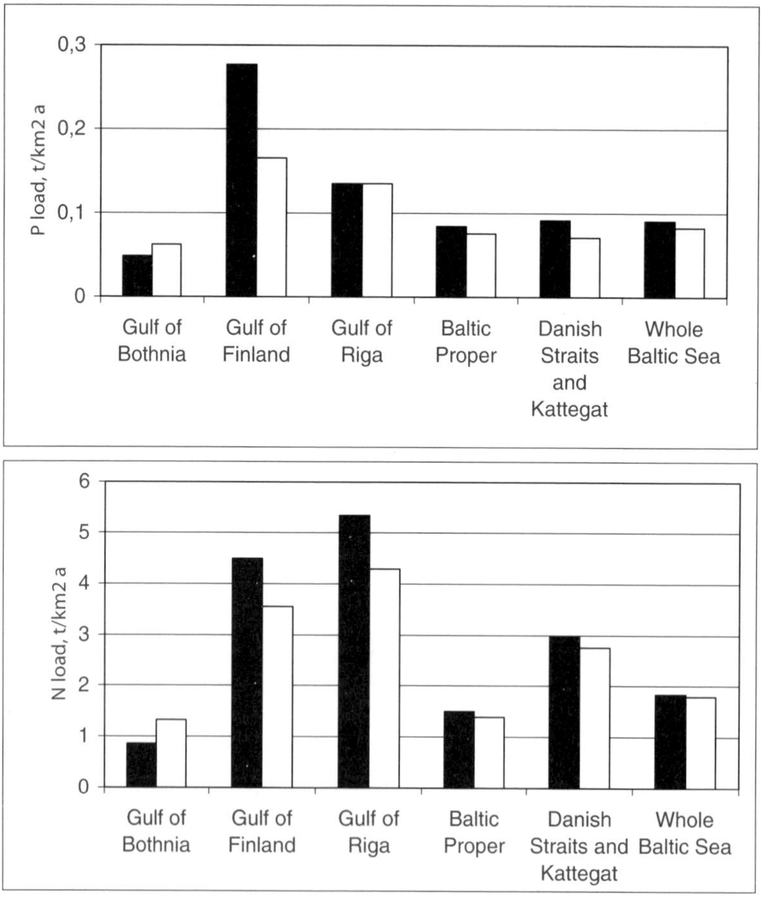

Fig. 13.2 The surface area-specific loads of total phosphorus (*P*) and total nitrogen (*N*) into the Baltic Sea and its sub-basins (t km^2 a^{-1}) according to values presented in HELCOM (1998, 2004). *Black bars* 1995, *white bars* 2000

of the N input (77,000 t a^{-1}) are readily bio-available for primary producers (Kiirikki et al. 2003; Fig. 13.3). The largest single source of bio-available P is the city of St. Petersburg, while the River Neva represents the largest single source of bio-available N. The mean ratio of bio-available N:P in the inputs is ca. 27 (w/w). An excess of N for primary producers is thus supplied, since the optimum Redfield ratio for primary production in the sea is only 7.2.

During the 1990s nutrient discharges into the Gulf of Finland were significantly reduced, by approximately 39% for total P and 36% for total N (Pitkänen and Räike 2004). The decrease was especially steep in the early 1990s, but it continued during the late 1990s and early 2000s. This development was only partly due to water protection measures. In the early 1990s the decrease was caused mainly by decreases in both industrial and agricultural production in Russia and Estonia after the collapse of the Soviet Union (Lääne et al. 2002), and the decrease in the nutrient load entering the eastern Gulf of Finland was thus especially steep. Regarding the northern (Finnish) coastal waters, no particular trend took place in the loading of total N and total P in the 1990s and the early 2000s,

13.5 Water Column Nutrients

13.5.1 Strong Vertical and Seasonal Variations

The Gulf of Finland is probably the single most nutrient-enriched basin in the whole Baltic Sea, although the values of inorganic N in the Gulf of Riga also are clearly elevated compared to other open sea areas of the Baltic Sea (Fig. 13.4). A peculiar feature of the Gulf of Finland is the strong variation in nutrients both inter-seasonally and vertically, which indicates prominent biological activity. In contrast, the coastal-open sea gradients in the horizontal direction usually remain relatively smooth, especially in winter. This is due largely to the elevated open sea concentrations, which partly counteract the effects of coastal nutrient sources.

For dissolved inorganic N (DIN), the average winter values of the Gulf vary from about 100 mg m^{-3} (7 mmol m^{-3}) in the west to about 300 mg m^{-3} (21 mmol m^{-3}) in the east (Pitkänen et al. 2001b; Lips et al. 2002; Table 13.1e, Fig. 13.4). In the coastal waters average values up to 400–500 mg m^{-3} (28–35 mmol m^{-3}) are generally found in restricted areas affected by direct coastal and catchment sources by the rivers entering the Gulf. In the open Baltic Proper, the average wintertime DIN varies from 30 to 90 mg m^{-3} (2 to 6 mmol m^{-3}) (Nausch and Lysiak-Pastuszak 2002; HELCOM 2003; Fig. 13.4), being in general less than one-half of the values found in the open Gulf of Finland.

The winter dissolved inorganic P (DIP) values of the mixed surface layer vary from approximately 20 mg m^{-3} (0.7 mmol m^{-3}) in the western Gulf to up to 40 mg m^{-3} (1.3 mmol m^{-3}) in the middle and eastern Gulf, with no particular coast–open sea gradient (Pitkänen et al. 2001b; Lips et al. 2002; Table 13.1e, Fig. 13.4). The

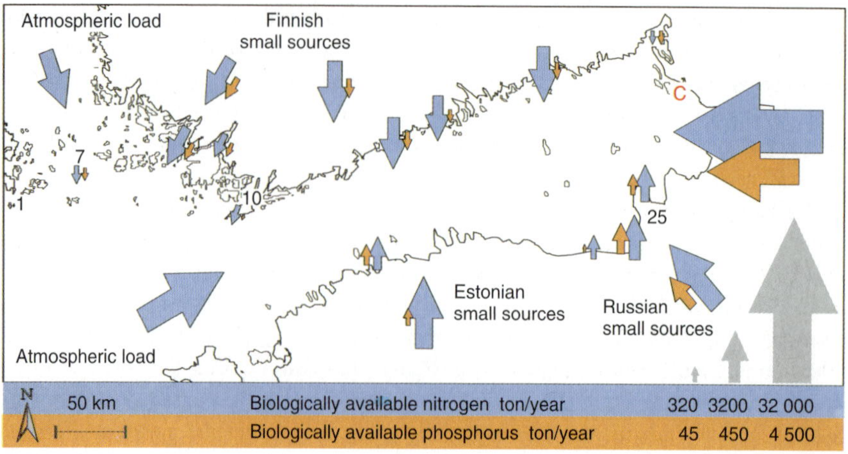

Fig. 13.3 The loads of N and P (t a^{-1}) biologically available to the Gulf of Finland and to the Archipelago Sea in 2000 (Kiirikki et al. 2003)

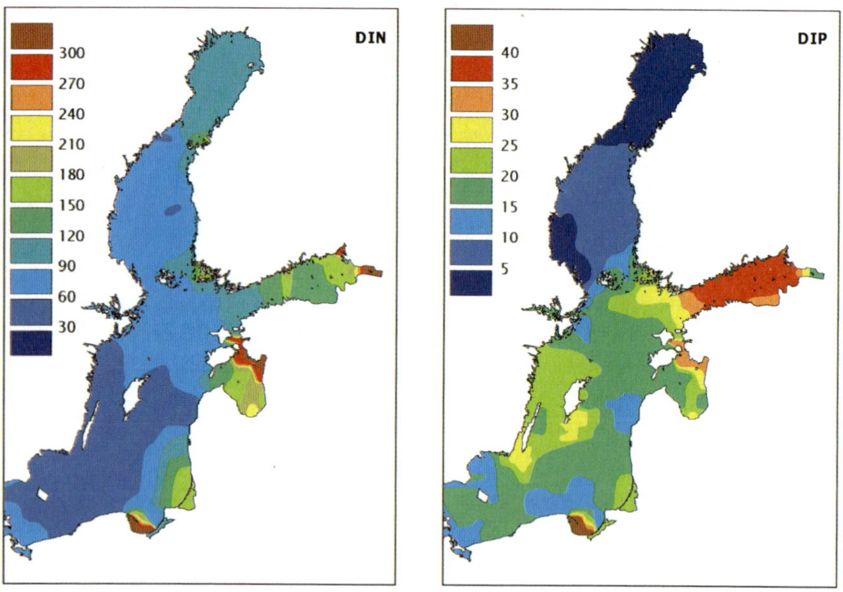

Fig. 13.4 Surface distributions of dissolved inorganic nitrogen (DIN, mg N m^{-3}) and dissolved inorganic phosphorus (DIP, mg P m^{-3}) concentrations in the Baltic Sea in the winter of 2002 according to data from 362 stations collected by the Finnish Institute of Marine Research, Finnish Environment Administration, Swedish Meteorological and Hydrological Institute and the City of Helsinki. Interpolation was carried out using the DAS-program developed at Stockholm University (data handling by Mikko Kiirikki and Paula Väänänen/SYKE)

loading from the River Neva and the city of St. Petersburg in Russia formerly elevated the concentrations of both DIP and DIN in the eastern Gulf. This east–west gradient was, however, weakened in the early 1990s due to decreased loading from the River Neva (Kiirikki et al. 2003) and, in the case of P, the voluminous internal loading taking place especially in the deeper middle parts of the Gulf.

A strong increase in winter DIP concentrations took place around the mid 1990s, despite the concomitant decrease in external P loading (Pitkänen et al. 2001a). In the late 1990s and the early 2000s the average wintertime surface levels of DIP in the middle and eastern Gulf (30–40 mg m^{-3}, 1–1.3 mmol m^{-3}) were almost twice those of the central Baltic (15–20 mg m^{-3}, 0.5–0.7 mmol m^{-3}) (Nausch and Lysiak-Pastuszak 2002; Fig. 13.4). This strongly increasing trend in P concentrations in the 1990s seems to be a unique feature of the Gulf of Finland. No such general increase has been recorded for the other main basins of the Baltic Sea (Nausch and Lysiak-Pastuszak 2002; HELCOM 2003).

In spring (late April/early May), both DIN and DIP are rapidly fixed by phytoplankton and transported towards deeper water layers and to the bottom by detritus (see also Chap. 12 by Telesh et al., this volume). Compared to winter conditions, biological production together with the prevailing hydrodynamics, benthic conditions, external nutrient loading and deep inflow from the northern Baltic Proper strongly modify the spatial and temporal distribution of inorganic nutrients.

In summer, the planktonic production of the open Gulf is limited by N or both N and P (Kivi et al. 1993; Pitkänen and Tamminen 1995). In the 1990s and the early 2000s the role of N became more and more prominent due to the increased concentration of DIP and declining concentration of DIN. As a result of the low winter N:P ratio, the spring bloom is limited by N. Inorganic P is thus available in the water also after the vernal bloom, fuelling nitrogen-fixing cyanobacteria later in the summer. In mid- and late summer, DIN- and DIP-values above analytical detection limits can be measured regularly only in the coastal waters and estuaries receiving nutrient inputs from rivers or coastal point-sources.

In contrast to the surface layer, DIP accumulates in the near-bottom layers of both the open Gulf and the coastal waters in summer (Fig. 13.5). Prior to the mid-1990s, the open sea DIP concentration usually varied between 30 and 60 mg m^{-3} (1 and 2 mmol m^{-3}) and only occasionally reached values of around 100 mg m m^{-3} (3–4 mmol m^{-3}) (Perttilä et al. 1995; Pitkänen et al. 2001b). Since the mid-1990s, open sea near-bottom DIP concentrations have reached values of over 200 mg m^{-3} (7 mmol m^{-3}), while values of ~100 mg m^{-3} (3–4 mmol m^{-3}) are common all over the Gulf (Pitkänen et al. 2001b, 2003). The highest measured values for the Gulf are in fact close to the maximum near-bottom values measured from the stagnant deep basins of the Baltic Proper in the late 1990s (Nausch and Lysiak-Pastuszak 2002). Even higher summer values, up to 800 mg m m^{-3} (27 mmol m^{-3}), have been measured in the enclosed basins near the Finnish coast at sites of intensive benthic P release (Pitkänen et al. 2003).

13.5.2 Temporal Trends: Increasing for P, Decreasing for N

The increase in the deep water P concentration of the open Gulf of Finland that has taken place since the mid-1990s is clearly connected to the strengthened salinity stratification, declining oxygen concentrations, and the various biological and chemical processes that lead to benthic P release (Lehtoranta 2003; Fig. 13.5a–d). Because poor oxygen conditions prevent effective nitrification and subsequent denitrification, nitrate values have generally decreased and ammonium values have increased (Fig. 13.5d). The basic reason for the increases in P concentration, i.e. benthic release as a result of reducing conditions at the sediment surface, is the same in coastal waters as in the open sea. However, in coastal waters the salinity stratification is weak or non-existent, and factors other than strengthened vertical stratification are needed to explain the observed phenomenon of intensified internal loading.

In contrast to the increasing trends in wintertime DIP, trends of surface layer DIN decreased in the late 1980s and 1990s, following an increasing trend since at least the early 1970s (Perttilä et al. 1995; Pitkänen et al. 2001b; Lips et al. 2002). This development is in accordance with the decrease of 36% in the total N input into the Gulf from the late 1980s to the year 2000 (Pitkänen and Räike 2004). According to

A

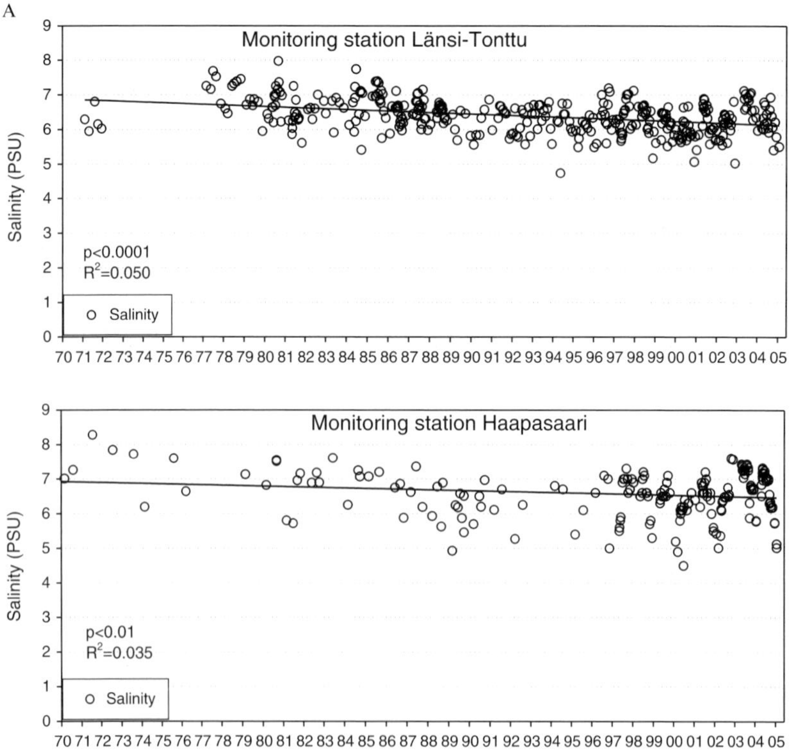

(continued)

B

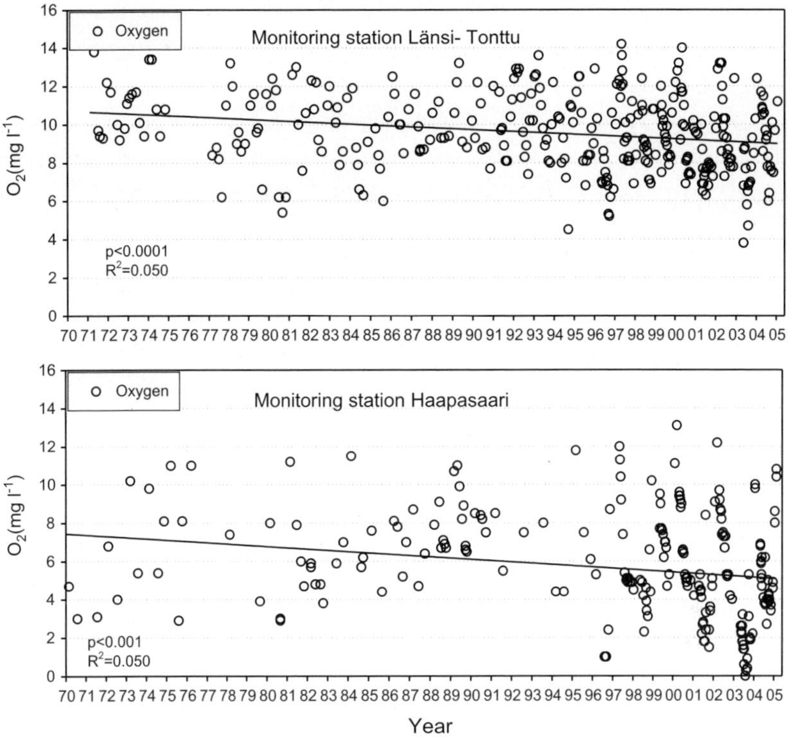

C

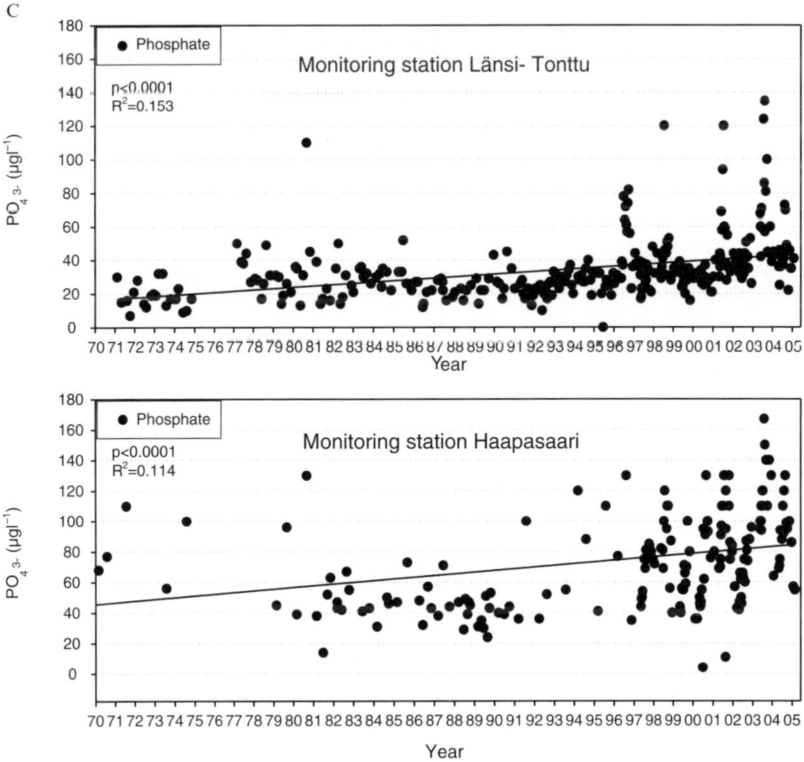

(continued)

D

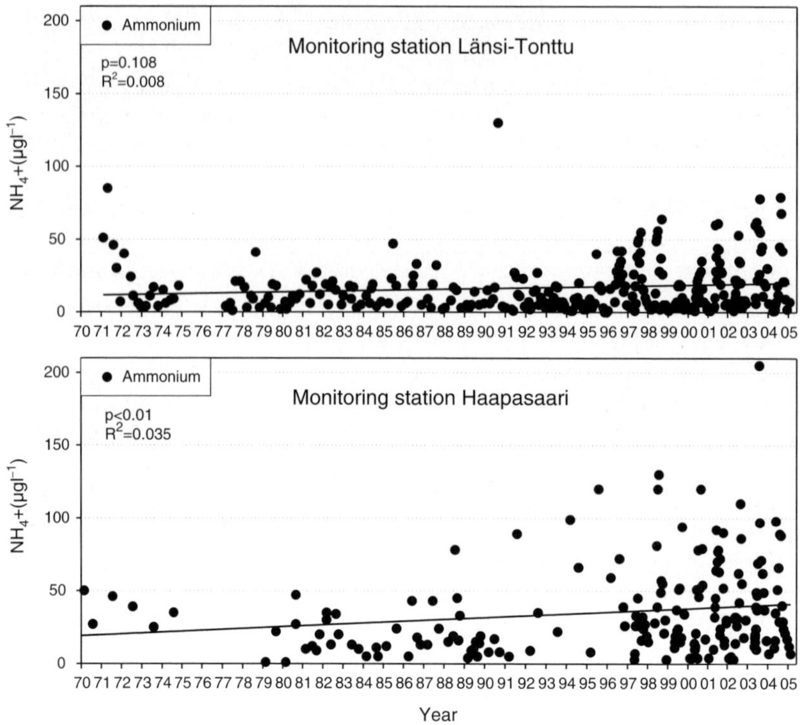

Fig. 13.5 Measurements of near-bottom **A** salinity (psu), **B** oxygen (mg l^{-1}), **C** phosphate-P (mg m^{-3}) and **D** ammonium-N (mg m^{-3}) at the off-coastal stations of Länsi-Tonttu (58 m) and Haapasaari (65 m) in 1970–2005. Data collected and analysed by Uusimaa and Kymi Regional Environment Centres, City of Helsinki and SYKE

loading estimates the present (year 2000) external N load is about 60,000 t a^{-1} smaller than in the late 1980s, which corresponds to as much as 15% of the estimated wintertime stock of total N (430,000 t; Perttilä et al. 1995) in the whole Gulf. Thus, it seems possible that the observed decrease in N concentration is connected to the decrease in loading rather than to internal processes. This is supported by the fact that decreasing deep water oxygen concentrations cause increasing (near-bottom ammonium-N concentration trends as conditions for denitrification deteriorate Fig. 13.5b,d).

13.5.3 Trends in Eutrophication

On the basis of chlorophyll-*a* concentrations, an increasing trend in summertime eutrophication has been seen in many parts of the Baltic Sea since 1980 (HELCOM

2003), despite the simultaneous decrease in external nutrient loading both from the catchment area and via the atmosphere (HELCOM 2004). Especially in the western part of the Gulf of Finland, eutrophication has continued since at least the late 1970s. This trend is very clear, especially after the late 1980s (Table 13.2).

In the eastern Gulf, which earlier received a higher nutrient load than at present, development has been gentler. This is also the case for the coastal waters (see also Chap. 12 by Telesh et al., this volume). However, despite some coastal exceptions caused by comprehensive improvements in loading conditions (e.g. the Helsinki capital region), summertime eutrophication has also continued in coastal waters (Kauppila and Lepistö 2001). The main reason for this development seems to be the general eutrophication development of the open Gulf rather than changes in the land-based nutrient load.

In the Gulf of Finland, the summertime eutrophication can be explained largely by the increase in DIP concentration and the cyanobacterial blooms induced by this increase (Kahru et al. 2000; Kauppila and Lepistö 2001; Raateoja et al. 2005). It is, however, unclear whether eutrophication development covers the whole growth season, since a major part of the annual primary production takes place during the vernal peak bloom in April/May. Most of the available datasets with frequent enough sampling in spring are not long enough to analyse changes prior to the 1990s. On the basis of a weekly to bi-monthly chlorophyll-*a* dataset starting in the early 1970s, Raateoja et al. (2005) suggest that the vernal blooms in the coastal western Gulf have shown a decreasing trend since the late 1980s. No increasing trend could be discerned in the whole growing season data for the period since the late 1980s, since spring decreases counteracted summer increases. The decreasing spring biomasses are in accordance with the decreased winter DIN concentrations, since N evidently limits vernal bloom production (Kivi et al. 1993).

To what extent can results concerning decreasing vernal blooms be extrapolated for the whole Gulf? On the basis of the available monitoring data, the winter decreases in N concentrations since the late 1980s cover at least the open middle and western Gulf, as well as the northern (Finnish) coastal waters (Pitkänen et al. 2001b; Lips et al. 2002). The hypothesis that spring blooms have also decreased in these parts of the Gulf is thus supported. The hypothesis is possibly also valid for

Table 13.2 The general late summer (July-September) levels of chlorophyll-*a* (mg m^{-3}) in the open sea areas around Finland in 1979–2001 (Pitkänen et al. 2004). The figures correspond to the basic level of the respective sea area during the each assessment period. The values refer only to Finnish territorial and international waters and do not include Estonian, Russian or Swedish territorial waters

Sea area	1979–1983	1986–1990	1991–1996	1997–2001	General change (1980s/1990s)[a]
Bothnian Bay	< 2	< 2	< 2	2–3	+
Bothnian Sea	< 2	< 2	< 2	2–3	+
Archipelago Sea	< 2	< 2	2–3	ca. 3	++
Western Gulf of Finland	2–3	2–3	3–5	ca. 5	++
Eastern Gulf of Finland	3–5	3–5	> 5	> 5	+

[a] + probable increase in Chl a, ++ clear increase

the easternmost Gulf, since the land-based inflow of N into the Russian waters of the Gulf has decreased strongly (Kiirikki et al. 2003).

13.6 Internal Nutrient Fluxes

13.6.1 Sediment–Water Exchange and Denitrification

Due to their large capacity for retaining and releasing nutrients, sediments play a significant regulatory role in shallow ecosystems such as the Gulf of Finland. In coastal marine waters, the oxidation state of the sediment surface depends on the content of oxygen-consuming organic matter in the sediments, and on the density stratification of the overlying water, which depends on salinity and temperature. In the Gulf of Finland, most new organic matter is settled on the surface of the sediment after the spring bloom (Heiskanen 1998).

The eutrophic state of the Gulf of Finland is reflected in the sediments. The surface sediment organic matter and nitrogen concentrations [loss-on-ignition (LOI) 17.6% dry matter (DM) and 8.6 mg N g^{-1} DM, Table 13.1e] are among the highest observed in the Baltic Sea and in other coastal sea areas (Lehtoranta 2003). The high organic matter content of the sediments influences the seasonal oxygen conditions in the near-bottom water, and the release of nutrients from sediment to water. The nutrients released to the water column in turn affect the seasonal primary production.

Microelectrode measurements have revealed that, despite high oxygen concentrations in the overlying water, the oxic sediment layer in the Gulf of Finland is only a few millimetres thick (J. Lehtoranta, unpublished data). In the Gulf of Finland the lowest oxygen concentrations are commonly measured in summer, when hydrodynamic conditions together with high organic matter content easily turn microbial mineralisation processes in surface sediments anaerobic. When the Neva Estuary is excluded (Fig. 13.1), reduced conditions have prevailed in large areas of both open and coastal organic accumulation bottoms in the Gulf since the mid-1990s (Pitkänen et al. 2001a, 2003). In addition, the deterioration of surface sediments is indicated by observations concerning the colonisation by white bacteria (filamentous sulphur oxidising *Beggiatoa* spp.) of vast areas of the Gulf. The presence of *Beggiatoa* indicates very low oxygen concentrations in the water layer immediately above the sediment surface.

The main mobile sediment P pools in the Gulf of Finland are iron-bound P and organic P, whereas calcium-bound P seems to be fairly constant with depth in sediments (Lehtoranta 2003). Phosphorus retention is favoured in bottom areas where the sediment surface is oxidised, i.e. where a brown surface layer indicating the presence of Fe(III) oxides is present. Extensive benthic P fluxes occur from bottom areas where the sediment surface is black, indicating the presence and formation of iron sulphides unable to capture P (Fig. 13.6). In the Gulf of Finland, brown surface

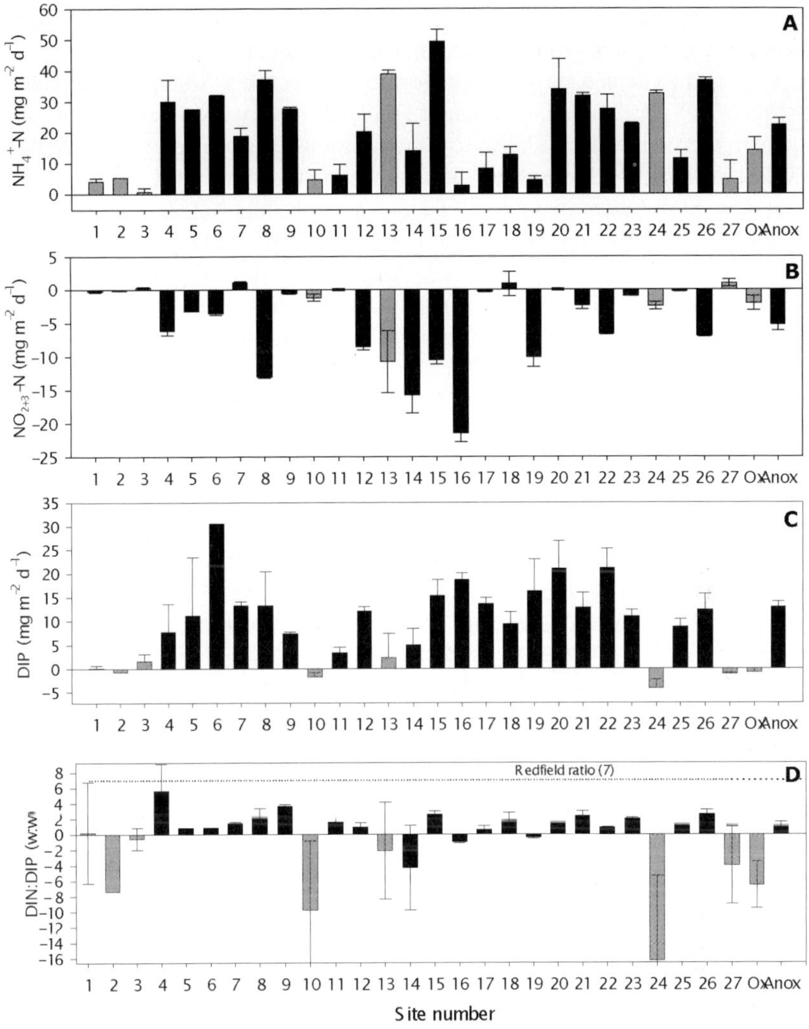

Fig. 13.6A–D Experimental results on sediment–water nutrient flux measurements in the Gulf of Finland in 1999–2002 according to Lehtoranta (2003). **A** Ammonium-N (mg m^{-2} day^{-1}); **B** nitrite + nitrate-N (mg m^{-2} day^{-1}); **C** phosphate-P (mg m^{-2} day^{-1}) and **D** DIN:DIP-ratio. *Black bars* denote reduced and *grey bars* oxidised sediment surfaces. Positive values denote sediment release. Values are the mean of three replicates ± SD. See Fig. 13.1 for the location of the sampling sites

layers are often found in winter and early spring but disappear in late summer. Accordingly, small P effluxes are measured in winter and the highest effluxes are measured in late summer (Lehtoranta and Heiskanen 2003).

Problems with the ability of the sediment to retain P occur mostly in summer, as evidenced by the high concentrations of P in the pore water and, consequently, as high benthic release rate of P at that time. The P efflux consists almost completely of

DIP, which is considered completely available for algae. The measured voluminous effluxes of DIP in summer are related to anoxic conditions, in which the cycling of iron may be inactivated by sulphides formed in microbial sulphate reduction and subsequent FeS formation. Formation of FeS leads to very low dissolved Fe concentrations in the pore waters relative to concentrations of dissolved P. New formation of Fe(III) oxides fuelled by upward diffusing Fe in the sediment is insufficient to bind all the upward diffusing dissolved P and, as a result, high release rates of P to the near-bottom water develop (Gunnars and Blomqvist 1997; Lehtoranta and Heiskanen 2003). Anaerobic microbial processes, such as dissimilatory Fe(III) oxide and sulphate reduction, evidently have a significant role in benthic P cycling, although only indirect evidence on the presence of these processes exists.

Oxic conditions at the sediment–water interface also favours anaerobic denitrification, since most denitrification is based on coupled nitrification–denitrification in the surface sediment. Only a small contribution to denitrification is made by nitrate diffusing from water to sediment (Kuparinen and Tuominen 2001). Usually, high effluxes of ammonium and small influxes of nitrate are observed at reduced bottoms, whereas outflux of nitrate prevails at oxidised bottoms (Lehtoranta 2003). Therefore, eutrophication and the subsequent anoxia/hypoxia largely inhibit benthic nitrification and accelerate the release of ammonium from sediment to water.

It has been estimated (Lehtoranta 2003) that the annual internal load of P in the Gulf of Finland varies between ca. 4,000 and 18,000 t annually, i.e. from 0.7 to 3 times the external annual loading from the catchment. Although the benthic ammonium efflux is also considerable, the inorganic N:P ratio (ranging from 1 to 3 by weight, Fig. 13.6) of the sediment nutrient efflux is strongly below the Redfield optimum of 7.2. This suggests that the intensified sediment release of P has been responsible for the observed summertime eutrophication and extensive cyanobacterial blooms in the Gulf of Finland in the late 1990s and the early 2000s.

13.6.2 The Role of Benthic Macrofauna

Bioturbation caused by macrofauna greatly increases the thickness of the oxidised sediment layer and affects the geochemistry of the sediment (e.g. Rosenberg 2001). If oxygen conditions are sufficient to allow the existence of macrofauna, this activity may both increase the binding of P (Hansen et al. 1998) and enhance denitrification (Tuominen et al. 1998; Gran and Pitkänen 1999). The abundance of benthic animals was exceptionally high in the open Gulf of Finland in 1991–1995 (Laine et al. 1997; Kotta et al. 2002) during the period of a weak or absent halocline, but since the increased salinity stratification in the mid-1990s, there has been a drastic decrease in benthic animal abundance (Pitkänen et al. 2003, and unpublished data of the Finnish Environment Institute and the Finnish Marine Research Institute).

In fact, comprehensive changes in benthic communities have been occurring during the whole of the past century. The coastal areas earlier dominated by the amphipod *Monoporeia affinis* are now dominated by the Baltic clam *Macoma baltica* (Laine et al. 2003). In shallow coastal areas the diverse *Corophium volutator–M. baltica*–chironomid community is dominated by *M. baltica* and by the polychaete *Marenzelleria viridis*. The invasive North American species *M. viridis* was observed for the first time at the entrance of the Gulf of Finland in 1990 and expanded into the eastern Gulf of Finland during 1990–1993 (Stigzelius et al. 1997).

13.7 Nutrient Dynamics are Reflected in the Phytoplankton

The intra-annual spatio-temporal variations in DIN and DIP are reflected clearly in the biomass and succession of phytoplankton. Two seasonal biomass peaks in the phytoplankton are evident: the first, in April/early May, is formed by diatoms and dinoflagellates, and the second, in July/August, is formed mostly by cyanobacteria [see Chaps.10 (Kotta et al.) and 12 (Telesh et al.), this volume].

The spring bloom is fuelled by the voluminous winter storage of inorganic nutrients. The average peak biomasses are generally around 5 mg l^{-1} wet weight (ww) (Jaanus et al. 2002). Individual samples can reach values of up to 20–30 mg l^{-1} (Kauppila and Lepistö 2001). In the late 1980s, maximum spring values as high as 40 mg l^{-1} [see Chaps. 7 (Kruk-Dowgiałło and Szaniawska) and 10 (Kotta et al.), this volume] were measured from the coastal eastern Gulf (Pitkänen et al. 1990). The dinoflagellate *Srcippsiella hangoei* and the diatoms *Skeletonema costatum*, *Chaetoceros* spp. and *Achnanthes taeniata* are the most dominant species (Table 13.1c). In the central and western Gulf *Scrippsiella* can, at its peak, be responsible for almost the whole phytoplankton biomass (Jaanus et al. 2002). After the vernal peak, the dinoflagellates *Peridinella catenata* and *Heterocapsa triquetra* dominate the strongly diminished biomass in June (Kauppila and Lepistö 2001).

In July the proportion of cyanobacteria starts to increase, and cyanobacteria may also form surface accumulations during favourable (calm and warm) weather conditions. The intensity of these surface blooms has clearly increased since the early 1990s (Kahru et al. 2000). The open sea blooms are usually dominated by the N_2-fixing *Nodularia spumigena* and *Aphanizomenon* spp. (Table 13.1c). According to semi-quantitative analysis of ship-of-opportunity data from the western Gulf, the relative abundance of *Nodularia* increased during the 1990s along with elevated DIP concentrations and increased bloom frequency (Jaanus et al. 2002). In the eastern Gulf, which is strongly affected by the N-rich waters from the River Neva, *Planktothrix agardhii* (not a N_2-fixer) is usually the dominant cyanobacteria in summer/early autumn (Kauppila et al. 1995; Kauppila and Lepistö 2001). Typical summer/early autumn biomasses vary between 0.5 and 1 mg l^{-1} (ww) in the western and middle Gulf, between 1 and 2 mg l^{-1} in the eastern Gulf, and between 2 and 4 mg l^{-1} in the Neva Estuary (Kauppila et al. 1995; Kauppila and Lepistö 2001; Jaanus et al. 2002).

13.8 Fish Assemblages and Environmental Conditions

The fish assemblage of the Gulf of Finland consists of a mixture of marine, freshwater and migratory species. About 50 species have been observed in this area (Koli 1994; Table 13.1c). Salinity strongly influences the composition of the fish assemblage in the Gulf of Finland, as marine fish species gradually disappear towards the less saline eastern part (Lappalainen et al. 2000). The most abundant species are herring (*Clupea harengus* L.) and sprat (*Sprattus sprattus* (L.), which dominate especially the pelagic areas (Peltonen et al. 2004), while in coastal habitats freshwater species are abundant (Lappalainen et al. 2000).

At least a fraction of the herring and sprat populations migrate seasonally from the Gulf of Finland to the Baltic Proper (Parmanne et al. 1994). While herrings spawn in a number of locations along the coasts, the hydrographic conditions, such as the low salinity in this area, do not favour sprat reproduction. Thus, sprat spawn mainly in the Baltic Proper, but also utilise the less saline areas of the Gulf of Finland for feeding. During recent decades sprat and herring stocks have shown considerable changes in both biomass and growth rates (ICES 2004; Rönkkönen et al. 2004). Several factors, especially fishing and changes in species interactions such as the predation by cod have influenced these dynamics (e.g. ICES 2004). However, environmental changes, such as changes in hydrography and eutrophication, may also influence the biomass of sprat and herring in the Gulf of Finland.

Of the marine species, cod is at present absent from the Gulf of Finland. During favourable environmental conditions, i.e. after major inflows of saline water from the North Sea, large year-classes may produce a considerable increase in Baltic Sea cod biomass, as seen in the 1970s and 1980s. In such years cod disperses to a wider geographic range, including the Gulf of Finland (Aro 2000). However, the intensive fishing in the Baltic Main Basin at present keeps the cod abundance at a low level (ICES 2004), and migration to the Gulf of Finland does not occur.

In the Northern Baltic Proper as well as in the Gulf of Finland, drastic changes have taken place in the growth of herring and sprat since the start of systematic sampling of commercial catches in the 1970s (ICES 2004). Weights-at-age peaked in the beginning of the 1980s, but in old age groups they decreased by ca. 50% during the last two decades of the twentieth century (Rönkkönen et al. 2004). The average weight-at-age of sprat was relatively stable during the 1980s, but during the 1990s a drastic decline occurred (ICES 2004). It seems that the same factors control growth rates in different parts of the Baltic Sea, because similar changes have taken place in different areas. In the Gulf of Finland, changes in prey resources have been especially studied (Flinkman et al. 1998; Rönkkönen et al. 2004). However, changes in prey populations may arise due to changes in hydrography (Viitasalo 1992; Flinkman et al. 1998; Hänninen et al. 2000), eutrophication (Viitasalo 1992) or depletion of zooplankton by dense clupeid fish stocks (Rudstam et al. 1994; Cardinale et al. 2002; Kornilovs et al. 2001). As selective predation may influence the size distribution of a prey community, it is also possible that changes in weights-at-age are linked with changes in cod abundance (Beyer and Lassen

1994). Cod abundance peaked in the northern Baltic Sea and in the Gulf of Finland in the beginning of the 1980s (Aro 2000; ICES 2004), i.e. the same time at which weights-at-age of clupeids were at their highest.

In the coastal waters of the Gulf of Finland the fish assemblage consists largely of freshwater species. Percid fishes such as perch (*Perca fluviatilis*), pikeperch (*Sander lucioperca*) and ruffe (*Gymnocephalus gernuus*) are abundant, as are cyprinids species, e.g. roach (*Rutilus rutilus*) and bream (*Abramis brama*). Other abundant species include three-spined stickleback (*Gastorosteus aculeatus*), pike (*Exos lucius*) and burbot (*Lota lota*) (e.g. Lappalainen et al. 2000). In the Baltic Sea the succession in the freshwater fish assemblages along productivity gradients resembles the succession patterns observed in European lakes, i.e. a shift in dominance from salmonids in oligotrophic conditions to dominance by percids at medium productivity and to dominance by cyprinids at high productivity (e.g. Persson et al. 1991). Evidently, the patterns within taxonomic groups resemble those observed in lakes. In percids, for example, perch dominates if productivity is low, but pikeperch increases in more productive conditions, and the dominance shifts to ruffe under still more eutrophic conditions (Leach et al. 1977). Roach is nowadays very abundant in the intermediate and outer archipelagoes of the Gulf of Finland, obviously due to eutrophication (Lappalainen 2002). However, the extremely eutrophic coastal bays rather favour bream and white bream (*Abramis bjoerkna*) over roach. Unlike the general trend in lakes, eutrophication may not increase fish biomass in coastal waters (Lappalainen 2002).

The rivers draining into the Gulf of Finland have previously supported salmon (*Salmo salar*) and sea trout (*Salmo trutta*) populations. At present, the fisheries of these species rely principally on the release of hatchery-reared smolts. Many original salmon and trout stocks have disappeared or their status is very poor, and this is due mostly to damming and eutrophication of rivers and to over-fishing (ICES 2006). Around 10 rivers support some salmon reproduction. The status of sea trout is classified as good in 7, satisfactory in 14 and poor in 31 rivers; in 10 rivers the status of the stock is not known (ICES 2006). However, due to intensive fishing, especially during feeding migration, the reproduction of these species is only a small fraction of the estimated carrying capacity of these rivers (ICES 2006). A substantial fraction of the Gulf of Finland salmon feed in the Baltic Proper and are also exploited there. On the other hand, salmon specimens from other Baltic Sea populations may migrate to the Gulf of Finland and contribute to catches in the Gulf of Finland (ICES 2006).

The M74 syndrome (Norrgren et al. 1993) causes high mortality of yolk-sac fry in Baltic Sea salmon and, to a lesser degree, in trout. The occurrence of this syndrome has been connected to changes in the food-web of the Baltic Sea, whereby salmon eggs and fry suffer from a maternally transferred deficiency of thiamin (vitamin B_1) (see Karlsson et al. 1999). M74 mortality of the fry was at its highest (56%) in 1997 among offspring of salmon ascending the River Kymijoki. The influence of this syndrome has since decreased, and in 2003–2005 the mortality was below 10% (ICES 2006). However, the decline of the annual salmon catches from the Gulf of Finland has continued in spite of lower M74 mortality and increasing release of hatchery-reared smolts. The annual salmon catçhes were more than 600 tons in 1991 and more than 400 tons in 1996–1997, but on average less that 100 tons in 2002–2005 (ICES 2006).

13.9 Conclusions – Is Recovery Possible?

The Gulf of Finland is a shallow, estuarial sea area and thus sensitive to external perturbations caused by nutrient loading from its large catchment, the Baltic Proper and the atmosphere, as well as by changes in climatic conditions affecting its physical regime. Since the mid-1990s the voluminous benthic release of nutrients, especially P, has strongly affected the state of the Gulf, and partly counteracted the potential positive effects of decreased nutrient loading into the Gulf.

The basic reason for the poor P-binding ability of the sediments in the Gulf of Finland is related to the high sedimentation and content of labile organic matter in relation to the prevailing mixing conditions and oxygen reserves of near-bottom waters. The lack of oxygen and the large reservoir of labile organic matter favour anaerobic Fe(III)-oxide and, especially, sulphate reduction in the top surface layers of the sediments. Due to the hypoxic/anoxic conditions at the sediment surface, either the abundance of benthic fauna is low or any such higher life is totally absent. This decisively decreases the ability of the sediments to retain and remove nutrients.

The ultimate recovery of the sediments and a reduction in sediment P release can happen only via decreased sedimentation of particulate organic matter caused by decreased pelagic primary production. According to present knowledge, the only way to permanently decrease autochthonous production is to further reduce external nutrient inputs into the Gulf of Finland. This especially concerns direct nutrient inputs to the Gulf, but also inputs to the Baltic Proper and inputs of N into the atmosphere. Recent results showing decreases in the vernal plankton bloom, most probably as a result of decreased DIN concentrations, suggest that in the long run, the content of labile organic matter in the surface sediment could start to decrease. This would also diminish the risk of internal loading. Due to the estuarine nature of the Gulf of Finland, deep water hypoxia/anoxia, benthic release of P and cyanobacterial blooms will, however, to some extent remain as typical features of the ecosystem, even if the level of external loading is lowered.

References

Alenius P, Myrberg K, Nekrasov A (1998) The physical oceanography of the Gulf of Finland: a review. Boreal Environ Res 3:97–125

Andersin A-B, Sandler H (1991) Macrobenthic fauna and oxygen deficiency in the Gulf of Finland. Memo Soc Fauna Flora Fenn 67:3–10

Aro E (2000) The spatial and temporal distribution patterns of cod (*Cadus morhua callarias* L.) in the Baltic Sea and their dependence on environmental variability – implications for fishery management. PhD thesis, University of Helsinki. 75 pp

Bartnicki J, Gusev A, Barrett K, Simpson D (2002) Atmospheric supply of nitrogen, lead, cadmium, mercury, and lindane to the Baltic Sea in the period 1996–2002. Final report for Helsinki Commission by EMEP (http://projects.dnmi.no/~emep/helcom2002/index.html)

Beyer JE, Lassen H (1994) The effects of size-selective mortality on the size-at-age of Baltic herring. Dana 10:203–234

Cardinale M, Casini M, Arrhenius F (2002) The influence of biotic and abiotic factors on the growth of sprat (*Sprattus sprattus*) in the Baltic Sea. Aquat Living Resour 15:273–281

Elken J, Matthäus W, Krzyminski W, Dubra J (2002) Meteorological, hydrological and hydrographic forcing. In: HELCOM. Environment of the Baltic Sea area 1994–1998. Baltic Sea Environ Proc 82B:39–45

Elken J, Golubkov S, Pitkänen H, Sarkkula J (eds) (2003) Special issue on the changing state of the Gulf of Finland ecosystem. Proc Estonian Acad Sci Biol Ecol 52:171–362

Flinkman J, Aro E, Vuorinen I, Viitasalo M (1998) Changes in northern Baltic zooplankton and herring nutrition from 1980s to 1990s: top-down and bottom-up processes at work. Mar Ecol Prog Ser 165:127–136

Gran V, Pitkänen H (1999) Denitrification in estuarine sediments in the eastern Gulf of Finland, Baltic Sea. Hydrobiologia 393:107–115

Gunnars A, Blomqvist S (1997) Phosphate exchange across the sediment-water interface when shifting from anoxic to oxic conditions – an experimental comparison of freshwater and brackish-marine systems. Biogeochemistry 37:203–226

Hänninen J, Vuorinen I, Hjelt P (2000) Climatic factors in the Atlantic control the oceanographic and ecological changes in the Baltic Sea. Limnol Oceanogr 45:703–710

Hansen K, Mouridsen S, Kristensen E (1998) The impact of *Chironomus plumosus* larvae on organic matter decay and nutrient (N,P) exchange in a shallow eutrophic lake sediment following a phytoplankton sedimentation. Hydrobiologia 364:65–74

Heinänen A (1992) Bacterioplankton in the open Baltic Sea. Finn Mar Res 260:1–32

Heiskanen A-S (1998) Factors governing sedimentation and pelagic nutrient cycles in the northern Baltic Sea. Monogr Boreal Environ Res, no 8, 80 pp

Haahti H, Kangas P (eds) (2004) State of the Gulf of Finland in 2003. Meri – Report Series of the Finnish Institute of Marine Research, no 51, 20 pp

Haapala J, Alenius P (1994) Temperature and salinity statistics for the northern Baltic Sea 1961–1990. Finn Mar Res 262:51–121

HELCOM (1996) Third periodic assessment of the state of the marine environment of the Baltic Sea, 1989–93; Background document. Baltic Sea Environment Proc, no 64B, 252 pp

HELCOM (1998) Third Baltic Sea pollution load compilation. Baltic Sea Environment Proc, no 70, 133 pp

HELCOM (2002) Environment of the Baltic Sea area 1994–1998. Baltic Sea Environment Proc, no 82B, 215 pp

HELCOM (2003) The Baltic marine environment 1999–2002. Baltic Sea Environment Proc, no 87, 46 pp

HELCOM (2004) The fourth Baltic Sea pollution load compilation (PLC-4). Baltic Sea Environment Proc, no 93, 188 pp

ICES (2004) Report of the Baltic Fisheries Assessment Working Group. ICES Advisory Committee on Fishery Management. ICES, Copenhagen, CM2004/ACFM:22, 513 pp

ICES (2006) Report of the Baltic Salmon and Trout Assessment Working Group (WGBAST). ICES Advisory Committee On Fishery Management. ICES CM 2006/ACFM:21. International Council for the Exploration of the Sea, Copenhagen, 210 pp

Jaanus A, Leppänen J, Hällfors S, Lepistö L, Kauppila P (2002) Phytoplankton. In: HELCOM Environment of the Baltic Sea area 1994–1998. Baltic Sea Environment Proc, no 82B, pp 78–80

Kahru M, Leppänen J-M, Rud O, Savchuk OP (2000) Cyanobacteria blooms in the Gulf of Finland triggered by saltwater inflow into the Baltic Sea. Mar Ecol Prog Ser 207:13–18
Karlsson L, Ikonen E, Mitans A, Hansson S (1999) The diet of salmon (*Salmo salar*) in the Baltic Sea and connections with the M74 syndrome. Ambio 28:37–42
Kauppila P, Lepistö L (2001) Changes in phytoplankton. In: Kauppila P, Beck S (eds) The state of Finnish coastal waters in the 1990s. The Finnish Environment, no 472, pp 61–70
Kauppila P, Hällfors G, Kangas P, Kokkonen P, Basova S (1995) Late summer phytoplankton species in the eastern Gulf of Finland. Ophelia 42:179–192
Kiirikki M, Rantanen P, Varjopuro R, Leppänen A, Hiltunen M, Pitkänen H, Ekholm P, Moukhametshina E, Inkala A, Kuosa H, Sarkkula J (2003) Cost effective water protection in the Gulf of Finland – focus on St. Petersburg. The Finnish Environment, no 632, 55 pp
Kivi K, Kaitala S, Kuosa H, Kuparinen J, Leskinen E, Lignell R, Marcussen B, Tamminen T (1993) Nutrient limitation and grazing control of the Baltic plankton community during annual succession. Limnol Oceanogr 38:893–905
Koli L (1994) Suomen kalaopas (Guide over fishes in Finland). WSOY, Porvoo, 160 pp
Kornilovs G, Sidrevics L, Dippner JW (2001) Fish and zooplankton interaction in the Central Baltic Sea. ICES J Mar Sci 58:579–588
Kotta J, Kotta I, Kangas P, Laine A, Andersin A-B (2002) Benthic conditions. Macrozoobenthos. In: HELCOM. Environment of the Baltic Sea area 1994–1998. Baltic Sea Environment Proceedings, no. 82B, pp 83–85
Kuparinen J, Tuominen L (2001) Eutrophication and self-purification: counteractions forced by large-scale cycles and hydrodynamic processes. Ambio 30:190–194
Laakkonen A, Mälkki P, Niemi Å (1981) Studies on the sinking, degradation and sedimentation of organic matter off Hanko peninsula, entrance to the Gulf of Finland, in 1979 (progress report). Inst Mar Res Meri 9:3–42
Laamanen M, Fleming V, Olsonen R (2004) Water transparency in the Baltic Sea between 1903 and 2004. http://helcom.navigo.fi/environment/indicators2004/secchi/en_GB/transparency/
Laine AO, Sandler H, Andersin AB, Stigzelius J (1997) Long-term changes of macrozoobenthos in the Eastern Gotland Basin and the Gulf of Finland (Baltic Sea) in relation to the hydrographical regime. J Sea Res 38:135–159
Laine AO, Luodekari K, Poikonen M, Viitasalo M (2003) A comparison between 1928 and 2000 indicates that major changes in the macrozoobenthos species composition and abundance on the SW coast of Finland (Baltic Sea). Proc Estonian Acad Sci Biol Ecol 52:3–16
Lappalainen A (2002) The effects of recent eutrophication on freshwater fish communities and fishery on the northern coast of the Gulf of Finland, Baltic Sea. PhD Thesis, University of Helsinki, 24 pp
Lappalainen A, Shurukin A, Alekseev G, Rinne J (2000) Coastal fish communities along the Northern Coast of the Gulf of Finland, Baltic Sea: responses to salinity and eutrophication. Int Rev Hydrobiol 85:687–696
Leach JH, Johnson MG, Kelso JRM, Hartmann J, Nümann W, Entz B (1977) Responses of percid fishes and their habitats to eutrophication. J Fish Res Board Can 34:1964–1971
Lehtoranta J (2003) Dynamics of sediment phosphorus in the brackish Gulf of Finland. Monogr Boreal Environ Res, no 24, 58 pp
Lehtoranta J, Heiskanen A-S (2003) Dissolved iron:phosphate ratio as an indicator of phosphate release to oxic water of the inner and outer coastal Baltic Sea. Hydrobiologia 492:68–84
Leppänen J-M, Tamelander G (1981) Composition of particulate matter and its relation to plankton biomass in the trophogenic layer off tvärminne, at the entrance to the Gulf of Finland. Inst Mar Res Meri 9:56–70
Lignell R (1993) Carbon flow supplied by algae and bacteria in the planktonic food web off the SW coast of Finland. PhD Thesis. Walter and Andree de Nottbeck Foundation Scientific Reports, no 9, 55 pp
Lips U, Poutanen E-L, Pitkänen H, Kauppila P, Basova S (2002) Hydrochemistry. Baltic Marine Environment Protection Commission. Fourth periodic assessment of the state of the marine environment of the Baltic Sea. Baltic Sea Environment Proc, no 82 B, pp 76–78

Lääne A, Pitkänen H, Arheimer B, Behrendt H, Jarosinski W, Sarmite L, Pachel K, Shekhovtsov A, Svendsen LM, Valatka S (2002) Evaluation of the implementation of the 1988 Ministerial Declaration regarding nutrient load reductions in the Baltic Sea catchment area. Finnish Environment Institute. The Finnish Environment, no 524, 195 pp

Nausch G, Lysiak-Pastuszak E (2002) Hydrochemistry. In: HELCOM. Environment of the Baltic Sea area 1994–1998. Baltic Sea Environment Proc, No 82 B, pp 42–45

Nielsen R, Kristiansen A, Mathiesen L, Mathiesen H (eds) (1995) Distributional index of the benthic macroalgae of the Baltic Sea area. Acta Bot Fenn 155:1–51

Norrgren L, Andersson T, Bergqvist PA, Björklund I (1993) Chemical, physiological and morphological studies of feral Baltic salmon (*Salmo salar*) suffering from abnormal fry mortality. Environ Toxicol Chem 12:2065–2075

Parmanne R, Rechlin O, Sjöstrand B (1994) Status of herring and sprat stocks in the Baltic Sea. Dana 10:29–59

Peltonen H, Vinni M, Lappalainen A, Pönni J (2004) Spatial feeding patterns of herring (*Clupea harengus* L.), sprat (*Sprattus sprattus* L.), and the three-spined stickleback (*Gasterosteus aculeatus* L.) in the Gulf of Finland, Baltic Sea. ICES J Mar Sci 61:966–971

Persson L, Diehl S, Johansson L, Andersson G, Hamrin SF (1991) Shifts in fish communities along the productivity gradient of temperate lakes – patterns and the importance of size-structured interactions. J Fish Biol 38:281–293

Perttilä M, Tulkki P, Pietikäinen S (1980) Mean values and trends of hydrographical and chemical properties of the Gulf of Finland. Finn Mar Res 247:38–50

Perttilä M, Niemistö L, Mäkelä K (1995) Distribution, development and total amounts of nutrients in the Gulf of Finland. Estuarine Coastal Shelf Sci 41:345–360

Pitkänen H (ed) (2004) The state of coastal and open sea waters at the turn of the new millennium (in Finnish with English summary). The Finnish Environment, no 669, 104 pp

Pitkänen H, Räike A (2004) Loading to waters from different sources. In: Pitkänen H (ed) The state of coastal and open sea waters at the turn of the new millennium (in Finnish with English summary). The Finnish Environment, no 669, pp 46–49

Pitkänen H, Tamminen T (1995) Nitrogen and phosphorus as production limiting factors in the estuarine waters of the eastern Gulf of Finland. Mar Ecol Prog Ser 129:283–294

Pitkänen H, Kangas P, Sarkkula J, Lepistö L, Hällfors G, Kauppila P (1990) Water quality and trophic status in the eastern Gulf of Finland (in Finnish with English summary). Publications of the Water and Environment Administration, Series A 50, 137 pp

Pitkänen H, Tamminen T, Kangas P, Huttula T, Kivi K, Kuosa H, Sarkkula J, Eloheimo K, Kauppila P, Skakalsky B (1993) Late summer trophic conditions in the north-east Gulf of Finland and the River Neva Estuary, Baltic Sea. Estuarine Coastal Shelf Sci 37:453–474

Pitkänen H, Lehtoranta J, Räike A (2001a) Internal nutrient fluxes counteract decreases in external load: the case of the estuarial Gulf of Finland. Ambio 30:195–201

Pitkänen H, Kauppila P, Laine Y (2001b) Nutrients. In: Kauppila P, Beck S (eds) The state of Finnish coastal waters in the 1990s. The Finnish Environment, no 472, pp 37–60

Pitkänen H, Lehtoranta J, Peltonen H, Laine A, Kotta J, Kotta I, Moskalenko P, Mäkinen A, Kangas P, Perttilä M, Kiirikki M (2003) Benthic release of phosphorus and its relation to environmental conditions in the early 2000s in the estuarial Gulf of Finland, Baltic Sea. Proc Estonian Acad Sci Biol Ecol 52:173–192

Pitkänen H, Kauppila P, Poutanen E-L (2004) Temporal changes in nutrients and trophic status. In: Pitkänen H (ed) The state of coastal and open sea waters at the turn of the new millennium (in Finnish with English summary). The Finnish Environment, no 669, pp 14–16

Raateoja M, Seppälä J, Kuosa H, Myrberg K (2005) Recent changes in trophic state of the Baltic Sea along SW coast of Finland. Ambio 34:188–191

Rosenberg R (2001) Marine benthic faunal successional stages and related sedimentary activity. Sci Mar 65 (Suppl 2):107–119

Rudstam L, Aneer G, Hildén M (1994) Top-down control the pelagic Baltic ecosystem. Dana 10:105–129

Rönkkönen S, Ojaveer E, Raid T, Viitasalo M (2004) Long-term changes in Baltic herring (*Clupea harengus membras*) growth in the Gulf of Finland. Can J Fish Aquat Sci 61:219–229

Sarkkula J (ed) (1997) Proceedings of the Final Seminar of the Gulf of Finland year 1996. Finnish Environment Institute. Suomen ympäristökeskuksen monistesarja, no 105

Seinä A, Peltola J (1991) Duration of the ice season and statistics of fast ice thickness along the Finnish Coast 1961–1990. Finn Mar Res 258:1–46

Stigzelius J, Laine A, Rissanen J, Andersin A-B, Ilus E (1997) The introduction of Marenzelleria viri-dis (Polychaeta, Spionidae) into the Gulf of Finland and the Gulf of Bothnia (northern Baltic Sea). Ann Zool Fenn 34:205–212

Tuominen L, Heinänen A, Kuparinen J, Nielsen LP (1998) Spatial and temporal variability of denitrification in the sediments of the northern Baltic Proper. Mar Ecol Prog Ser 172:13–24

Vallius H (1999) Recent sediments of the Gulf of Finland: an environment affected by the accumulation of heavy metals. Academic dissertation. Åbo Akademi University, Department of Geology and Mineralogy

Velner H-A (ed) (2005) Water protection of the Gulf of Finland and Estonian water bodies. Tallinn University of Technology. Land and Water Technology Foundation, Finland, 165 pp

Viitasalo M (1992) Mesozooplankton of the Gulf of Finland and northern Baltic Proper a review of monitoring data. Ophelia 35:147–168

Chapter 14
The Stockholm Archipelago

C. Hill and K. Wallström

14.1 Environmental Characteristics

The Stockholm archipelago is a brackish-water archipelago that extends along the Swedish east coast, just south of the border between the Bothnian Sea and the northern Baltic Proper. The Stockholm archipelago stretches about 200 km from Singö in the north to Nynäshamn in the south (Fig. 14.1). A nearly continuous belt of islands stretches eastwards from the Stockholm archipelago across to the Åland islands and the Åbo archipelago in Finland.

The Stockholm archipelago has about 30,000 large and small islands spread over an area of more than 35,000 km^2, which makes it the largest connected archipelago in Swedish waters. The archipelago is part of the Precambrium peneplane and is characterised by joint–valley landscapes that give rise to a network of cliffs and rifts. Its waters consist of a series of basins and inlets that are connected by passages. The marine biotopes include open waters, rocky coasts (bedrock, mostly granite and gneiss), long, narrow fjord-like bays and sheltered inlets. The archipelago is widest at the latitude of Stockholm, where it stretches about 100 km out to the east (Fig. 14.1).

The inner archipelago is characterised by shallow bays and fjords that are sheltered from the open sea. The islands are relatively large and are covered by pine and spruce forests and some agricultural land. The inner waters are strongly influenced by freshwater and land-based substances from streams and coastal catchments. The inner archipelago is usually covered by ice during winter (Table 14.1).

The central archipelago is influenced by wind and waves and the fjords are larger. There are more deciduous trees, bushes and bare rocky shores on the islands. The eastern parts of the outer archipelago are very exposed to wind and waves, so there is extensive water exchange with the open sea. These outer areas are seldom covered by ice during winter. Large expanses of open water are sprinkled with groups of small, barren islands and skerries. There are often large colonies of seabirds in the outer archipelago.

The land in the region is still rebounding after the retreat of the last icesheet. Land uplift in the region is about 5 mm per year, and has a very visible effect on the lowest parts of the archipelago. As the land rises, seashores expand and new islands

U. Schiewer (ed.), *Ecology of Baltic Coastal Waters. Ecological Studies 197.*

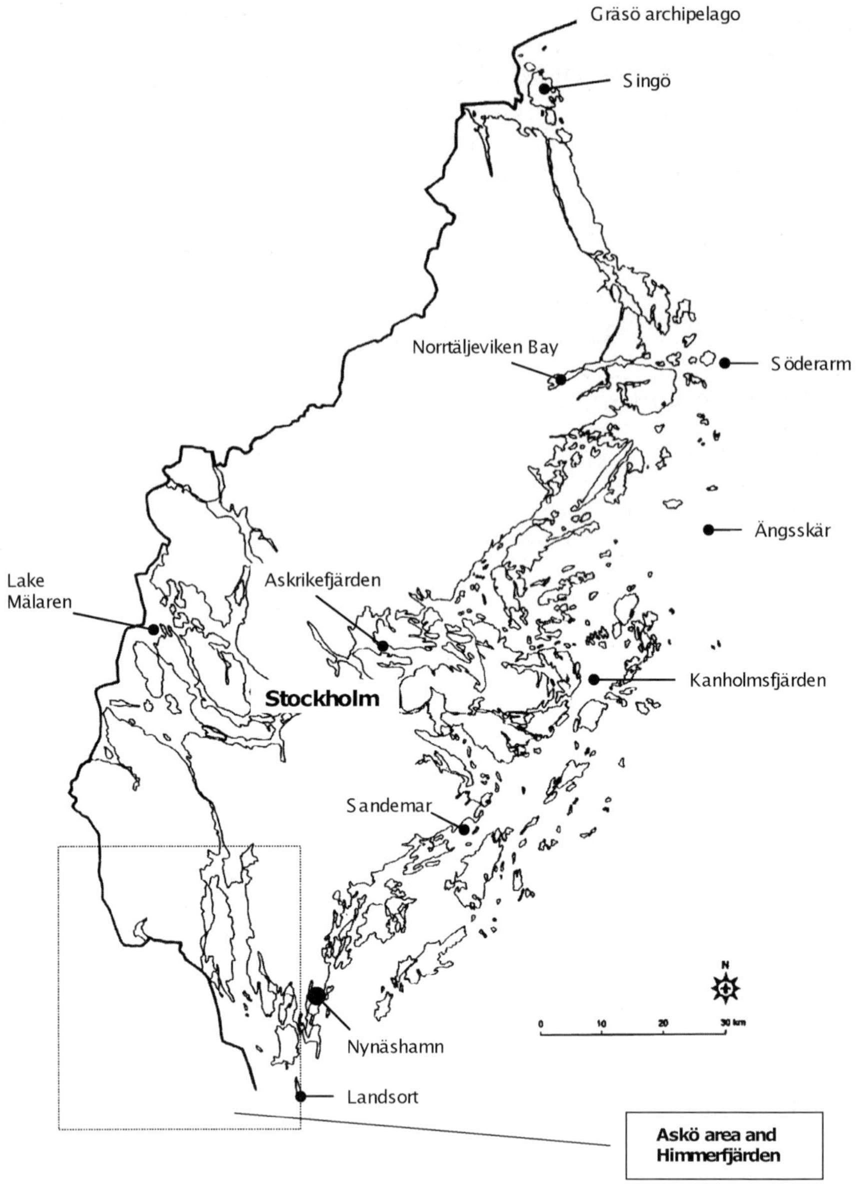

Fig. 14.1 The greater Stockholm region; Stockholm archipelago and the Askö–Himmerfjärden area

emerge above the water. Shallow inlets that have narrow openings and protective sills become cut off from the sea by the surrounding land. This gradual isolation results in a succession from flads (small, shallow brackish water bodies that are still connected to the sea) to glo lakes (water bodies isolated from the sea) (Munsterhjelm 1997). New lakes are thus continually being formed in the coastal area.

In the outer parts of the archipelago the surface water flows mainly from north to south (Fig. 14.2), which gives a salinity gradient from 5.5 psu in the north to

Table 14.1 Main characteristics of the Stockholm archipelago. *dw* Dry weight, *ww* wet weight, *Chl a* chlorophyll *a*, *DIN* dissolved inorganic nitrogen, *DIP* dissolved inorganic phosphorus

	Stockholm archipelago
a. **Climate and hydrology**	
Air temperature (°C)	−3.0 to +17.2 (monthly mean)[a]
Water temperature (°C)	+ 0.5 to + 16.1 (monthly mean)[a]
Ice covering (days a^{-1})	30–80[b]
Annual global radiation (kWh m^{-2})	975 (northern parts); 1,000 (southern parts)[a]
Average precipitation (mm a^{-1})	540[a]
Mean freshwater inflow ($10^6 m^3 a^{-1}$)	6,300[c]
Mean inflow ($10^6 m^3 a^{-1}$) (from Baltic)	47,300[d]
Mean total outflow ($10^6 m^3 a^{-1}$)	54,000
Residence time (days)	<10 (outer areas) >40 (inner areas)[e]
b. **Morphometry and sediments**	
Area (km^2)	5,010[f]
Volume ($10^6 m^3$)	109.6[f]
Mean depth (m)	14[f]
Maximum depth (m)	241[f]
Catchment area (km^2)	27,265[f]
Area/catchment relation	0.18
Organic matter content (% dw)	14–25[g]
C/N ratio (0–10 cm depth)	8.5–11.2[h]
Phosphate release (mg P m^{-2} day^{-1})	No data
Ammonia release (mg N m^{-2} day^{-1})	No data
Denitrification rate (mg N m^{-2} day^{-1})	No data
Sediment distribution	Soft bottoms ~50% Rocky bottoms ~20%[i]
c. **Pelagic biological components**	
Secchi depth (m)	4 (inner areas); 7 (outer areas)[j]
Bacteria ($10^6 ml^{-1}$)	<0.05 (outer areas); >0.1 (inner areas)[j]
Phytoplankton biomass (μg Chl *a* l^{-1})	5,7; spring max: ~20[j]
Dominating phytoplankton species	**Spring**: *Perediniella catenata, Thalassiosira baltica, Pauliella taeniata, Chaetoceros wighamii, Skeletonema costatum*; additional in inner areas: *Diatoma tenuis, Asterionella formosa, Tabellaria fenestrata*[j] **Autumn**: *Woronichinia compacta, Microcystis* spp, *Aphanizomenon cf. baltica, Thallassiosira baltica, Chaetoceros wighamii*[j]

(continued)

Table 14.1 (continued)

	Stockholm archipelago
c. Pelagic biological components	
Flagellates	
Dominating species	Cryptomonads[j]
Ciliates	
Dominating species	*Lohmaniella stellaris*, *Tinntinnopsis tubulosa*, *T. brandti*, *Vorticella* spp, *Myrionectra rubra*[j]
Mesozooplankton	
Dominating species	*Synchaeta* spp, *Asplanchna* spp, *Keratella cochlearis*, *K. quadrata*, *Bosmina coregoni*, *Daphnia longispina*, *Podon polyphemoides*, calanoid and cyclopoid copepods[j]
Fish	
Number of species	50[k]
Dominating species	*Clupea harengus*, *Sprattus sprattus*, *Abramis bjoerkna*, *Abramis brama*, *Alburnus alburnus*, *Leuciscus idus*, *Rutilus erythrophthalmus*, *Rutilus rutilus*, *Osmerus eperlanus*, *Salmo trutta*, *Esox lucius*, *Perca fluviatilis*, *Sander lucioperca*, *Platichthys flesus*[l]
Fish catches ($t\ a^{-1}$)	325[m]
d. Benthic biological components	
Macrophytes	
Dominating species	Rocky substrates: *Cladophora glomerata*, *Fucus vesiculosus*, *Furcellaria lumbricalis*, *Rhodomela confervoides*, *Polysiphonia fucoides*, *Ceramium* spp, *Sphacelaria arctica* Soft bottoms: *Potamogeton pectinatus*, *P. perfoliatus*, *P. filiformis*, *Ceratophyllum demersum*, *Myriophyllum spicatum*, *Najas marina*, *Chara aspera*, *C. tomentosa*[l]
Macrozoobenthos	
Biomass ($g\ m^{-2}$)	No data
Dominating species	Rocky substrates: *Theodoxus fluviatilis*, *Lymnea peregrea*, *Idothea baltica*, *Gammarus* spp, *Hydrobia* spp, *Mytilus trossulus*, *Balanus improvisus* Soft bottoms: *Cerastoderma* sp, *Mya arenaria*, *Crangon crangon*, *Corophium volutator*, *Chironomus* spp, *Macoma balthica*, *Monoporeia affinis*, *Harmothoe sarsi*, *Saduria entomon*, *Halicryptus spinulosus*, *Marenzelleria neglecta*[l]
e. Water chemistry, trophic status and pollution	
Salinity (psu)	0–7 (surface)[l]
pH	No data
Total nitrogen ($\mu mol\ l^{-1}$)	Annual mean: 20 (outer parts) Annual mean: 60 (inner parts)[j]

(continued)

Table 14.1 (continued)

	Stockholm archipelago
	Winter max: 24 (outer parts)
	Winter max: 80 (inner parts)[j]
DIN (μmol l^{-1})	no data
	Annual mean: 0.6 (outer parts)
Total phosphorus (μmol l^{-1})	Annual mean: 1.2 (inner parts)
	Winter max: 0.8 (outer parts)
	Winter max: 2.0 (inner parts)[j]
DIP (μmol l^{-1})	No data
Annual nitrogen input (t a^{-1})	11,500[n]
Annual phosphorus input (t a^{-1})	295[n]
Limitation of PP and period	Phosphorus limitation summer[j]
Main pollutants	Nutrients[l]
Trophic level	Very high–high (inner areas)
	Low–very low (outer areas)[o]

[a]SMHI (2005); [b]Cederwall (2000); [c]A. Engqvist (personal communication), [d]Engqvist and Andrejev (2003); [e]Swedish Environmental Protection Agency (1999); [f]SMHI (2003); [g]Jonsson (2003); [h]Jönsson et al. (2005); [i]Mattisson (2005); [j]Stockholm Water Co (2004); [k]Olburs (2002); [l]see main chapter text for references; [m]Andersson (2003); [n]Kautsky et al. (2000); [o]Hagström and Pansar (2003)

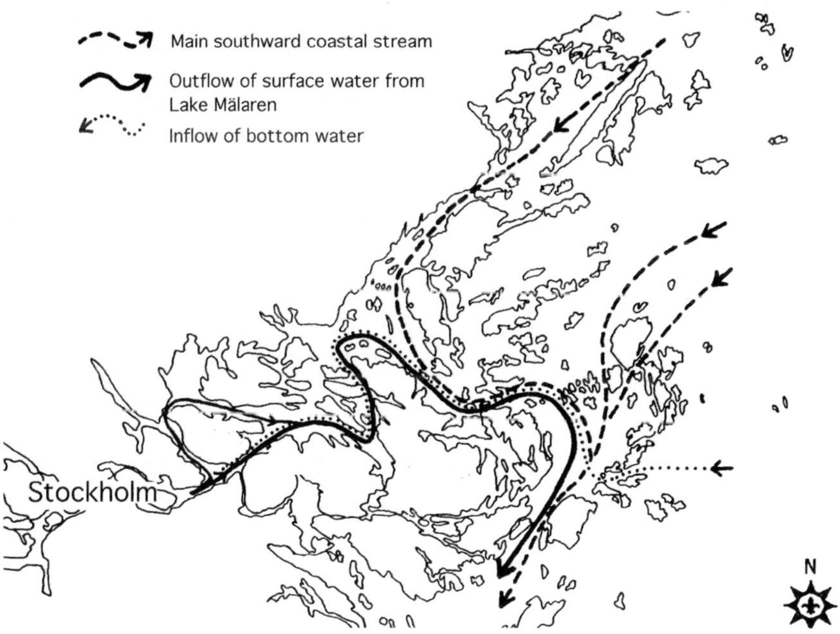

Fig. 14.2 The outward surface flow of freshwater from Lake Mälaren to the Stockholm archipelago (which creates the "nutrient gradient" referred to in the text), and the countercurrent of saltier bottom water from the open Baltic Sea. A southerly surface current carries less saline water from the Gulf of Bothnia along the Swedish coast to the northern parts of the Stockholm archipelago. Redrawn from Kautsky et al. (2000), with permission from the County Administrative Board of Stockholm

about 7 psu in the south. The salinity also increases from west to east, due to the outflow of fresh water from coastal catchments.

In the archipelago east of Stockholm, salinity is heavily influenced by the outflow of Lake Mälaren. The mean outflow of freshwater from Lake Mälaren was 4,800 Mm^3 per year in 1968–2003, and varied considerably both within and between years. Surface salinity was 1–2.5 psu 10 km east of the outflow of Lake Mälaren in 1998, when the outflow was relatively high and even (total flow was 5,500 Mm^3). The corresponding salinity in the same area was 2–4 psu in 2003, when the freshwater outflow was low (2,600 Mm^3) (data from Stockholm Water Company 2004).

The surface outflow of freshwater from Lake Mälaren causes a deep inward counter-current that transports saltier water from the northern Baltic Proper into the archipelago (Fig. 14.2). At 10 km east of the outflow of Lake Mälaren, the mean salinity at 50 m depth was 5.2 psu during 1994–2003. The salinity of the bottom water in the inner archipelago has decreased gradually by about 1 psu during the last 20 years (Stockholm Water Company 2004).

A weak halocline develops due to the differences in salinity between the fresh surface water and the underlying saltier water. In summer, a thermocline usually strengthens the separation of surface and bottom waters. When it is windy, upwelling bottom water may mix the stratified layers, especially in the outer parts of the southern archipelago (Gidhagen 1987).

The water exchange in different sub-areas of the archipelago depends on several factors, such as the location and size of a certain basin as well as the influence of freshwater outflow and/or the presence of protective sills. In general, water exchange increases the more contact there is with the open sea. Water exchange is more than 40 days in the innermost sheltered areas, 10–40 days in the central archipelago, and less than 10 days in open coastal areas (Swedish Environmental Protection Agency 1999).

In shallow areas, the inflow and outflow of water is driven by changes in water level, which depend mainly on air pressure and strong winds (Kautsky et al. 2000). The maximum amplitude in water level during 1990–2003 was 1.9 m (Stockholm Water Company 2004). There are essentially no tides in the Stockholm archipelago, and tidal water levels usually vary by only a few centimetres daily.

During a normal winter, the Stockholm archipelago is ice-covered for 30 days. In severe winters, the ice cover may last as long as 70–80 days. Completely ice-free winters occur about once every 10 years (Cederwall 2000).

14.2 The Pelagic Zone

14.2.1 Nutrient Cycles

The main inputs of nutrients to the Stockholm archipelago are discharges from sewage treatment plants, the outflow from Lake Mälaren in central Stockholm, and atmospheric deposition on the sea surface. The nutrient load to the archipelago is greatest in the inner archipelago near Stockholm (Aneer and Arvidsson 2003).

In the 1990s, discharges from sewage treatment plants and the outflow of Lake Mälaren contributed an annual load of 5,960 t nitrogen and 160 t phosphorus to the inner archipelago. Most of the nitrogen came from sewage treatment plants, while most of the phosphorus came from Lake Mälaren (Stockholm Water Company 2004). Moreover, sediments provide an internal load of 30–60 t phosphorus each year in the inner areas (Stockholm Water Company op. cit.).

The heavy load of nutrients and the eastward transport of water results in a nutrient gradient from Stockholm to the outer archipelago. From 1995 to 2003, the annual mean concentration (0–4 m depth) of total nitrogen was 60 $\mu mol\ l^{-1}$ in the inner areas and 20 $\mu mol\ l^{-1}$ in the outer areas. The corresponding concentrations of total phosphorus were 1.2 and 0.6 $\mu mol\ l^{-1}$, respectively (data from Stockholm Water Company 2004).

According to Swedish water-quality criteria (Swedish Environmental Protection Agency 1999), summer concentrations of nitrogen and phosphorus in surface waters (0–10 m) of the inner archipelago were "high" or "very high" (26– >32 μmol tot-N l^{-1}; 0.77– >1.0 μmol tot-P l^{-1}) in 1996–2002. In the outer archipelago, corresponding nutrient concentrations were classified as "low" or "very low" (≤18–22 μmol tot-N l^{-1}; ≤0.48–0.60 μmol tot-P l^{-1}) (Hagström and Pansar 2003). Nutrient concentrations were slightly lower in the northern archipelago than in areas south of Stockholm (Aneer and Arvidsson 2003).

Concentrations of nutrients that are available for primary production in spring (i.e. nitrate, nitrite, ammonia and phosphate) also decreased along the nutrient gradient from Stockholm to the outer archipelago in 1995–2003. After the spring phytoplankton bloom, there was still more inorganic nitrogen in the inner areas (18–20 μmol inorg-N l^{-1}) than in the outer areas (<4 μmol inorg-N l^{-1}). However, phosphate concentrations usually sank to below 0.2 μmol PO_4-P l^{-1} in the whole archipelago in late spring (data from Stockholm Water Company 2004).

Silica, which is an essential nutrient for diatoms, is also heavily utilised during the spring bloom. Concentrations of available inorganic silica were low (<0.5 μmol inorg-Si l^{-1}) in the central archipelago for a short period in May–June 2003 (Stockholm Water Company 2004).

14.2.2 Plankton

The biomass and species composition of the phytoplankton depend mainly on prevailing nutrient conditions. The nutrient gradient from Stockholm to the outer archipelago is reflected in phytoplankton biomasses in summer: phytoplankton biomass is classified as "very high" [>5.0 μg chlorophyll *a* (Chl *a*) l^{-1}] in the inner archipelago and "medium high" (2.2–3.2 μg Chl *a* l^{-1}) in the outer areas (Stockholm Water Company 2004).

The phytoplankton community changes from spring (February–June) to summer (July–September) to autumn (October–November). Long-term investigations of phytoplankton along the west–east nutrient gradient (net samples with mesh size > 25 μm) showed that typical spring species were the dinoflagellate *Peredinella catenata* and the diatoms *Thallassiosira baltica*, *Pauliella taeniata* (syn. *Achnanthes taeniata*) and *Chaetoceros wighamii*. Freshwater diatoms such as *Diatoma tenuis*, *Asterionella formosa* and *Tabellaria fenestrata* occurred in the innermost areas, while *Skeletonema costatum* was common in the saltier outer archipelago (Stockholm Water Company 2000, 2004).

Cyanobacteria characterise the summer phytoplankton community (see Chap. 15 by Kautsky, this volume). During the 1980s the species composition of Cyanobacteria shifted, from a dominance of *Planktothrix agardhii* to a mixture of *Woronichinia compacta*, *Microcystis* spp and *Aphanizomenon* cf. *baltica* (see Sect. 14.4). A quantitative study (where samples included plankton < 25 μm) in 2003 showed that cryptomonads constituted up to half of the total biomass in summer. In summers when there were heavy blooms of Cyanobacteria in the open Baltic Sea, *Nodularia spumigena* was common in the outer Stockholm archipelago (Stockholm Water Company 2000, 2004; Lindahl et al. 1980).

Cyanobacteria are still abundant in the Stockholm archipelago in autumn, together with diatoms such as *Thallassiosira baltica* and *Chaetoceros wighamii*, which occur all year round (Stockholm Water Company 2000, 2004).

Investigations of the zooplankton communities in 1994–1998 (Stockholm Water Company 2000) indicate which species are likely to occur in different seasons and areas of the Stockholm archipelago. In spring, ciliates (i.e. *Lohmaniella stellaris, Tinntinnopsis tubulosa, Mesodinium rubrum*), rotifers (i.e. *Synchaeta* spp.) and early developmental stages of copepods (nauplii and copepodites) occur in the whole archipelago.

In summer and autumn the species composition of the zooplankton shifts to other species of ciliates (i.e. *Tinntinnopsis brandti*, *Vorticella* spp), rotifers (i.e. *Asplanchna* spp, *Keratella cohlearis*, *K. quadrata*) and cladocerans (*Bosmina coregoni*, *Daphnia longispina*, *Podon polyphemoides*). All developmental stages of calanoid and cyclopoid copepods, including adults, were present. In the outer archipelago some additional species were found during summer and autumn, e.g. the ciliates *Leprotintinnus bothnicus* and *Helicostomella subulata* and the cladoceran *Evadne nordmanni* (Stockholm Water Company 2000).

There are only a few measurements of pelagic primary production in the Stockholm archipelago, and no measurements of secondary production. During the period June to August in 1973 and 1975, daily carbon assimilation varied between 400 and 900 mg C m^{-2} day^{-1} in the outermost parts of the archipelago (Brattberg 1977; Lindahl et al. 1980).

14.2.3 Fish

About 50 fish species live in the Stockholm archipelago (Table 14.2) but there are few quantitative investigations of their abundance and geographical distribution (Olburs 2002). Herring (*Clupea harengus*) is the dominant species in the marine

Table 14.2 Fish species characteristic of the inner, central and outer areas of the Stockholm archipelago

Fish species		Inner archipelago	Central archipelago	Outer archipelago
Eel	*Anguilla anguilla*	x		
Herring	*Clupea harengus*		x	x
Sprat	*Sprattus sprattus*		x	x
Silver bream	*Abramis bjoerkna*	x	x	
Bream	*Abramis brama*	x	x	
Vimba	*Abramis vimba*	x		
Bleak	*Alburnus alburnus*	x	x	
Ide	*Leuciscus idus*	x	x	
Dace	*Leuciscus leuciscus*	x		
Minnow	*Phoxinus phoxinus*	x		
Rudd	*Rutilus erythrophthalmus*	x	x	
Roach	*Rutilus rutilus*	x	x	
Tench	*Tinca tinca*	x		
Smelt	*Osmerus eperlanus*	x	x	
Whitefish	*Coregonus* sp.		x	
Salmon	*Salmo salar*	x		
Brown trout	*Salmo trutta*	x	x	
Pike	*Esox lucius*	x	x	
Cod	*Gadus morhua*			x
Burbot	*Lota lota*	x		
Three-spined stickleback	*Gasterosteus aculeatus*			x
Sand goby	*Pomatoschistus minutus*		x	
Common goby	*Pomatoschistus microps*		x	
Two-spotted goby	*Gobiusculus flavescens*		x	
Black goby	*Gobius niger*		x	
Perch	*Perca fluviatilus*	x	x	
Ruffe	*Gymnocephalus cernuus*		x	
Pikeperch	*Sander lucioperca*	x	x	
Viviparous blenny	*Zoarces viviparus*		x	
Turbot	*Psetta maxima*			x
Flounder	*Platichthys flesus*		x	x

fishery in Stockholm County, with a catch of about 180t in 1999 (Andersson 2003). About 125t cod (*Gadus morhua*) was landed in the county in 1999, but these catches were mainly from the southern Baltic Sea. About 20t eel (*Anguilla anguilla*) was caught in Stockholm County in 1999.

Freshwater species, mainly cyprinids, dominate in the sheltered inner archipelago, where salinity is low and the water warm (Table 14.2). Salmonids migrate in towards the coast to spawn, and salmon (*Salmo salar*) can be seen leaping in Stockholm harbour in late autumn. Freshwater species also dominate in the central archipelago, but small marine fish, e.g. gobies, also occur (Kautsky et al. 2000; Andersson 2003). Marine species such as cod are found in the outer archipelago.

14.2.4 Seabirds

About 170,000–180,000 pairs of seabirds breed in the Stockholm archipelago (Kautsky et al. 2000), including about 50% of the Swedish populations of eider duck (*Somateria mollissima*), velvet scoter (*Melanitta fusca*) and arctic skua (*Stercorarius parasiticus*). About 25% of the Swedish populations of razor bill (*Alca torda*), goosander (*Mergus merganser*) and mute swan (*Cygnus olor*) live in the archipelago.

Three areas are especially important for breeding seabirds: Gräsö archipelago in the north, the outer Stockholm archipelago, and Sandemar in the south (Skov et al. 2000).

The Gräsö archipelago is important for gulls (Laridae), the great cormorant (*Phalacrocorax carbo sinensis*), the white-tailed sea eagle (*Haliaeetus albicilla*), the Caspian tern (*Sterna caspia*) and the common tern (*Sterna hirundo*).

The outer archipelago is an important breeding area for eider duck, velvet scoter, red-breasted merganser (*Mergus serrator*), goosander, redshank (*Tringa totanus*), turnstone (*Arenaria interpres*), arctic skua, common gull (*Larus canus*), lesser black-backed gull (*Larus fuscus*), great black-backed gull (*Larus marinus*), Caspian tern, arctic tern (*Sterna paradisaea*), guillemot (*Uria aalge*), razor bill and black guillemot (*Cepphus grylle*).

Sandemar is an important stopover site and breeding area for species such as the white-tailed sea eagle, whooper swan (*Cygnus cygnus*) and tufted duck (*Aythya fuligula*) (Skov et al. 2000).

The Stockholm archipelago is also important as a wintering area for great cormorant; mute swan, mallard (*Anas platyrhynchos*), tufted duck, long-tailed duck (*Clangula hyemalis*), common goldeneye (*Bucephala clangula*), goosander, common coot (*Fulica atra*) and black guillemot (Durinck et al. 1994).

14.2.5 Marine Mammals

Numbers of grey seals (*Halichoerus grypus*) have increased from a low of about 300–400 animals in the 1970s and 1980s (Helander and Sjöåsen 1985) to more than 2,500 animals in the outer archipelago in 2002 (Karlsson and Helander 2003). The populations have increased by 7.6% yearly since 1990.

Until the 1950s, harbour porpoises (*Phocoena phocoena*) were common in the Stockholm archipelago. Today, the harbour porpoise is very rare (A. Roos, personal communication).

14.3 Benthic Communities

14.3.1 Bottom Substrate

According to a recent classification based on the European Nature Information System (EUNIS), most of the marine substrates in Stockholm County consist of soft bottoms (23% sublittoral sediments and 28% glacial clay), while rocky bottoms (11% infralittoral rock and 9% circalittoral rock) are less common (Mattisson 2005).

Accumulation bottoms, where fine material is continuously deposited, generally prevail in shallow, sheltered areas and at depths below 15 m. The mean rate of sediment accumulation is as high as 19 mm a^{-1} in the Stockholm archipelago, compared to 1–4 mm a^{-1} in offshore areas of the northern Baltic proper. The carbon content of the sediments is similar throughout the archipelago, and ranges from 4.5–6.5% in sediment layers from the 1950s to 5–8% in layers from the 1990s (Jonsson 2003). About 39% of the carbon in sediments in the innermost archipelago near Stockholm originates from the outflow of Lake Mälaren. The relatively low C:N ratio (8.5–11.2 mol ratio) in sediments along a transect 30 km eastwards from Stockholm indicates that the organic matter comes from plankton rather than from terrestrial sources (Jönsson et al. 2005).

14.3.2 Vegetation of Hard Bottoms

On rocky substrates in the outer archipelago, the macroalgal belt extends from the surface down to a depth of about 20 m. The maximum depth distribution of algae is shallower in the inner archipelago, where high turbidity limits light penetration. At the outflow of Lake Mälaren in central Stockholm, the algal zone extends only to about 2.5 m depth (Kautsky et al. 2000).

The species composition of the upper sublittoral zone (0.5–1 m) varies seasonally. After cold winters, when ice has ripped off the uppermost zone of the previous year's algae and water levels are high, benthic diatoms colonise the bare cliffs and rocks. Later in spring, annual filamentous algae such as *Pylaiella littoralis* and *Cladophora glomerata* take over. In summer, *C. glomerata* usually dominates in the upper sublittoral zone. In late fall, the red alga *Ceramium* becomes abundant (Kautsky et al. 2000).

In the inner archipelago, near the outflow of Lake Mälaren, the freshwater alga *Cladophora aegagropila* grows in patches below the belt of *C. glomerata* (Wærn 1948; Kautsky 1995). In more saline waters, the marine brown alga *Fucus vesiculosus* (bladderwrack) usually dominates below the belt of *C. glomerata*. Regional differences in the light regime are reflected in the depth distribution of bladderwrack. In the outer archipelago where light penetrates deeper, bladderwrack grows down to a maximum depth of 8.7 m, whereas it grows closer to the water surface (1–2 m depth) in the murky inner archipelago (Kautsky 1995).

Marine red algae take over below the bladderwrack belt. In the outer archipelago east of Stockholm, the red algal belt consists mainly of *Furcellaria lumbricalis*, *Coccotylus truncatus*, *Rhodomela confervoides* and *Polysiphonia fucoides* (Kautsky et al. 2000). In the northernmost parts of the archipelago, about 100 km north of Stockholm, the brown alga *Sphacelaria arctica* occurs as undergrowth in the red algal belt and becomes totally dominant from 10 to 20 m depth (Eriksson et al. 1998).

14.3.3 Macrozoobenthos of Hard Bottoms

Fucus vesiculosus is considered to be a key species in the Baltic food web, as it forms a species-rich biotope that supports about 30 species of macrofauna and epiphytes, and provides spawning, foraging and nursery areas for fish (Haage and Jansson 1970; Jansson 1972; Kautsky et al. 1992, 2000).

On shallow hard bottoms in the Stockholm archipelago, the species number is highest in sheltered bays, and includes many freshwater species, such as the gastropods *Theodoxus fluviatilis* and *Lymnaea peregra* and insect larvae (e.g. Trichoptera) (Kautsky et al. 2000). Common brackish-water and marine species include isopods (e.g. *Idothea baltica*), amphipods (e.g. *Gammarus* spp.), mysids, gastropods (e.g. *Hydrobia* spp.), and bivalves (e.g. *Cerastoderma edule* and *Mytilus edulis*). The marine barnacle *Balanus improvisus* is common in exposed areas.

The blue mussel *Mytilus edulis* forms dense mussel beds on rocky bottoms that are not covered by macroalgae. Blue mussels are an important link between the pelagic and benthic zones. Mussels filter phytoplankton from the water, and produce nutrients and faeces that are assimilated by phytoplankton, algae and macrobenthos (Kautsky and Evans 1987). Mussel larvae are eaten by small fish and herring, and adult mussels are eaten by eider ducks and flatfishes (Kautsky 1981).

In the warm summer of 1994, large colonies of the bryozoan *Electra crustulenta* were common on shallow rocky bottoms, especially in the exposed outer archipelago (Kautsky 1995). In the same period, mass mortality occurred in blue mussels in the Stockholm archipelago and in many other areas of the Baltic Proper.

14.3.4 Vegetation of Soft Bottoms

Aquatic phanerogams and charophytes of freshwater, brackish water or marine origin grow on shallow soft sediments in the archipelago. Species composition varies with depth and bottom characteristics. In exposed areas with coarse sediment, sparse stands of *Potamogeton perfoliatus* and *Ranunculus baudotti* occur from 2.5 to 8 m depth, while *Potamogeton filiformis* and *Chara aspera* dominate in shallower waters (Dahlgren 1997). Eelgrass (*Zostera marina*) occurs sparsely on sandy bottoms in the central and outer parts of the southern archipelago (Kautsky 1995).

The phanerogams *Ceratophyllum demersum*, *Myriophyllum spicatum*, *Najas marina* and *Potamogeton pectinatus* grow in sheltered bays with thick soft sediment. Charophytes such as *Chara aspera* and *C. tomentosa* (Fig. 14.3) are also characteristic of these biotopes, especially in the northernmost archipelago (Schreiber 2003; Wallström et al. 2000). *Chara horrida*, which is characterised as "Vulnerable" on the Swedish Red List of threatened species (Gärdenfors 2005), occurs at two localities in the outer archipelago (Dahlgren 1997; Schreiber 2003).

Ongoing land uplift results in a gradual change in species composition and plant density in shallow coastal waters. The vegetation is usually most dense during the

Fig. 14.3 The stonewort *Chara tomentosa*, a charophyte typical of undisturbed shallow waters in the Stockholm archipelago. Drawing by Johanna Fredenberg

late stages of succession, when sheltered bays are nearly cut off from the sea (e.g. flads). When the nutrient load from the catchment area is low, these bays develop into spectacular water bodies with clear water and luxuriant vegetation (Schreiber 2003; Wallström et al. 2000).

Shallow soft bottoms with dense vegetation generally house a rich invertebrate fauna, and provide spawning and nursery grounds, feeding areas and shelter for fish. The shallow coastal zone of the archipelago is of particular importance for the recruitment of freshwater fish such as perch, pike and cyprinids (Karås 1999).

14.3.5 Macrozoobenthos of Soft Bottoms

On shallow soft bottoms (< 10–15 m depth), typical marine species are bivalves (e.g. *Cerastoderma* sp. and *Mya arenaria*), the brown shrimp *Crangon crangon* and the amphipod *Corophium volutator*. Freshwater species include insect larvae (e.g. *Chironomus* spp.) and gastropods (e.g. *Bithynia* sp.) (Kautsky et al. 2000).

Deep muddy bottoms are inhabited by only a few species of macrobenthic animals, e.g. the marine bivalve *Macoma balthica*, the deposit-feeding amphipods *Monoporeia affinis* (freshwater origin) and *Pontoporeia femorata* (marine origin), the marine polychaete *Harmothoe sarsi*, the predatory isopod *Saduria entomon* and the marine priapulid *Halicryptus spinulosus* (Kautsky et al. 2000). A recent invader is the North American marine polychaete *Marenzelleria viridis*. Deep sandy bottoms are uncommon in the archipelago, but provide important spawning grounds for flatfish, e.g. turbot.

14.4 Effects of Eutrophication and Other Anthropogenic Disturbances

The wide variety of biotopes in the Stockholm archipelago provides considerable natural variation in the prerequisites for pelagic and benthic communities. Human activities also have a major impact on the environmental status of the archipelago. The most obvious is the input of nutrients from Stockholm County and the Baltic Sea region. According to Lindahl et al. (1993) the Stockholm archipelago is the most eutrophic area along the entire Swedish coast.

The pelagic ecosystem near Stockholm has shown signs of eutrophication for almost 100 years (Johansson and Wallström 2001). Sewage discharges from Stockholm and a high nutrient load from Lake Mälaren caused extensive summer blooms of Cyanobacteria in the inner archipelago in the early 1900s.

In the late 1960s, nutrient concentrations were higher in the vicinity of Stockholm than in the surrounding coastal areas (Wærn and Pekkari 1973). In the summer of 1970 (June–September) mean surface concentrations of nutrients along a transect 30 km eastwards from Stockholm were 56 μmol tot-N l^{-1} and 4.2 μmol tot-P l^{-1} (data from Wærn and Hübinette 1973). At that time, sewage treatment plants contributed about 70% and Lake Mälaren about 30% of the total phosphorus load to the archipelago (Brattberg 1986).

Better sewage treatment was introduced in 1968–1973 to remove phosphorus from wastewater, and summer concentrations of phosphorus in surface waters declined to about 1.3 μmol tot-P l^{-1} by the early 1980s (Brattberg 1986). The resulting transport of excess nitrogen to the outer archipelago was undesirable, as nitrogen is the most limiting nutrient in the Baltic Proper (Granéli et al. 1990).

Further measures to improve sewage treatment, e.g. efficient nitrogen reduction in the late 1990s, have resulted in a gradual decline in both nitrogen and phosphorus concentrations in surface waters. In 1995–2003 mean summer concentrations of nitrogen and phosphorus were 43 μmol tot-N l^{-1} and 0.7 μmol tot-P l^{-1} in the inner 30 km of the archipelago (data from Stockholm Water Company 2004).

Although nutrient conditions have improved, parts of the inner archipelago are still very eutrophic. According to Swedish water-quality criteria (Swedish Environmental Protection Agency 1999), nitrogen concentrations close to Stockholm are classified as "very high" and phosphorus concentrations have been "high" since the mid 1990s. In other areas, concentrations of both nutrients have dropped by one classification level since 1996, and now range from "high" in the inner archipelago to "very low" in the outer archipelago. One exception is Norrtäljeviken Bay in the north, where levels of nitrogen and phosphorus have increased steadily since the mid 1980s (Hagström and Pansar 2003).

The total biomass and species composition of phytoplankton have changed in response to changing nutrient conditions along the gradient eastwards from Stockholm. For example, following the removal of phosphorus from sewage in the 1970s, the summer phytoplankton biomass declined substantially in the inner and central archipelago (Brattberg 1986). The dominant species of the cyanobacterial blooms since the 1900s, *Planktothrix agardhii* (Oscillatoriaceae), was replaced by species of *Woronochinia* and *Microcystis* (Chroococcales) in the mid-1980s. Since 1995, different species of *Aphanizomenon* (Nostocales) have increased in abundance (Stockholm Water Company 2000, 2004).

In the outer archipelago, where there is a surplus of phosphorus and deficiency of mineralised nitrogen, nitrogen-fixing species of *Aphanizomenon* and *Nodularia* dominate summer blooms (Lindahl et al. 1980). Species of *Aphanizomenon* that occur in the inner areas near Stockholm probably do not fix nitrogen due to a surplus of nitrogen relative to phosphorus.

Unique studies of the depth distribution of *Fucus vesiculosus* during the 1900s illustrate the eutrophication process in the Stockholm archipelago (Kautsky et al. 1986; Eriksson 2002). In the northernmost, outer archipelago the depth distribution of bladderwrack decreased by 2.5 m from the 1940s to the 1980s, and the shallower distribution persisted into the 1990s (Fig. 14.4). The main cause is probably reduced water transparency (Kautsky et al. 1986). Other factors that relate to eutrophication may also contribute, e.g. overgrowth by epiphytic algae and a greater sediment load (Pekkari 1973; Eriksson 2002).

In the central archipelago, the maximum depth distribution of bladderwrack was about 1–1.5 m shallower in the 1990s than in the 1960s (Kautsky et al. 2000). The decline in the depth distribution was less severe in the inner archipelago, and a reversal of the trend from the late 1960s to 1990 indicated that conditions had improved (Lennmark and Strömberg 1990).

The benthic vegetation of soft, shallow bottoms is threatened mainly by poor light conditions, which may be caused by excessive nutrient inputs or mechanical

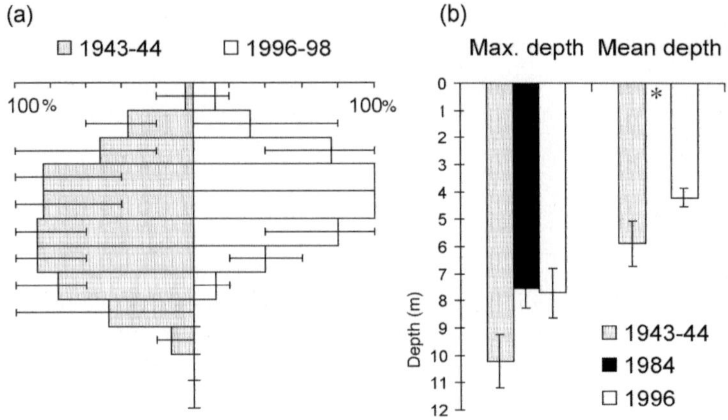

Fig. 14.4a,b Changes in the depth distribution of bladderwrack (*Fucus vesiculosus*) in the northern Stockholm archipelago from the 1940s to 1996. The depth distributions are based on divers' measurements along five transects in the Öregrund area. **a** Percentage cover of *F. vesiculosus* at different depths in 1943–1944 and 1996–1998 (based on Eriksson et al. 1998; Eriksson and Bergström 2005). *Bars* show mean, minimum and maximum cover along the transects. **b** Lower distribution limits of *F. vesiculosus* (maximum depth) and mean depths in 1943–1944, 1984 and 1996 (based on Kautsky et al. 1986; Eriksson et al. 1998). No data were available for mean depth in 1984 (*). *Error bars* One standard deviation of the mean. The depth penetration of *F. vesiculosus* decreased significantly both in terms of lower distribution limit and mean depth (*t*-test, n = 5 sites, $P < 0.05$). By courtesy of Britas Klemens Eriksson

disturbance of plants and sediment (e.g. due to dredging). Charophytes (stoneworts) can be used as an indicator of anthropogenic disturbance in shallow waters, as they are less common where light conditions are poor (Schubert and Blindow 2004).

Charophytes are more common in the northern parts of the Stockholm archipelago than in the southern parts (see Sect. 14.3), probably due to the impact of human activities that increase water turbidity in the southern archipelago. Eriksson et al. (2004) found that vegetation cover and species richness decreased and species dominance changed in inlets that were disturbed by small boat marinas or ferry traffic (Fig. 14.5). Plants such as *Myriophyllum spicatum* and *Ceratophyllum demersum* were characteristic of inlets with marinas, while *Chara* spp and *Ruppia* spp were more abundant in pristine inlets.

Such changes in the species composition of plants impair the recruitment of fish (e.g. pike, *Esox lucius*) that depend on benthic vegetation for spawning and feeding (Sandström et al. 2005).

Although environmental conditions have deteriorated in many shallow coastal areas, there are still some relatively pristine bays with clear water and dense benthic vegetation (Schreiber 2003). Such undisturbed shallow areas should be targeted for nature conservation, due to their high biodiversity and importance as recruitment areas for coastal fish.

Eutrophication has caused anoxic conditions on deep bottoms in many parts of the Stockholm archipelago. Healthy sediments should be well mixed due to bioturbation

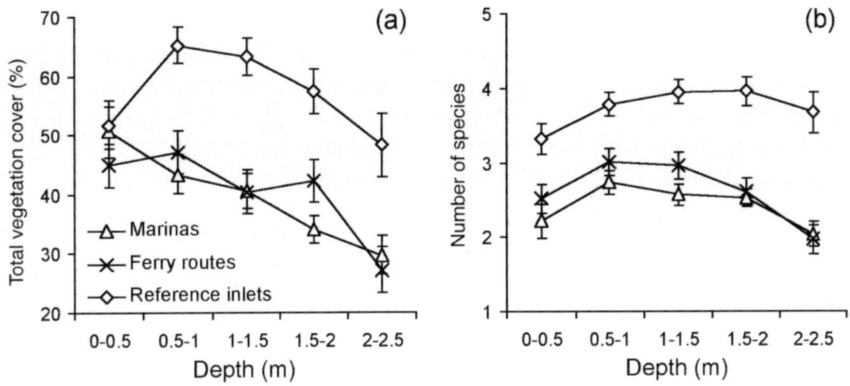

Fig. 14.5a,b The vegetation cover and species richness of aquatic macrophytes in shallow inlets in inner, central and outer parts of Stockholm archipelago. Some inlets had marinas for small boats, some inlets were close to ferry routes for medium-sized ferries, and reference inlets were similar in size and morphology to the disturbed inlets. Least square means are shown for **a** total vegetation cover and **b** species richness. Least square means were calculated for depth intervals of 0.5 m, using the exposure index WI_{mean} as a co-variable. *Error bars* 95% confidence intervals. Reprinted from Eriksson et al. (2004), with permission from Elsevier

by benthic fauna, but where anoxia has killed the macrofauna, sediments become laminated. Mats of the sulphur bacteria *Beggiatoa*, which indicate anoxia, covered extensive areas of soft-sediment below 20 m depth in the inner archipelago in 1991 (Rosenberg and Diaz 1993).

A study of 23 fjords in the Stockholm archipelago showed that the proportion of laminated sediments increased from 1910 until the 1990s, and then levelled off. The increase was greatest in the 1970s, even though better sewage treatment measures had been introduced. Laminated sediments occurred in the central archipelago in the 1950s, but did not appear in the outer archipelago until 1970–1980. This suggests that nutrient conditions in the offshore Baltic are the main cause of eutrophication in the outer archipelago (Jonsson 2003).

It is hard to distinguish the effects of eutrophication on fish communities from the effects of natural variation, overfishing and top predators such as seals. However, eutrophication may have contributed to greater catches of pikeperch (*Sander lucioperca*) and smaller catches of whitefish (*Coregonus* sp.) in the Stockholm archipelago (Hansson and Rudstam 1990).

14.5 Environmental Pollutants

Sediments of the Stockholm archipelago are heavily contaminated by metals and organic pollutants (Meili et al. 2000; Sundelin et al. 2003). Levels of contaminants are generally highest in sediments close to Stockholm and in the inner archipelago, which indicates that they come from local sources. Sediment cores from central

Stockholm show that levels of metals and polyaromatic hydrocarbons (PAHs) from industrial activities and coal firing were already high in the early 1900s. Concentrations of metals peaked in 1960–1980, while concentrations of polychlorinated biphenyl (PCB) increased until the 1970s. Levels of metals and hydrocarbons have declined in the last 10–20 years (Östlund et al. 1998).

On anoxic bottoms, heavy metals occur as insoluble metal sulphides. However, if the sediments become reoxygenated and are recolonised by macrofauna, there is a risk that amphipods such as *Monoporeia affinis* will accumulate lead (Pb) and mercury (Hg) (Sundelin et al. 2003).

Levels of heavy metals in macrophytes are generally highest in the inner archipelago. An exception is cadmium in bladderwrack, where concentrations were highest at Ängsskär and Söderarm in the outer archipelago (11 mg Cd kg^{-1} dw^{-1}) (Kautsky et al. 2000).

There are a number of hotspots in the Stockholm archipelago where levels of organic pollutants [e.g. PAHs, polybrominated diphenyl ethers (PBDEs), tributyltin (TBT), PCB and dichloro-diphenyl-trichloroethane (DDT)] and heavy metals are particularly high in sediments. Hotspots include harbours and boat marinas, the Östra Askrikefjärden fjord in the inner archipelago, a fjord near a military base in the southern archipelago and Södra Kanholmsfjärd fjord near a deepwater dumping site in the outer archipelago (Kautsky et al. 2000; Berglind et al. 2003; Sternbeck et al. 2003).

In the outer archipelago, the main source of PCB and PBDEs in sediments is long-range transport of contaminants from the open Baltic Sea (Meili et al. 2000; Sternbeck et al. 2003). PCB and DDT from the open sea may also end up in sediments of the inner archipelago (Jonsson et al. 2000).

The filtering activities of blue mussels (*Mytilus edulis*) facilitate the transport of organic contaminants through the food chain. Blue mussels from the Stockholm archipelago accumulated PCB, and sedimentation of PCB increased by 50% above mussel beds (Björk et al. 2000). Benthic animals that ingest sediment and diving ducks that feed on blue mussels may thus accumulate PCB and other organic contaminants.

Top marine predators have been severely affected by organic pollutants. Populations of white-tailed sea eagle (*Haliaeetus albicilla*) along the Swedish Baltic coast were previously threatened by hunting in combination with organochlorine pollutants such as DDT, which caused reproductive disorders such as eggshell abnormalities and embryo mortality (Helander et al. 2002).

Sea eagle reproduction declined from the 1950s onwards, and reached its lowest level in 1965–1985 (Helander op.cit.). The populations began to recover in the 1980s, largely thanks to a special feeding programme, and in 2002 there were 225 pairs along the Swedish Baltic coast (Helander 2003). There are nearly 50 pairs of sea eagle in the Stockholm archipelago today (B. Helander, personal communication).

Hunting and environmental pollutants have also been a serious threat to Baltic Sea seals. High levels of DDT and PCB caused a complex of pathological symptoms and reproductive disorders. The symptoms were most severe in the 1970s, and included uterine lesions and tumours, uterine occlusion, skeletal deformities and erosion, skin and claw lesions, colonic ulcers and damage to the kidneys and adrenal gland (Bergman and Bäcklin 1999; Ambio 1992; Olsson et al. 1992).

Levels of PCB and DDT have decreased in the Baltic Sea, and the health of seals has now improved. The prevalence of uterine obstructions has decreased and the proportion of pregnant female grey seals increased from 9% in the 1970s to 60% in the late 1990s. However, colonic ulcers are very common. Researchers suspect that new pollutants such as organobromines (e.g. PBDE) affect the seals' immune systems (Bergman and Bäcklin 1999).

Harbour porpoises (*Phocoena phocoena*) used to be common along the Swedish coast of the Baltic Sea, including the Stockholm archipelago. Populations have crashed since the Second World War, due to hunting in combination with bycatches in fishing gear (Berggren 1995).

The Baltic population of harbour porpoise is estimated to have declined 100- to 400-fold from the 1950s to about 600 animals today (excluding Polish waters) and is now classified as "Critically Endangered" on the Swedish Red List of threatened species (Gärdenfors 2005). Bycatches are the major threat, but organochlorine pollutants may also threaten the health and reproduction of porpoises (Berggren et al. 1999).

14.6 Physical Exploitation and Erosion of Shorelines

Many shorelines and shallow bottoms in the Stockholm archipelago have been cut off and fragmented by coastal development. Aerial photographs of the density of jetties, harbours, buildings and urban centres show that the inner and central parts of the Stockholm archipelago are extensively exploited. More than 30% of the shoreline is very heavily exploited in the inner archipelago, and about 50% of the coast is exploited to some degree in the central archipelago. The outer archipelago is still relatively unexploited (Mattisson 2004).

Boat traffic produces waves that erode shorelines and resuspend sediments, with negative impacts on plant and animal communities. Shoreline erosion was previously associated with the large ferries that run between Stockholm and Finland (Cederwall 2000; Kautsky et al. 2000). Today, a growing number of large, heavy recreational boats produce particularly high wave energies that are eroding shorelines at alarming rates along some shipping lanes (Granath 2004).

Resuspension of sediments by boat traffic also increases the turbidity of coastal waters, particularly along shipping lanes. Sedimentation of particles and organic material limits the depth distribution of algae by blocking light and preventing plants from colonising hard bottoms, particularly along shipping lanes, where stands of bladderwrack have become sparser (Rönnberg and Östman 2000).

14.7 The Spread of Alien Species

A number of alien (non-indigenous) species have colonised the Stockholm archipelago, including freshwater invertebrates (zebra mussel *Dreissena polymorpha*), marine invertebrates (mysid shrimp *Hemimysis anomala*), birds (Canada goose *Branta canadensis*) and mammals (American mink *Mustela vison*). American mink preys on eggs and fledglings, and poses a serious threat to seabirds in the northern and outer Stockholm archipelago (Skov et al. 2000). Some new invaders are feared to have wide-ranging ecological effects, e.g. the pelagic cladoceran (water flea) *Cercopagis pengoi* and the benthic polychaete *Marenzelleria neglecta* [see Chaps. 12 (Telesh et al.) and 15 (Kautsky), this volume].

14.8 Recruitment Failure in Coastal Fish

Several coastal fish species, e.g. herring (*Clupea harengus*), pike (*Esox lucius*), perch (*Perca fluviatilus*), and whitefish (*Coregonus* sp.) have declined in abundance in the Stockholm archipelago. Several factors may be responsible: eutrophication of shallow bays, which are important spawning and nursery areas for fish; physical exploitation of nursery areas (due to jetties, boat traffic, dredging etc.); and overfishing (Andersson 2003).

During the last 10 years, recruitment has failed in several species of freshwater fish along the Swedish coast of the Baltic Proper, mainly pike and perch, but also roach (*Rutilus rutilus*), bream (*Abramis brama*), silver bream (*Abramis bjoerkna*), rudd (*Rutilus erythrophthalmus*), and tench (*Tinca tinca*). The outer areas of the Stockholm archipelago are severely affected, and there is little or no recruitment of pike and perch (Ljunggren et al. 2005). Recruitment failure is especially severe in shallow bays with relatively high water exchange (juvenile flads), compared to semi-enclosed bays (glo-flads) where abundances of perch fry are higher.

Recruitment failure in pike and perch is correlated to a low abundance of zooplankton, which is an important food for juvenile fish. Bays that lacked pike and perch had high abundances of three-spined sticklebacks (*Gasterosteus aculeatus*). Sticklebacks may be better adapted to low food availability, or may be able to exploit alternative food sources (Ljunggren et al. 2005).

14.9 Abnormal Mortality in Seabirds

Abnormal mortality has been observed in seabirds along the Swedish coast of the Baltic Proper, including the Stockholm archipelago, since 2000, but the cause is still unknown. The symptoms (paralysis, desiccation, necrosis and inflammation of the kidneys, necrosis in the intestines, and brain damage) affect mainly the herring gull (*Larus argentatus*), but also other species of gulls, geese, swans and ducks.

14.10 Nature Conservation

Sweden is now applying a holistic approach to the management and protection of river basins and coastal waters, as part of the European Union's Water Framework Directive. The Stockholm archipelago belongs to the new Swedish water district "Northern Baltic Sea".

One of the main challenges is to introduce cost-effective measures to reduce eutrophication and its negative effects on coastal ecosystems. It will be a difficult task to limit the diffuse runoff of nutrients from agricultural land, and prevent the release of nutrients that are already stored in coastal sediments. We also need to survey coastal marine biotopes and monitor plants and animals in the phytobenthic zone.

There are as yet no marine reserves in the Stockholm archipelago. Some marine habitats are protected within terrestrial nature reserves, Natura 2000 areas, or bird and seal reserves. The Helsinki Commission lists three areas as potential Baltic Sea Protected Areas: Gräsö/Singö archipelago in the north; Storö/Bockö/Stora Nassa/Svenska Högarna/Svenska Björn in the outer archipelago; and Landsort/Askö/Hartsö/Landsort Deep in the south (HELCOM 1996).

In 2004, a working group that includes professional fishermen, sport fishermen, coastal water owners, regional authorities and fisheries scientists established 17 no-take zones in the Stockholm archipelago, to protect spawning and nursery areas for fish. All fishing is prohibited in these zones from April to mid-June, and the effects of the closure are now being evaluated.

14.11 Summary and Conclusions

Stockholm archipelago is the largest connected archipelago in Swedish waters. It extends 200 km along the Swedish east coast, just south of the border between the Bothnian Sea and the northern Baltic proper. The archipelago is widest off Stockholm, where it stretches about 100 km to the east. The inner parts of the archipelago are characterised by relatively large islands and shallow bays and fjords that are sheltered from the open sea. The central and outer parts are more exposed, and water exchange with the open sea increases successively towards the east. The shallow parts of the archipelago are clearly affected by land uplift of about 5 mm per year.

A varied topography and a wide variety of biotopes including open waters, rocky coasts, long, narrow fjord-like bays and sheltered inlets give prerequisites for a rich flora and fauna in the archipelago. The species composition of plant and animal communities depends greatly on the salinity, which varies from freshwater in the inner areas to 5–7 psu in the outer archipelago.

The inner archipelago is strongly influenced by runoff from streams and catchments, in particular the outflow from Lake Mälaren in central Stockholm. The runoff from land results in higher water turbidity and nutrient concentrations in the inner archipelago, compared to the outer areas. The nutrient load is greatest in the inner archipelago near Stockholm, where there are also several large sewage treatment plants.

The Stockholm archipelago is possibly the most eutrophic coastal area in Sweden. According to Swedish water-quality criteria, summer concentrations of nitrogen and phosphorus are classified as "high" or "very high" in the inner archipelago near Stockholm. In the outer archipelago, nutrient concentrations are classified as "low" or "very low". Nutrient concentrations are slightly lower in the northern parts of the archipelago, than in the south.

Excessive input of nutrients has resulted in anoxic conditions on extensive areas of deep soft bottoms in the archipelago. Anoxic sediments that lack macrofauna are laminated (varved) and are covered by sulphur bacteria, especially in the inner archipelago.

The sediments in the archipelago have been heavily contaminated by metals and organic pollutants since the early 1900s. Pollutant levels are particularly high in the inner archipelago, but there are also hotspots in the central and outer areas. Long-range transport of organic pollutants from the open Baltic Sea contaminates sediments in the outer archipelago.

The abundance of phytoplankton reflects the nutrient gradient from Stockholm eastwards to the outer archipelago. Summer phytoplankton abundance is characterised as "very high" in the inner archipelago and "medium high" in the outer archipelago. Eutrophication caused summer blooms of Cyanobacteria in the inner archipelago near Stockholm already in the early 1900s. Conditions in the inner archipelago have gradually improved since the 1980s, mainly as a result of better sewage treatment measures. The Cyanobacteria *Planktothrix agardhii* dominated the summer phytoplankton community for most of the twentieth century, but species diversity is higher today.

The depth distribution of bladderwrack, *Fucus vesiculosus*, depends on light conditions, which are influenced partly by natural variation in water transparency but also by eutrophication. The maximum depth distribution of bladderwrack is 8.7 m in the outer archipelago, but only 1–2 m in the murky inner archipelago. As a result of eutrophication, the lower limits of bladderwrack belts became shallower from the 1940s onwards, in both the central and outer archipelago. In the inner archipelago near Stockholm, the decline in depth distribution was less severe, and a reversal of the trend indicated that light conditions improved after the late 1960s.

In the low-lying parts of the archipelago, shallow flads and glo-flads with bottoms covered by dense vegetation are highly productive and particularly important as recruitment areas for freshwater fish. Environmental conditions have deteriorated in many of these shallow areas due to human activities, e.g. dredging and boating, that reduce light conditions and cause mechanical disturbance. Some undisturbed shallow areas with nearly pristine conditions still occur in the Stockholm archipelago, mainly in the northern parts. However, physical exploitation of shorelines and shallow bottoms is a growing problem in the archipelago.

The fish communities are characterised by a mixture of freshwater and marine species. Freshwater species, especially cyprinids, dominate in the inner and central archipelago, while marine species are found in the outer archipelago. The Stockholm archipelago is an important breeding and wintering area for many species of seabirds. Populations of top predators such as grey seals and white-tailed sea eagle are now recovering from reproductive disturbances caused by organic pollutants such as PCB and DDT. Populations of harbour porpoise have been decimated by hunting in combination with bycatches in fishing gear, and have still not recovered.

14.11.1 Final Remarks

The inner Stockholm archipelago suffered from severe eutrophication for most of the twentieth century, but conditions have gradually improved during the last 10–20 years. However, concentrations of nitrogen are still high, and further measures must be introduced to diminish the nutrient load. We still lack quantitative studies of many of the biotopes in the archipelago, including the phytobenthic zone, the zooplankton and the fish communities. There is an urgent need to protect vulnerable and pristine shallow areas that house threatened species and are important for spawning and feeding of fish. Recruitment failure of pike and perch in the outer archipelago is alarming and needs to be further investigated.

Acknowledgements We would like to thank the following persons for providing material and commenting on drafts of the chapter on the Stockholm archipelago: Gunnar Aneer, Gunilla Brattberg, Anders Engqvist, Britas Klemens Eriksson, Johanna Fredenberg, Michael Gilek, Björn Helander, Hans Kautsky, Nils Kautsky, Annelie Mattisson, Anna Roos, Sonja Råberg and Alfred Sandström.

References

Ambio (1992) Special issue: Seals and seal protection. Ambio 21(8)

Andersson HC (2003) Fish and fishermen in the county of Stockholm: status report for 2002 (in Swedish with English summary). Stockholm County Administrative Board, Report 2003:19

Aneer G, Arvidsson D (2003) Inputs of nutrients (nitrogen and phosphorus) to coastal waters of Svealand, northern Baltic Sea, in 1997 (in Swedish with English summary). Stockholm County Administrative Board, Report 2003:17

Berggren P (1995) Stocks, status and survival of harbour porpoises in Swedish waters. PhD thesis, Department of Zoology, Stockholm University. ISBN 91-87272-41-5

Berggren P, Ishaq R, Zebühr Y, Näf C, Bandh C, Broman D (1999) Patterns and levels of organochlorines (DDTs, PCBs, non-*ortho* PCBs and PCDD/Fs) in male harbour porpoises (*Phocoena phocoena*) from the Baltic Sea, the Kattegat-Skagerrak seas and the west coast of Norway. Mar Pollut Bull 38:1070–1084

Berglind R, Larsson R, Leffler P, Ragnvaldsson D, Sjöström J, Wingfors H (2003) Ecotoxicological investigation of exposure levels in fish and mussels and toxic effects of sediment from the aquatic vicinity of the Muskö base. Swedish Defense Research Agency, Report FOI-R-0844-SE, ISBN 1650–1942

Bergman A, Bäcklin BM (1999) The health of grey seals – better but not good. In: Tidlund A, Brenner U (eds) (2000) Baltic '99. Swedish marine monitoring in the Baltic proper August 2000. Stockholm Marine Research Centre, Stockholm University, pp 10–14

Björk M, Gilek M, Kautsky N, Näf C (2000) *In situ* determination of PCB biodeposition by *Mytilus edulis* in a Baltic coastal ecosystem. Mar Ecol Prog Ser 194:193–201

Brattberg G (1977) Nitrogen fixation in a polluted brackish water archipelago. Ambio Special Report 5:27–42

Brattberg G (1986) Decreased phosphorus loading changes phytoplankton composition and biomass in the Stockholm archipelago. Vatten 42:141–152

Cederwall K (2000) The archipelago landscape – physical water environment, shipping and building in water. In: von Numers M (ed) The environment of archipelagos – present situation, problems and opportunities. Nordic Council of Ministers' collaborative project on archipelagos. Kirjapaino Grafia Oy, Åbo. ISBN 951-53-2099-2, pp 123–130

Dahlgren S (1997) Aquatic vegetation in shallow semi-enclosed bays in Stockholm County (in Swedish). County Administrative Board of Stockholm, Report U:33

Durinck J, Skov H, Jensen FP, Pihl S (1994) Important marine areas for wintering birds in the Baltic Sea. Report to the European Commission. EU DG IX research contract no. 22452/90-09-01. Ornis Consult Report 1994

Engqvist A, Andrejev O (2003) Water exchange of the Stockholm archipelago – a cascade framework modelling approach. J Sea Res 49:275–294

Eriksson BK (2002) Long-term changes in macroalgal vegetation on the Swedish coast. An evaluation of eutrophication effects with special emphasis on increased organic Sedimentation. Acta Universitatis Upsaliensis. Comprehensive Summaries of Uppsala Dissertations from the Faculty of Science and Technology No 746

Eriksson BK, Bergström L (2005) Local distribution patterns of macroalgae in relation to environmental variables in the northern Baltic Proper. Estuarine Coastal Shelf Sci 62:109–117

Eriksson BK, Johansson G, Snoeijs P (1998) Long-term changes in the sublittoral zonation of brown algae in the southern Bothnian Sea. Eur J Phycol 33:241–249

Eriksson BK, Sandström A, Isæus M, Schreiber H, Karås P (2004) Effects of boating activities on aquatic vegetation in the Stockholm archipelago, Baltic Sea. Estuarine Coastal Shelf Sci 61:339–349

Gärdenfors U (ed) (2005) The 2005 Red List of Swedish species. Swedish Species Information Centre, Swedish University of Agricultural Sciences, Uppsala. ISBN 91-88506-30-4

Gidhagen L (1987) Coastal upwelling in the Baltic Sea. Satellite and in situ measurements of sea-surface temperatures indicating coastal upwelling. Estuarine Coastal Shelf Sci 24:449–462

Granath L (2004) Boat traffic and erosion of shorelines in the Stockholm archipelago (in Swedish with English summary). County Administrative Board of Stockholm, Report 2004:19

Granéli E, Wallström K, Larsson U, Granéli W, Elmgren R (1990) Nutrient limitation of primary production in the Baltic Sea area. Ambio 19:142–151

Haage P, Jansson B-O (1970) Quantitative investigations of the Baltic *Fucus* belt macrofauna. 1. Quantitative methods. Ophelia 8:187–195

Hagström J, Pansar J (2003) Environmental status and trends in lakes, streams and coastal areas impacted by urban pollution in the County of Stockholm (in Swedish with English summary). Stockholm County Administrative Board, Report 2003:23

Hansson S, Rudstam LG (1990) Eutrophication and Baltic fish communities. Ambio 19:123–125

Helander B (2003) Sea eagles are found all along the Swedish coastline (in Swedish). In: Tidlund A (ed) Swedish report on the environmental state of the Baltic proper including English summaries. Stockholm Marine Research Centre, Stockholm University, pp 20–21

Helander B, Sjöåsen T (1985) Seal populations on the Swedish south and east coasts 1975–1984 (in Swedish). Seal information 1985:2. Seal Project, Swedish Museum of Natural History, Stockholm

Helander B, Olsson A, Bignert A, Asplund L, Litzén K (2002) The role of DDE, PCB, coplanar PCB and eggshell parameters for reproduction in the white-tailed sea eagle *Haliaeetus albicilla* in Sweden. Ambio 31:386–403

HELCOM (1996) Coastal and marine protected areas in the Baltic Sea region. Helsinki Commission. Baltic Sea Environment Proceedings No. 63

Jansson BO (1972) Ecosystem approach to the Baltic problem. Bulletins from the Ecological Research Committee/NFR. Swedish Natural Science Research Council 16:1–82, ISBN 91-546-0001-4

Johansson L, Wallström K (2001) Urban impact in the history of water quality in the Stockholm archipelago. Ambio 30:277–281

Jönsson A, Lindström M, Carman R, Mörth C-A, Meili M, Gustafsson Ö (2005) Evaluation of the Stockholm Archipelago sediments, northwestern Baltic Sea Proper, as a trap for freshwater runoff organic carbon. J Mar Syst 56:167–178

Jonsson P (ed) (2003) The seabed of the archipelago. A summary of sediment investigations carried out in 1992-1999 in archipelagos along the Swedish east coast (in Swedish with English summary). Swedish Environmental Protection Agency, Report 5212

Jonsson P, Eckhéll J, Larsson P (2000) PCB and DDT in laminated sediments from offshore and archipelago areas of the NW Baltic Sea. Ambio 29:268–276

Karås P (1999) Recruitment areas for stocks of perch, pike and pikeperch in the Baltic Sea (in Swedish with English summary). Swedish National Board of Fisheries, Report 6:31–65

Karlsson O, Helander B (2003) Grey seals – lack of ice gave a better count (in Swedish). In: Tidlund A (ed) Swedish report on the environmental state of the Baltic proper. Stockholm Marine Research Centre, Stockholm University, pp 22–23

Kautsky H (1995) Divers' survey of the vegetation of shallow bottoms in the Stockholm archipelago 1994 (in Swedish). Pilot study for regional environmental monitoring. Technical Report no. 18, Institute of Systems Ecology, Stockholm University, ISSN 1104–8298

Kautsky H, Kautsky L, Kautsky N, Kautsky U, Lindblad C (1992) Studies on the *Fucus vesiculosus* community in the Baltic sea. Acta Phytogeographica Suecica 78:33–48

Kautsky L, Norberg Y, Aneer G, Engqvist A (2000) Under the surface of the Stockholm Archipelago (in Swedish). County Administrative Board of Stockholm. Published by Lenanders Tryckeri, Stockholm

Kautsky N (1981) On the trophic role of the blue mussel (*Mytilus edulis* L.) in a Baltic coastal ecosystem and the fate of the organic matter produced by the mussels. Kieler Meeresforsch Sonderh 5:454–461

Kautsky N, Evans S (1987) Role of biodeposition by *Mytilus edulis* in the circulation of matter and nutrients in a Baltic coastal ecosystem. Mar Ecol Prog Ser 38:201–212

Kautsky N, Kautsky H, Kautsky U, Wærn M (1986) Decreased depth penetration of *Fucus vesiculosus* (L.) since the 1940s indicates eutrophication of the Baltic Sea. Mar Ecol Prog Ser 28:1–8

Lennmark I, Strömberg I (1990) Bladderwrack in the Stockholm archipelago: a photographic survey (in Swedish). Study commissioned by the Stockholm County Administrative Board, Stockholm County Council and Stockholm Water Company

Lindahl AH, Persson G, Olsson H (1993) Eutrophication of Swedish coastal areas and surrounding seas: status, trends, causes and effects (in Swedish). Swedish Environmental Protection Agency, Report 4151

Lindahl G, Wallström K, Brattberg G (1980) Short-term variations in nitrogen fixation in a coastal area of the Northern Baltic. Arch Hydrobiol 89:88–100

Ljunggren L, Sandström A, Johansson G, Sundblad G, Karås P (2005) Recruitment failure in coastal fish populations in the Baltic Sea (in Swedish with English summary). Swedish National Board of Fisheries, Report Finfo 2005:05

Mattisson A (2004) Physical exploitation of coastal areas in the County of Stockholm – an analysis of the shorelines of Lake Mälaren and the Baltic Sea (in Swedish with English summary). County Administrative Board of Stockholm, Report 2004:05

Mattisson A (2005) Mapping marine habitats. Pilot study for the coastal areas of the Stockholm County. Stockholm County Administrative Board, Report 2005:21

Meili M, Jonsson P, Carman R (2000) PCB levels in laminated coastal sediments of the Baltic Sea along gradients of eutrophication revealed by stable isotopes ($\delta^{15}N$, $\delta^{13}C$). Ambio 29:282–287

Munsterhjelm R (1997) The aquatic macrophyte vegetation of flads and gloes, S coast of Finland. Acta Bot Fennica 157:1–68

Olburs C (2002) Fish, fisheries and aquaculture in the Svealand archipelago (in Swedish). Management and environmental objectives. Uppsala County Administrative Board, Report 2002:11

Olsson M, Karlsson B, Ahnland E (1992) Seals and seal protection: a presentation of a Swedish research project. Ambio 21:494–496

Östlund P, Sternbeck J, Brorström-Lundén E (1998) Metals, PAH, PCB and total hydrocarbons in sediments around Stockholm – flows and concentrations (in Swedish). Swedish Environmental Research Institute, IVL Report B1297, ISBN 91-630-6738-2

Pekkari S (1973) Effects of sewage water on benthic vegetation. Nutrients and their influence on the algae in the Stockholm Archipelago during 1970. Oikos 15[Suppl]:185–188

Rönnberg O, Östman M (2000) Effects of waves from boat traffic in archipelagos. In: von Numers M (ed) The environment of archipelagos – present situation, problems and opportunities. Nordic Council of Ministers' collaborative project on archipelagos. Kirjapaino Grafia, Åbo. ISBN 951-53-2099-2, pp 111–121

Rosenberg R, Diaz RJ (1993) Sulfur bacteria (*Beggiatoa* spp.) mats indicate hypoxic conditions in the inner Stockholm archipelago. Ambio 22:32–36

Sandström A, Klemens Eriksson B, Karås P, Isæus M, Schreiber H (2005) Boating and navigation activities influence the recruitment of fish in a Baltic Sea archipelago area. Ambio 34:125–130

Schreiber H (2003) Shallow areas with high conservation values in the archipelagos of Svealand (in Swedish). Stockholm County Administrative Board, Report 2003:5

Schubert H, Blindow I (eds) (2004) Charophytes of the Baltic Sea. Baltic Marine Biologists Publ No 19. Gantner, Ruggell

Skov H, Vaitkus G, Flensted KN, Grishanov G, Kalamees A, Kondratyev A, Leivo M, Luigujõe L, Mayr C, Rasmussen JF, Raudonikis L, Scheller W, Sidlo PO, Stipneice A, Struwe-Juhl B, Welander B (2000) Inventory of coastal and marine important bird areas in the Baltic Sea. BirdLife International, Cambridge. ISBN 9986-756-97-9

SMHI, Swedish Meteorological and Hydrological Institute (2003) Depth data for sea areas 2003 (in Swedish). SMHI Oceanography No 73

SMHI, Swedish Meteorological and Hydrological Institute (2005) Climate data available at http://www.smhi.se

Sternbeck J, Brorström-Lundén E, Remberger M, Kaj L, Palm A, Junedahl E, Cato I (2003) WFD Priority substances in sediments from Stockholm and the Svealand coastal region. IVL Swedish Environmental Research Institute, Report B1538

Stockholm Water Company (2000) Results of the control programme for discharges of treated wastewater to the Stockholm Archipelago 1999 (in Swedish). Stockholm Vatten, Report MV-00115, 2000-03-27

Stockholm Water Company (2004) Results of the control programme for discharges of treated wastewater to the Stockholm Archipelago 2003 (in Swedish). Stockholm Vatten, Report MV-04168, 2004-03-25

Sundelin B, Eriksson Wiklund AK (2003) Contaminated sediments – what happens if they become oxygenated? (in Swedish with English summary) In: Tidlund A (ed) Swedish report on the environmental state of the Baltic proper. Stockholm Marine Research Centre, Stockholm University, pp 10–15

Swedish Environmental Protection Agency (1999) Environmental quality criteria – coasts and seas (in Swedish with English summary). Report 4914

Wærn M (1948) Algal vegetation of the rocky shores of the province of Uppland (in Swedish). In: Hörstadius S, Curry-Lindahl K (eds) Natural history of the province of Uppland. Svensk Natur, Stockholm

Wærn M, Hübinette L (1973) Phosphate, nitrate and ammonium in the archipelago during 1970. Nutrients and their influence on the algae in the Stockholm Archipelago during 1970. Oikos 15[Suppl]:164–170

Wærn M, Pekkari S (1973) Outflow studies. Nutrients and their influence on the algae in the Stockholm Archipelago during 1970. Oikos 15[Suppl]:155–163

Wallström K, Mattila J, Sandberg-Kilpi E, Appelgren K, Henricson C, Liljekvist J, Munsterhjelm R, Odelström T, Ojala P, Persson J, Schreiber H (2000) Environmental status of shallow coastal bays. Description of shallow coastal bays in the counties of Uppsala and Stockholm, the Åland islands and the Västnyland archipelago in Finland. Development of methods for studies of benthic plant and animal communities (in Swedish). The Uppland Foundation, Stencil no 18

Chapter 15
Askö Area and Himmerfjärden

H. Kautsky

15.1 Askö Area

This chapter is focused on the different types of benthic, plant and animal communities found in the photic zone of the Askö area. The description of these phytobenthic communities is also relevant for the entire Swedish coastline of the Baltic Proper. Only a few species disappear from the southern coast up to the Askö area and a few new species of freshwater origin occur at the northern border of the Stockholm Archipelago. For the Askö area, a comparison is made with the 1970s, and trends over the last 10 years are described. Some original data are presented. The deeper benthic and pelagic systems are more briefly described. Only a fraction of the results of the extensive research carried out in the Askö area since the 1960s can be presented in this review.

In the Himmerfjärd area (see Sect. 15.2), the emphasis is more on the pelagic and deeper benthic communities, but findings there are usually relevant also for the Askö area.

15.1.1 Environmental Characteristics

Starting already in the County of Blekinge archipelago (latitude 56°), the Swedish coast is characterised by its abundance of islands and skerries forming archipelagos. The archipelagos continue northward, via the Askö area to the Archipelago of Stockholm (lat. 60° 30′), where they also extend eastward over the Åland Sea to the Archipelago Sea of the Finnish coast. North of the Island of Öland (lat. 57° 30′) the coastline is steeper and hard substrates usually dominate. These relatively steep shorelines, and the abundance of islands, give rise to a characteristic more sheltered, inner archipelago, an intermediate part and an outer, wave exposed part.

The main characteristics of the Askö and Himmerfjärden area are given in Table 15.1. The Askö area is characterised by a relatively open archipelago with good water exchange and a small catchment area. The catchment area of 95 km^2 land is dominated by forests (http://www.mare.su.se) (Fig. 15.1). The annual

U. Schiewer (ed.), *Ecology of Baltic Coastal Waters. Ecological Studies 197.*

Table 15.1 Main characteristics of the greater Stockholm region: Askö and Himmerfjärden area. *dw* Dry weight, *ww* wet weight, *Chl a* chlorophyll *a*, *DIN* dissolved inorganic nitrogen, *DIP* dissolved inorganic phosphorus

	Askö area and Himmerfjärden
a. Climate and hydrology	
Air temperature (°C)	–3,0 to +17.2 (monthly mean)[a]
Water temperature (°C)	+0.5 to 16.1 (monthly mean)[a]
Ice covering (days a^{-1})	30–80[b]
Annual global radiation (kW m^{-1})	975 (northern parts); 1,000 (southern parts)[a]
Average precipitation (mm a^{-1})	540[a]
Mean freshwater inflow (10^6 m^3 a^{-1})	No data
Mean inflow (10^6 m^3 a^{-1})	No data
Mean total outflow (10^6 m^3 a^{-1})	No data
Residence time (days)	No data
b. Morphometry and sediments	
Area (km^2)	365[c]
Volume (10^6 m^3)	No data
Mean depth (m)	No data
Maximum depth (m)	No data
Catchment area (km^2)	1,381[c]
Area/catchment relation	0.26
Organic matter content (% dw)	No data
C/N-ratio (0–10 cm depth)	No data
Phosphate release (mg P m^{-2} day^{-1})	12–120[c]
Denitrification rate (mg N m^{-2} day^{-1})	0–23[c]
Ammonia release (mg N m^{-2} day^{-1})	10–110[c]
Sediment distribution	No data
c. Pelagic biological components	
Secchi depth (m)	No data
Bacteria (10^6 ml^{-1})	No data
Phytoplankton biomass (μg Chl *a* l^{-1})	No data
Dominating phytoplankton species	No data
Flagellates	
Dominating species	No data
Ciliates	
Dominating species	No data
Mesozooplankton	
Dominating species	No data
Fish	
Number of species	No data
Dominating species	No data
Fish catches (t a^{-1})	No data
d. Benthic biological components	
Macrophytes	
Dominating species	Rocky substrates: *Ceranium tenuicorne, Furcellaria lumbricalis, Pilayella/Ectocarpus, Cladophora glomerata, Phyllophora* sp., *Fucus versiculusus, Rhodomela confercvoides*

(continued)

Table 15.1 (continued)

	Askö area and Himmerfjärden
	Soft bottoms: *Patomageton pectinatus, P. perfoliatus, Zannichellia palustris, Ruppia* spp., *Myriophyllum* sp.[d]
Macrozoobenthos	
Biomass (g m^{-2})	<600 (soft substrate, ww); <2,000 (hard substrate, ww)
Dominating species	Rocky substrates: *Mytilis trossulus, Macoma balthica, Idotea baltica, Gammarus* spp., *Theodoxus fluviatilis, Balanus improvisus, Electra crustelenta*
	Soft bottoms: *Macoma balthica, Monoporeia affinis, Harmothoe sarsi*[e]
e. Water chemistry, tropic status and population	
Salinity (psu)	0.12–6.8[e]
pH	ca. 8
Total nitrogen (µmol l^{-1})	17
DIN phosphorus (µmol l^{-1})	5
Total phosphorus (µmol l^{-1})	0.5
DIP (µmol l^{-1})	0.6
Annual nitrogen input (t a^{-1})	500.6–969.9[f]
Annual phosphorus input (t a^{-1})	19–28.9[f]
Limitation of PP and period	No data
Main pollutants	No data
Trophic level	No data

[a]SMHI (2005); [b]Cederwall (200); [c]Elmengren and Larsson (1997); [d]Engqvist (personal communication); [e]see Chap. 14 by Hill and Wallström, this volume, 2.4.2 for references; [f]Aneer and Arvidsson (2003)

temperature in the water is over 20 °C in mid-July and August and can be −0.5 °C in late January and February. Ice cover usually occurs from mid-January until March, at least in the inner parts of the archipelago and close to shore. The annual freshwater inflow to the area is 0.7 m^3 s^{-1} by land runoff and 2.2 m^3 s^{-1} as direct precipitation over the sea (sea area of estimate is 232 km^2). Annual precipitation over Sweden has increased constantly, from 550 mm in the year 1900 to 750 mm in 2000 (Fig. 15.2). The Askö area is a fairly open system with no sills and the water residence time is usually but one to a few days in the area. Currents are mainly wind driven. Over the last 10 years, the Secchi depth in spring, which had been relatively constant since the 1970s, has tended to increase, but summer values are more or less the same or slightly less (Fig. 15.3). A number of attached algae have increased their depth extension, which would indicate an overall increase in Secchi depth in the area. The oxygen contents of the Askö have not been reported as critically low. The total nutrient content of the area over the year is around 30 mg tot-P m^{-3} and 300 mg tot-N m^{-3}. Trends in the discharge rates from the largest sewage treatment plant are illustrated in Fig. 15.13, discussed below.

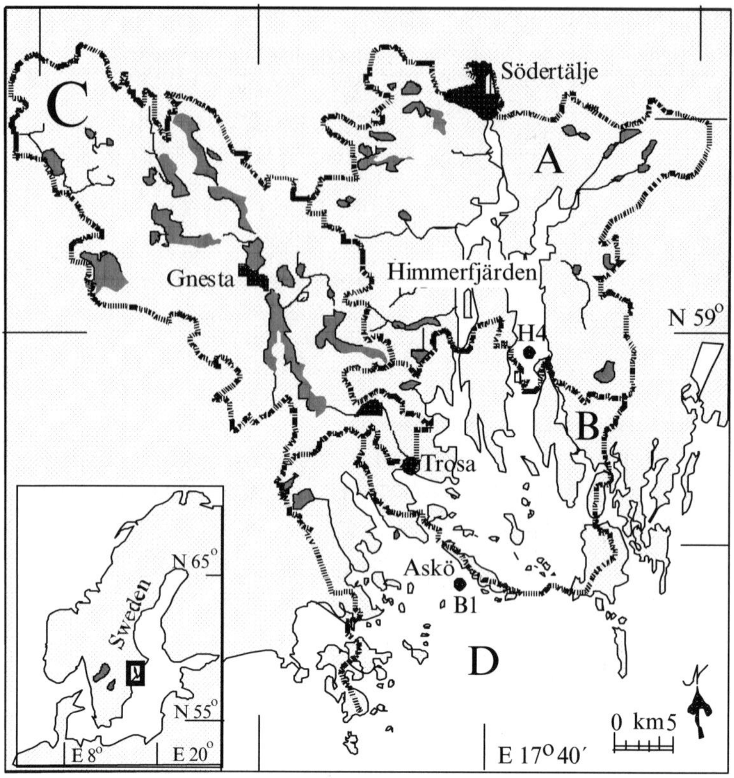

Fig. 15.1 The receiving area of the Askö archipelago (*D*) and the Himmerfjärden area (*A–C*)

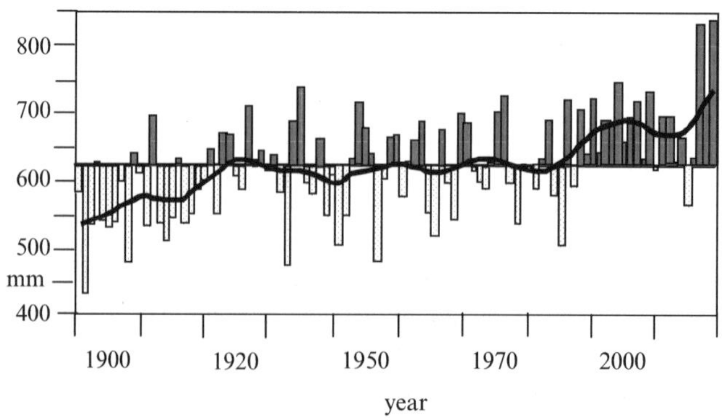

Fig. 15.2 Annual precipitation in Sweden during the twentieth century. Mean values from over 20 stations (from Alexandesson 2000)

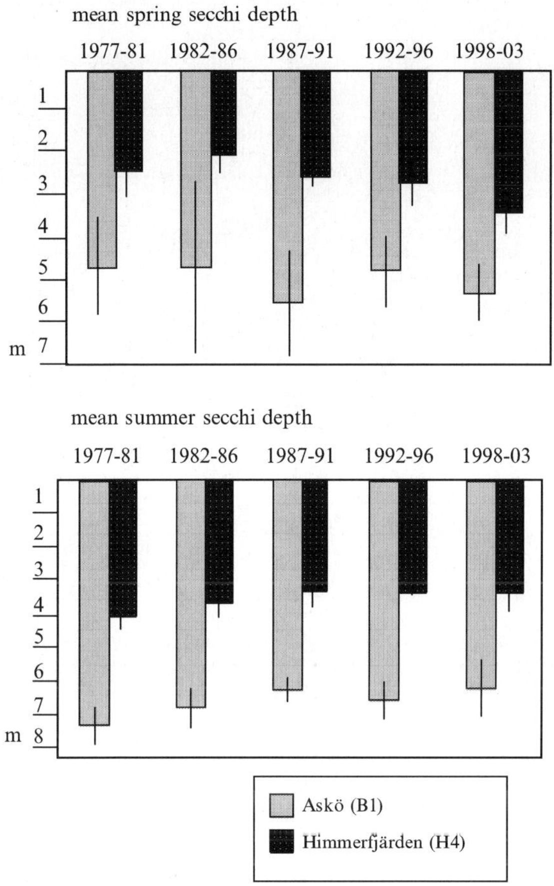

Fig. 15.3 Five-year mean spring and summer Secchi depths in the Askö archipelago (B1) and Himmerfjärden (H4) area (R. Elmgren and U. Larsson, unpublished)

15.1.2 Planktonic Communities

The pelagic system in the Askö area has been monitored since the mid-1970s (Fig. 15.4). There is a characteristic spring bloom, consisting mainly of diatoms, set off by temperature just after or during the melting of winter ice cover (see Chap. 14 by Hill and Wallström, this volume). In late summer a more or less pronounced Cyanobacteria-bloom occurs, especially after periods of hot, calm weather. After the spring bloom, zooplankton (mostly the mesozooplankton Rotatoria, Cladocera and Copepoda) occur in larger numbers (Fig. 15.5) (Johansson 1992), but bacteria and nano- and picoplankton are also present. Recently, Johansson et al. (2004) have studied nano- and picoplankton, especially the ciliates, focusing on their seasonal variation as well as their importance for the recycling of nutrients in the pelagic system. Their ability to regenerate nutrients can explain some of the primary

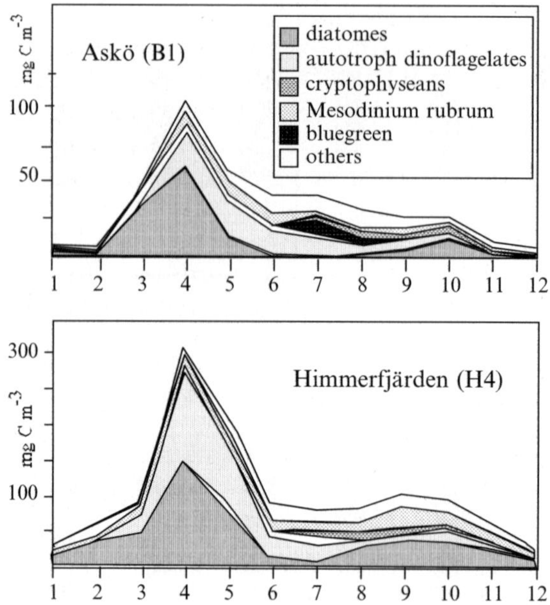

Fig. 15.4 Phytoplankton seasonal dynamics of the Askö archipelago (B1) and Himmerfjärden (H4) area. Major plankton groups are indicated. Data are based on mean values from 1977 to 1988 (redrawn from Hajdu et al. in Elmgren and Larsson 1997)

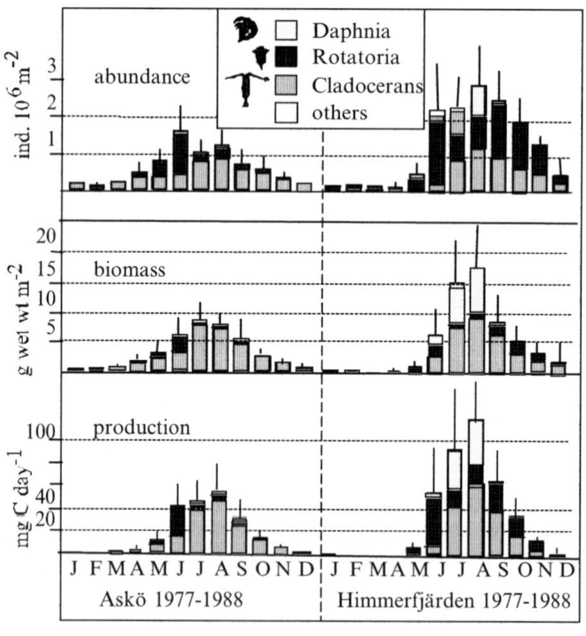

Fig. 15.5 Zooplankton seasonal dynamics in the Askö archipelago (B1) and the Himmerfjärden area (H4). Major zooplankton groups are indicated (redrawn from Johansson in Elmgren and Larsson 1997)

production potential during spring despite the depletion of nutrients in the water. During summer they are a food source for mesozooplankton. The mean annual chlorophyll *a* (Chl *a*) content is approximately 1.5 µg l^{-1} (Fig. 15.13).

The population dynamics of the copepod fauna was described in more detail by Adrian et al. (1999). Based on 12 years of field observations and experimental studies, they concluded that the dominant calanoid copepods (*Acartia* spp. *bifilosa/ A. longiremis* and *Eurytemora affinis*) in the Askö area and the eutrophicated Himmerfjärden area were equally abundant at both sites (see Chap. 14 by Hill and Wallström, this volume). However, *Eurytemora* was more abundant in the eutrophied area while *Acartia* spp. dominated in the non-eutrophied reference area outside Askö. There was, however, no significant difference between the two areas in biomass dynamics (biomass specific rate of change) of the two groups of copepods. However, in spring and early summer the population biomass increased faster for *Eurytemora* as compared to *Acartia* spp.

The nutrient load to the ecosystem from direct anthropogenic sources has decreased successively since the 1970s by, e.g., the construction of sewage treatment plans. However, land runoff still is fluctuating due to seasonal precipitation. In the years 2004 and 2005 unusually high concentrations of phosphorous were attributed to up-welling of offshore deep water from the reduced part of the Baltic Proper below the halocline. This was predicted to favour the growth of Cyanobacteria in summer, and in fact an extreme growth of the genera *Nodularia* and *Aphanizomenon* was observed in the area in 2005. Their ability to fix nitrogen directly was substantial (Larsson et al. 2001). The potential occurrence of the toxic *Chrysochromulina* species was described by, among others, Hajdu et al. (1996). She found potentially toxic species in high abundance but no toxic bloom has yet been recorded.

Rudstam et al. (1986) studied mysids and their role in the ecosystem, and revealed that e.g. *Mysis mixta* could consume up to 50% of the zooplankton at a given site.

Pelagic fish have traditionally been investigated by net fishing, but in recent years more and more by echo sounding and other non-destructive methods. The occurrence and reproduction success of herring (*Clupea harengus membras*) (Aneer 1975, 1980, 1985; Aneer and Nellbring 1982), as well as the top-down versus bottom-up mechanisms determining the structure of the system have been studied (Hansson 1993; Hansson et al. 1990, 1997a, 1997b). The choice of spawning ground of herring was studied in an extensive SCUBA investigation; divers observed that herring spawn close to the surface (not deeper than 11 m depth) and prefer to place their eggs on filamentous algae (Aneer and Nellbring 1982). Later experimental studies also showed that hatchery success was higher when the eggs were placed on more coarse algae (e.g. *Furcellaria lumbricalis* and *Fucus vesiculosus*) compared to filamentous brown algae (e.g. *Pilayella/Ectocarpus*) or areas with a recent oil spill (Aneer and Nellbring 1982; Aneer 1985). There is no commercial fishing in the Askö area.

The sand goby, *Pomatoschistus minutus*, is by far the most common fish in the area, and is seen on almost all substrates except soft substrates, where *P. microps* is more common. However, its contribution to the total fish biomass is small. Nellbring used a fish drop-net technique to establish the species composition and dynamics in shallow soft substrates, following the seasonal changes over the year (Nellbring 1985) and observing that competition for suitable breeding sites below shells etc. restricted the population size (Nellbring 1986). Using the SCUBA census technique in vegetation-covered substrates, Jansson et al. (1985) revealed that the gobies *Pomatoschistus minutus, P. microps, Gobius niger, Cottus gobio, Zoarches viviparus Pholis gunnellus* and *Platichys flesus* are the most common species present (see Chap. 14 by Hill and Wallström, this volume). *Gobius niger* together with *Z. viviparus* are commonly seen on substrates with boulders and stones, where *P. gunnellus* is also occasionally seen.

15.1.3 Benthic Communities

Early descriptions of the phytobenthic zone in the Askö area recorded the plant and animal species biomass depth distribution (Jansson and Kautsky 1977), the upper half-meter zone of the filamentous algae, especially *Cladophora glomerata* (Jansson 1974), the seasonal animal species composition in *Fucus vesiculosus* communities (Haage 1975a, 1975b), the plants occurring and their distribution along a eutrophication gradient (Wallentinus 1972), and the biomass composition and seasonal fluctuations of the deeper red algae communities (Kautsky 1974) [see Chaps. 10 (Kotta et al.) and 14 (Hill and Wallström), this volume]. Subsequently, factors structuring the distribution of species have been discussed, e.g. the major importance of environmental factors for the ecosystem (Kautsky and van der Maarel 1990), but biotic interactions have also been addressed, mainly in experimental studies. Comparisons of phytobenthic communities in the Askö area with those in other parts of the Baltic Sea indicate that the general trends described below are of general relevance for the Swedish coast of the Baltic Proper (Kautsky 1995a).

The plant and animal communities described below are found along the entire coast of the Swedish part of the Baltic Proper, with a few exceptions. For example, *Fucus serratus* occurs frequently in the southern parts up to just north of the Island of Öland. It grows together with *F. vesiculosus* but also penetrates a few metres deeper down. Sometimes, the distance from inner to exposed archipelago is a mere 1–2 km, although, as in the Stockholm Archipelago, it can be over 80 km. In the Askö area, this distance is approximately 15 km at its widest (see Chap. 14 by Hill and Wallström, this volume).

15.1.3.1 Inner Sheltered Archipelago

The inner, sheltered archipelago has a notable complement of freshwater species, e.g. the reed *Phragmites australis* grows along the shoreline, and different *Potamogeton* species are found in the water body, with *P. pectinatus* and *P. perfoliatus* being the most common. The blue mussels *Mytilus edulis, Macoma balthica* and, more recently, *Cerastoderma glaucum* occur in low biomass, and Hydrobidae as well as the fresh water snails *Theodoxus fluviatilis, Bithynia tentaculata* and *Lymnaea* spp. can be found. In the characteristic, soft substrate, shallow enclosed bays, various phanerogames and characeans are frequent. Here, close to the water surface down to a depth of one or a few metres, hard substrates may occur with growth of a compressed algal belt dominated by annual, filamenous algae for the first half metre. Thereafter, a narrow belt of *Fucus* and scattered *Furcellaria*, which are usually overgrown by filamentous brown algae for most of the year, may occur. The epiphytes are dominated by *Pilayella*/ *Ectocarpus* during spring and early summer, and *Dictyosiphon foeniculaceus* in late summer and autumn. The typical floral epiphyte on *Furcellaria* is *Ceramium tenuicorne*.

The bryophyte *Electra crustulenta* and the limpet *Balanus improvisus* are characteristic faunal epiphytes, occurring mainly on *Fucus* but also on *Furcellaria* in the inner sheltered areas, especially if the area receives increased nutrient load.

15.1.3.2 Shallow Enclosed Bays

The shallow, enclosed bays have attracted much attention in recent years as these habitats are decreasing due to exploitation by man, e.g. for marinas. Their natural communities are also destroyed by dredging. The natural development of such bays is dictated by the land uplift since the ice age, which tends to close the inlet, turning the bays into enclosed, coastal freshwater lakes. This process is described, and different stages of the closing bays are called "glo" and "glan" in the Finnish literature (Munsterhjelm 1997). The change eventually prevents the passage of boats into the bays and dredging is carried out with or without permission from the authorities, and thus these unique ecosystems are lost.

15.1.3.3 Intermediate Wave-Exposed Archipelago

On hard substrates in the intermediate wave-exposed archipelago, the surface has a characteristic belt of filamentous algae down to a depth of 0.5 m. The algae found depend not only on the season but also on whether there has been a preceding low water timed with their seasonal growth. The northern Baltic Proper has no tides, but irregular water level fluctuations due to weather conditions (both wind and pressure) are frequent, especially in early summer. This dries out the first 0.5 m depth and causes a succession of filamentous species. In early spring, filamentous brown algae such as *Pilayella/Ectocarpus* and *Scytosiphon lomentaria* dominate. The latter also occurs in autumn. These algae are usually preceeded by monosiphonal green algae (mostly *Ulothrix* spp. and *Urospora* spp.). Usually, after a low water period, the brown algae are replaced by, predominantly, the green alga *Cladophora glomerata* and in autumn by the red alga *Ceramium tenuicorne*, which then dominates throughout the winter until spring (e.g. Waern 1952; Wallentinus 1979, 1991; Qvarfordt et al. 2006). This succession does not always happen, depending on the preceding weather conditions; Waern (1952) described brown, green and red years depending on which algae managed to remain. Occasionally, the uppermost metre may be scraped by ice in spring, but this occurs only at certain, ice-exposed locations and not every year. Below this upper zone with its filamentous algae, the only large kelp-like, canopy-forming perennial brown alga – the bladder wrack (*Fucus vesiculosus*) – forms more or less dense belts down to 3–4 m, disappearing at 4–8 m depth depending where in the archipelago we are. *Fucus* is found deepest in the outer areas, whereas in more sheltered areas the lack of suitable substrate (hard) sets the lower limit of this species.

Close to the surface, *Ceramium, Furcellaria, Pilayella/Ectocarpus, Cladophora glomerata, C. rupestris, Phyllophora, Polysiphonia* are the most abundant species. There is also a seasonality deeper than 0.5 m depth, with some brown and green algae usually occurring during the winter period or in early spring or late autumn (e.g. *Monostroma greveli, Dictyosiphon chordaria, D. foeniculaceus, Stictyosiphon tortilis*) (Wallentinus 1979).

The bladder wrack is, in general, totally covered by filamentous brown algae in spring and more coarse brown algae (*Dictyosiphon foeniculaceus*) together with *Ceramium tenuicorne* in autumn. The common understory algae are – besides the crust-forming red alga *Hildenbrandia* and some brown crusts (mostly *Lithoderma*) – the same species that occur as epiphytes but also *Furcellaria lumbricalis, Phyllophora* and *Coccothylus*. The most common red alga is *Ceramium tenuicorne* but in later years *Polysiphonia fucoides* has also become more abundant. *Cladophora rupestris* and *Polysiphonia violacea* occur. The red algae dominate deeper down, where *C. rubrum, Rhodomela conefervodes* and the brown alga *Sphacelaria arctica* grow together with *Furcellaria* and *Ceramium tenuicorne*. At all depths except the first half metre at the surface, the animal biomass is dominated by the blue mussel *Mytilus edulis*. On vertical faces these mussels form almost monocultures with high biomass [1–2 kg dry weight (dw) m^{-2}]. Other common sessile animals are the barnacle *Balanus improvisus*, the bryozoan *Electra pilosa* and the hydrozoan *Cordylophora*. The latter is found mainly deeper down and under "overshootings".

Among the vagile fauna within the vegetation, different snails (*Theodoxus fluviatilis, Bithynina tentaculata* and Hydrobidae) are common as well as *Gammarus* spp. and the isopodes *Jaera* spp., *Idothea baltica, I. granulosa* and *I. chelipes*. Next to *Mytilus edulis, Macoma balthica* and *Cerastoderma glaucum* are the main contributors to the biomass.

The animal biomass usually follows the plant biomass well, being high when plant biomass is high. However, in some cases *Mytilus* reverse this pattern, especially deeper down, e.g. recent studies have revealed a vigorous competition for space between filamentous plants and *Mytilus*.

15.1.3.4 Wave-Exposed Skerries

In the outer, wave-exposed skerries, hard substrates extend deeper down, to 20 m depth and sometimes even deeper. Here, red algal communities are characteristic of the area. *Fucus vesiculosus* can be observed only at sites protected from direct wave action, are reduced in size and usually lack bladders. When *Fucus* is lacking, *Ceramium tenuicorne* dominates instead. Red algae and their associated fauna, mainly *Mytilus edulis, Macoma balthica, Cerastoderma glaucum*, Hydrobidae, Gammaridae and *Idothea* spp., dominate.

15.1.3.5 The Deeper Benthic Ecosystem

The deeper benthic ecosystem was described by Ankar and Elmgren (1976). Although few dominating macroscopic animal species are found in this ecosystem (the Baltic Sea mussel *Macoma balthica*, the amphipods *Monoporeia affinis, Pontoporeia femorata*, the polychetes *Harmotoe sarsi*, the occasional *Hediste diversicolor*, and, in later years, also *Marenzelleria neglecta*), the interaction between the species present and their environment is quite intriguing. Changes in the deeper benthic communities over the last 30 years were reported by Cedervall (e.g. http://www.smf.su.se/nyfiken/ostersjo/arsrapp/ostersjo00/djurliv.pdf). He observed how the two amphipodes were to a large extent replaced by *Macoma balthica* (and *Mytilus edulis*) starting at the end of the 1970s. As *Macoma* increased, *Harmotoe sarsi* disappeared and *Saduria entomon* occurred in the 1980s. Thereafter, *Saduria* decreased and *Halicrypus spinulosa* increased. Although the number of species is low, the species that are presetn seem to interact quite substantially. The intriguing interactions between the few species occurring in the benthic system has been studied experimentally. The interaction between the two most common amphipods on soft substrates in the area (*Monoporeia affinis* and *Pontoporeia femorata*) showed that the former occurred closer to the sediment surface (Hill 1991, 1992). These mud-living amphipods feed on juvenile *Macoma balthica*, thus influencing the recruitment of their population (Ejdung and Elmgren 1998; Ejdung et al. 2000).

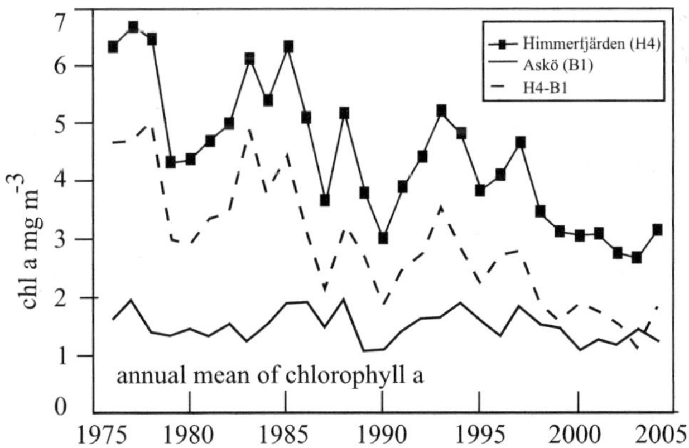

Fig. 15.6 Trends in the annual mean chlorophyll *a* (Chl *a*) concentration in the Himmerfjärden area (H4) and the Askö archipelago (B1), as well as a between-year comparison, from 1975 to 2004 (R. Elmgren and U. Larsson, unpublished)

The meiofauna in *Fucus* communities was described by (Kangas 1978). He came to the conclusion that they play a minor role compared to the soft benthic communities. Highest abundance was found among the filamentous algae. One-third of the biomass of 0.07–0.95 g m^{-2} was composed of temporary meiofauna. Nematoda, Halacarida, Chironomidae, Rotatoria and Ostracoda were observed in his samples.

Quantitative studies of the meiofauna of the deeper benthic communities revealed their importance not only as direct recyclers of nutrients but also for concentrating energy for larger organisms at higher trophic levels (Elmgren 1976). Elmgren estimated the production of meiofauna to be 110 kJ m^{-2} a^{-1}. Nematoda dominate the meiofauna in the deeper parts of the sediment, but the kinorhynch *Echinoderes levanderi*, the ostracod *Paracyprideis fennica* and Harpacticoida are also abundant closer to the sea floor. Depending on the season, a number of temporary meiofauna also occur. Experimental studies on the influence of *Macoma balthica* on the benthic meiofauna came to the conclusion that harpacticoids seem to compete for food resources with clams, while *Macoma* had no effect on the densities and distribution of other meiofauna in muddy habitats (Olafsson et al. 1993).

15.1.3.6 Biomasses, Succession and Productivity

The total biomass depth distribution of the plants and animals of the Askö area shows a distinctive plant biomass maximum from 0.5 m depth down to 4 m, after which it decreases with decreasing light towards deeper depths (Fig. 15.7). The bladder wrack *Fucus vesiculosus* contributes most to the biomass, but phanerogames are also common as well as red algae deeper down. The animal biomass is totally dominated by *Mytilus edulis*, which has its highest biomass coinciding with the maximum plant biomass. However, *Mytilus* is also found in high biomass deeper down. *Mytilus* constitutes over 90% of the total animal biomass. This is a characteristic feature of the low diversity Baltic Sea area, where the main natural predators of *Mytilus* are lacking and no extensive competition for space occurs, at least on more vertical faces and deeper down.

This absolute dominance of the marine blue mussel *M. edulis* has attracted special attention (see, e.g. Kautsky 1981). The importance of *Mytilus* is highlighted by the abundance of studies describing its role in the natural ecosystem as well as its role for toxicants. In the Baltic Sea, *Mytilus* lives on the edge of its salinity tolerance, thus any additional stress (e.g. toxicants) has a higher impact on this community compared to communities found on the more marine Swedish west coast. *Mytilus* filters approximately 5 l water g dw^{-1} h^{-1} and thus is a major contributor of nutrients within the phytobenthic communities in the Askö area (Kautsky and Evans 1987; Kautsky and Wallentinus 1980), recirculating 10–20% of the pelagic nutrient demand. Settlement studies have revealed the ability of *Mytilus* to attach to any hard substrate and that they have the high ability to move along substrates (Littorin and Gilek 1999a, 1999b). Coupled to toxic substances, protein levels and physiology change more in the Baltic populations than in populations from the more marine Swedish west coast, and the Baltic populations perform less well, indicating higher sensitivity to additional stress (Tedengren and Kautsky 1986, 1987; Björk et al. 2000).

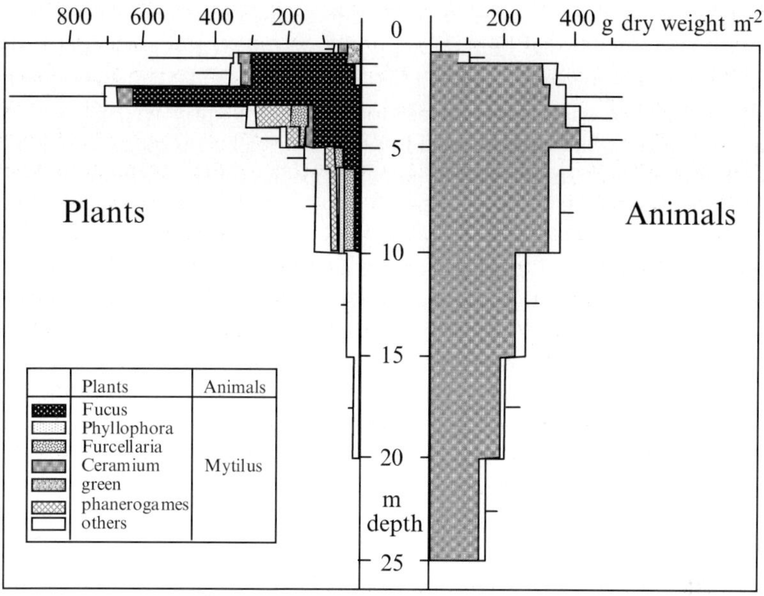

Fig. 15.7 The phytobenthic total depth distribution of plants and animals in the Askö archipelago. Major plant groups are indicated

In recent years, the succession within the phytobenthic plant and animal communities has been studied more intensely. A clear seasonal pattern of the settlement of germlings was observed, which repeated in a cyclic manner over the 2 years of study (Qvarfordt et al. 2006; Qvarfordt and Kautsky, manuscript submitted). Empty granite plates placed in the water at different seasons of the year had, after 3 months, an overgrowth of the species spawning at the time of submersion. After a few years, the differences decrease and the community develops towards the same annual species that occurs on plates submerged for more than 1 year. The perennial species occur in slow succession and after a few years the system probably looks the same as the surrounding area. It seems that the onset of the system is ruled by chance, i.e. the season, and which species happen to have spores in the water column at that moment. The succession then seems to go towards a given goal, determined mainly by the environmental conditions of the site.

The intriguing reproduction of *Fucus vesiculosus* has been discussed by Andersson et al. (1994). *Fucus* reproduces either in spring or autumn, with spawning synchronised to the phase of the moon (mainly during full moon) and taking place when the water is relatively calm. This makes spawning more or less successful. Furthermore, the zygotes are short-lived and spread mostly just below the mother plant, precluding geographically wide dispersion. Both spring- and autumn-reproducing *Fucus* plants are found but it seems that spring spawning dominates. One hypothesis is that the autumn spawners should dominate as eutrophication hampers the establishment of new germlings. This has not yet been observed.

The species found in the Askö area are influenced by its low salinity. Several marine species found in the southern part of the Baltic Sea (e.g. Denmark and the German coast) are missing (Wallentinus 1991; Kautsky 1995a) but, due to the vicinity of the field station, the species occurring are fairly well known. A frequency plot of the plant species found in the area, based on quantitative samples ($n = 932$), is characterised by a few dominating species that occur in more than 25% of the samples and a tail of more or less rare species (Fig. 15.8). The contribution of these species to the total biomass is quite low. A more detailed description of the most frequently occurring plant species follows below. Although a high number of samples have been collected, several species described from the area were not found.

15.1.4 Long-Term Change

Fucus vesiculosus was more common in the 1970s than at the beginning of the 1990s (Fig. 15.9), but since the mid-1990s it has again increased. *Fucus vesiculosus* has a patchy occurrence in the area and seems to come and go more or less stochastically from a given restricted area. *Ceramium tenuicorne* occurs in almost all samples collected. There is a weak tendency towards decrease in *Ceramium* at the same time as *Polysiphonia fucoides* tends to increase. *Ceramium* is the typical red alga found along the coast, whereas *Polysiphonia* is the dominating alga along the more clean coasts of the Island of Gotland and surrounding the Island of Öland. *Furcellaria lumbricalis* has increased in the last 15 years whereas other red algae expected in clean areas, *Phyllophora* spp and *Rhodomela confervoides*, seem to be decreasing. It seems that several plant species indicate a positive trend in the area, while other species indicate the opposite trend. However, today we observe, for example, the perennial *Fucus vesiculosus* deeper down than in the 1970s and 1980s and, together with the positive trends mentioned above, I see a positive development of the coastal system.

Animal species show a distribution pattern similar to that of the plants, but here *Mytilus edulis* totally dominates the mean biomass in the samples. Besides *Mytilus, Hydrobia* spp, *Macoma, Gammarus, Cerastoderma, Balanus, Hediste, Idothea baltica, Jaera* and *Chironomidae* are the most frequent species found in the area.

15.1.5 No Recent Increase in Eutrophication

The ongoing eutrophication of the Baltic Sea has caused changes in the phytobenthic zone, mainly by replacing the depth extension of attached plants towards the surface (Kautsky et al. 1986). Although increasing growth of ephemeral, filamentous algae has been observed, this is hard to observe in the quantitative material collected in the Askö area since 1993 and including samples from the same sites collected in

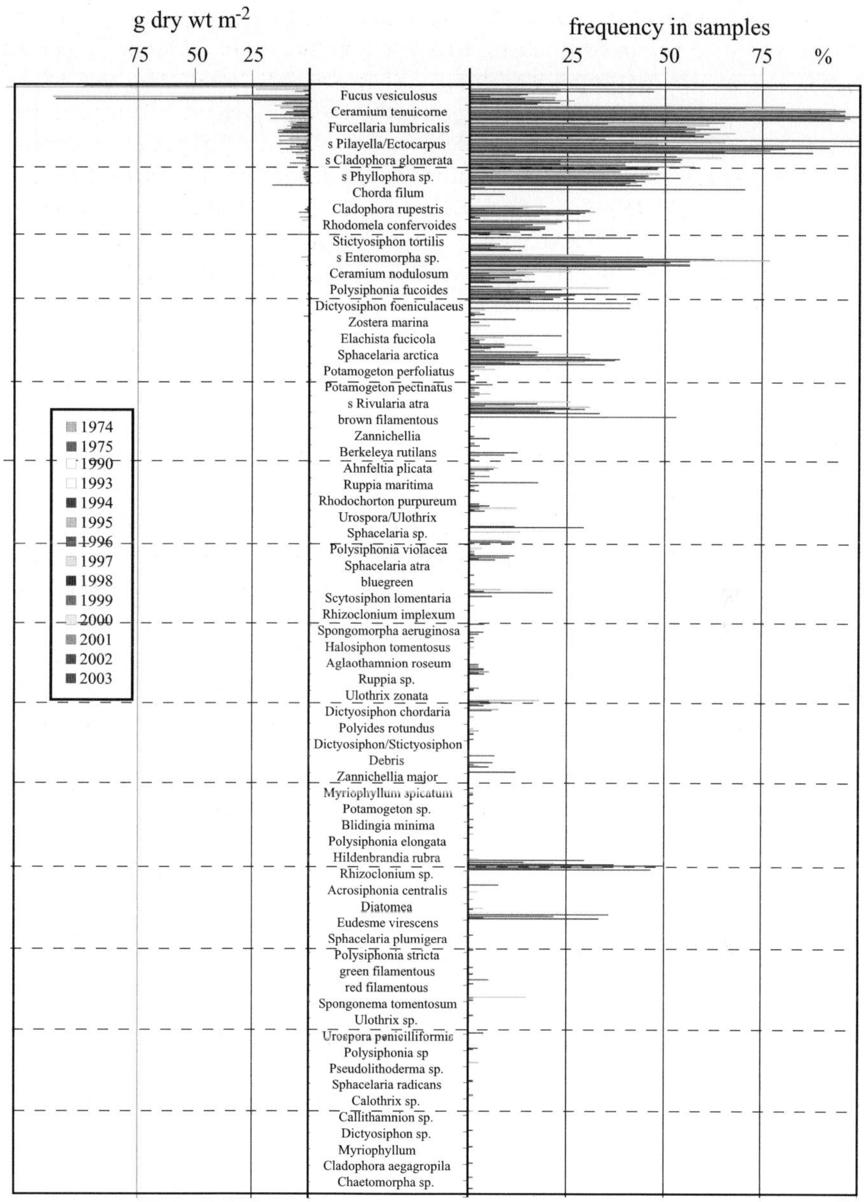

Fig. 15.8 The frequency of occurrence and mean biomass of plants in samples collected in the Swedish monitoring programme of vegetation-covered substrates in the Baltic Proper

the years 1974 and 1975 (Fig. 15.10). The proportion of annual and perennial plants has changed towards a dominance of annual plants, but this is not due to an increase in annual species but rather a decrease in perennials. Also, changes in animal biomass since the 1970s do not clearly indicate eutrophication. Herbivores, which

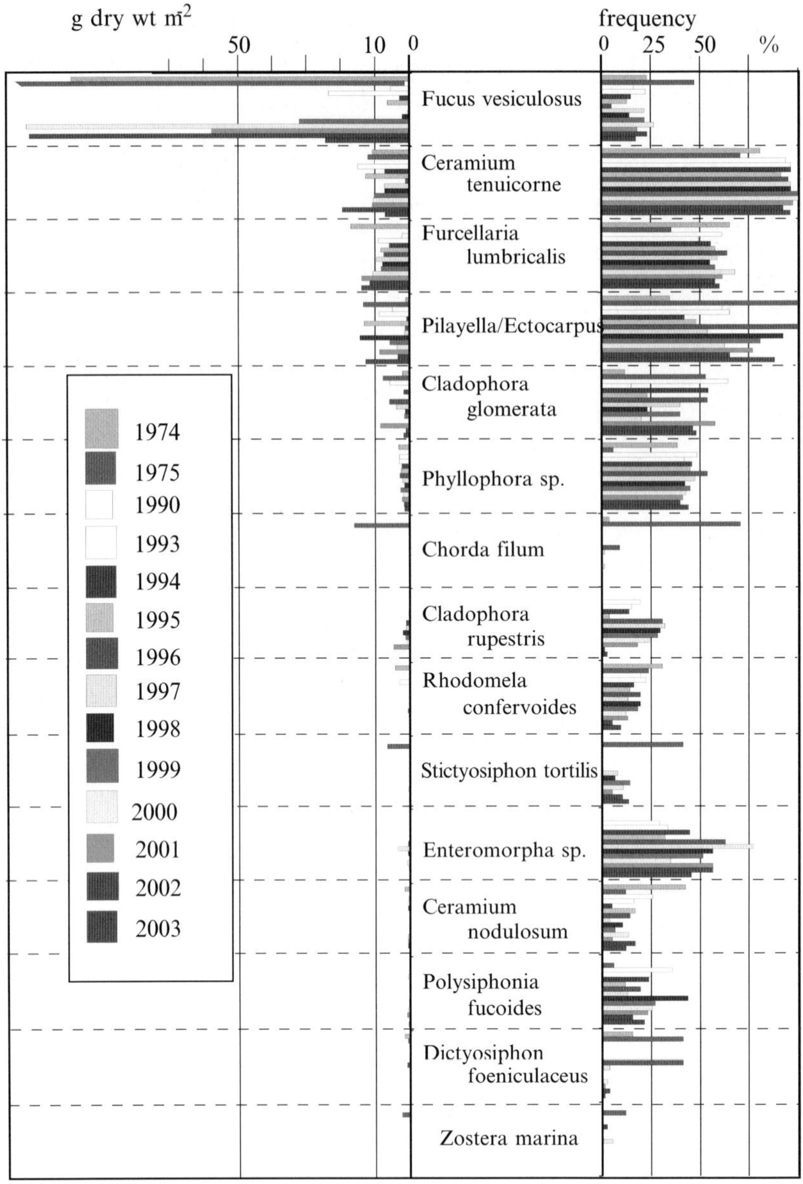

Fig. 15.9 The frequency of occurrence and mean biomasss of the most common plants found in samples collected in the Swedish monitoring programme of vegetation-covered substrates in the Baltic Proper from 1974/1975 and from 1993 to 2003

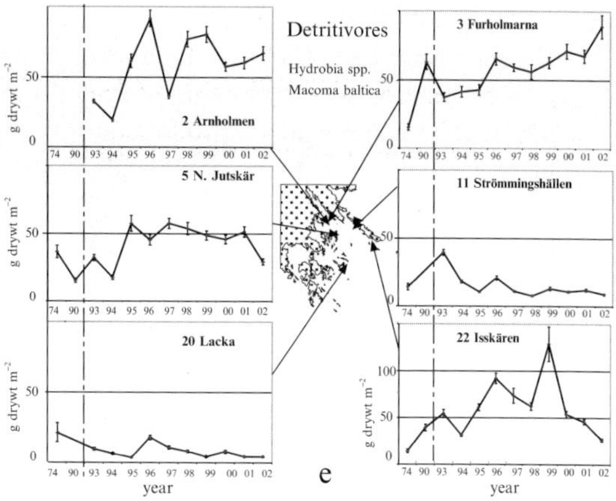

Fig. 15.10 Trophical group long-term change in the Askö archipelago, plotted as mean biomass (g dry weight m^{-2}) per station. Changes over the years 1974/1975 and from 1993 to 2003 in annual perennial plants (**a**), the biomass of annual and perennial plants (**b**), herbivores (**c**), filter feeders (**d**) and detritivores (**e**) are shown

supposedly should increase in eutrophicated waters (Kangas et al. 1982), have not increased. Filter feeders doubled from the 1970s to the beginning of the 1990s but then, due to unusually warm summers, decreased again to a mean biomass of the same magnitude as that seen in the 1970s. Detritivores show a clear increasing trend in biomass, and thus are the only group actually reflecting eutrophication. Other species show no, or even a trend towards less, eutrophication in the area. Each individual station has its own wildlife, and, for example, the maximum depth distribution of *Fucus* in the area has increased at some stations, is the same as in the 1970s in others, and at some stations *Fucus* has disappeared. The general trend, however, is that the maximum depth of *Fucus* in the area today is about 1 m deeper compared to the 1970s. From several places in the Stockholm archipelago (see Chap. 14 by Hill and Wallström, this volume), and also further south along the coast, there are observations that *Fucus* is nowadays more common, and that it has returned to places from which it was missing in the 20 years before the new millennium.

The occurrence of *Fucus vesiculosus* in the area seems not as dependent on eutrophication. This somewhat contradicts the view of Kangas et al. 1982 [see also Schramm and Nienhuist (1996) for a review]. Trends show a stabilised concentration of nutrients in the last 10 years. At several stations, *Fucus vesiculosus* has increased its maximum depth of occurrence, while at other stations no change or a decrease in depth distribution was observed compared to results from the 1970s. The deepest findings of *Fucus* have increased by 1 m compared to the 1970s. The occurrence of *Fucus vesiculosus* at a given site seems to depend on a combination of its population dynamics and the difficulties of becoming re-established in areas where it has disappeared. Pollution is an obvious factor reducing the distribution of *Fucus* (Kautsky 1992), with eutrophication first of all reducing the depth distribution (Kautsky et al. 1986; Eriksson et al. 1998). However, the observed disappearance in the Askö area is local and patchy, sometimes affecting areas of only a few square metres, indicating some local influence rather than regional factors such as eutrophication and pollution. Few grazers could cause this type of patchy disappearance; although a single plant in a depleted area is bound to attract hundreds of grazers, this does not explain the occurrence of holes in the canopy that is frequently observed. A *Fucus* stand tends to be of equal age at a given site, indicating recruitment success as being of major importance for the occurrence of this species with its complicated reproduction strategy (Andersson et al. 1994).

The conclusion is that the phytobenthic system is healthy, and that no major negative changes have occurred since the 1970s. Indeed, some data even indicate improved conditions in the ecosystem, with a tendency to find plants deeper than in the 1970s. Also, there is a shift in dominance towards species that usually occur in less eutrophied areas, and these species have increased their biomass. In spring, there is a tendency towards reduced nutrient load and increased Secchi depth in the area, as well as reduced pelagic spring bloom compared to the early 1990s (see also below, Sect. 15.2 Himmerfjärden)

15.1.6 Production and Energy Flow

Assessment of the production of the plant and animal communities in the Askö area was performed early using an ecosystem approach (Jansson and Wulff 1977; Jansson et al. 1982), where the total production of an enclosed sound was studied intensely. A programme monitoring pelagic production has continued since then (see e.g. http://www2.ecology.su.se/dbbm/index.shtml for details). Kautsky (Kautsky 1995b; Kautsky and Kautsky 1995) discussed the production of the coastal area and recalculated it using two different approaches to estimate the respiration of the system. The traditional calculation resulted in a net export of organic matter from the coastal to the deeper and outer areas (Fig. 15.11). In contrast, if respiration was actually measured, net production was seen to have been traditionally overestimated, being in fact only one-tenth of the amount previously calculated, although production varies according to the season. This reduces the contribution from the pelagic system substantially [see line indicating high respiration alternative (HIR-case) in Fig. 15.12]. Kautsky concluded that not only the

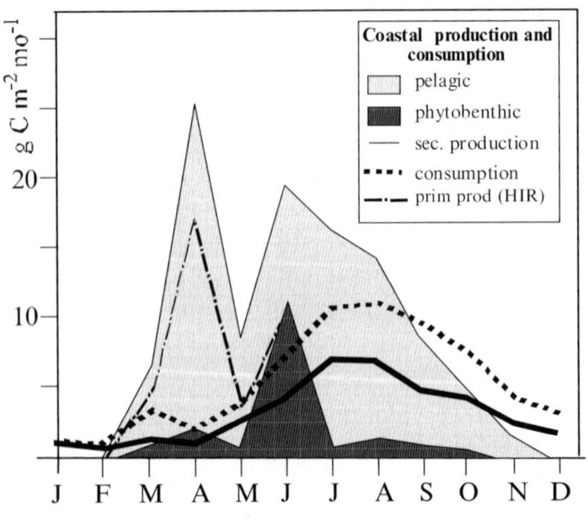

Fig. 15.11 Annual (months indicated on *x*-axis) coastal net primary production (*prim prod*) of the pelagic and phytobenthic system and the benthic consumption as well as faecal and secondary production (*sec. production*). The high respiration case (HIR) for the pelagic production is indicated by a *dashed line* (redrawn from Kautsky 1995b)

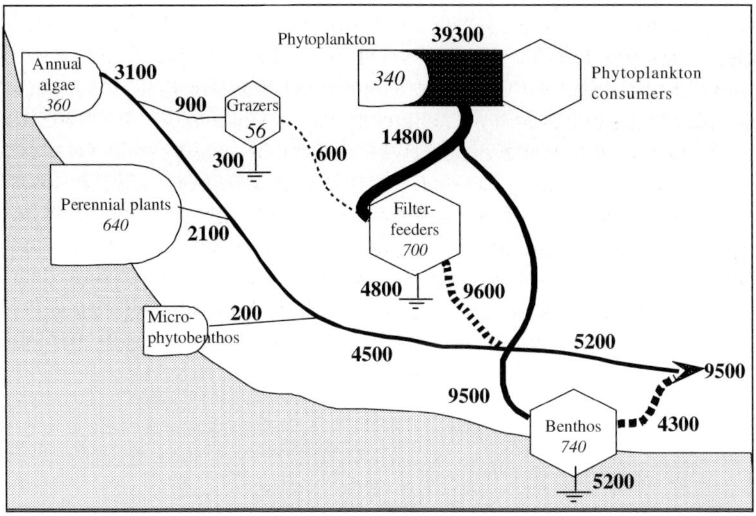

Fig. 15.12 Standing stock and flow of carbon through the Baltic Proper coastal region (area down to 25 m depth). *Bullets* Primary producers, *hexagons* secondary producers. Phytoplankton consumers include both zooplankton and the microbial loop. Numbers in *italics* are standing stock (10^5 kg C), and the *bold* numbers are carbon flow (10^5 kg C a^{-1}). *Solid lines* Flow of primary produced matter, *dashed lines* flow of secondary production. Respiration is indicated by *downward-pointing arrows* (modified from Kautsky and Kautsky 1993)

pelagic system but also the phytobenthic zone contribute substantially (50%) to the food supply of deeper benthos during parts of the summer. According to the HIR-case, pelagic production is highly overestimated. That this is actually true has yet to be proven, although the measured carbon dioxide balance in the pelagic system indicates that it is the case. Also, since internal rates are higher than the exchange with land and offshore areas, Kautsky claims that the coastal system should be regarded as an ecosystem in itself and not merely a transition zone (Kautsky 1995b).

Figure 15.12 illustrates the energy flow for the ecosystem down to 25 m depth in the Baltic Proper (from Kautsky and Kautsky 1995). The fluxes are based on biomass data collected along the Baltic Sea coast (Kautsky 1995a) and production measurements taken in the Askö area (for references see Kautsky and Kautsky 1995). Phytobenthic production takes place during summer when the nutrient content in the water is low and therefore production measurements taken in summer might be an underestimate compared to pelagic production based on annual calculations. The dominance in production by annual algae is due to their higher productivity per unit biomass compared to perennial algae (see, e.g. Wallentinus 1978). The annual production of the phytobenthic community of the Baltic Proper coastal area down to 25 m depth is 5.4×10^{11} g C a^{-1}. Compared to the pelagic production of 39×10^{11} g C a^{-1} this represents 12% of the total annual production. Filter feeders demand 14×10^{11} g C if we assume that the consumption rate is three times the respiration rate. This means that 35% of the pelagic production goes through the *Mytilus* population of the phytobenthic zone. According to Larsson (1986), a large fraction of the pelagic production is used by the microbial loop and zooplankton, and only 33–42% is available for benthic consumers. This is close to the demand of the filter feeder *Mytilus edulis*. A large fraction of the mussel consumption is deposited as faeces, which in turn can be used by the decomposers. The deeper benthic communities have a carbon demand comparable to the phytobenthic production or the faecal production of *Mytilus*. The excess carbon coming from the coastal zone is 9×10^{11} g C a^{-1} and is composed of detritus (faeces and dead organisms) and, at most, 75% of the primary produced matter from the phytobenthos. The excess equals one-half of the amount of sequestered carbon in the sediments of the entire Baltic Sea. This in turn means that the breakdown of faecal pellets and dead organisms in the coastal zone is probably underestimated. This should explain the high estimates of export of particulate organic carbon (POC) from the coastal system.

15.1.6.1 Interactions and Alien Species

The phanerogames have been especially studied by Idestam-Almqvist (1998a), who found a rare, for Baltic conditions, biotic interaction, namely the suppression of (mainly) *P. pectinatus* by waterfowl, not by grazing on shoots, but by eating subterranean tubers between late April and early June. His study suggests that waterfowl herbivory can constrain submerged macrophyte populations in shallow sheltered areas (Idestam-Almquist 1998b). In experimental studies of the interaction between different phanerogames in situ, Kautsky (1991) found that interspecific relationships were generally weaker than intraspecific relationships.

Some of the alien species introduced to the Baltic Sea have established themselves and are a “natural” part of the whole ecosystem, e.g. *Mya arenaria*, introduced already 1,000 years ago, and *Balanus improvisus*, introduced approximately 100 years ago, are now found up to the northernmost part of Bothnian Bay. *Balanus* is well established but seems to be less competitive, although mass occurrence on below boat hulls and submerged substrates may cause problems with fouling. On natural substrates it seems to be pushed to less favourable sites, i.e. overhangs below stones and boulders. A nice zonation pattern can be seen on boulders, with algae on top then *Mytilus edulis* and finally *Balanus* and *Cordylophora caspia* (also a late alien). More recently, the polychete *Marenzelleria neglecta* (formerly *M. viridis)* has been observed in large numbers on soft substrates but also among algae. A study of its influence on the ecosystem indicated a competitive ability, especially in deeper parts of the sediment (Kotta and Olafsson 2003; Kotta et al. 2003, 2006). The mysid *Hemimysis anomala*, frequently observed in the area, may have an influence on pelagic zooplankton through predation. Recently, the mass occurrence of the zooplankton *Cercopagis pengoi* has been observed, e.g. in the Finnish Bay, but it has also occurred in high numbers especially in the Himmerfjärden area. It competes with herring larvae for food and clogs the nets of fishermen due to its elongated spine. The voracious consumer and aggressive goby *Neogobius melanostomus* has been observed in the Åland Sea and in the Gulf of Finland and is well established in Pucks Bay, Poland [see Chaps. 7 (Kruk-Dowgiałło and Szaniawska) and 13 (Pitkänen et al.), this volume]. It seems only a matter of time before it arrives in the Askö area.

This may eventually change the system as, for example, *Mytilus* will decrease through predation. The Asian red alga *Gracilaria vermiculophylla* has also had mass occurrences on the Swedish west coast, and has been seen in the Kiel Bight. As it lives loose on soft substrates it competes with *Zostera marina*. Again, it may just be a matter of time before we find it also in the Askö area.

15.2 Himmerfjärden Area

The bay of Himmerfjärden has provided an opportunity to study how the ecology of a well defined, previously fairly unpolluted area changes with the establishment of a large sewage treatment plant. The effluents into the receiving area are substantial but have changed in character since 1974. The investigations started in 1976 and are still ongoing (http://www2.ecology.su.se/dbhfj/hfjsmall2.shtml).

15.2.1 Environmental Characteristics

The area of the Himmerfjärden is 174 km^2 and the mean depth is 17 m (Fig. 15.1). The drainage basin is 1,286 km^2, of which about 20% is agricultural land. Several smaller municipalities with a total of 17,000 inhabitants are served by smaller sewage treatment plants, whereas the newly established Himmerfjärdverken served a population of 240,000 in 1993. The nitrogen load increased gradually from 1976 to 1983, whereas the phosphorus load decreased (Fig. 15.13). Since 1985, the nitrogen load has dropped continuously with a short recession due to a deliberate large-scale experiment in the mid-1990s where nitrogen was allowed to increase for a few years; thereafter it has again decreased to values similar to those found in the Askö

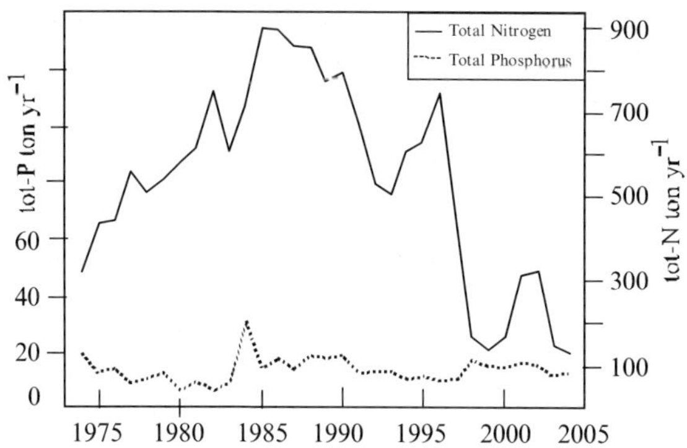

Fig. 15.13 Annual nutrient input to the Himmerfjärden area from the sewage treatment plant. Comparison of discharges of nitrogen and phosphorous from 1974 to 2004 (R. Elmgren and U. Larsson, unpublished)

area. This led to a reduction in the autumn bloom of Cyanobacteria to the same level as found outside the receiving area. The mean water exchange of the bay is 30 days, with a maximum of 10 days.

15.2.2 Eutrophication

The following lessons have been learned from the almost 30 years of intensive study of the influence of eutrophication in the bay of Himmerfjärden.

The nutrient content was higher in the receiving area compared to reference sites. However, the elevated nutrient level was not only due to sewage power plant effluents, although these were the major source, but also, as the large-scale experiment showed, to the weather, which influences both the water exchange in the coastal area, and also changes in regional land run off with local rainfall. The Himmerfjärden area also has an extensive exchange of nutrients with the open sea.

The ecological effects in the archipelago depend not only on the nutrient load but also on the weather, i.e. the temperature, solar radiation and wind. A study in bladder wrack (*Fucus vesiculosus*) of the content of stable 15N-isotopes, which are overrepresented in effluents, showed that nitrogen discharges were greatly diluted when reaching the reference area. The production of the phytoplankton is determined by the nutrient contents. The limiting nutrient was deduced using several methods, e.g. inorganic nutrient concentration in the water, nutrient elements in the macroalgae, phosphatase activity in macroalgae and phytoplankton, chlorophyll contents and algal growth potential. They all showed mutually consistent results. Initially, nitrogen was always limiting, except for the normally phosphorous-limited nitrogen-fixing Cyanobacteria. As phosphorous discharges decreased in the beginning of the 1980s and nitrogen increased, phosphorous became limiting in spring and early summer in the inner parts of the receiving area. An excess of dissolved nutrients was then exported to the outer areas.

This phosphorous experiment in the years 1983/1984 revealed a decrease in the period of phosphorous limitation, but this did not increase phytoplankton primary production (Fig. 15.6). However, the 1985 spring bloom was exceptionally high due to excess phosphorus in the water in combination with a large nitrogen-loaded spring flood. This was followed in 1985 by a 40% increase in nitrogen due to an increase in the population served by the plant. Finally, from 1988 to 1993 the nitrogen load was reduced by the construction of facilities for nitrogen reduction in the sewage treatment plant. This reduced the nitrogen load in the recipient area to levels of the order of those found in the mid-1970s (Fig. 15.6).

The efficient nitrogen reduction in the effluents, starting in the early 1990s, reduced and later even eliminated the spring–early summer excess of nitrogen in the inner receiving area (Fig. 15.6). In summer, the phytoplankton biomass was twice as high in the receiving area compared to the reference sites, and four times as high during the spring and autumn bloom. The bloom also started earlier and ended later in the receiving area. The species composition was similar and changed more with annual changes in the weather. However, potentially toxic Cyanobacteria and Prymnesiophyceae (e.g. *Chrysochromulina polylepis*) were more abundant in the receiving area. In 2004, the annual mean Chl *a* content in the Himmerfjärden area had decreased to one-half of what it was in 1976 (Fig. 15.13).

The same species of zooplankton were found in both the receiving area and adjacent areas, but in different proportions. The annual cycle showed a peak in July–August (see Fig. 15.3). Rotifers and cladocerans were more abundant in the receiving area. The decrease in zooplankton in late summer and autumn seemed to be caused by grazing from mysid shrimp and that year's young herring.

Fish were more abundant in the receiving area and, together with zooplankton, contributed to 10–30% of the nutrient supply to the phytoplankton. The importance of grazers for nutrient recirculation was tested by the introduction of the predatory fish pikeperch into the area, but with no visible success.

The biomass of benthic macrofauna increased with the increase of primary production in the first years after the mid-1970s. Oxygen deficiency eradicated the benthic fauna in 1976 and 1980. Since the mid-1980s, stormy summers have again oxygenised the benthos and the biomass has increased, especially *Macoma balthica.*

As demonstrated in a mass balance model, water exchange in the Himmerfjärden is driven by density differences caused by freshwater input and by wind-driven up- and down-welling in the coastal zone outside the area. Water exchange is extremely weather dependent and large changes in residence time vary with depth, between seasons and between years. Therefore, most measurements have to be taken over several seasons to obtain meaningful results. There was a net export of nutrients from the system. The exchange of nutrients across the sediment–water interface showed a high release of phosphorus and, to a lesser extent, nitrogen. Denitrification was a considerable nitrogen sink. Nitrogen export and denitrification increased with larger local loads. Although a net export of phosphorus from the system was observed, import and export alternated depending on the season.

Mass balance calculations showed that the archipelago was a net nutrient sink and that only a fraction reached the open sea. Thus, the coastal zone functions as an extra reduction zone, protecting the open Baltic Sea.

Following the successful reduction in nutrient release from the sewage power plant since 1993, nitrogen is once more the limiting nutrient in the receiving area, and all nitrogen released into the area is assimilated into algal biomass. An increase in nitrogen reduction will increase the potential bloom of nitrogen-fixing Cyanobacteria.

References

Adrian R, Hansson S, Sandin B, Stasio Bd, Larsson U (1999) Effects of food availability and predation on a marine zooplankton community: a study on copepods in the Baltic Sea. Int Rev Hydrobiol 6:609–626

Alexandesson H (2000) Var det växthuseffekten? http://www.smf.su.se/nyfiken/ostersjo/arsrapp/ostersjo00/vaxthus.pdf

Andersson S, Kautsky L, Kalvas A (1994) Circadian and lunar gamete release in the *Fucus vesiculosus* in the atidal Baltic Sea. Mar Ecol Prog Ser 110:195–201

Aneer G (1975) Composition of food of the Baltic herring (*Clupea harengus* v. *membras* L.) fourhorn sculpin (*Myxocephalus quadricornis* L.) and eel-pout (*Zoarces viviparus* L.) from deep soft bottom trawling in the Askö-Landsort area during two consecutive years. Merentutimuslait. Julk/Havsforskningsinst Skr 239:146–154

Aneer G (1980) Estimates of feeding pressure on pelagic and benthic organisms by Baltic herring (*Clupea harengus* v. *membras* L.). Ophelia 1 [Suppl]:265–275

Aneer G (1985) Some speculations about the Baltic herring in connection with the eutrophication of the Baltic Sea. Can J Fish Aquat Sci 42:83–90

Aneer G, Nellbring S (1982) A SCUBA-diving investigation of Baltic herring (*Clupea harengus* v. *membras* L.) spawning grounds in the Askö-Landsort area, northern Baltic proper. J Fish Biol 21:433–442

Ankar S, Elmgren R (1976) The benthic macro- and meiofauna of the Askö-Landsort area (northern Baltic proper). Contrib Askö Lab 11:1–115

Björk M, Gilek M, Kautsky N, Näf C (2000) In situ determination of PCB biodeposition by *Mytilus edulis* in a Baltic coastal ecosystem. Mar Ecol Prog Ser 194:193–201

Ejdung G, Byrén L, Elmgren R (2000) Benthic predator–prey interactions: evidence that adult *Monoporeia affinis* (Amphipoda) eat postlarval *Macoma balthica* (Bivalvia). J Exp Mar Biol Ecol 253:243–251

Ejdung G, Elmgren R (1998) Predation on newly settled bivalves by deposit-feeding amphipods: a Baltic Sea case study. Mar Ecol Prog Ser 168:87–94

Elmgren R (1976) Baltic benthos communities and the role of the meiofauna. Contrib Askö Lab 14:1–31

Elmgren R, Larsson U (1997) Himmerfjärden- förändringar i näringsbelastat kustekosystem i Östersjön. Naturvårdsverket, Report 4565:1–197

Eriksson BK, Johansson J, Snoeijs P (1998) Long-term changes in the sublittoral zonation of brown algae in the southern Bothnian Sea. Eur J Phycol 33:241–249

Haage P (1975a) Quantitative investigations of the Baltic *Fucus* belt macrofauna 3. Seasonal variation in biomass, reproduction and population dynamics of the dominant taxa. Contrib Askö Laboratory 10:1–84

Haage P (1975b) Quantitative investigations of the Baltic *Fucus* belt macrofauna. Quantitative seasonal fluctuations. Contrib Askö Lab 9:1–88

Hajdu S, Larsson U, Moestrup O (1996) Seasonal dynamics of *Chrysochromulina* species (Prymnesiophyceae) in a coastal area and a nutrient-enriched inlet of the northern Baltic proper. Bot Mar 39:281–295

Hansson S (1993) Variation in hydroacoustic abundance of pelagic fish. Fish Res 16:203–222

Hansson S, Arrhenius F, Nellbring S (1997a) Food web interactions in a Baltic Sea coastal area. Monograph: Forage fishes in marine ecosystems. In: Proceedings of the International Symposium on the Role of Forage Fishes in Marine Ecosystems, 13–16 November 1996, Anchorage, Alaska, University of Alaska, Fairbanks, pp 281–291

Hansson S, Hobbie JE, Elmgren R, Larsson U, Fry B, Johansson S (1997b) The stable nitrogen isotope ratio as a marker of food-web interactions and fish migration. Ecology 78:2249–2257

Hansson S, Larsson U, Johansson S (1990) Selective predation by herring and mysids, and zooplankton community structure in a Baltic Sea coastal area. J Plankton Res 12:1099–1116

Hill C (1991) Mechanisms influencing the growth, reproduction and mortality of two co-occuring amphipod species in the Baltic Sea. PhD Thesis, Stockholm University, ISBN: 91-87272-27-X, http://www.baltic.vtt.fi/pdfs/C_Hill.pdf

Hill C (1992) Interactions between year classes in the benthic amphipod *Monoporeia affinis*: effects on juvenile survival and growth. Oecologia 91:157–162

Idestam-Almqvist J (1998a) Temporal and spatial variation of submersed aquatic plants in the Baltic Sea. PhD Thesis, Stockholm University

Idestam-Almqvist J (1998b) Waterfowl herbivory on *Potamogeton pectinatus* in the Baltic Sea. Oikos 81:323–328

(1997 and 2003), and hence the lowest freshwater discharge. This is because the longer residence time of water in these years favours denitrification when nitrate concentrations are high in winter and spring, but also favours the exhaustion of the oxidised iron pool and the subsequent phosphorus release during summer due to lower oxygen availability.

16.2.4 Phytoplankton

The levels of chlorophyll-*a* (Chl *a*) are highest in the inner Odense Fjord and, as expected, decrease along the estuarine nutrient gradient (Fig. 16.2). Typically for fjords, the highest biomass is found during early summer, whereas a more traditional pattern is found in the open sea outside the fjords with biomass peaks in early spring and autumn (the very high winter chlorophyll peak in the shallow inner fjord is an "artefact", originating from macrophytes detritus and resuspended material). A more pronounced seasonal pattern is found for area-based primary production, the rates being lowest in the inner fjord due to the shallow depth (Fig. 16.2). The annual means of Chl *a* and primary production were about 16 and 20% lower in 2003 compared to 2002 (data not shown), due to the 2- to 3-times lower run-off and nutrient loads to the fjord in 2003 compared to 2002 (Fig. 16.2). This is a clear indication of the impact of the nutrient load on production patterns in Odense Fjord.

The phytoplankton community in Odense Fjord is dominated by diatoms [see Chaps. 3 (Schiewer), 7 (Kruk-Dowgiałło and Szaniawska), 10 (Kotta et al.) and 12 (Telesh et al.), this volume]. On an annual scale, 50–75 % of the phytoplankton carbon biomass and most blooms are due to diatoms, which is typical of eutrophicated fjords. Bloom-forming (e.g. >200 μg C l^{-1}) species are e.g. *Skeletonema costatum*, *Proboscis alata* and potentially toxic species belonging to the *Pseudonitzschia* group complexes. The second largest group in terms of carbon biomass are normally nanoflagellates. However, blooms of dinoflagellates (e.g. *Prorocentrum minimum*) or more opportunistic species belonging to *Euglenophyceae* occasionally occur. Zooplankton has not been monitored regularly in Odense Fjord, but a study conducted by Fyn County in 2002 suggests that the most important grazers on phytoplankton [see Chaps. 3 (Schiewer), 5 (Radziejewska and Schernewski), 8 (Chubarenko and Margoński) and 12 (Telesh et al.), this volume] are calanoid copepods (e.g. *Acartia tonsa*) and heterotrophic ciliates (e.g. *Strombidium* spp., *Balanion* spp.). However, the biomass level of pelagic phytoplankton in Odense Fjord is relatively low, despite the high nutrient loads and high concentrations in the fjord. Furthermore, the seasonal development of the phytoplankton biomass is very dynamic, with rapidly shifting concentrations of Chl *a*. Besides the dynamic water exchange, a high biomass of zoobenthic filter-feeders, which may efficiently filter the phytoplankton from the water column, especially when the water is efficiently mixed by the wind (Riisgård et al. 2004), is a major factor responsible for the shifting concentrations of Chl *a* (see below). Thus, there is no clear (statistical) trend in Chl *a* in the fjord over a longer time scale despite the fact that the nutrient load reductions (Fig. 16.2) have caused an increasing number of days where nutrients are potentially limiting for pelagic phytoplankton production (Fyn County 2003).

16.2.5 Macrophytes

In the shallow Odense Fjord, with its abundant filter-feeding zoobenthos, the biomass of phytoplankton is determined by nutrients and grazing forces. Because grazing forces apparently prevail, this means that high nutrient loads result in only a limited response by phytoplankton biomass. Instead, a surplus of nutrients may increase the biomass of ephemeral macroalgae, such as the sea lettuce *Ulva lactuca* and horsehair seaweed *Chaetomorpha linum*, which are abundant in Odense Fjord during the summer [see Chaps. 3 (Schiewer; Schlei) and 4 (Schiewer; Salzhaff), this volume]. A surplus of nutrients is generally known to change the balance among autothrophic components from dominance of perennial algae and sea grasses toward dominance of ephemeral algae (Borum 1996; Schramm 1996; Nienhuis 1996; Pedersen and Borum 1997).

In the 1980s, *Ulva lactuca* especially, but also *Chaetomorpha linum*, appeared in large quantities in Odense Fjord. *U. lactuca* appeared in the inner part of the fjord with extremely high biomasses [about 1 kg dry weight (dw) m^{-2}] and with an annual production of about 1,000 t carbon, or twice the annual phytoplankton production in the same area. In the outer part of the fjord, filamentous algae, primarily *Cladophora* sp. and *Ectocarpus siliculosus*, appeared in quantities of 150–300 g dw m^{-2} in the 1980s (Fyn County 1991). The very high sea lettuce biomass caused large fluctuations in oxygen conditions and pH in the water, both on a daily basis and over longer periods during the summer. When the sea lettuce biomass decomposed, this was frequently accompanied by oxygen deficit and the release of hydrogen sulphide (HS^-) (Fyn County 1991), even in very shallow water. Oxygen deficiency and the presence of HS^-, together with the shading and physical disturbance from ephemeral algae, is known to reduce the viability of rooted macrophytes (Bonsdorff et al. 1997; Holmer and Bondgaard 2001; Plus et al. 2003; Greve et al. 2004). Accordingly, rooted macrophytes were almost absent from the inner part of the fjord in the 1980s (Fyn County 2003). Since then, especially the phosphorus and the summer nitrogen loads (and accordingly, the nutrient concentrations in the water column) have decreased (Fig. 16.2), coinciding with a decrease in the abundance of ephemeral macroalgae in the fjord. In particular, the biomass and coverage of sea lettuce have been markedly reduced in the inner Odense Fjord, and the biomass of filamentous algae in the outer fjord has been reduced to a low level, i.e. below 20 g dw m^{-2} (Fyn County 2004), while the abundance of perennial macroalgae such as bladder wrack (*Fucus vesiculosus*) and of rooted macrophytes such as eelgrass (*Zostera marina*) and widgeon grass (*Ruppia maritima*) have increased (Fyn County 1991, 2003). The macrophyte community in the fjord is still very unstable, and still exhibits large inter-annual variation in coverage. Eelgrass has virtually disappeared in the northwestern part of the outer Odense Fjord. Whereas in the mid-1990s there were dense stands, the overall eelgrass coverage there is now at an all-time low, at less than 7% of suitable bottom area (Fyn County 2003).

Historical data show a reduction of eelgrass-depth limit in the outer Odense Fjord from 6.7 m (Ostenfeld 1908) to 2.5 m during the past ca. 100 years, which is most likely linked to the general deterioration of the light climate in the fjord caused by eutrophication, in parallel to the general development observed in Danish and European shallow coastal waters (Schramm 1996; Nienhuis 1996; Nielsen et al. 2002). The decreasing sea lettuce biomass has provided more stable oxygen conditions in the inner Odense Fjord, and oxygen deficit and release of HS^- are now rare or absent in this area. In a study of the ability of sedimentary metal oxides to buffer against release of HS^- in an Odense Fjord muddy sand (Kristensen et al. 2003), it was found that oxygen depletion could last for at least a month before sulphide was released into the water, during which time about 25% of the total pools of manganese and iron disappeared.

The outer Odense Fjord still experiences occasional oxygen deficit, primarily as a result of a density-driven circulation caused by the intrusion of hypoxic bottom water from the area outside the fjord via the excavated fairway.

16.2.6 Benthic Filter-Feeders and Grazing Impact

Odense Fjord is characterised by a large biomass of filter feeding polychaetes (*Nereis diversicolor*), clams (*Mya arenaria*) and cockles (*Cerastoderma glaucum*), which together make up about 70% of the total animal biomass (Fyn County 2001b). Other species of bivalves in Odense Fjord are *Mytilus edulis*, *Macoma balthica*, *Scrobicularia plana*, and *Ensis* sp. (see Chap. 3 by Schiewer, this volume; Schlei).

The density of the facultative filter-feeder *Nereis diversicolor* (cf. Riisgård 1991; Riisgård and Kamermans 2001) especially is unusually high in the inner part of Odense Fjord, and therefore it has been assumed that *N. diversicolor* along with *Mya arenaria* and *Cerastoderma glaucum* play an essential role in the regulation of the biomass of phytoplankton in the inner part of the fjord. A recent study conducted by Riisgård et al. (2004) has thrown light on the grazing impact of the filter-feeding zoobenthos in Odense Fjord. Thus, the filtration rate of each population (F_{tot}) of *N. diversicolor*, *M. arenaria* and *C. glaucum* was related to the total water volume (V_{tot}) in the different areas of Odense Fjord and expressed as 'the grazing impact':

$$Q = V_{tot}/F_{tot} \tag{1}$$

which is related to the half-life time for phytoplankton by:

$$t_{1/2} = Q \times \ln 2 \tag{2}$$

representing the time to reduce the concentration by a factor of 2 in the water column above the filter-feeding populations, assuming efficient vertical mixing.

Nereis diversicolor was distributed evenly in the inner part of Odense Fjord, and the estimated area-specific population filtration rates (F_{pop}) of *N. diversicolor* are shown in Fig. 16.3. The mean value in the inner part was found to be $F_{pop} = 2.74 \pm 1.80\,m^3\,m^{-2}\,day^{-1}$. Further, it was found that $Q = 0.29$ days, and $t_{1/2} = 0.2$ days. In other words, *N. diversicolor* is able to filter a volume of water equivalent to the whole water mass in the inner part of the fjord about three times per day, and – under conditions of efficient vertical mixing – the worm may reduce the phytoplankton biomass by 50% within less than 5 h, assuming 100% retention. *Mya arenaria* was found on most stations in the inner part of the fjord, with higher densities in a belt running north–south along a channel extending from the Odense River and out through the fjord. *Cerastoderma glaucum* was found on 50% of the stations in the inner part of Odense Fjord, with highest densities along the channel extending northwards from the outlet of Odense River (Fig. 16.3). The total potential grazing impact of these three filter-feeding species combined, using $F_{tot} = \Sigma F_{pop}$, was estimated to be $Q = 0.17$ days, and $t_{1/2} = 0.12$ days, or less than 3 h.

However, a dense population of the amphipod *Corophium volutator* may further help to exert a pronounced grazing impact. *C. volutator* lives in a U-shaped semicircular tube in the sediment in shallow water. Like *Nereis diversicolor*, *C. volutator* has two feeding modes: (1) surface deposit feeding, and (2) filter feeding (cf. review by Riisgård and Kamermans 2001). During the latter, particles larger than about 7 µm are retained. The ability of *C. volutator* to filter suspended particles from the ambient water has recently been studied by Møller and Riisgård (2006), who made experimental measurements of water pumping and particle retention efficiency and used the data obtained to assess the potential grazing impact and the area-specific population filtration rate (F_{pop}), of a dense population (up to about 18,000 ind. m^{-2} in August) of this burrowing amphipod in the inner part of Odense Fjord. The potential grazing impact values of F_{pop} were 1 and $20.6\,m^3\,m^{-2}\,day^{-1}$ in January and July, respectively, and the estimated grazing impact varied between Q = 20 and 1 h, whereas the potential half-time varied between 14.5 and 0.7 h. These values indicated that *C. volutator* is able to filter a volume of water equivalent to the whole column about 25 times per day, and – under conditions of efficient vertical mixing – the amphipod may reduce the phytoplankton biomass by 50% within less than 0.7 h in June and July. However, it must be strongly emphasised that the estimated grazing impacts are potentials that may be realised only if a decisive prerequisite is fulfilled, namely that the filter-feeding animals on the bottom are exposed to the whole water column by effective vertical mixing of the water.

In the shallow inner part of Odense Fjord with a mean depth of only 0.8 m, the water may often be well mixed by wind action, but so far no systematic studies focusing on this aspect have been performed. On the other hand, it has previously been clearly demonstrated that, on calm days a phytoplankton-depleted near-bottom layer only 5–10 cm thick is established above a dense population of *N. diversicolor* (Vedel et al. 1994; Riisgård et al. 1996c; Vedel 1998). The thickness of this algal depleted layer is determined by the ability of the worms sitting in

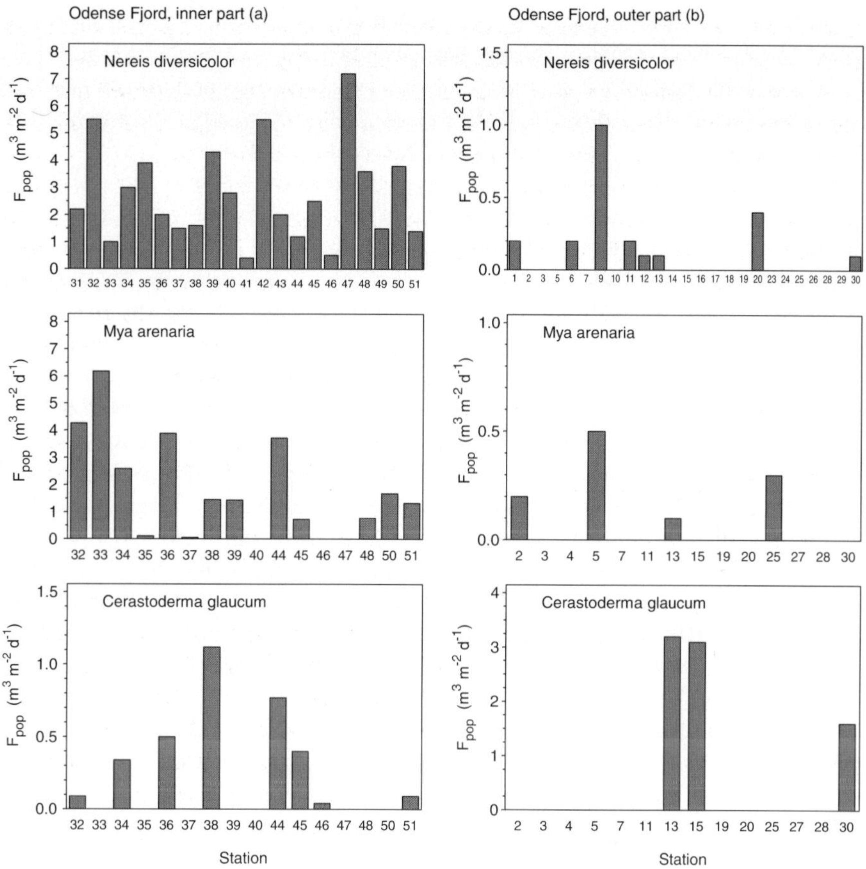

Fig. 16.3 Area-specific population filtration rates (F_{pop}) estimated for *Nereis diversicolor*, *Mya arenaria* and *Cerastoderma glaucum* in Odense Fjord. *Left panels* Inner part (stations 31 to 51); *right panels* outer part of fjord (stations 1–30). From Riisgård et al. (2004)

their U-shaped burrows to 'biomix' the overlaying water by means of their inhalant and exhalant feeding currents (Larsen and Riisgård 1997). The short residence time for phytoplankton can be as little as 3 h (with a generation time of about 1 day for larger species and ca. 6 h for very small species), showing that the filter-feeding zoobenthos may, on especially windy days, control the phytoplankton biomass in the inner part of Odense Fjord. It may therefore be predicted that, depending on the actual wind speed, the biomass of phytoplankton measured as Chl *a* may fluctuate strongly. Figure 16.4 illustrates that this is actually seen to be the case in the inner part of Odense Fjord, whereas the variation in Chl *a* at 17 m water depth outside the fjord is much less conspicuous (apart from the phytoplankton spring bloom in March). When the wind is weak, an 'uncoupling' of the benthic filter-feeders in the shallow Odense Fjord is likely to occur, which may contribute to triggering a rapid algal bloom in the overlying water. This phenomenon may be the reason for the very noticeable peak of Chl *a* in May during a long preceding period with generally decreasing wind speed (Fig. 16.4). In contrast, in windy weather the benthic filter will again be coupled, soon resulting in clear water.

As a consequence of intense zoobenthic grazing, surplus nitrogen released as ammonium in urine may subsequently accumulate in the ambient water, in particular if a grazing-reduced phytoplankton biomass simultaneously leads to a reduced assimilation of ammonium (NH_4^+) in the algal cells. To examine the possibility of increased grazing impact due to wind mixing subsequently leading to reduced Chl *a* and elevated ammonium concentrations, the actual concentrations of these substances (measured routinely by Fyn County in Odense Fjord) have been plotted as a function of wind speed (Riisgård et al. 2004). The slopes of the regression lines support the suggested causal connection, although the correlations are not strong because the actual fate of ammonium and Chl *a* are also determined by other factors.

More extensive field studies in Odense Fjord, involving time series of simultaneous measurements of vertical profiles of Chl *a*, wind-mixing of the water column above benthic filter feeders, and water current speed and direction, are now been carried out by the Marine Biological Research Centre in order to further investigate the suggested causal connections. Field studies are supplemented

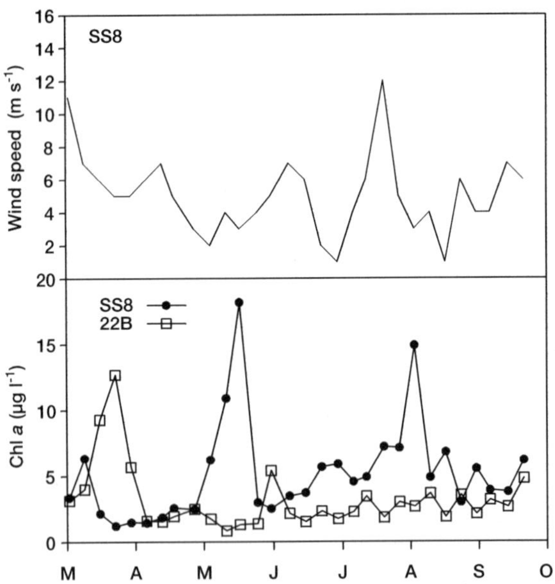

Fig. 16.4 *Upper panel* Variation in wind speed at station SS8 (water depth 0.8 m) in Odense Fjord from March to October 2000. *Lower panel* Variation in Chl *a* at station ODF22B (water depth 17 m) outside Odense Fjord compared to station SS8 (see Fig. 16.1). From Riisgård et al. (2004)

with underwater video observations of the filter-feeding zoobenthos. Preliminary results show that when the filter feeders are uncoupled, with an algal depleted near-bottom layer created soon after, this results in switching from a filter-feeding mode of nourishment to surface deposit feeding in *Nereis diversicolor* and *Corophium volutator* (Riisgård and Kamermans 2001; Møller and Riisgård 2006), whereas filter-feeding bivalves withdraw their siphons and close their valves (Riisgård et al. 2003).

16.2.7 Fish

In addition to Odense River, five streams flow into Odense Fjord, and several of these are important spawning and maturation sites for sea trout (*Salmo trutta*). During the last 25 years great efforts have been made to improve conditions for fish life, and fishing restrictions have been imposed. Recreational fishing of, primarily, eel (*Anguilla anguilla*) and flounder (*Platichthys flesus*) takes place by means of net, trap and fishing rod, but few professional fishermen operate in the fjord. Other species of fish in Odense Fjord are: *Belone belone* (garfish), *Gadus morhua* (cod), *Gobius niger* (black goby), *Pomatoschistus minutus* (spotted goby), *Zoarces viviparus* (viviparous blenny), *Syngnathus typhle* (greater pipefish), *Psetta maxima* (turbot), *Crenimugil labrosus* (thick-lipped grey mullet), *Salmo salar* (salmon), *Clupea harengus* (herring), *Trachurus trachurus* (horse mackerel), *Limanda limanda* (dab), *Pleuronectes platessa* (plaice), *Pungitius pungitius* (nine-spined stickleback), and *Gasterosteus aculeatus* (three-spined stickleback).

16.2.8 Impacts of Hazardous Substances

In Odense Fjord, hazardous substances occur in high and potentially toxic concentrations in the sediment, and for some compounds in such quantities that they may impact the flora and fauna in certain areas of the fjord (Fyn County 2004). This is true for compounds such as tributyltin (TBT) – for which concentrations in common mussels *Mytilus edulis* from Odense Fjord are 3–35 times higher than the ecotoxicological criteria endorsed by the Oslo-Paris Commission (OSPARCOM) – as well as for certain PAHs (polyaromatic hydrocarbons), PCBs (polychlorinated biphenyls) and others, but may also be true to a lesser extent for heavy metals such as copper, nickel and lead.

TBT may especially affect the biota in Odense Fjord. The Lindø Shipyard, located in the eastern part of the outer fjord, and traffic in the shipping lane are the major sources of TBT (Fyn County 2001a). Thus, Jensen et al. (2004) demonstrated a likely impact on *Ruppia maritima* in the inner Odense Fjord at ambient sediment concentrations of TBT. A conspicuously unfortunate effect of TBT is malformation of the genital systems in certain marine snails, e.g. the common periwinkle *Littorina littorea*, which seriously affects their ability to reproduce. In females, male sexual characteristics can develop (intersex), and in males, the penial glands are affected. Investigations have revealed that a large number of common periwinkles show such signs of being affected by TBT (Fyn County 2001a).

16.3 Case Study: Kerteminde Fjord/Kertinge Nor

16.3.1 Basic Characteristics

The fjord-system consisting of Kerteminde Fjord and Kertinge Nor covers an area of 8.5 km^2 and has a mean water depth of approximately 2 m and a maximum depth of 8 m (Fig. 16.1). The fjord has a sill at its mouth to the open sea (Great Belt). Discharge over the sill is forced by a diurnal tide with an average amplitude of approximately 20 cm. The tide gives rise to maximum discharges at the fjord entrance of 100–200 $m^3\ s^{-1}$. The catchment areas to Kerteminde Fjord and Kertinge Nor are limited (1% of the island of Fyn), at 18.6 and 17.4 km^2, respectively, most of which is agricultural land and forests. The freshwater input of <0.05 $m^3\ s^{-1}$ is negligible with respect to the water exchange of the fjord system. The salinity in the central part of the system varies typically between 15 and 21 psu over the year, and the monthly mean temperature ranges between 2 and 18°C (Fig. 16.5).

Water exchange in the fjord-system is governed by density-driven circulation. The salinity in the Great Belt outside the fjord varies as a result of changing flow situations (Jürgensen 1995; Møller 1996). Outflow of water from the Baltic Sea gives salinities down to less than 10 psu whereas inflow to the Baltic Sea gives salinities up to 27 psu in the upper layer of the Great Belt. Because saline water is more dense than fresh water the salinity variations cause longitudinal density variations from the inner part of the fjord system to the mouth, and density-driven vertical circulation occurs. When dense water is flushed over the sill by tidal forcing it will flow down below the fjord water and give rise to a density-driven circulation system within the entire fjord system. When, on the other hand, lighter water is forced into the fjord the circulation is in the opposite direction. On an annual time scale the two circulation directions have equal probability. Because of the dynamics of the exchange processes, the term 'residence time' is somewhat dubious in this fjord system. An approximate time-scale for residence time of water in the central areas of the system lies between 1 week and a few months [see Chaps. 3 (Schiewer), 5 Radziejewska and Schernewski), 10 (Kotta et al.) and 12 (Telesh et al.), this volume], with an average of approximately 6 weeks (Jürgensen 1995). In Kertinge Nor – and similar localities characterised by low current velocities caused by density-driven currents and moderate tides – it should be emphasised that the alternating tidal current does not give rise to a net transport of water whereas the density-driven circulation can potentially flush the entire water mass within 10 days (Jürgensen 1995).

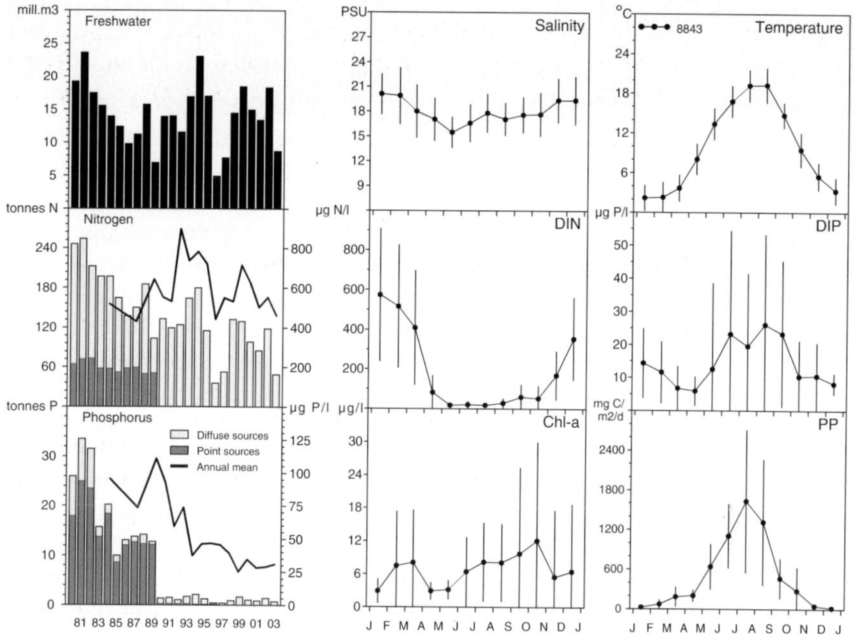

Fig. 16.5 *Left panels* Freshwater, nitrogen and phosphorus run-off to Kertinge Nor/Kerteminde Fjord and annual means of total nitrogen (*TN*) and phosphorous (*TP*) at station 8843 in the central part of Kertinge Nor, 1980–2003. The nitrogen and phosphorus run-off are source-apportioned into point sources (mainly industrial and municipal waste water) and diffuse sources (mainly run-off from agricultural land). *Central and right panels* Temperature, salinity (psu), dissolved inorganic nitrogen (DIN), dissolved inorganic phosphate (DIP), Chl-*a* and depth-integrated primary production (PP) at station 8843 (2.7 m); data are monthly means ± SD (1989–2003) in surface water, except primary production (1998–2003)

Sandy sediments dominate the Kerteminde Fjord/Kertinge Nor system (Miron and Kristensen 1993). The silt and clay fraction (mean particle diameter <63 µm) is usually under 7%, and percentages of particulate organic carbon (POC) are also low (<1.5%). Differences between localities are most conspicuous with respect to the pore water HS^- concentration, with the mussel-bed area near the mouth of Kerteminde Fjord having the highest HS^- concentrations.

16.3.2 Nutrients and Phytoplankton

The stoppage of the sewage outfall to the Kertinge Nor/Kerteminde Fjord system during the late 1980s meant that land-based nutrient loads fell markedly (Fig. 16.5). Thus, annual land-based discharges of nitrogen and phosphorus were reduced by 43% and 92%, respectively, as compared to mean values for 1976–1989 and 1990–

2003. Since 1990, the nutrient loads have been due almost solely to diffuse sources, which imply that they co-vary with the freshwater run-off (Fig. 16.5). The significant load reductions, down to 110t N and 1.2t P (mean 1990–2003), made the fjord system suitable for studying the effects of nutrient reduction on its recovery from eutrophication, as the reduction obtained almost fulfilled the objectives of the first Danish Action Plan of the Aquatic Environment, the so-called APAE 1 from 1987 (Fyn County 2001a), of a 49% reduction in N and an 80% reduction in P.

The dramatic decrease in the P-load is clearly reflected in the annual means of phosphate in Kertinge Nor (Fig. 16.5). The decrease discernable for the N-load is less pronounced, and the actual annual load varies by more than 100% depending on the freshwater run off. The annual means of nitrogen in Kertinge Nor show large fluctuations and no temporal trend, but to some extent seem to follow the annual fluctuations in the N-load. This may reflect the fact that both the relative importance and the temporal changes in load are secondary. Thus, the inter-annual variations in N-load are still considerable, and the magnitude of the mean reduction is less than that of the P-reduction.

The mean seasonal variation in inorganic nutrients in Kertinge Nor is shown in Fig. 16.5. The seasonal variation in inorganic nitrogen is typical, with high concentrations during winter (up to ca. 600 μg N l^{-1}), and a progressive depletion of the pool with low concentrations, often potentially limiting for the phytoplankton, during summer. The inorganic phosphate concentrations peak during summer at about 25 μg P l^{-1}, when release from the sediment is high due to depletion of the oxidised iron pool to which the phosphate is bound.

There are generally two biomass peaks of phytoplankton in Kertinge Nor, as reflected in the seasonal Chl *a* cycle (Fig. 16.5): one in spring and one in late summer/autumn (see Chap. 4 by Schiewer, this volume). The seasonal primary production pattern is typically more distinct, with a clear peak in mid-summer (Fig. 16.5). This peak is displaced in relation to the later biomass peak, presumably reflecting grazing patterns (see below). There are no systematic recordings of phytoplankton species in the Kerteminde Fjord/Kertinge Nor fjord system, but sporadic blooms of e.g. dinoflagellates such as *Prorocentrum minimum*, and Prymnephyceae such as *Chrysochromulina parkae*, have been observed after nutrient release events triggered by oxygen-depletion.

16.3.3 Biological Structure and Nutrient Dynamics in Kertinge Nor

The combination of exceptionally clear water, a varied vegetation of rooted macrophytes (*Zostera marina*, *Potamogeton pectinaus*, *Zannichellia palustris* and *Ruppia* sp.) and periodic thick mats of filamentous algae (primarily *Chaetomorpha linum* and *Cladophora sericea*) with biomasses of 50–100 g dw m^{-2} (Fyn County 1991), high densities of small jellyfish (*Aurelia aurita*), and a dense population of benthic ascidians (*Ciona intestinalis*) makes the cove Kertinge Nor interesting for studying the dynamics of the biological structure in an eutrophic ecosystem, in which nutrient fluxes and suspension-feeding organisms play a decisive role.

Plankton dynamics are at times governed by the interplay between filter feeders and hydrodynamics, especially density-driven currents. This is especially true in the Kertinge Nor/Kerteminde Fjord system (Riisgård et al. 1996a; Riisgård 1998). Attention must be paid to these phenomena in order to understand the dynamic behaviour of Danish fjords and coastal areas that are influenced by the water exchange between the Baltic Sea and the North Sea.

16.3.3.1 'Normal Years'

The biological structure of the Kertinge Nor ecosystem in a 'normal year' is summarised in Fig. 16.6. The water column is clear, allowing sufficient light penetration to the bottom where a significant benthic primary production of filamentous algae and eelgrass may be found (see Chap. 4 by Schiewer, this volume; Salzhaff). The algal mat is important for the control of nutrient flux from the sediment to the water column, and in the spring and summer period N and P concentrations often drop below potentially limiting levels for phytoplankton production (i.e. about 2 µm DIN and 0.2 µmol DIP) (Fyn County 2004). Below the algal mat, the sediment is black and sulphidic due to anoxic conditions and lack of living animals. A dense population of filter-feeding ascidians (*Ciona intestinalis*) keep the water clear, and a large number of small, maximum diameter 4–6 cm, umbrella jellyfish (*Aurelia aurita*), dominate the water column.

During summer, the water processing capacity of the many (up to ca. 250–300 ind. m^{-3} in April–June) small *Aurelia aurita* is very high, with a maximum rate attained in early September, where the jellyfish population can daily process a water volume corresponding to approximately 10–15 times the whole water volume of Kertinge Nor (Olesen et al. 1994). This indicates that *A. aurita* controls zooplankton in the cove during summer and autumn. Laboratory experiments have proved that the medusae are food-limited at in situ zooplankton concentrations. Moreover, *A. aurita* is presumably growing in excess of its food source, as zooplankton densities in the water column of Kertinge Nor during the day cannot explain the observed growth of *A. aurita*. It has been observed, however, that the density of harpacticoids in the water column at night can exceed the density during the day by a factor of 20, and night-swimming harpacticoids may therefore be an important food source for jellyfish in Kertinge Nor (Olesen et al. 1994; Olesen 1995; Nielsen et al. 1997).

The dense population of the filter-feeding *Ciona intestinalis* exerts a high grazing pressure on phytoplankton, which may partly explain the low observed phytoplankton biomass. In particular, during late summer and autumn, the *Ciona* population can reach densities of approximately 250 ind m^{-2}. During fall, the population of *C. intestinalis* usually has the potential capacity to filter the total water volume of Kertinge Nor 0.2–1.2 times daily, and the mean residence time of an algal cell in the water column in September may only be about 7 h (Petersen and Riisgård 1992). From previous data and observations, the 'normal' summer situation for the

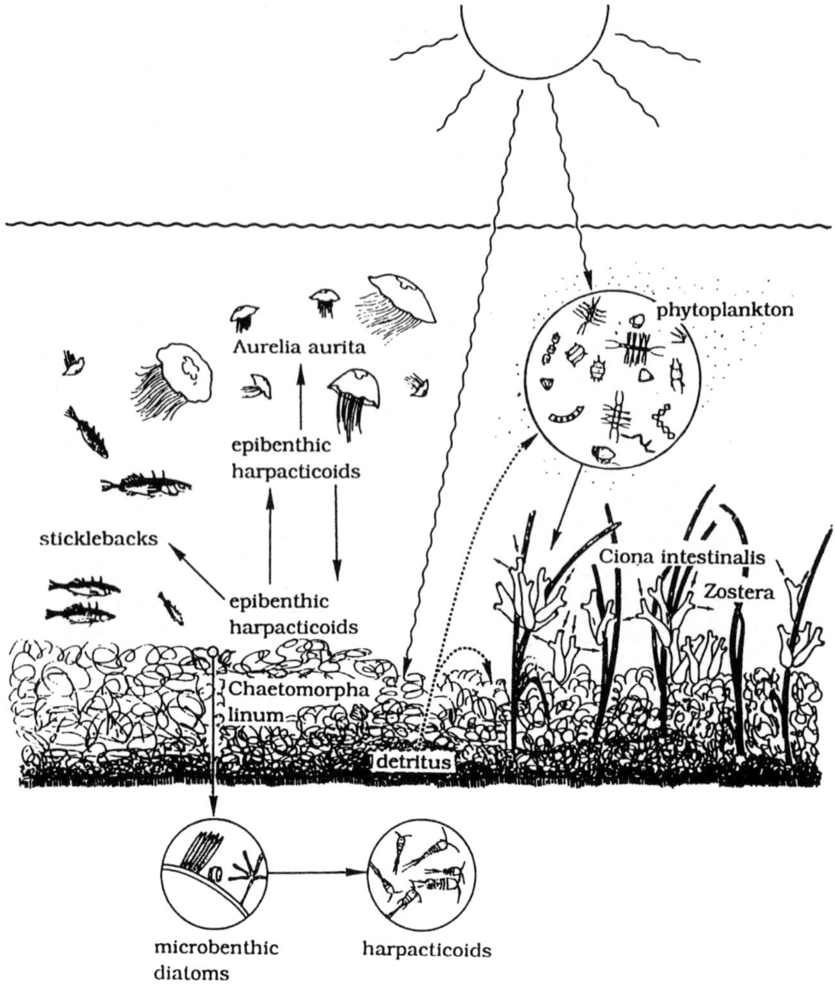

Fig. 16.6 Biological structure in 1991 ('normal year') in Kertinge Nor. Three food-chains were identified: (1) phytoplankton → ascidians (*Ciona intestinalis*); (2) epiphytic diatoms → epibenthic harpacticoids → jellyfish (*Aurelia aurita*) + sticklebacks; (3) macrophytes → detritus → decomposing microorganisms. From Riisgård et al. (1994b)

biological structure and dynamic of Kertinge Nor outlined above prevails, but the ecosystem is inherently unstable.

16.3.3.2 The 'Unusual Year': 1992

As explained above, the water column of Kertinge Nor is usually very clear, and Chl *a* concentrations are low throughout the growth season, but this situation can change dramatically as observed in 1992 where the 'normal' situation existed only until June.

A high biomass of the filamentous macroalga *Chaetomorpha linum* was recorded on the bottom throughout the year (mats up to 40 cm thick during the summer). The filamentous algal production was limited by the availability of N, and the algal mat therefore controlled the release of inorganic N from the sediment (Christensen et al. 1994). This 'normal' situation existed for Kertinge Nor until June, but a long period of calm clear weather with high insolation began in the middle of May and persisted for about 2 months. This caused a high primary production of the filamentous algal mat, which led to the formation of large oxygen bubbles that caused parts of the whole algal mat to float up and away from the sediment to the water surface (Christensen et al. 1994). In this position the filamentous algal mat could no longer act as an absorbing filter for the nutrient flux from the underlying anoxic sediment and large amounts of nutrients were released to the water column. The phytoplankton population immediately responded to the elevated nutrient concentrations. Within a week, in areas where the algal mat was more or less separate from the sediment surface, the phytoplankton biomass increased 80 times due to photosynthetic activity (Christensen et al. 1994). During June 1992, the phytoplankton biomass increased almost exponentially, which reduced the light penetration depth into the water to only 30 cm. This massive shading effect caused a total collapse of the remaining benthic algal mats and, from August, *Chaetomorpha linum* disappeared from Kertinge Nor (Fig. 16.7). The growth of eelgrass also decreased (see Chap. 3 by Schiewer, this volume). This enabled a constant high flux of nutrients from the sediment, and extremely high Chl *a* concentrations were measured in October. During June 1992, the parent generation of *Ciona intestinalis* died off (due to its natural life cycle) and

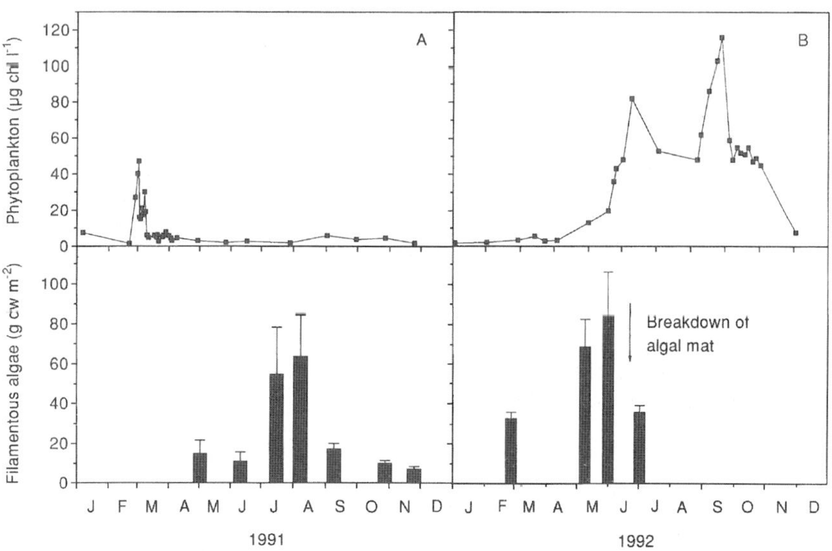

Fig. 16.7 Chlorophyll-*a* concentration and biomass of filamentous algae in Kertinge Nor during 1991 (**A**) and 1992 (**B**). From August and throughout the remaining part of 1992, filamentous algae were absent in Kertinge Nor. From Riisgård et al. (1995)

the filtration capacity of the ascidian population was low at the end of June to the beginning of July because only newly settled specimens were present. The decline in filtration capacity coincided with the increasing phytoplankton biomass. In July, algal cell concentrations exceeded optimum concentrations for filtration, and *C. intestinalis* became saturated and subsequently reduced its filtration rate to a minimum of its potential capacity (Petersen and Riisgård 1992). *C. intestinalis* did not, therefore, control phytoplankton during the summer and autumn of 1992 in Kertinge Nor. However, the marked change in primary producers had no significant influence on the jellyfish population, and the high predation rate prevented the meso-zooplankton from increasing in numbers in response to the increased phytoplankton biomass. Thus, even during extreme situations with Chl *a* concentrations of up to 120 µg l^{-1}, the *Aurelia aurita* population had an important controlling impact on the zooplankton density.

The pelagic biomass levels and the successions of plankton species were similar during the spring periods of 1991 and 1992, but developments were very different during the two summer and autumn periods (Fig. 16.7). In 1991, auto- and heterotrophic biomasses were low and dominated by diatoms and dinophyceans as well as ciliates, rotifers and epibenthic harpacticoids. In 1992 the biomass of phytoplankton was exceptionally high. At the beginning of that period the autotrophic biomass was dominated by the diatoms *Skeletonema costatum* and *Stephanodiscus hantzschii*. The diatom bloom was succeeded by a bloom of small Cyanobacteria that lasted until the end of November. The heterotrophic biomass was dominated by ciliates at the beginning of the period, but was later succeeded by heterotrophic dinoflagellates appearing in very high biomasses (Riisgård et al. 1996a).

16.3.3.3 Impact of Chaetomorpha linum Mats

Krause-Jensen et al. (1996) studied the productivity of dense mats of *Chaetomorpha linum* in Kertinge Nor in relation to the vertical gradients of light and nutrient availability created within the mats. This was done by incubating a 15 cm dense mat of *C. linum* in the laboratory at low and high surface irradiance, and simultaneously simulating the nutrient efflux from the anoxic sediment by pumping nutrients up through the mat. The algal activity resulted in steep vertical gradients in O_2 and NH_4^+ concentration profiles within the mats. In the light, O_2 production caused supersaturation in the surface layers, and algal assimilation significantly reduced the flux of nutrients to the water column. The depth gradients of decreasing light and increasing nutrient availability within the mat suggested light limitation in the bottom of the mat, and progressive N limitation towards the mat surface. Algal productivity declined with depth in the mats, reflecting a pronounced self-shading, and the depth of 1% surface irradiance was only 8 cm. Krause-Jensen et al. (1996) found that filamentous macroalgal mats can switch from being net productive to a status where consumption exceeds production, and it was concluded that reduced irradiance and high water temperature may trigger such shifts in a macroalgal-dominated ecosystem. In similar studies, McGlathery et al. (1997) found that the patterns of ammonium uptake within dense mats of actively growing *C. linum* can efficiently restrict benthic nutrient inputs to the overlying water and thus reduce nutrient availability to a level that may limit phytoplankton production.

16.3.3.4 Ecosystem Stability

It may be concluded that Kertinge Nor is an inherently unstable eutrophic ecosystem in which interactions between suspension-feeding organisms and mobilisation of nutrients from the sediment determine the dynamics of the biological structure. The instability of the system blurs the possible effects of the significant reduction in nutrient discharge to the fjord system by the end of 1989. Since 1992, mats of filamentous algae (*Chaetomorpha linum* and *Cladophora sericea*) have frequently been observed in Kertinge Nor, but not to the same extent as in 1992 (Fyn County 2004), while the number of both ascidians and jellyfish do not seem to have not changed since then. During summer periods with calm and warm weather, sporadic incidents of oxygen depletion and macroalgae up-floating and subsequent decay followed by nutrient release from the bottom and subsequent local algal blooms may still – at least for a certain period of time – change the 'normal' clear water into green water, as e.g. observed in 2001 (Fyn County 2002).

16.3.4 Implications of Density-Driven Currents for Predation Impact by Jellyfish

In Kertinge Nor, the maximum diameter of the umbrella of *Aurelia aurita* is usually only a few centimetres, although high abundances of such small jellyfish (up to several hundred per cubic metre of water) control the zooplankton biomass (Olesen et al. 1994; Riisgård et al. 1995; Frandsen and Riisgård 1997). The implications of density-driven currents for the interaction between *Aurelia aurita* and zooplankton, and thus the phytoplankton, were studied in the Kertinge Nor by Nielsen et al. (1997). It was found that the local population of jellyfish is highly influenced by the density-driven circulation created by the frequent salinity changes in the adjacent Great Belt. When new water of either higher or lower salinity enters the fjord, changes in the jellyfish distribution are apparent. The time it takes for the jellyfish to enter a new water mass of higher or lower salinity has been found to be dependent on the degree of changes in salinity. In laboratory experiments, Nielsen et al. (1997) found that the adaptation time for equilibrium buoyancy and normal swimming of the jellyfish is directly proportional to the salinity difference. Osmo-conforming times of 2–4 h were found for the salinity gradients of 2–4 psu typically measured in the fjord. Further, the number and distribution of zooplankton was found to be highly influenced by the presence of jellyfish. The disappearance of incoming holoplanktonic copepods from the Great Belt occurred simultaneously with the conquest of the new water mass by the jellyfish (Fig. 16.8). The filtering activity of the *A. aurita* population was able to keep the zooplankton biomass low (mean residence time of a zooplankton organism varied between 10 and 20 h), and negligible zooplankton grazing explains why the Chl *a* concentration can increase markedly during a few days of rapid growth when the benthic filter feeding *Ciona intestinalis* is uncoupled due to stratification (see Riisgård et al. 1996b, 1998). The occurrence of *A. aurita* in different hydrographical situations showed that the density of jellyfish was always highest in the "old" fjord water (Nielsen et al. 1997).

16.3.5 Density-Driven Currents and Grazing Impact of Filter-Feeding Benthos

The realisation of the benthic filter feeders grazing potential is highly dependent on currents and mixing of the overlying water. Grazing impact is enhanced by the turbulent mixing of the water mass due to wind-, wave- and current action, coupling the benthic filter feeders to the pelagic biomasses. However, the grazing of *Ciona intestinalis* in Kertinge Nor may be restricted by the stratification of the water body caused by salinity differences, as frequently seen in the Great Belt. The stratification

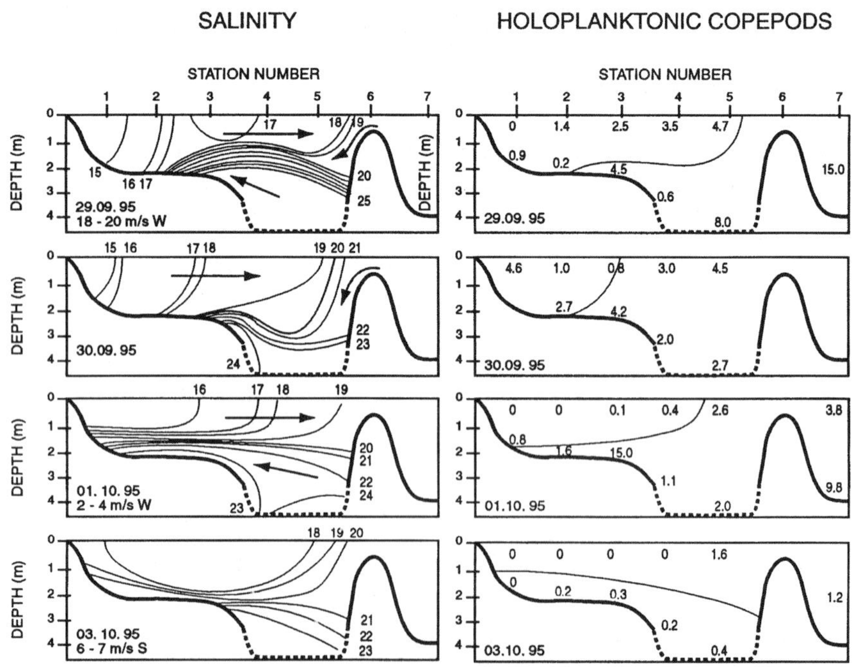

Fig. 16.8 Hydrographical situations (sketches on *left*) and the biomasses (μg C l^{-1}) of holoplanktonic copepods (*right*) on 4 days between 29 September 1995 and 3 October 1995. Wind speed and direction is indicated, and *arrows* show the direction of density-driven water circulation in the fjord. From Nielsen et al. (1997)

reduces, or even prevents, the supply of algae across the interface, decoupling the benthic filter feeders from the pelagic biomass. Thus, a key to the understanding of spatial and time dependent variations in pelagic biomasses is a knowledge of the circumstances under which coupling/decoupling of zoobenthic filter feeding take place. The hydrodynamic conditions in Kertinge Nor have been studied in detail by Riisgård et al. (1996b, 1998). The main focus of these studies was on the importance of density-driven currents in determining the grazing impact of *C. intestinalis*. Thus, it was observed that the grazing behaviour of *C. intestinalis* may at times be restricted to a boundary-layer flow created by these currents, which are in turn controlled by varying hydrographic conditions in the surrounding open sea (Fig. 16.9). It was shown that the grazing impact caused by ascidians on horizontally flowing water could be described by means of a simple numerical model. A satisfactory agreement between observations and modelled predictions emphasised the importance of filter-feeding benthos. For the idealised case, the model is expressed by the equation (Riisgård et al. 1996b):

$$Cx = C_0 e^{--(fx/Y)} \qquad (3)$$

where C_x = algal concentration at a distance downstream x, C_0 = initial concentration, $f = F/v_c$; F = area specific population filtration rate of filter-feeding benthos, v_c = current velocity; and Y = depth of mixed bottom layer. The above equation illustrates how the algal concentration decreases as a function of dimensionless scales for velocity and length.

The velocity of density-driven currents, as well as the importance of tidal currents and wind-mixing, for the supply of phytoplankton to the filter feeding *Ciona intestinalis* in Kertinge Nor was studied by Riisgård et al. (1998). The aim was to examine the dynamic interactions between the water column and benthic filter

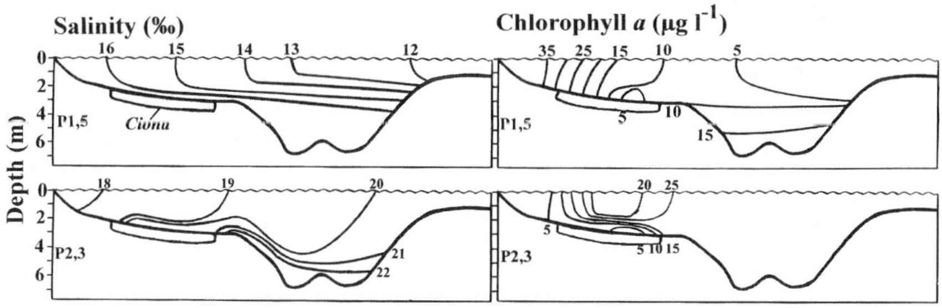

Fig. 16.9 Iso-lines for salinity and Chl *a* in Kerteminde Fjord/Kertinge Nor on two separate days (P1,5 and P2,3) in late summer 1994. The area populated with *Ciona intestinalis* in Kertinge Nor is indicated. On the first day, salinity observations (*upper left*) suggest that a 0.5 m-thick near-bottom density-driven current was flowing northwards along the bottom. The upstream Chl *a* concentration, measured to be 36 µg l^{-1}, was reduced to a downstream concentration of 5 µg l^{-1} by passing over the *C. intestinalis* bed. On the 2nd day (*lower panels*), no density-driven current was detected. The stratification, which uncoupled the benthic filter feeders, caused low Chl-*a* concentrations in the near-bottom water. However, above the halocline, a phytoplankton bloom was developing. From Riisgård et al. (1996b)

feeders in order to account for the variability of phytoplankton biomass seen in the fjord. The moderate tidal- and density-driven near-bottom currents were both estimated and directly measured by means of an underwater-video technique, and the downstream reduction in algal concentration over the ascidian bed was both measured and modelled by means of Eq. 3.

Riisgård et al. (1998) undertook a number of cruises in Kerteminde Fjord/Kertinge Nor during 1995 in order to measure salinity profiles. Over the whole investigation period, the number of clockwise and anti-clockwise circulations were almost the same, and the incidence of salinity stratifications (i.e. salinity differences > 0.2 psu between surface and bottom) in Kertinge Nor was about 50%. In September 1995, during a pronounced density-driven circulation, 'extended' cruises were performed, including collection of water samples for Chl *a* measurement. The Chl *a* concentrations in the near-bottom water above the *Ciona* bed were lower than in the upper layer, and Riisgård et al. (1998) suggested that the relatively high algal concentrations developing in the surface layers during stratification were available to the benthic grazers only if a density-driven circulation carried these algae down to the bottom. It was concluded that, although wind is a crucial force for vertical mixing and transport of phytoplankton to the bottom, the frequent salinity changes in the Great Belt give rise to density-driven currents of considerable importance for the nourishment of the filter-feeding benthos in Kertinge Nor.

Food consumption of benthic filter feeders is, in general, dependent on currents in the benthic boundary layer. In some cases, for example in calm weather or in waters characterised by very moderate tidal currents (e.g. the diurnal tide is only 0.2 m in Kertinge Nor), the near-bottom current speed may be slow: <0.001 m s^{-1}. Near-bottom vertical profiles of phytoplankton caused by a dense population of filter-feeding *Nereis diversicolor* in the southernmost part of Kertinge Nor have been observed in the field by Riisgård et al. (1996c). Water samples were collected simultaneously at different heights above the bottom where *N. diversicolor* were present, and it appeared that a phytoplankton-reduced near-bottom water layer of 0.05–0.1 m in thickness developed on calm days. That such a depletion of phytoplankton in near-bottom waters plays a significant role for this worm was demonstrated in field-growth experiments performed with worms transferred to glass tubes placed at different heights above the bottom (Riisgård et al. 1996c). A reduction in growth rate of bottom dwelling *N. diversicolor* of around 10 times compared with that of worms elevated just 0.1 m above the sediment surface indicates that extremely meagre food conditions are prevalent near the seafloor.

16.3.6 Biogeochemistry, Microphytobenthos and Macrozoobenthos

The distribution of the three nereid polychaetes *Nereis virens*, *N. diversicolor* and *N. succinea* between the shoreline and 1 m water depth in Kerteminde Fjord/Kertinge Nor was studied during the autumn of 1991 by Miron and Kristensen

(1993). The results suggested that pore water sulphide produced during sulphate reduction – the major anaerobic mineralisation process in the sediment – acts on the distribution of these species. *N. virens* was confined to low-sulfidic areas (<50 µM) and was found throughout most of the fjord system, with highest densities (>500 ind. m^{-2}) in the western part of Kertinge Nor. *N. succinea* was found in high-sulfidic sediments (from 50 to 2,000 µM) of the dense bed of mussels (*Mytilus edulis*) near the mouth of Kerteminde Fjord; *N. diversicolor* showed a broader distribution with respect to pore water sulfide and was found mainly in the southern and eastern part of Kertinge Nor.

The impact of the non-suspension-feeding *Nereis virens* and the facultative suspension-feeding *N. diversicolor* on C and N dynamics in organic-poor sediment from the innermost part of Kertinge Nor was investigated in the laboratory by Christensen et al. (2000). It was found that the oxygen consumption and DIN release were increased by a factor of 3 in sediment with *N. diversicolor*, but only by a factor of 1.5 in sediment with *N. virens*, and the deposition of particulate C and N to the sediment was up to 30 times higher in sediments inhabited by the much more actively water pumping *N. diversicolor* (see also review by Riisgård and Larsen 2005).

Rysgaard et al. (1995) studied seasonal variations in oxygen and nutrient fluxes, and denitrification in the shallow (0.5 m) southernmost part of Kertinge Nor, which has sandy sediment colonised by benthic microalgae and burrowing zoobenthos. Oxygen dynamics in the upper sediment layers were controlled by the microalgae and there was a net flux of O_2 out of the sediment during spring and autumn and a reduced efflux of NH_4^+ and PO_4^{3-} from the sediment to the water during daytime. Denitrification based on NO_3^- from the water column (D_w) occurred only in winter and spring – when NO_3^- was present in the water column – and activity was proportional to the water column NO_3^- concentration. Coupled nitrification–denitrification (D_n) in the sediment was stimulated by O_2 production during winter and spring, at which times NO_3^- and NH_4^+ were present in the water column in high concentrations. In contrast, during summer, when the concentration of NO_3^- and NH_4^+ in the water column was low, benthic microalgae inhibited D_n by competing with nitrifying bacteria for NH_4^+. D_w accounted for 80% of the total denitrification during winter, while on an annual basis, D_w and D_n each accounted for 50% of the total denitrification activity. Oxygen consumption, D_w and D_n were linearly correlated with the density of *Corophium volutator*, and all the processes studied were stimulated by the water-pumping zoobenthos. But, because the concentration of inorganic nitrogen in the overlying water and the sediment nitrification potential were both low during summer when zoobenthic density is high, Rysgaard et al. (1995) concluded that the stimulatory effect of burrowing, water-pumping zoobenthos is of minor importance to the annual denitrification budget in Kertinge Nor. Impacts of bioturbating zoobenthos (*Nereis diversicolor*, *Corophium volutator*, *Hydrobia* spp.) on sediment metabolism and nutrient fluxes in different Kertinge Nor sediments were also studied by Hansen and Kristensen (1997).

16.3.7 Fish

Around 16 species of fish have been recorded in Kerteminde Fjord/Kertinge Nor (Larsen et al. 1994). A number of fish species are found only, or mainly, in Kerteminde Fjord: *Clupea harengus* (herring), *Salmo trutta* (sea trout), *Syngnathus typhle* (pipefish), *Solea solea* (sole), *Platichtys fleses* (flounder), *Merlangius merlangus* (whiting), *Sprattus sprattus* (sprat). Other species of fish are found mainly in Kertinge Nor: *Gasterosteus aculeatus* (three-spined stickleback), *Pungitius pungitus* (nine-spined stickleback), *Myoxocephalus scorpius* (short-horn sculpin), *Perca fluviatilis* (perch), and some fish species occur in the whole fjord system, e.g. *Anguilla anguilla* (eel), *Hyperoplus lanceolatus* (sand eel), *Zoarces viviparus* (viviparous blenny), *Pomatoschistus microps* (clay goby), *Gobis niger* (black goby).

The dominating species in Kertinge Nor are three- and nine-spined stickleback. Due to their very high numbers, especially of three-spined sticklebacks, they have been suggested to play an important ecological role in Kertinge Nor, where they compete with jellyfish for zooplankton (Fig. 16.6). Only a few part-time fisherman catch fish, mainly eel (and shrimp), in the fjord system but no official fishing statistics exist.

16.4 Odense Fjord and Kerteminde Fjord/Kertinge Nor: Protection Measures and Environmental Objectives

16.4.1 International Protection Measures

The European Union (EU) Water Framework Directive encompasses all surface waters, stipulating that all water bodies will have to fulfill "good ecological status" by 2015, meaning that only slight deviations from undisturbed conditions will be acceptable. The whole of the inner Odense Fjord and the western part of the outer fjord has been designated as an international nature protection area – Natura 2000 site – according to the EU "Habitats" and the "Birds" directives. In these areas, "good preservational status" must be achieved for the species and habitat types selected following the directives. In Denmark, the whole country has been designated as nutrient-sensitive area due to the "Nitrates" and "Urban Waste Water" directives, meaning that outlets from point sources and diffuse sources to all surface waters are basically regulated by national measures pursuant to these directives, i.e the Aquatic Action Plans I, II and III. No areas of the fjord are highlighted for protection following the "Bathing Water" or "Shellfish" directives. The international Helsinki Convention encompasses the Baltic Sea and adjacent fjord areas, and thus includes Odense Fjord and Kerteminde Fjord/Kertinge Nor. The ministerial declaration of the Convention aims at a 50% reduction in nutrient outlets and a 'generation target' of reducing outlets of hazardous substances to near zero.

16.4.2 National Protection Measures

In Denmark, the water quality of coastal waters is administered at the regional political and administrative level, i.e. the counties, through a Regional Plan that is revised every 4 years. In this plan, the objectives of water quality, and measures to protect and enhance water quality, are decided politically. Due to the natural qualities of Odense Fjord and Kertinge Nor, the northwestern part of Odense Fjord and Kertinge Nor have been designated as "reference areas for scientific studies" (stringent quality objective) in Fyn County's Regional Plan for 2001–2013. The quality objective for the remainder of the fjord is "fish waters for angling and/or fishery" and, where natural conditions permit, "spawning and/or nursery grounds for fish" (general/basic quality objective). The fjord is also encompassed by a number of other national and regional protection regulations, focusing especially on the impact of public access, hunting and fishing.

Denmark as a nation has been progressive with respect to the introduction of legislation designed to reduce eutrophication. In 1987, measures were adopted to reduce nitrogen and phosphorus input to the sea by 50 and 80%, respectively, within the following 5 years (Richardson 1996). The goals with respect to phosphorus reduction and nitrogen reduction from wastewater treatment were indeed met on time (see Fig. 16.2). But the desired reduction in runoff of nitrogen from agricultural land and the ecological quality objectives defined for the coastal waters have not yet been fulfilled (Rask et al. 1999; Iversen et al. 1998; Conley et al. 2000; Ærtebjerg et al. 2003). New measures are thus being considered for the purpose of further reducing, primarily, the run off from diffuse sources (agriculture and scattered dwellings), in order to also fulfill international requirements pursuant to e.g. the Water Framework directive, and the Birds and Habitat directives. This reduction demand on nitrogen is considered to be important for the marine environment because nitrogen is the primary limiting nutrient in Danish coastal waters, as is most pronounced in open areas (Ærtebjerg et al. 2003). A close connection between nitrogen concentration and phytoplankton biomass may therefore be expected. The median phytoplankton biomass in Danish fjords and coastal waters is about 5 μg Chl *a* l^{-1} and the biomass is coupled to the total nitrogen concentration (Sand-Jensen et al. 1994). During the summer period, total nitrogen concentrations can account for about 60% of the variability in Chl *a* concentrations (Sand-Jensen et al. 1994). The remaining 40% variation, however, is likely to be due to the complicated interplay between hydrography and filter-feeding zoobenthos, and nutrient competition with ephemeral macroalgae, etc. as demonstrated for Odense Fjord and Kertinge Nor in the present case study. Knowledge about the interactions between currents, wind- and biomixing, density-driven circulation, macrophytes and filter-feeding animals, may explain the otherwise unaccountable large variations in phytoplankton biomass. Such insight is of great importance for marine monitoring programmes using Chl *a* as a parameter in the assessment of environmental conditions.

Acknowledgements Thanks are due to Inga Housted Holm, Flemming Nørgaard, Birte Vindt, Mogens Michael Møller, Elsebeth Glob, and Nikolaj Holmboe, Fyn County, for help with data treatment, figures and maps. Poul Scheel Larsen and Marianne Holmer are acknowledged for constructive comments on the manuscript.

References

Ærtebjerg G, Andersen JH, Hansen OS (eds) (2003) Nutrients and eutrophication in Danish marine waters. A challenge for science and management. National Environmental Research Institute, Copenhagen

Borum J (1996) Shallow waters and land/sea boundaries. In: Jørgensen BB, Richardson K (eds) Eutrophication in coastal marine ecosystems. Coastal and estuarine studies, vol 52. American Geophysical Union, Washington DC, pp 179–203

Bonsdorff E, Blomquist EM, Mattila J, Norkko A (1997) Coastal eutrophication: causes, consequences and perspectives in the archipelago areas of the northern Baltic Sea. Estuarine Coast Shelf Sci 44[Suppl A]:63–72

Christensen B, Vedel A, Kristensen E (2000) Carbon and nitrogen fluxes in sediment inhabited by suspension-feeding (*Nereis diversicolor*) and non-suspension-feeding (*N. virens*) polychaetes. Mar Ecol Prog Ser 192:203–217

Christensen PB, Møhlenberg F, Krause-Jensen D, Jensen HS, Rysgaard S, Clausen P, Schluter L, Josefsen SB, Andersen FØ, Thomassen J, Thomsen MS, Jürgensen C, Møller MM (1994) Stoftransport og omsætning i Kertinge Nor/Kerteminde Fjord (in Danish with English summary). Havforskning fra Miljøstyrelsen. Nr. 43 1992. Miljøstyrelsen

Conley DJ, Kaas H, Møhlenberg F, Rasmussen B, Windolf J (2000) Characteristics of Danish Estuaries. Estuaries 23:820–837

Fyn County (1991) Eutrophication of coastal waters. Coastal water quality management in the county of Funen, Denmark, 1976–1990. Funen County Council

Fyn County (2001a) Aquatic environment of Fyn, Denmark, 1976–2000. Streams and lakes, coastal waters, groundwater, environmental impact of wastewater and agriculture. Fyn County, Odense, Denmark

Fyn County (2001b) Kystvande 2000 (in Danish). Vandmiljøovervågning. Natur- og Vandmiljøafdelingen, Fyns Amt

Fyn County (2002) Kystvande 2001 (in Danish). Vandmiljøovervågning. Natur- og Vandmiljøafdelingen, Fyns Amt

Fyn County (2003) Odense Pilot River Basin. Provisional Article 5 report pursuant to the Water Framework Directive. Fyn County

Fyn County (2004) Kystvande 2003 (in Danish). Fyn County Council

Frandsen K, Riisgård HU (1997) Size dependent respiration and growth of jellyfish (*Aurelia aurita*). Sarsia 82:307–312

Greve TM, Borum J, Pedersen O (2004) Meristematic oxygen variability in eelgrass (*Zostera marina*). Limnol Oceanogr 48:210–216

Hansen K, Kristensen E (1997) Impact of macrofaunal recolonization on benthic metabolism and nutrient fluxes in a shallow marine sediment previously overgrown with macroalgal mats. Estuarine Coastal Mar Sci 45:613–628

Holmer M, Bondgaard EJ (2001) Photosynthetic and growth response of eelgrass (*Zostrea marina)* to low oxygen and high sulfide concentrations during hypoxic events. Aquat Bot 70:29–38

Iversen TM, Grant R, Nielsen K (1998) Nitrogen enrichment of European inland and marine waters with special attention to Danish policy measures. Environ Poll 102:771–780

Jensen HF, Holmer M, Dahllöf I (2004) Effect of tributyltin (TBT) on Seagrass (*Ruppia maritima*). Mar Pollut Bull 49:564–573

Jürgensen C (1995) Modelling of nutrient release from sediment in a tidal inlet, Kertinge Nor, Funen, Denmark. Ophelia 42:163–178

Krause-Jensen D, McGlathery K, Rysgaard S, Christensen PB (1996) Production within dense mats of the filamentous macroalga *Chaetomorpha linum* in relation to light and nutrient availability. Mar Ecol Prog Ser 134:207–216

Kristensen E, Jensen MH, Aller, RC (1991). Direct measurement of dissolved inorganic nitrogen exchange and denitrification in individual polychaete (*Nereis virens*) burrows. J Mar Res 49:355–377

Kristensen E, Kristiansen KD, Jensen MH (2003) Temporal behavior of manganese and iron in a sandy coastal sediment exposed to water column anoxia. Estuaries 26:690–699

Larsen GR, Jürgensen C, Christensen, PB, Olesen NJ, Petersen JK, Jensen JN, Mortensen E, Sortkjær O, Andersen P (1994) Kertinge Nor/Kerteminde Fjord - status og udvikling (in Danish with English summary). Havforskning fra Miljøstyrelsen Nr. 44 1994. Miljøstyrelsen

Larsen PS, Riisgård HU (1997) Biomixing generated by benthic filter-feeders: a diffusion model for near-bottom phytoplankton depletion. J Sea Res 37:81–90

McGlathery KJ, Krause-Jensen D, Rysgaard S, Christensen PB (1997) Patterns of ammonium uptake within dense mats of the filamentous macroalga *Chaetomorpha linum*. Aquat Bot 59:99–115

Miron G, Kristensen E (1993) Factors influencing the distribution of nereid polychaetes: the sulfide aspect. Mar Ecol Prog Ser 93:143–153

Møller JS (1996) Water masses, stratification and circulation. In: Jørgensen BB, Richardson K (eds) Eutrophication in a coastal marine ecosystem. Coastal and Estuarine Studies vol 52. American Geophysical Union, Washington DC, pp 51–66

Møller LF, Riisgård HU (2006) Filter feeding in the burrowing amphipod *Corophium volutator*. Mar Ecol Prog Ser 322:213–224

Nielsen AS, Pedersen AW, Riisgård HU (1997) Implications of density driven currents for interaction between jellyfish (*Aurelia aurita*) and zooplankton in a Danish fjord. Sarsia 82:297–305

Nielsen SL, Sand-Jensen K, Borum J, Geertz-Hansen O (2002) Depth colonization of Eelgrass (*Zostrea marina*) and macroalgae as determined by water transparency in Danish coastal waters. Estuaries 25:1025–1032

Nienhuis PH (1996) The North Sea coasts of Denmark, Germany and the Netherlands. In: Schramm W, Nienhuis PH (eds) Marine benthic vegetation. Recent changes and the effects of eutrophication. Ecological Studies, vol 123. Springer, Berlin, pp 187–221

Olesen NJ (1995) Clearance potential of the jellyfish *Aurelia aurita*, and predation impact on zooplankton in a shallow cove. Mar Ecol Prog Ser 124:63–72

Olesen NJ, Frandsen K, Riisgård HU (1994) Population dynamics, growth and energetics of jellyfish *Aurelia aurita* in a shallow fjord. Mar Ecol Prog Ser 105:9–18

Ostenfeld CH (1908) On the ecology and distribution of the grasswrack (*Zostera marina*) in Danish waters. Report of the Danish Biological Station to the Board of Agriculture XVI:1–62

Pedersen MF, Borum J (1997) Nutrient control of estuarine macroalgae. Growth strategy and the balance between nitrogen requirements and uptake. Mar Ecol Prog Ser 161:155–163

Petersen JK, Riisgård HU (1992) Filtration capacity of the ascidian *Ciona intestinalis* and its grazing impact in a shallow fjord. Mar Ecol Prog Ser 88:9–17

Plus M, Deslous-Paoli J-M, Dagault F (2003) Seagrass (*Zostera marina* L.) bed recolonisation after anoxia-induced full mortality. Aquat Bot 77:121–134

Rask N, Pedersen SE, Jensen MH (1999) Response to lowered nutrient discharges in the coastal waters around the island of Fynen, Denmark. Hydrobiologia 393:69–81

Rasmussen B, Josefson A (2002). Residence times of micro-tidal estuaries. Estuarine Coast Shelf Sci 54:65–73

Richardson K (1996) Conclusion, research and eutrophication control. In: Jørgensen BB, Richardson K (eds) Eutrophication in a coastal marine ecosystem. Coastal and Estuarine Studies, vol 52. American Geophysical Union, Washington DC, pp 243–267

Riisgård HU (1991) Suspension feeding in the polychaete *Nereis diversicolor*. Mar Ecol Prog Ser 70:29–37

Riisgård HU (1998) Filter feeding and plankton dynamics in a Danish fjord: a review of the importance of flow, mixing, and density-driven circulation. J Environ Manage 53:195–207

Riisgård HU, Banta GT (1998) Irrigation and deposit feeding by the lugworm *Arenicola marina*, characteristics and secondary effects on the environment. A review of current knowledge. Vie Milieu Period Ecol Gen 48:243–257

Riisgård HU, Kamermans P (2001) Switching between deposit and suspension feeding in coastal zoobenthos. In: Reise K (ed) Ecological comparisons of sedimentary shores. Ecological studies series. Springer, Berlin, pp 73–101

Riisgård HU, Larsen PS (2005) Water pumping and analysis of flow in burrowing zoobenthos: an overview. Aquat Ecol 39:237–258

Riisgård HU, Christensen PB, Olesen NJ, Petersen JK, Møller MM, Andersen P (1995) Biological structure in a shallow cove (Kertinge Nor, Denmark) – control by benthic nutrient fluxes and suspension-feeding ascidians and jellyfish. Ophelia 41:329–344

Riisgård HU, Jürgensen C, Andersen FØ (1996a) Case study: Kertinge Nor. In: Jørgensen BB, Richardson K (eds) Eutrophication in coastal marine ecosystems. Coastal and Estuarine Studies, vol 52. American Geophysical Union. Washington DC, pp 205–221

Riisgård HU, Jürgensen C, Clausen T (1996b) Filter-feeding ascidians (*Ciona intestinalis*) in a shallow cove: implications of hydrodynamics for grazing impact. J Sea Res 35:293–300

Riisgård HU, Poulsen L, Larsen PS (1996c) Phytoplankton reduction in near-bottom water caused by filter-feeding *Nereis diversicolor* – implications for worm growth and population grazing impact. Mar Ecol Prog Ser 141:47–54

Riisgård HU, Jensen AS, Jürgensen C (1998) Hydrography, near-bottom currents and grazing impact of benthic filter-feeding ascidians (*Ciona intestinalis*) in a Danish fjord. Ophelia 49:1–16

Riisgård HU, Kittner C, Seerup DF (2003) Regulation of opening state and filtration rate in filter-feeding bivalves (*Cardium edule*, *Mytilus edulis*, *Mya arenaria*) in response to low algal concentration. J Exp Mar Biol Ecol 284:105–127

Riisgård HU, Seerup DF, Jensen MH, Glob E, Larsen PS (2004) Grazing impact of filter-feeding zoobenthos in a Danish fjord. J Exp Mar Biol Ecol 307:261–171

Rysgaard S, Christensen PB, Nielsen LP (1995) Seasonal variation in nitrification and denitrification in estuarine sediment colonized by benthic microalgae and bioturbating infauna. Mar Ecol Prog Ser 126:111–121

Sand-Jensen K, Nielsen SL, Borum J, Geertz-Hansen O (1994) Fytoplankton- og makrofytudviklingen i danske kystområder (Phytoplankton and macrophyte development in Danish coastal waters) (in Danish with English summary). Havforskning fra Miljøstyrelsen nr. 30, Miljøministeriet, Copenhagen

Schramm W (1996) The Baltic Sea and its transition zones. In: Schramm, W, Nienhuis PH (eds) Marine benthic vegetation. Recent changes and the effects of eutrophication. Ecologial Studies 123. Springer, Berlin, pp 131–163

Vedel A (1998) Phytoplankton depletion in the benthic boundary layer caused by suspension-feeding *Nereis diversicolor* (Polychaeta): grazing impact and effect on temperature. Mar Ecol Prog Ser 163:125–132

Vedel A, Andersen BB, Riisgård HU (1994) Field investigations of pumping activity of the facultatively filter-feeding polychaete *Nereis diversicolor* using an improved infrared phototransducer system. Mar Ecol Prog Ser 103:91–101

Chapter 17
Synthesis

U. Schiewer

17.1 Common Features and Differences

Due to their geology, Baltic coastal areas are highly diverse. Thus the northern and northeastern Baltic is shaped by rock basement of the Scandinavian plate, whereas moraine landscapes and soft soils are typically found throughout the southern, southwestern and southeastern Baltic. Furthermore, constantly changing conditions occur due to the rising northeastern edge and trailing southern shore, in addition to the successive formation of mature shorelines throughout the south and southeast. This results in at least ten different coastal forms (see Chap. 2 by Schiewer, this volume; Fig. 2.1), which can comprise a variety of soil formations ranging from silt, sand and gravel to various types of hard soil.

On the basis of case studies a rough geographic subdivision can be made into southern and eastern coastal waters on the one hand and northern and western Baltic waters on the other.

Some main features of the southern and eastern Baltic coastal waters are:

- Heavier eutrophication as compared to the northern and western Baltic.
- Higher chlorophyll *a* (Chl *a*) concentrations.
- Nutrient limitations, particularly of N. With heavy eutrophication light limitation occurs, at least temporarily.
- Noticeably higher phytoplankton diversity than in the open Baltic.
- Frequent loss of diatom dominance.
- In extreme cases, all-season dominance of Cyanobacteria and green algae.
- Higher zooplankton diversity than in the open sea. Dominant groups are copepods, rotifers and cladocerans.
- With high eutrophication, decreased importance of copepods and increased importance of rotifers and protozoa.
- With greater river influx, a characteristic increase of rotifers and protozoa.
- Due to increased sedimentation of organic material a reduced amount of oxygen in the sediment is typical. This is particularly distinct for deeper waters.
- Composition of submersed macrophytes is subject to strong variability; fast-growing species are favoured.
- Decreased species diversity of submersed macroalgae with temporary absence.

- Decreased species diversity of macrozoobenthos, particularly of bivalve molluscs. Increased importance of oligochaetes and chironomid larvae.

Exceptions to this pattern are coastal waters with:

- Stronger exchange with the Baltic.
- Very strong fresh water influx from rivers.

In contrast, the northern and western Baltic waters are characterised by:

- Less eutrophication.
- Lower Chl *a* amount, but still noticeably higher than in the Baltic Proper.
- Permanent dominance of diatoms and summer dominance of dinoflagellates and N_2-fixating Cyanobacteria.
- Distinct seasonal succession of diatoms, green algae/Cyanobacteria, diatoms.
- Noticeably less oxygen limitation in the sediment.
- Frequent occurrences of hard soil, providing better opportunity for colonisation for marine submersed macroalgae.
- Higher species diversity of macrozoobenthos.
- Permanent problems caused by heavy metals.
- Frequent exposure to polychlorinated biphenyls (PCBs) and other pollutants.

Exceptions to this pattern occur only in:

- Inner waters with low exchange with the Baltic.
- Stronger influx from lakes: e.g. from the Mälar Lake to the Stockholm Archipelago.

The main reasons for these differences are the different structure and outputs of the discharge areas, the shallowness of the southern and eastern coastal zones and their often reduced exchanges with the Baltic Sea.

A further differentiation occurs through the hydrographical characteristics of coastal waters, which influence retention times and thereby water exchange modes. Thus we find open coastal areas such as the Greifswalder Bodden alongside series connected subsystems such as the Darß-Zingst boddens; and freshwater-dominated gulf systems such as the Gulf of Finland adjacent to a girdle of skerries, the so-called 'skerry gardens', such as the Stockholm Archipelago.

Coastal ecosystems also differ by their varying catchment areas, by the size ratio between surface of coastal water and catchment area, and by their mean depth.

The degree of eutrophication and pollution of inner coastal waters is influenced by the exchange dynamics with the Baltic: the higher the exchange rate with the Baltic, the lesser the pollution of coastal water ('dilution effect'), but the stronger the impact on the Baltic Sea. This is particularly evident in Neva Bay. Due to the high freshwater influx of $77.6\,km^3\,a^{-1}$, the entire bay is inhabited by freshwater flora and fauna (see Chap. 12 by Telesh et al.). The capacity for self-purification arises, in this case, from a pure flow effect; the actual eutrophication by nutrients occurs only at a reduced water flow rate. If original outlets are degraded or partly eliminated, so that the retention time is prolonged, water quality can deteriorate dramatically. The reasons for this are stronger sedimentation as well as increased eutrophication caused by nutrient utilisation under prolonged retention time (see Chap. 12 by Telesh et al.).

Considering the water exchange dynamics with the Baltic, five main types of coastal areas can be distinguished, as detailed in the following sections.

17.1.1 *Type 1: Open Coast/Near Shore*

Representatives of this type are Tromper Wiek and Koserow, the western Polish shore, the Lithuanian outer coast, and the Askö area. The differences between them arise from their position relative to Baltic basins, their wind exposure, coastal morphometry, sediment type and the inflow from inner coastal waters. The influence of their position relative to the deep basin is already evident near shore when comparing the coastal zones Tromper Wiek and Koserow. For the Tromper Wiek, its proximity to the Arkona basin and the low influx from the catchment area leads to a clear shift in the dissolved organic carbon/particulate organic carbon (DOC/POC) ratio in favour of DOC (Fig. 17.1).

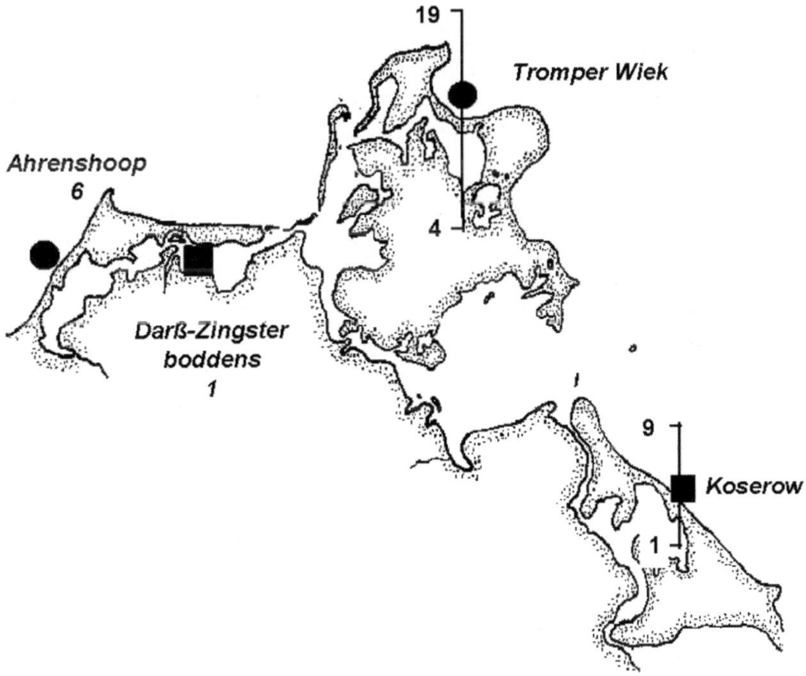

Fig. 17.1 Dissolved organic carbon / particulate organic carbon (DOC/POC) ratios at Koserow and Tromper Wiek, fitted in order with values of neighboring inshore waters (Görs et al. 2000, figure modified from Estrum-Yousef 2001) depending on the expected trophic level with the highly eutrophic Darss-Zingst bodden chain and Ahrenshoop (2 m water depth, Baltic Sea coast without pronounced river run-off). *Bars* Whole range of values during the investigation period in 1997

17.1.2 Type 2: Big Bays

Type 2 is represented by four main areas: the Szczecin Lagoon, the Gulfs of Gdansk and Riga, and the Gulf of Finland.

These big bays generally possess characteristic features. The decisive factors are:

- morphology
- river influx
- exchange opportunity with the Baltic Sea, which influences the water retention time
- water shallowness or deepness and thus the absence or presence of a stable thermocline. However, the formation of the latter is a rare occurrence. A stable halocline never occurs near shore.
- position and connection to open Baltic basins
- size of catchment area and extent of urban colonisation of the shore
- divergent climate: in south–north temperature gradient and in west–east climate variations
- variable anthropogenic exposure.

In general, deeper big bays possess a diverse macrozoobenthos. The individual water bodies exhibit the following features:

- Szczecin Lagoon: medium freshwater input ($18\,km^3\,a^{-1}$); very shallow; average retention time 55 days; no submersed plants; high Chl *a* concentrations; diatoms still in spring; *Microcystis* blooms in summer; typical freshwater macrofauna as well as mass occurrence of oligochaetes and chironomid larvae. Heavy metal exposure (Pb, Zn) and high organic load.
- Gulf of Gdansk: high river input ($35\,km^3\,a^{-1}$); very short retention time (7 days), good connection with Gdansk Deep; stratified water body offshore; lack of oxygen in deep zones; accumulation of faecal bacteria leading to temporary bathing ban in coastal regions; increase in diatom biomasses; N_2-fixing Cyanobacteria and flagellates in summer; dominance of *Acartia*, *Synchaeta* and *Keratella*. Still diverse macrophytobenthos; increase in macrofauna: crustaceans, bivalvia, gastropods. Heavy metal exposure: Pb, Cd, Hg. Puck Bay: distinctly higher eutrophication.
- Gulf of Riga: high river input ($31\,km^3\,a^{-1}$); very long retention time: 2–4 years; mean depth 27 m; still <3–>5 m Secchi depths; boreal diatoms, dinoflagellates; N_2-fixing Cyanobacteria in summer; assumed transition from N- to P-limitation; dominant zooplankton genera *Eurytemora*, *Acartia*, *Bosmina*; medium macrophytobenthos diversity (64 species) but highly productive; diverse macrozoobenthos with biomass increase between 1970 and 1980; *Macoma balthica* as major producer.
- Gulf of Finland: very high freshwater input ($114\,km^3\,a^{-1}$); long retention time (2–3 years); unstable salinity gradient; occurrence of thermocline; ca. 2–3 times more external N- and P-loading compared to the Baltic Sea; diatoms, dinoflagellates; N_2-fixing Cyanobacteria in summer; O_2 lack in deeper zones; diverse macrophytobenthos (97 species); rich macrofauna with *Macoma balthica* as major representative.

17.1.3 Type 3: Good Water Exchange with the Baltic

Water bodies with good exchanges with the Baltic Sea include Kerteminde Fjord/ Kertinge Nor, Wismar-Bucht, Greifswalder Bodden, Haapsalu and Matsalu Bays and the Stockholm Archipelago. The strong exchange with the Baltic makes these water bodies relatively compatible with the adjacent Baltic coast (allochthonal). Their main characteristics are:

- lower risk of eutrophication
- distinct submerse macrophytobenthos
- high oxygen level in the sediment
- diverse macrozoobenthos.

17.1.4 Type 4: Impeded Exchange with the Baltic

Type 4 water bodies are largely autochthonal. Representatives are the Darß-Zingst Bodden (DZB) chain, the Northern Rügener Bodden (NRB), the Usedomer Achterwasser, the Vistula Lagoon, the Schleimündung. Bodden and Haffs belonging to this type are characteristic patterns of the western, southern and southeastern Baltic. Their characteristics are:

- well-defined but unstable gradients
- dominance of physical factors
- short-circuited nutrient recycling
- high sensitivity to nutrient exposures
- increased carbon turnover rates due to dominance of microbial food web (MFW) as a consequence of eutrophication.

A subdivision of a coastal water body into two or more clearly separated subsystems can be beneficial (see Chap. 3 by Schiewer; NRB). Due to frequently alternating transport directions, however, negative influences on the near Baltic sections are inevitable in nutrient-overloaded water bodies (see Chap. 3 by Schiewer; DZB). Such transport fluctuations can also distinctly diminish the chances of successful remediation.

17.1.5 Type 5: Coastal Waters Dominated by River Inflow

This type of coastal waters are allochthonous and have limnic character. Parts of the Szczecin Lagoon, the inner Gulf of Gdansk, the inner Gulf of Riga and the Neva Bay in the Gulf of Finland fall into this category. Their features are:

- high nutrient input
- high flow rates, short water retention times
- little conversion of the available production potential in the transition zone

- dominance of limnetic organisms
- high sensitivity to natural and artificial changes in morphology.

17.1.6 Characteristics of Baltic Coastal Waters

The above-mentioned differentiation simplifies to a certain extent the complexity of coastal ecosystems, but is beneficial for generalised reviews. However, every coastal water has to be considered as unique once the diversity of influencing factors is taken into account. Thus, considering their specificity, Baltic coastal waters may be characterised briefly as follows:

- Schleimündung: Fjord character; decreased exchange with the Baltic Sea ; relatively high salinity near the Baltic coast; dominance of marine organisms in the Outer Schlei; very high eutrophication of the Inner Schlei.
- Salzhaff: small haff with relatively high freshwater input and very good exchange with the Baltic; relatively high salinity in the coastal zone; dominance of submersed macrophytes; well developed macrozoobenthos; mesotrophic.
- DZB: Bodden system divided into four sub-basins connected by narrow and/or very shallow straits; low freshwater input; very high eutrophication due to their shallowness and restricted exchange with the Baltic; increased phytoplankton taxonomic diversity compared to the Baltic; light limitation of phytoplankton; dominance of small Cyanobacteria and green algae; important role of the "fluffy-sediment-layer" (FSL); macrozoobenthos dominated in abundances and biomasses by oligochaetes and chironomid larvae; little pollution; no heavy metals; main alien *Marenzelleria neglecta* (formerly *M. viridis*).
- Northern Rügener Bodden: Bodden system with partial medium exchange opportunities with the Baltic (except Kleiner Jasmunder Bodden); low freshwater input, low nutrient load; increased phytoplankton diversity compared with the Baltic; no heavy metals and low pollution.
- Greifswalder Bodden: Bodden with very good exchange with the Baltic; low pollution through river input; medium nutrient load; eutrophic; normal annual periodicity of phytoplankton; reduced macrophytobenthos, but still relatively rich in submersed macrophytes; well developed macrofauna; very important spawning ground for herring; intensive herring fishing (catches of up to 20,000 t a^{-1}).
- Szczecin Lagoon: Lagoon with temporarily extensive river input; very shallow water; restricted exchange with the Baltic; high eutrophication; partly with light limitation; mass development of *Microcystis aeruginosa* during summer; macrozoobenthos is dominated in abundances and biomasses by oligochaetes and chironomid larvae; well developed beds of *Dreissena polymorpha*; high heavy metal loads and pollution; main aliens are *Eriocheir sinensis*, *Gammarus tigrinus*, *Marenzelleria neglecta*, *Dreissena polymorpha*.
- Near-shore zones:
  - Koserow: dominated by erosion processes; eutrophication by inputs from the Odra River: eutrophic/mesotrophic water body; nutrient limitation (mainly

nitrogen) of phytoplankton; dominance of the MFW and pronounced FSL; greater influence of soft-bottom fauna; large fraction of carbon net-production exported into the Baltic basins: typical transition zone.
- Tromper Wiek: equilibrium between erosion and accumulation; low nutrient input; nutrient limitation (mixture of nitrogen and phosphorus); meso- /oligotrophic water body; grazer-controlled food chain and less developed FSL; balanced net-carbon ecosystem.

- Gulf of Gdansk: A system of estuaries; Gulf with high river input; good exchange with the Baltic; very short retention times (<7 days); direct transition into Gdansk Deep; thermocline, partly halocline; very strong nutrient load; N_2-fixing Cyanobacteria during summer; increased mesozooplankton that has much in common with that of the open Baltic Sea; very strong loads of heavy metals (Pb, Cd, Hg) and organic material; development of faecal bacteria, e.g. *Escherichia coli*, in coastal zones; increase of phyto- and zooplankton; increased abundances of macrozoobenthos; O_2 deficiency in deep zones and decrease of macrozoobenthos; pollution; main aliens *Gammarus tigrinus*, *Marenzelleria neglecta*, *Cercopagis pengoi*.
- Vistula Lagoon: Lagoon with north–south gradient of abiotic parameters due to very narrow connection to the Baltic; small freshwater input since 1915; extensive nutrient load; organic load in the north; crucial role of wind-induced mixing and water transport; medium retention time; high proportion of N_2-fixing Cyanobacteria; dominance of rotifers and freshwater taxa of macrozoobenthos in the western part; decrease of species diversity of macrozoobenthos; eutrophic with a tendency to mesotrophy; main aliens *Marenzelleria neglecta*, *Gammarus tigrinus*, *Cercopagis pengoi*.
- Curonian Lagoon: Lagoon with strong fresh water input in the middle section; restricted exchange with Baltic (wind driven), higher only in northern part; medium nutrient load; important role of wind-induced mixing; dominance of freshwater phyto-and -zooplankton; eutrophic (45 μg Chl *a* l^{-1}); grazing and export of nutrients; macrozoobenthos controlled by salinity; sand- and mud bottom substrates; in Klaipeda area impact of shipping activities (oil, tributyltin, heavy metals); main aliens *Dreissena polymorpha*, *Marenzelleria neglecta*, *Cercopagis pengoi*.
- Gulf of Riga: Gulf of medium depth; strong fresh water inputs; interfering exchange with Baltic; long residence time (2–4 years); high nutrient input; very high numbers of phytoplankton taxa in the Pärnu Bay; phytoplankton dominance (medium Chl *a* amounts between 13–77 μg l^{-1}); very high numbers of phytoplankton species in Pärnu Bay; main zooplankton genera *Eurytemora*, *Acartia*, *Bosmina*; temporary thermocline in some parts; rich macrofauna with biomass increase between 1970/1980; main aliens *Dreissena polymorpha*, *Marenzelleria neglecta*, *Cercopagis pengoi*.
- Haapsalu and Matsalu Bays: Outer coast with good exchange with the Baltic; Haapsalu Bay more differentiated, both morphologically and biotically; low to moderate freshwater inputs; shallow water; moderate nutrient load; low Chl *a* (10–15 μg l^{-1}); normal phytoplankton seasonality; diverse macrozoobenthos; marine as well as fresh water species; ongoing re-mesotrophication since 1990.

- Neva Estuary: Shallow bay with freshwater influx dominating ($79\,km^3\,a^{-1}$); good exchange with the Gulf of Finland, susceptible to decrease in exchange; high load of nutrients and heavy metals (Hg, Cd, Pb, etc.), PCBs, phenols, oils, chlororganic substances; freshwater phyto- and zooplankton; dominance of microzooplankton; mesozooplankton dominated by rotifers; emersed and submersed macrophytes (during last years *Cladophora glomerata* dominance); diverse macrozoobenthos, dominance of oligochaetes and increasing role of chironomid larvae; main aliens *Dreissena polymorpha*, *Pontogammarusd robustoides*, *Marenzelleria neglecta*, *Gmelinoides fasciatus*, *Cercopagis pengoi*.
- Gulf of Finland: Gulf with direct connection to Baltic Proper; 2–3 times higher N- and P-load than in the Baltic Sea; residence time 2–3 years; temperature gradient; instable salinity gradient; two phytoplankton peaks: diatoms, dinoflagellates in spring, N_2-fixing Cyanobacteria in summer; O_2-deficiency in sediment causes P-release and denitrification; rich macrofauna with *Macoma balthica* (formerly *Monoporeia affinis*) dominance.
- Stockholm Archipelago: Subdivided archipelago with direct connection to Baltic; moderate retention time <10–>40 days; low Chl *a* ($2–3\,\mu g\,l^{-1}$); dinoflagellates, diatoms in spring/summer; Cyanobacteria in autumn; species-rich zooplankton including ciliates; hard- and soft soil with species-rich macroflora and -fauna. Declining exposure to PCBs, dichloro-diphenyl-trichloroethane (DDT), Pb, Hg and Cd; main aliens *Cercopagis pengoi*, *Marenzelleria neglecta*, *Dreissena polymorpha*, *Hemimysis anomala*.
- Askö area and Himmerfjärden: Open outer shore or fjord; Baltic water dominance; 2–6 m Secchi-depth; very low Chl *a* ($1.5\,\mu g\,l^{-1}$); diatoms; N_2-fixing Cyanobacteria in summer; dominance of rotifers, cladocerans and copepods; very rich macrophytobenthos and macrozoobenthos; mesotrophic.
- Odense Fjord: Shallow fjord; exchange through tides; 9–17 day retention time; distinct freshwater input and nutrient load: eutrophic (Chl *a* ca. $5\,\mu g\,l^{-1}$, max. 23); diatoms, nanoflagellates, dinoflagellates; ciliates; *Acartia*, copepods; well developed macrophythobenthos and zoobenthos; pollutants: tributyltin (TBT), polycyclic aromatic hydrocarbons (PAHs), PCBs and heavy metals.
- Kerteminde Fjord/Kertinge Nor: Clear tidal rhythmic; low freshwater input; average retention time 1.5 months; mesotrophic (Chl *a* ca. $6\,\mu g\,l^{-1}$); blooms of dinoflagellates *Prymnesium*; main grazer *Aurelia aurita* and *Ciona intestinalis*; alternate dominance of phytoplankton and macrophytes (*Chaetomorpha*, *Cladophora sericea*). Macrozoobenthos: *Nereis virens*, *N. diversicolor*, *Corophium volutator*, *Mytilus edulis*.

The above mentioned basic ecosystems are modified further by abiotic factors. As estuaries are ecotones, i.e. transition zones between adjacent limnic, terrestrial and marine ecosystems, their most characteristic features are gradients. These gradients level off the transition between ecosystems or within the ecosystem itself. Among the abiotic factors, salinity has the strongest effect on the biocoenosis, constituting a relatively sharp barrier between marine and limnic biotopes (Remane and Schlieper 1958). Salinity ranges from 25–15 psu in the surface waters of western

Baltic outlets (see Chap. 16) to 2–0 psu in the pure limnic northeastern parts (see Chap. 12). Salinity accounts for a species minimum in the Baltic (see Fig. 17.2), at which the salinity range of 6–8 psu – characteristic of the central Baltic (Baltic Proper) – is considered to be least hostile ('horohalinikum').

Salinity gradients are much lower in the coastal waters of the central Baltic regions, where they specifically affect only limnic communities. The effect of salinity is enhanced by its chaotic fluctuations in coastal zones, e.g. rapidly varying between > 8 to 0 psu, causing additional metabolic stress to organisms.

Analysis of the influence of salinity on the above-mentioned Baltic coastal waters provides the following results, which are in agreement with Remane's (Remane and Schlieper 1958) statement:

- Higher number of plankton species, particularly phyto- and zooplankton, due to increase of freshwater species diversity in coastal waters compared to marine waters.
- Impact of salinity on macrophytobenthos, which is however compounded by additional negative eutrophication effects.
- Strong influence of salinity on macrozoobenthos composition and species number.

The results for zooplankton are different. In the Baltic Sea, L. Postel (personal communication) found a clear "Remane-relationship" (Bothnian Sea and Bay excluded). In contrast, in the coastal zones the minimum of marine zooplankton is compensated, or overcompensated in most cases, by the introduction of freshwater species, accompanied by a change in the community structure (e.g. see Chaps. 9 and 12).

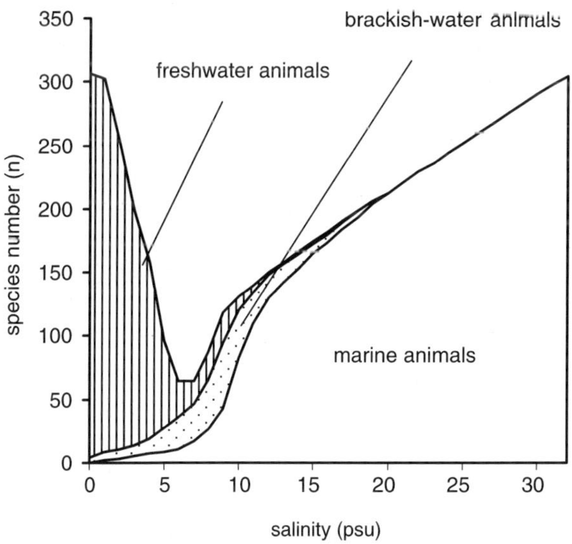

Fig. 17.2 Salinity dependent species minimum according to Remane and Schlieper (1958)

Exceptions are found in coastal waters with low freshwater inputs like Northern Rügener boddens, Greifswalder Bodden and Darß-Zingster boddens. In the latter case the number of species is additionally reduced by the high eutrophication level.

Other important abiotic factors affecting coastal ecosystems are temperature, wind and nutrients. The south–north extent of the Baltic Sea of more than 1,000 km implies a pronounced temperature gradient. Furthermore, there is a west–east influence with the transition from Atlantic climate in the west to continental climate in the east. The relevant variation in the length of the growing season in coastal waters influences species occurrence, their development and metabolic rates.

In contrast, temporary temperature gradients in coastal waters are less important due to their low spatial range. Here, the existence of improved heat resistance for shallow waters seems more important.

Wind force and direction are of considerable importance, causing different rates of mixing and exchange of water masses within the estuary and with the Baltic (e.g. Vistula Lagoon, Curonian Lagoon, Chaps. 8 and 9). In shallow coastal waters, wind-induced Langmuir spirals play an important role in phytoplankton productivity (see Chap. 3).

Tidal effects rarely play a role. Exceptions are the Western Baltic Odense Fjord and the Kerteminde Fjord/Kertinge Nor.

Nutrient gradients are often mesoscalic and depend strongly on catchment areas, the structure of coastal waters and their exchange rate with the Baltic. While limitations of nitrogen (and in spring also temporarily of phosphorus and silica) are common, the Bothnian Sea and Bothnian Bay are phosphor-limited owing to the character of their catchment areas. An increased eutrophication risk exists primarily for shallow coastal waters (e.g. see Chaps. 3, 5, 8 and 16), where steady exchange between sediment and pelagic zone is amplified, thereby assuming the exceptional role of the so-called 'internal' eutrophication of these waters (see Chaps. 3 and 7, etc.). In highly eutrophic coastal waters (e.g. DZB), the FSL plays an important role in this respect. In contrast, temporary "decoupling" between the pelagic zone and the sediment occurs in deeper, more or less stratified costal waters (see Chaps. 13 and 14). Water exchange in these areas proceeds by saltwater inflow and outflow near the bottom. The "dilution effect" in these waters increases, and thus the pollution of the adjacent Baltic waters is enhanced.

Near the shore line, and particularly in the inner costal waters, allochthonous input of terrestrial material (humic substances) is important. In extreme cases a notable saprobisation of water bodies occurs. With the persistence of humic substances, this may cause a shift in the DOC/POC ratio and the occurrence of corresponding gradients (see above; Fig. 17.1). Related to this is a clear reduction in O_2 supply in deeper water layers and particularly in the sediment. Invertebrate soil organisms are primarily affected; their existence in these zones is guaranteed only if their euryhalinity is coupled with resistance to O_2 deficiency (von Oertzen 1988).

17.2 Eutrophication

Eutrophication is a central problem for the Baltic (see Chap. 1). The same goes for its coastal waters, where the process of eutrophication is coupled with characteristic ecosystem changes. Judging from the data available, certain behaviour patterns and relevant processes can be described for different groups of organisms and structural components of coastal ecosystems under eutrophication stress (see Chap. 3):

- Phytoplankton
- Summer blooms of Cyanobacteria are typical for all Baltic costal waters; however, there are clear distinctions depending on the degree of eutrophication of these waters
- N_2-fixing Cyanobacteria occur in the Baltic and in less eutrophicated coastal zones
- Differences in species composition are obvious: *Nodularia spumigena, Aphanizomenon baltica, A. flos-aquae* in the Baltic, *Anabaena elenkinii, A. spiroides, Anabaenopsis arnoldii* and *Aphanizomenon flos-aquae* in coastal waters
- Dominance of chroococcal Cyanobacteria in eutrophic costal waters (*Microcystis aeruginosa, Chroococcus* spp., e.g. Odra Lagoon)
- Changes to nano- and pico-Cyanobacteria like *Aphanothece clathrata* etc. in polytrophic coastal water.

The first effect of eutrophication is a rise in phytoplankton production. Initially, this will be buffered by higher feeding activity and increased zooplankton development. As eutrophication increases further this "buffer" no longer functions. Due to the greater occurrence of green algae and filamentous Cyanobacteria, and in summer of N_2-fixing Cyanobacteria, dominance of diatoms is lost. At the same time, the number of species determining primary production decreases, although there is no loss of species. Massive phytoplankton blooms interfere with the development of submersed macrophytes and finally cause their loss. In this respect it is interesting that phytoplankton populations are subsequently subject to light limitation, and restructure accordingly. As expected, representatives of nano- and pico-phytoplankton prevail under conditions of light deficiency. The expected dominance of green algae does not occur. On the contrary, the ratio between the biomasses of green algae and Cyanobacteria remains almost constant. This is because Cyanobacteria are better adapted to short-term light stress (Schubert 1996). Such light stress also arises in highly eutrophic waters due to the focusing effects of the waves, with short-term light intensities of up to 8 times solar radiation attained on the water surface. Due to the presence of Langmuir spirals, the algae are exposed in turn to very high and very low light intensities. As the circulation time for Langmuir spirals ranges, e.g. in the DZB, between 10 and 20 min, Cyanobacteria are more efficient at using these short recreation phases in low light compared to green algae. The latter are better at compensating light stress only after prolonged light exposure (Fig. 17.3).

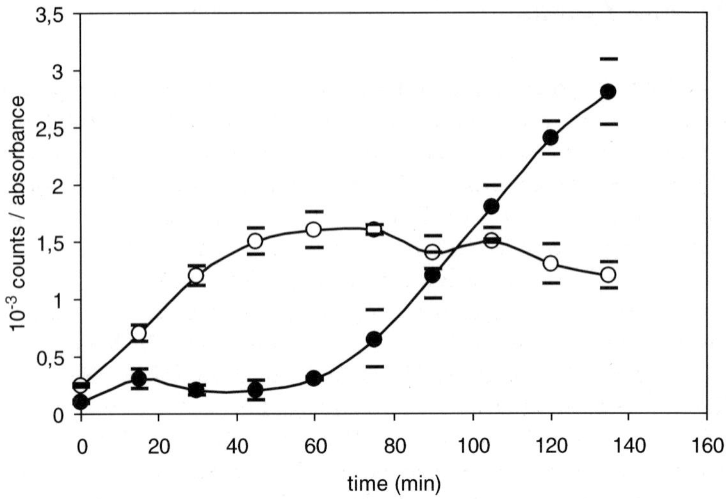

Fig. 17.3 Kinetics of radioactively labelled phaeophytin content after increasing the irradiance from 25 to 260 μmol photons m^{-2} s^{-1}. The concentration of radioactively labelled phaeophytin at different time points after the beginning of the light stress is shown (disintegrations per absorption unit), with standard deviations for four samples. The algae were marked with $NaH^{14}CO_3$ 6 h before the start of the experiment. *Triangles* Green alga (*Chlorella*); *squares* Cyanobacterium (*Synechocystis*). For further details, see Schubert (1996)

17.2.1 Zooplankton

The restructuring of the phytoplankton community goes hand in hand with a structural shift in zooplankton (Schiewer 1998a). This results in a decrease in the numbers of larger zooplankton (*Eurytemora affinis*, larger rotifers) and an increase in small rotifers and particularly in protozooplankton). During the last 20 years, the species number in certain coastal areas has been reduced, but regular annual cycles still exist formed by the succession of already existing species as well as invaders (see Chap. 3).

17.2.2 Protozooplankton

The protozooplankton is always present. Its role becomes more important in the coastal areas with higher river inflow (e.g. Chap. 12). A marked increase in the biological importance of planktonic protists variety of taxa, abundances and production occurs only at higher eutrophication due to the switch of planktonic communities from a grazing food web to a MFW (see Chap. 3).

17.2.3 Fish Populations

A diversity of fish species is typical for almost all described coastal waters. Their role for fisheries is, however, variable. For instance, it is significant in the Greifswalder Bodden because it is a local herring breeding area; in contrast, it is currently relatively unimportant in the Gulf of Riga as a result of over-fishing. Dominant commercial species are *Clupea harengus*, *Sprattus sprattus*, *Plathichthys flesus*, *Belone belone* and *Gadus morhua*. For freshwater fish the catch of *Sander lucioperca*, *Abramis brama* and *Rutilus rutilus* predominates.

The first steps of eutrophication usually favour fish population development due to increasing food availability (see Chap. 4, Greifswalder Bodden: *Esox lucius, Sander lucioperca*; cf. Chaps. 9 and 10). In contrast, Lappalainen (2002) found that "…unlike the general trend in lakes, eutrophication may not increase fish biomasses in coastal waters" of the Gulf of Finland (see Chap. 13). Further stages of eutrophication have a negative impact on fish development. Fish communities lose their most vulnerable populations (e.g. Chap. 3, Schleimündung: *Anguilla anguilla*; Chap. 3: *Sander lucioperca*, *Perca fluviatilis*; Chap. 4, Greifswalder Bodden: *Acipenser sturio*, *Alosa fallax*; Chap. 7: *Acipenser sturio*, *Spinachia spinachia*, *Alosa falax*, *Coregonus lavaretus*, Chap. 13: *Gadus morhua*, partly *Clupea harengus* and *Sprattus sprattus*; *Salmo salar*, *Salmo trutta*, *Coregonus* spp). As this process usually affects edible fishes, eutrophication leads to commercial losses (see Chap. 3, etc). The causes are diverse, but include the negative impact of high pH values (up to pH 11.0); deterioration of fish spawning and nursery grounds (see Chap. 3, DZB: decrease of *Esox lucius*; the same processes, however, favour occurrence of *Sander lucioperca*); and unfavourable food availability due to lack of O_2. Moreover, over-fishing leads to a reduction in the stock of certain species (see Chap. 11: *Perca fluviatilis*). Particular reasons for the temporary decrease of *Gadus morhua* and *Plathichthys flesus* include over-fishing in the Baltic and eutrophication in the Baltic Proper (see Chap. 4, Wismar-Bucht, Salzhaff).

17.2.4 Submersed Macrophytes

Usually, submersed macrophytes are dominant in shallow Baltic coastal waters. They are, however, very sensitive to even the very early stages of eutrophication. The main reason lies in the decrease in light availability due to increasing phytoplankton production, and the increase in the concentration of detritus in the water. The causes

- a decline in colonisation depth
- an increase of epiphytic growth
- a loss of slow-growing macrophytes as seaweed (*Fucus*) and some red algae
- an increase of fast-growing algae, e.g. *Cladophora* and *Pilayella*
- disappearance of Charophytes and the increased appearance of *Potamogeton pectinatus*.

At the end of this development there is extensive loss of submersed macrophytes.

The transition from dominance of submersed macrophytes to dominance of phytoplankton therefore occurs more or less abruptly. It is prepared by a chronic exposure and finally triggered by a short-term combination of unfavourable factors (see Chap. 3). In the case of the Darß-Zingster boddens described previously, high rainfalls, strong freshwater influx and high turbidity coincided in the spring of 1981.

17.2.5 Sediments

The decline of submersed macrophytes leads to a simultaneous destabilisation of the sediments. The higher sediment movement associated with this intensifies the eutrophication. It significantly affects the transition from nutrient- to light-limitation thus initiated, as well as the phytoplankton production. Furthermore, it causes a change of the phytoplankton to low-light-adapted species. Following the loss of submersed macrophytes, a dominance of nano-/picophytoplankton (green algae and Cyanobacteria) and an increasing importance of the MFW was observed in the polytrophic/hypertrophic Darß-Zingster boddens (see Chap. 3). The degradation of organic material was thereby facilitated, inducing an additional increase in nutrient release. Such a development can lead to hypertrophy of a water body and with it to a total loss of its buffer function for the Baltic (see Chap. 3, Kleiner Jasmunder Bodden).

17.2.6 Microbial Food Webs and Aggregates/Fluffy-Sediment-Layer

Since the original report of Azam et al. (1983), the important role of bacteria in the turnover of organic material in the water column of oligotrophic marine waters has been proved many times. Subsequent investigations established a MFW that was organised in aggregates (e.g. Azam 1998; Heissenberger et al. 1996). This was also true for the oligotrophic northern Baltic Sea (e.g. Wikner and Hagström 1988; Samuelsson and Andersson 2003). The reasons for this are

- high DOM/POM rates
- nutrient limitation
- available light.

In 1983, the results of mesocosm experiments in the eutrophic Darß-Zingster boddens (Zingster Strom) supported the existence of an important bacterial influence in the turnover of organic matter (Schiewer et al. 1986). The expression of MFWs took place accompanied by increasing eutrophication. Finally, 90–95% of the turnover of organic matter in the Darß-Zingster boddens (see Chap. 3) was connected with the MFW.

The occurrence and function of aggregates in the coastal waters of the Baltic have barely been investigated. The first comprehensive findings have resulted from the BMBF-funded joint project "ÖKOBOD" (Meyer-Reil 1999), proving aggregate occurrence in the pelagic zone. However, these results also showed that aggregate structuring in eutrophic shallow waters was not as pronounced as in marine areas due to the high proportions of phytoplankton and strong wind-shearing activities. Aggregates play an important role as FSL. Depending on the production within the water body and its depth they formed layers from one to several millimeters thick.

In sediment layers of deeper water zones, smaller aggregates combine to form larger ones involving diatoms and bacteria. Thus, a more or less stable sediment layer can be formed. This causes a partial re-stabilisation of the sediment (see Chap. 3, GB). In contrast, in highly eutrophic shallow waters, sediment layers remain highly mobile and only slightly differentiated. They are thereby able to immediately bind the nutrients released from the sediment and to use them for the production of new biomass (see Chap. 3, DZB). Exchange processes in the Darß-Zingster boddens are thus primarily bound to these particles (Fig. 17.4). With aggregates also being the preferred sites of colonisation of bacteria and protozoa, short nutrient cycles exist, which contribute significantly to the "internal" nutrient load in these waters.

In contrast to the oligotrophic Baltic Proper, the background for establishing MFWs in the polytrophic coastal waters is different, i.e.:

- nearly equal DOM/POM ratios but very high concentrations of both parts
- no nutrient limitation
- light limitation, which favours the development of eukaryotic and prokaryotic nano- and prokaryotic pico-phytoplankton
- reduced abundances and biomasses of mesozooplankton
- increased concentration of protozooplankton, e.g. heterotrophic nanoflagellates (HNF) and ciliates.
- development of an unstable FSL instead of more stable aggregates.

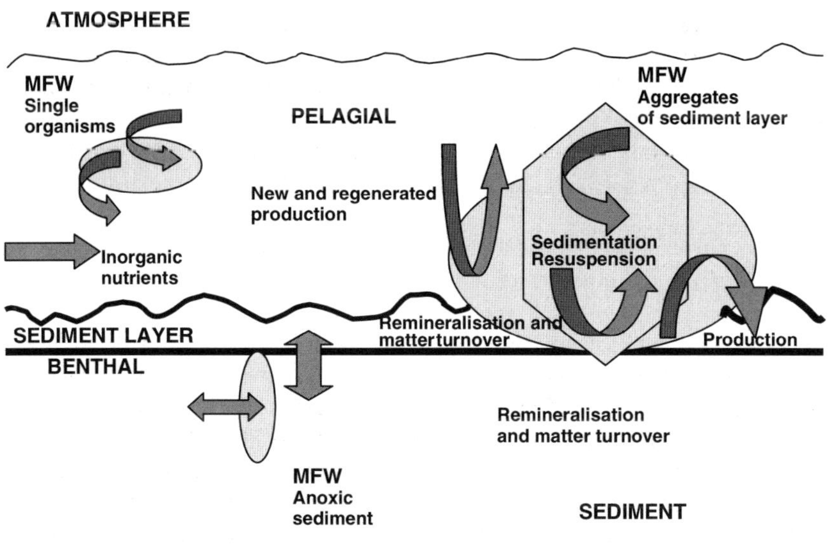

Fig. 17.4 Matter turnover in pelagial and sediment: impact of sediment layer (fluffy sediment layer; FSL). *MFW* Microbial food web

17.2.7 *Benthos*

Increasing eutrophication leads to improved feeding conditions for macrozoobenthos. Due to intensive benthic–pelagic coupling, high quality planktonic material is available to these organisms, which results in higher biomasses of macrozoobenthos. If continued, this process leads to selection of r-strategists and this implies a reduction in biodiversity. A dominance of oligochaetes and chironomid larvae often occurs. As higher organic content boosts activity of microorganisms, the share of O_2 supplied to the sediment decreases. This leads to a reduction in the depth of settlement by macrozoobenthos. Because of the temporary occurrence of H_2S, living conditions deteriorate with progressing eutrophication. Only specialists, which invade the sediment to limited extent only, frequently at high abundances, survive. As a result, only epibenthic organisms remain (Rumohr 1996). This latter step is usually not reached in the shallow coastal waters of the Baltic. It can occur however in the highly eutrophic deeper layers of coastal waters, temporarily stratified by a halocline (see Chap. 7), and is characteristic of the deeper sinks of the Baltic Sea.

Despite repeated efforts, the control of eutrophication in Baltic coastal waters has so far failed. Only targeted reductions of nutrient inputs on the coasts of Sweden and Denmark have led to the first positive results (see Chaps. 14 and 16).

The reduction in nutrient inflow from the mainland that followed the political and economic changes in Middle and Eastern Europe beginning in the early 1990s has at least produced a certain stabilisation. Locally, the first signs of relief can be observed. It remains open, however, whether these signs will lead to long-term ecosystem improvement. The renewed boom of agriculture and tourism in coastal regions of the Baltic Sea could reverse this development.

17.3 Other Threats

Apart from eutrophication, pollution from oil and oil products (particular in relation to increased shipping), heavy metals, rising saprobisation and pesticide entry play an important role in the coastal waters of the Baltic Sea. However, for these factors, stronger differentiation is needed.

Within estuarine regions of large rivers and along coastal stretches with big cities, pollution by oil and heavy metals as well as saprobisation naturally play a considerable role (see Chap. 12). In the last few years, saprobisation was reduced significantly through wastewater treatment (see Chap. 14). Pesticide loads occur predominantly in agricultural regions, and their frequently diffuse entry is difficult to control.

A new kind of danger arises from the increasing inflow of non-indigenous species (aliens) (see Chap. 12). There are extensive studies on the entry of aliens into the Baltic by Leppäkoski and Olenin (2000), Leppäkoski (2002), Orlova et al. (2006) and other authors. The occurrence of aliens is not directly linked to eutrophication.

It can, however, foster the arrival of new species through an excess of food and/or loss of other species. General migrants into all brackish eastern coastal waters are *Dreissena polymorpha*, *Marenzelleria neglecta* and *Cercopagis pengoi.* Microorganisms and viruses have so far attracted little attention. Arguably they present an increasing danger in ever-cleaner harbour waters.

17.4 Buffer and Filter Capacity

By connecting rivers and sea, coastal waters perform a formal role as both a buffer and a filter, based on flow-driven sedimentation and the physical, chemical and biological transformation of these ecosystems. The underlying processes are affected in multiple ways by coastal zone morphology, soil topology, water circulation, freshwater outflow and saltwater inflow, as well as by other abiotic factors and biological components and processes. Hence, high filter and buffer capacity cannot be postulated a priori for all coastal waters. In addition, anthropogenic influences affect not only internal conditions but also exchange processes. A number of these aspects have been discussed earlier. A further distinction has to be made between shallow and deeper coastal waters.
(1) Shallow coastal waters (e.g. GB, DZB, Szczecin Lagoon)

- These waters are polymictic and thereby preside over an intensive pelagic–benthic coupling. At the same time the fixing of nutrients in the sediment is inhibited by permanent re-suspension.
- The permanent oxygen supply accelerates the degradation of organic substances. The supply of allochthonal terrestrial material offsets this process.
- The high re-mineralisation rates associated with degradation of organic matter give rise to water internal eutrophication.
- In the winter season biotic processes play a minor role.

During the vegetation period organic materials are barely enriched even in highly eutrophic coastal waters. It could be shown in mesocosm experiments that formation and degradation rates balance each other even under strong algal blooms (Schiewer 1998b). It is not surprising that so far sediment accumulations of only 1 mm a^{-1} have been found, even in the highly eutrophic Szczecin Lagoon. All in all, the turnover in the pelagic zone is significantly higher than that by benthic organisms. Even under the relatively optimal conditions of Pomeranian Bay (Kube 1996), decomposition of organic matter by zoobenthos amounted only to one-third of primary production.

Through water exchange with the open Baltic Sea, considerable shares of primary production are exported even under natural conditions. The latter can be applied for instance to the Greifswalder Bodden, the river inputs of which are very small. Under these conditions horizontal exchange processes dominate over the internal material cycles. In contrast, the Darß-Zingster boddens represents a water type with far larger 'autonomy' due to its subdivided structure and the very narrow exit to the Baltic. Remarkably, during summer, nutrients are transported into the Baltic predominantly as organic loads. Inorganic nitrate transports are dominant during winter due to the high NO_3 content in the water. In contrast to other coastal waters, inorganic phosphorus concentrations in the pelagic waters of Darß-Zingster boddens are generally very low, so that the transport of the organic form is predominant. Lack of NO_3 supply in late spring and summer reduces N_2 release by denitrification. Promotion of this process via aggregate formation was not clearly detectable. For larger river mouths (e.g. Szczecin Lagoon), considerable freshwater inflow is released directly into the Baltic Sea. Patches formed in this way have a variable life span. Currents and wind conditions are particularly crucial, but the prevailing water density differences are also important. In extreme cases, they initiate toxic algal blooms. A classic example for the Baltic Sea was the *Chrysochromulina*-bloom in 1988 (Maestrini and Granéli 1991).

(2) Deeper coastal waters

- Temporary occurrence of a thermocline
- Formation of an unstable halocline
- Temporary decoupling of pelagic and benthic interactions
- Temporary sink-function of the benthal
- Dimictic behaviour as in lakes, but near-bottom transport of organic material from coastal zones into the Baltic and finally into its deep basins.

17.5 Ecosystem Stability

If external stress exceeds the threshold biotic parameters for individual organisms or communities, self regulation processes can be initiated. At the ecosystem level, such a process can be demonstrated by the transition from dominance of submersed macrophytes to dominance of phytoplankton in shallow waters (see Chaps. 3 and 16). This apparently represents the adjustment of the ecosystem from one stable state to another (cf. Scheffer 1998 for limnic waters). The Kerteminde Nor (see Chap. 16) seems to be an ecosystem switching between the two extremes – dominance of submersed macrophytes versus dominance of phytoplankton. Due to the initiating effect of external nutrient release, it is obviously changing to phytoplankton dominance under certain conditions. My own studies prove for coastal waters that intermediate steps can also be stabilised (see Chap. 3). On the one hand, this concerns the oligo-mesotrophic conditions – for example at the initial stages of formation of the Darß-Zingster boddens. The status of the water body remained stable as long as the island cores were not connected by silting processes, and the exchange with Baltic water was almost undisturbed. In terms of phytoplankton, this stage is characterised by the dominance diatoms. On the other hand, there is the eutrophic intermediate stage, which can be distinguished by a reduced submersed macrophytes population but also by dominance of green algae–Cyanobacteria–biocoenosis in phytoplankton. Another intermediate stage is the polytrophic state, which is characterised by a loss of submerse macrophytes, dominance of low-light-adapted Cyanobacteria and green algae, and dominance of the MFW with a large fraction of protozoa.

Regarding the stability of the intermediate stages, Kalbe (1996) obtained comparable results when analysing the eutrophication process of Brandenburgian lakes.

17.6 Knowledge Gaps

The overview of existing results reveals a range of fundamental deficiencies in our knowledge of the Baltic coastal ecosystems:

- Water exchange models are often lacking.
- Long-term monitoring is frequently missing or is performed only with a very limited number of parameters.
- There are only limited propositions regarding the autecology of species.
- Population dynamics have rarely been analysed with regard to the interaction between species. There is a general lack of studies on the role of the MFW and microzooplankton communities in the ecosystem functioning.
- Our understanding of how changes in external forcing are mediated within the coastal ecosystems is at best fragmentary. In particular, it is still necessary to identify, define and quantify thresholds, switches, bottle-necks and key species/processes in the ecosystem response to changes in external forcing. Moreover, there are but few experimental analyses of coastal ecosystems regarding the variability, stability and elasticity of their population dynamics. Micro-scale analyses of the coastal ecosystems are the exception, knowledge of sediment structure and function remains patchy and the interaction between sediment and water remains scarcely analysed.
- It is unknown how the sediment mirrors system changes by proxy-variables. The invasion of microorganisms and viruses into coastal ecosystems remains unexplored, and there are only a few propositions regarding the palaeoecology of coastal water bodies.

17.7 Protection Measures

The coastal ecosystems are the last filters and buffers for the Baltic Proper. Usually, these capacities are high, but they are now threatened by eutrophication, pollution, aliens, etc.

Eutrophication is one of the most important threats to the health and productivity of the open Baltic Sea as well as to the coastal ecosystems. It is stimulated by nutrient import from point and non-point sources by discharges from agriculture, touristic and industrial landscapes and atmospheric deposition. During the past 50 years, the tideless estuaries of the Baltic Sea have often been overloaded by anthropogenic activities, mainly through high inputs of inorganic nutrients inducing eutrophication. Meanwhile, in some estuaries, e.g. Darß-Zingster boddens, the eutrophication is so extreme that they have not only lost their filter and buffer capacity, but have themselves become loading sources for the Baltic Sea. The existing Baltic Sea Joint Comprehensive Action Programs are oriented primarily on the point sources. The current priorities will not ensure the eventual improvement of environmental conditions of the Baltic. The existing action programs have produced some initial positive results, but they should be modified. Alternatively, additional, non-point source-oriented, action programs must be developed. They should include as priorities:

- Identification of "hot catchment areas", where most excessive nutrient losses takes place.
- Restoration of the discharge areas that impact eutrophic aquatic ecosystems.
- Construction of three-step sewage treatment plants for "Hot Spots".
- Creation of low-tech sewage treatment technology for small point sources in the entire catchment area.
- Recommendations on improvement of land use and drainage systems to minimise nutrient loss.
- Implementation of the best available agricultural methods, e.g. proper circulation of crops, no-till agriculture, rational use of artificial fertilisers, proper management and storage of manure to prevent nutrient discharge into water bodies and atmosphere.
- Activities to control NO_x discharge from transportation and combustion of fossil fuel.
- Controlling volatilisation of NH_4 from agricultural wastes.
- Application of a "concept of atmospheric origin areas", considering that most of the atmospheric discharges affecting the Baltic Sea originate in the industrial and agricultural regions of Western Europe are far removed (sometimes by hundreds of kilometres) from the immediate drainage basin of the Baltic Sea.

In the future, there will be increasing scope for conflicts between nature conservation, tourism, different types of use in the coastal and near-shore areas, and coastal protection. Integrated Coastal Zone Management (ICZM) is one way to provide suitable concepts for a sustainable development (European Commission 1999b), but joint approaches are still missing. It is necessary to select areas for combined and comparative natural scientific and socio-economic studies. The main goal should be to support the function of natural rather than anthropogenically disrupted capital by feedback from society.

Large amounts and various kinds of data about the Baltic Sea have been collected to date. Ongoing monitoring programs and projects – often applying very traditional and insufficient methods – increase the amount of data. It is necessary to develop more general monitoring and classification systems with more integrated assessments of processes in coastal ecosystems. New impetus will come from the implementation of the European Water Frame Work Directive (European Commission 1999a). At present, there is nearly no application of bio-indication techniques. Complex aquatic microbiological test systems are lacking and the development of eco-technology for coastal waters is rather low.

The insufficient availability of these data, and lacking knowledge, is a problem. Further attempts are needed to link existing data and information in interdisciplinary, inter-sectoral, regional and national databases, and to provide easy access to these data sets.

Legislation and competence in respect to the coastal area problems are very different, e.g. they vary considerably even between the different federal states of Germany. Moreover, there is so far no official definition of the coastal zone, and joint terminology, principles and outlined main topics do not exist. The coastline is the main boundary within the coastal zone. In Germany, for example, laws, responsibilities, stakeholders and approaches differ between terrestrial areas and coastal waters. The Planning Programs cover the terrestrial region up to the shoreline, but coastal waters in Germany are not subject to spatial planning. In this way, a problematic overlap of uses already exists.

Public participation is needed, but it is doubtful whether it holds the key to problem-solving in the coastal zone. Questions to be answered include: how will the stakeholders be chosen, how can the continuity of the process be ensured, who will be the moderator, and how will the large number of various topics be dealt with?

17.8 Conclusions and Outlook

The Baltic coastal ecosystems are highly diverse. Based on spatial distribution, 12 typical coastal waters can be distinguished according to Klug (1985), whereas Lampe (1996) refers to at least 10 basic types. However, considering both human impact and ecosystem reactions, it is not far fetched to state that every system is unique!

Therefore we need much more comprehensive data for each subsystem, regarding the past and current ecological situation and the development, state and fate of eutrophication, pollution etc., as well as knowledge of the role of the introduction of alien species such as micro-organisms and viruses.

In future, pressure on the Baltic coastal ecosystems will increase. Therefore it is necessary

- to develop more general monitoring programs and new classification systems with more integrated assessments of coastal processes and systems
- to ensure the multivalent use of estuaries, to establish a mosaic of co-evolving socio-economical and ecological systems
- and to study the self-regulation of ecosystems, especially with regard to life-support functions.

The coastal waters have a limited capacity for maintaining the nutrient balance under the increasing nutrient loads. Because of their shallowness, an increasing nutrient load leads to rapid eutrophication. Interaction with other factors (e.g. salinity) and processes (e.g. exchange with the Baltic Proper, freshwater entries) imply differences with regard to both the degree of eutrophication and its temporal pattern. One advantage of further research will be that results obtained in the Baltic coastal waters could be useful for the handling of other coastal waters elsewhere in the world.

The recent socio-economic changes in the countries of Central and Eastern Europe have resulted in considerable alterations in terms of their impacts on the environment, especially due to declines in industrial and agricultural production. Several unexpected outcomes (e.g. changes in the nutrient dynamics in the Gulf of Riga, Chap. 10) have occurred along with the general decrease of anthropogenic discharges. It is of vital importance to ensure the environmentally sound re-industrialisation and economic re-structuring of these countries. Environmental scientists in transition countries must receive support from developed countries to be able to follow and predict rapid environmental changes. The social demands dictating political priorities of countries in transition differ somewhat from those of the developed countries. Thus, environmental protection is a field of low priority level in many Baltic countries! Therefore, policy-makers must receive encouragement from their western colleagues to re-prioritise environmental protection activities.

For the sustainable future of the Baltic coastal waters it is necessary to develop:

- legally based mechanisms to transfer scientific research results to appropriate governmental authorities
- environmental education programs that bring research results to the public in a comprehensible form
- mechanisms for public involvement in the development of national environmental policy. The public should be educated and informed about these mechanisms and relevant issues
- environmental management programs that are geographically and ecologically suitable. For example, to manage better water quality, a river-basin approach should be considered. Intergovernmental collaboration will be necessary in many instances
- management programs that are legally sound and defensible. Support law and policy with a solid scientific foundation. This is particularly important in cases that could affect the economic value of private properties. The degree to which the value of property is protected by constitutional and other forms of law will likely vary in different countries
- environmental mediation tools as a way of addressing conflicts without always having to enter an adversarial juridical system
- examination of alternative management programs and approaches in regions with comparable ecosystems and systems of government. Understanding successes and failures from these programs should be helpful for the implementation of the above mentioned goals.

References

Azam F (1998) Microbial control of oceanic carbon flux: the plot thickens. Science 280:694–606

Azam F, Fenchel T, Field JG, Gray JS, Meyer-Reil LA, Thingstad F (1983) The ecological role of the water-column microbes in the sea. Mar Ecol Progr Ser 10:257–263

Estrum-Yousef S (2001) Untersuchungen zur partikulären organischen Fracht küstennaher Gewässer in der Pommerschen Bucht, südliche Ostsee. Diss. Univ Rostock, Rostock

European Commission (1999a) European Water Framework Directive. European Commission, Brussels

European Commission (1999b) Toward an European Integrated Coastal Zone Management (ICZM). European Commission, Brussels

Görs S, Schumann R, Rentsch D, Schiewer U, Schlungbaum G (2000) Untersuchungen zum gelöstem und partikulärem organischen Material im Trophiegradienten der Darß-Zingster Boddenkette unter besonderer Berücksichtigung des Aminosäurepools. DGL-Tagungsbericht 1999, pp 394–398

Heissenberger A, Leppard GG, Herndl GJ (1996) Relationship between the intercellular integrity and the morphology of the capsular envelope in attached and free-living marine bacteria. Appl Environ Microbiol 62:4521–4528

Kalbe L (1996) Zur Stabilität von limnischen Ökosysteme. Limnologica 26:281–291

Klug H (1985) Küstenformen der Ostsee. In: Newig J, Theede H (eds) Die Ostsee: Natur und Kulturraum. Heide, Husum, pp 70–78

Kube J (1996) Spatial and temporal variations in the population structure of the soft-shell clam, *Mya arenia*, in the Pomeranian Bay (Southern Baltic Sea). J Sea Res 35:335–344

Lampe R (1996) Küstentypen. In: Rheinheimer G (ed) Meereskunde der Ostsee, 2nd edn, Springer, Berlin, Heidelberg, pp 17–25

Lappalainen A (2002) The effects of recent eutrophication on freshwater fish communities and fishery on the northern coast of the Gulf of Finland, Baltic Sea. PhD thesis, University of Helsinki

Leppäkoski E (2002) Harmful non-native species in the Baltic Sea – an ignored problem. In: Schernewski G, Schiewer U (eds) Baltic coastal ecosystems: structure, function and management. Springer, Berlin, pp 253–275

Leppäkoski E, Olenin S (2000) Non-native species and rates of spread: lessons from the brackish Baltic Sea. Biol Invasions 2:151–163

Maestrini S, Granéli E (1991) Environmental conditions and ecophysiological mechanisms which led to the 1988 *Chrysochromulina polylepis* bloom and hypothesis. Oceanol Acta 14:397–413

Meyer-Reil L-A (1999) Verbundprojekt ÖKOBOD – Beitrag zum wissenschaftlichen Verständnis der Ökologie der Boddengewässer. Bodden 8:87–105

Oertzen J-A von (1988) Das Leben im Brackwasser – Konfrontation oder Opportunismus? Biol Rundsch 26:197–212

Orlova MI, Telesh IV, Berezina NA, Antsulevich AE, Maximov AA, Litvinchuk LF (2006) Effects of nonindigenous species on diversity and community functioning in the eastern Gulf of Finland (Baltic Sea). Helgoland Mar Res 60:98–105

Remane A, Schlieper C (1958) Die Biologie des Brackwassers. In: Thienemann A (ed) Die Binnengewässer 22:1–348

Rumohr H (1996) Veränderungen des Lebens am Meeresboden. In: Lozán JL, Lampe R, Matthäus W, Rachor E, Rumohr H, Westernhagen H von (eds) Warnsignale aus der Ostsee. Parey, Berlin, pp 162–168

Samuelsson K, Andersson A (2003) Predation limitation in the pelagic microbial food web in an oligotrophic aquatic system. Aquat Microb Ecol 30:239–250

Scheffer M (1998) Ecology of shallow lakes, 1st edn. Chapman and Hall, London

Schiewer U (1998a) 30 years eutrophication in shallow brackish waters – lessons to be learned. Hydrobiologia 363:73–79

Schiewer U (1998b) Hypertrophy of a Baltic estuary – changes in structure and function of the planktonic community. Verh Int Verein Limnol 26:1503–1507

Schiewer U, Arndt H, Baader G, Ballin G, Börner R, Evert F-K, Georgi F, Heerkloss R, Jost G, Kell V, Krüger B, Walter T (1986) The bounds and potential effects of NH_4^+ (loading) on the pelagic system of a Baltic estuary. Limnologica (Berlin) 17:7–28

Schubert H (1996) Ökophysiologie der Lichtanpassung des Phytoplanktons eutropher Flachgewässer. Habil. Univ Rostock, Rostock

Wikner J, Hagström Å (1988) Evidence for a tightly coupled nanoplanktonic predator-prey link regulating the bacterivores in the marine environment. Mar Ecol Progr Ser 50:137–145

Index

O

P

R

Ecological Studies

Volumes published since 2003

Volume 165
Population Viability in Plants: Conservation, Management, and Modeling of Rare Plants (2003)
C.A. Brigham, M.W. Schwartz (Eds.)

Volume 166
North American Temperate Deciduous Forest Responses to Changing Precipitation Regimes (2003)
P. Hanson and S.D. Wullschleger (Eds.)

Volume 167
Alpine Biodiversity in Europe (2003)
L. Nagy, G. Grabherr, Ch. Körner, D. Thompson (Eds.)

Volume 168
Root Ecology (2003)
H. de Kroon and E.J.W. Visser (Eds.)

Volume 169
Fire in Tropical Savannas: The Kapalga Experiment (2003)
A.N. Andersen, G.D. Cook, and R.J. Williams (Eds.)

Volume 170
Molecular Ecotoxicology of Plants (2004)
H. Sandermann (Ed.)

Volume 171
Coastal Dunes: Ecology and Conservation (2004)
M.L. Martínez and N. Psuty (Eds.)

Volume 172
Biogeochemistry of Forested Catchments in a Changing Environment: A German Case Study (2004)
E. Matzner (Ed.)

Volume 173
Insects and Ecosystem Function (2004)
W.W. Weisser and E. Siemann (Eds.)

Volume 174
Pollination Ecology and the Rain Forest: Sarawak Studies (2005)
D. Roubik, S. Sakai, and A.A. Hamid (Eds.)

Volume 175
Antarctic Ecosystems: Environmental Contamination, Climate Change, and Human Impact (2005)
R. Bargagli

Volume 176
Forest Diversity and Function: Temperate and Boreal Systems (2005)
M. Scherer-Lorenzen, Ch. Körner, and E.-D. Schulze (Eds.)

Volume 177
A History of Atmospheric CO_2 and its Effects on Plants, Animals, and Ecosystems (2005)
J.R. Ehleringer, T.E. Cerling, and M.D. Dearing (Eds.)

Volume 178
Photosynthetic Adaptation: Chloroplast to Landscape (2005)
W.K. Smith, T.C. Vogelmann, and C. Chritchley (Eds.)

Volume 179
Lamto: Structure, Functioning, and Dynamics of a Savanna Ecosystem (2006)
L. Abbadie et al. (Eds.)

Volume 180
Plant Ecology, Herbivory, and Human Impact in Nordic Mountain Birch Forests (2005)
F.E. Wielgolaski (Ed.) and P.S. Karlsson, S. Neuvonen, D. Thannheiser (Ed. Board)

Volume 181
Nutrient Acquisition by Plants: An Ecological Perspective (2005)
H. BassiriRad (Ed.)

Volume 182
Human Ecology: Biocultural Adaptations in Human Cummunities (2006)
H. Schutkowski

Volume 183
Growth Dynamics of Conifer Tree Rings: Images of Past and Future Environments (2006)
E.A. Vaganov, M.K. Hughes, and A.V. Shashkin

Volume 184
Reindeer Management in Northernmost Europe: Linking Practical and Scientific Knowledge in Social-Ecological Systems (2006)
B.C. Forbes, M. Bölter, L. Müller-Wille, J. Hukkinen, F. Müller, N. Gunslay, and Y. Konstantinov (Eds.)

Volume 185
Ecology and Conservation of Neotropical Montane Oak Forests (2006)
M. Kappelle (Ed.)

Volume 186
Biological Invasions in New Zealand (2006)
R.B. Allen and W.G. Lee (Eds.)

Volume 187
Managed Ecosystems and CO_2: Case Studies, Processes, and Perspectives (2006)
J. Nösberger, S.P. Long, R.J. Norby, M. Stitt, G.R. Hendrey, and H. Blum (Eds.)

Volume 188
Boreal Peatland Ecosystem (2006)
R.K. Wieder and D.H. Vitt (Eds.)

Volume 189
Ecology of Harmful Algae (2006)
E. Granéli and J.T. Turner (Eds.)

Volume 190
Wetlands and Natural Resource Management (2006)
J.T.A. Verhoeven, B. Beltman, R. Bobbink, and D.F. Whigham (Eds.)

Volume 191
Wetlands: Functioning, Biodiversity Conservation, and Restoration (2006)
R. Bobbink, B. Beltman, J.T.A. Verhoeven, and D.F. Whigham (Eds.)

Volume 192
Geological Approaches to Coral Reef Ecology (2007)
R.B. Aronson (Ed.)

Volume 193
Biological Invasions (2007)
W. Nentwig (Ed.)

Volume 194
***Clusia*: A Woody Neotropical Genus of Remarkable Plasticity and Diversity** (2007)
U. Lüttge (Ed.)

Volume 195
The Ecology of Browsing and Grazing (2008)
I.J. Gordon and H.H.T. Prins (Eds.)

Volume 196
Western North American *Juniperus* Communites: A Dynamic Vegetation Type (2008)
O. Van Auken (Ed.)

Volume 197
Ecology of Baltic Coastal Waters (2008)
U. Schiewer (Ed.)

Printing: Krips bv, Meppel, The Netherlands
Binding: Stürtz, Würzburg, Germany